STRUCTURE REPORTS

for 1990

Volume 57A

Structure Reports is prepared under the guidance of a Commission of the International Union of Crystallography. The members of the Commission sometime concerned with the preparation of this volume are listed below.

STRUCTURE REPORTS

for 1990

Volume 57A

METALS AND INORGANIC SECTIONS

General editor

G. Ferguson

Section editor

J. Trotter

Published for the

INTERNATIONAL UNION OF CRYSTALLOGRAPHY

SPRINGER-SCIENCE+BUSINESS MEDIA, B.V.

First published in 1992

ISSN 0166–6983

Printed on acid-free paper

ISBN 978-94-017-2251-3   ISBN 978-94-017-2249-0 (eBook)
DOI 10.1007/978-94-017-2249-0

TABLE OF CONTENTS

INTRODUCTION

The present volume continues the aim of Structure Reports to present critical accounts of all crystallographic structure determinations of metals and of inorganic compounds. Details of the arrangement in the volumes, symbols used etc. are given in volume 53A, pages v and vi.

University of Guelph,   G. FERGUSON
Guelph, Ontario, Canada

4 January 1992

STRUCTURE REPORTS

SECTION I

METALS

Edited by

J. Trotter

(University of British Columbia)

STRUCTURE REPORTS

SECTION I

METALS

ARRANGEMENT

The metals reports in this volume are arranged under the classifications:
binary alloys, ternary alloys (within each of these classifications, the entries
are sorted alphabetically by formula), hydrides, borides, carbides, silicides,
pnictides (N, P, As), chalcogenides (S, Se, Te) (within each of these classifications, the entries are sorted alphabetically by formula on the elements other
than that of the classification). See 47A, VII for further details. To find
particular substances the metals formula index may be used.

AlAu$_4$

J. Less-Common Metals, 61, 347-354.

P2$_1$3, 6.9227, Z = 4, R = 0.042. Au in 12(b): 0.1328,0.2007,0.4625; Au(2) in 4(a): x,x,x, x = 0.0673; Al in 4(a): x = 0.6903. Ordered variant of β-Mn structure (2, 3; 44A, 78).

Al$_{11}$Au$_6$, AlAu

J. Less-Common Metals, 160, 143-152.

Al$_{11}$Au$_6$, Fm3m, 5.9988, Z = 2/3, R = 0.012. 4 Au in 4(a): 0,0,0; 7.3 Al in 8(c): 1/4,1/4,1/4. CaF$_2$-type structure (1, 148), with partially occupied Al site; previously described as Al$_2$Au (3, 20, 314; 29, 96).

AlAu, P2$_1$/m, 6.3998, 3.3286, 6.3254, 93.140, Z = 4, R = 0.071. Atoms in 2(e): x,1/4,z, x = 0.4737, 0.0076, 0.8215, 0.3180, z = 0.3092, 0.1871, 0.5492, 0.9211, for Au(1), Au(2), Al(1), Al(2), respectively. Distorted NiAs-type, with increased coordination numbers: Al = 8, Au = 9.

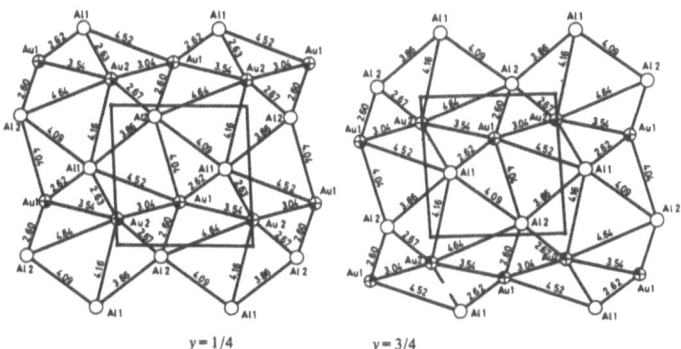

y = 1/4   y = 3/4

AlCe$_3$

Physica B, 163, 587-590.

Above 520K, cubic AuCu$_3$ type (1, 486).

115-520K, P6$_3$/mmc, 7.008, 5.422 A, at 310K, Z = 2, neutron powder data. Al in 2(c): 1/3,2/3,1/4; Ce in 6(h): 0.8208,0.6416,1/4. Ni$_3$Sn-type (5, 7).

Below 115K, P2$_1$/m (c unique), 6.824, 12.458, 5.336, γ = 89.69°, at 15K, Z = 4, neutron powder data. Distortion of the Ni$_3$Sn type, isostructural with α-Pu (21, 165; 28, 38; 56A, 58).

$Au_{51}U_{14}$

J. Less-Common Metals, 160, 171-180.

P6/m, a = 12.6521, 12.615, 12.615, c = 9.1381, 9.118, 9.118 A, at 295, 30, 11K, Z = 1, neutron powder data. Structure as previously described (55A, 5).

$Be_{13}U$

Acta Cryst., C46, 1579-1580.

Fm3c, 10.268, Z = 8, R = 0.022. U in 8(a): 1/4,1/4,1/4; Be(1) in 8(b): 0,0,0; Be(2) in 96(i): 0,0.1151,0.1765. Structure as previously described (16, 27; 52A, 17), with U surrounded by eight $Be(2)_{12}$ dodecahedral cages, each containing a Be(1) atom. U-Be(2) = 3.013, Be(1)-Be(2) = 2.163 A.

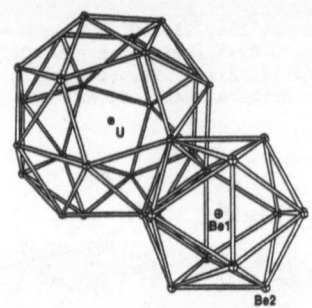

$Ce_{4.8}Ga_{3.2}$ 1.6 x Ce_3Ga_2

J. Less-Common Metals, 160, 229-235.

P4/ncc, 8.066, 14.495, Z = 4, R = 0.035. Ce(1) in 16(g): 0.9124,0.4307, 0.0993; Ce(2) (0.8Ce + 0.2Ga) in 4(c): 3/4,3/4,0.2766; Ga(1) in 8(f): 0.1330,0.6330,1/4; Ga(2) in 4(c): 3/4,3/4,0.0012. Ba_5Si_3-type (31A, 25). Coordination numbers are: Ce = 13, 16, Ga = 9; shortest interatomic distances: Ce-Ce = 3.396, Ce-Ga = 3.180, Ga-Ga = 2.670 A.

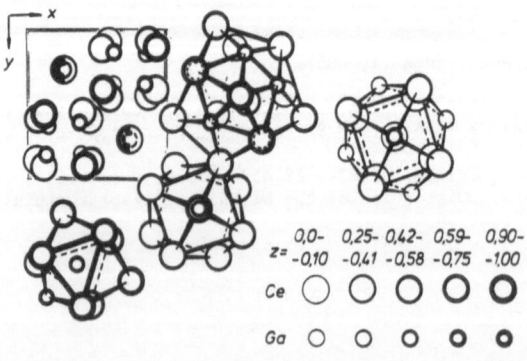

CeGe$_{1.6}$

J. Less-Common Metals, 167, 53-63.

I4$_1$, a = 3 x 4.223, c = 14.208 A, Z = 36, R = 0.075 and neutron powder data. Superstructure of the α-ThSi$_2$-type (9, 121), with Ge vacancies.

	x	y	z
Ce1	0.3395	0.0006	-0.0035
Ce2	0.0005	0.3421	-0.0034
Ce3	0.3370	0.3361	0.0055
Ce4	0.3372	0.6647	0.0057
Ce5	0.000	0.000	0.000
Ge1	0.3338	-0.0001	0.4151
Ge2	0.000	0.000	0.5503
Ge3	-0.0011	0.3189	0.4131
Ge4	0.3329	0.6778	0.5742
Ge5	-0.0010	0.6366	0.5969
Ge6	0.3338	-0.0024	0.5910
Ge7	0.3358	0.3197	0.5763
Ge8 a	0.320	0.360	0.412
Ge9 a	0.318	0.641	0.403

ᵃOccupancy factor of 0.25

Co$_5$U

J. Less-Common Metals, 158, 287-294.

R$\bar{3}$m, a = 4.80, 4.78, c = 36.43, 36.34 A, at 293, 15K, Z = 9, neutron powder data. 9-Layer derivative of the CaCu$_5$ type (11, 59).

Cu$_6$Ln (Ln = La, Ce, Pr, Nd)

J. Solid State Chem., 84, 93-101.

High-temperature phases, Pnma, a ∿ 8.1, b ∿ 5.1, c ∿ 10.2 A, Z = 4, neutron powder data. Structures as previously described (24, 99; 52A, 39; 53A, 4; 54A, 7).

Low-temperature phases (below 455, 214, 206, 151K for the four compounds, respectively), P2$_1$/c, a ∿ 5.1, b ∿ 10.2, c ∿ 8.1 A, β ∿ 91°, Z = 4, neutron powder data. Structures as previously described (52A, 39; 53A, 4; 54A, 7).

Dy$_3$Ge$_5$ (I), DyGe$_{1.9}$ (II)

J. Less-Common Metals, 163, 319-330.

I, Fdd2, 5.729, 17.190, 13.678, Z = 8, neutron powder data. Dy(1) in 8(a): 0,0,0; Dy(2) in 16(b): 0.756,0.082,0.261; Ge(1) in 8(a): 0,0,0.428; Ge(2) in 16(b): 0.818,0.073,0.653; Ge(3) in 16(b): 0.776,0.082,0.827. Superstructure of the tetragonal ThSi$_2$-type (9, 121).

II, Cmmm [cf. Cmc21 in 56A, 6], a = 4.091, 4.072, b = 29.807, 29.766, c = 3.987, 3.964 A, at 293, 4.2K, Z = 8, X-ray and neutron powder data. Isostructural with Ge$_2$Tb (55A, 7).

		Neutron diffraction at 293 K			Neutron diffraction at 4.2 K		
		x	y	z	x	y	z
Dy(1)	4i	0.0	0.4247	0.0	0.0	0.4245	0.0
Dy(2)	4j	0.0	0.3101	0.5	0.0	0.3093	0.5
Ge(1)	2a	0.0	0.0	0.0	0.0	0.0	0.0
Ge(2)	2c	0.5	0.0	0.5	0.5	0.0	0.5
Ge(3)	4i	0.0	0.1581	0.0	0.0	0.1592	0.0
Ge(4)*	4i	0.0	0.2362	0.0	0.0	0.239	0.0
Ge(5)	4j	0.0	0.1049	0.5	0.0	0.1047	0.5

*This position is only half occupied.

$Er_{26}Ge_{23-x}$

Dopov. Akad. Nauk Ukr. RSR, Ser. A: Fiz-Mat. Tekh. Nauki , No.8, 78-82 (1989).

P4/nmm, 14.576, 10.228, Z = 2, R = 0.036. The structure is similar that of $Ce_{26}Li_5Ge_{23+x}$ (54A, 10).

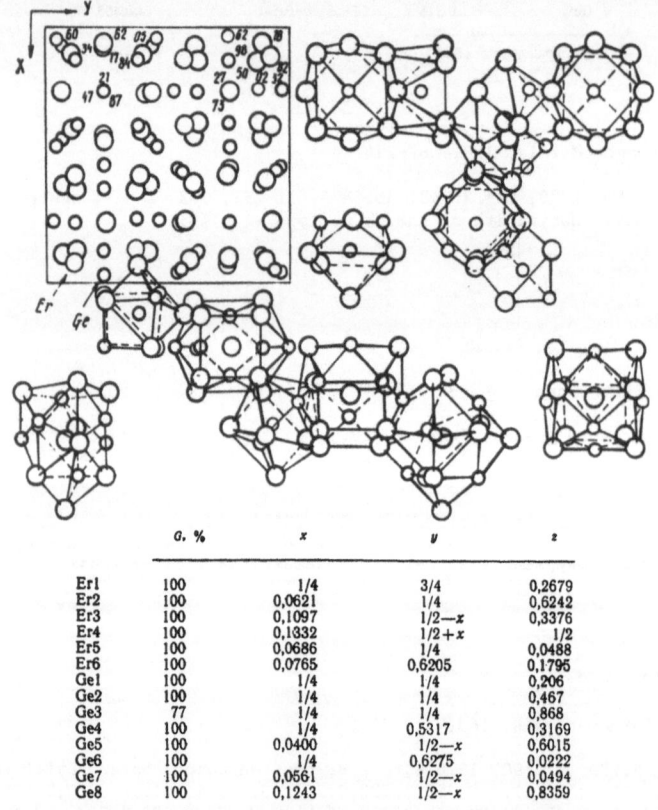

	a. %	x	y	z
Er1	100	1/4	3/4	0,2679
Er2	100	0,0621	1/4	0,6242
Er3	100	0,1097	1/2—x	0,3376
Er4	100	0,1332	1/2+x	1/2
Er5	100	0,0686	1/4	0,0488
Er6	100	0,0765	0,6205	0,1795
Ge1	100	1/4	1/4	0,206
Ge2	100	1/4	1/4	0,467
Ge3	77	1/4	1/4	0,868
Ge4	100	1/4	0,5317	0,3169
Ge5	100	0,0400	1/2—x	0,6015
Ge6	100	1/4	0,6275	0,0222
Ge7	100	0,0561	1/2—x	0,0494
Ge8	100	0,1243	1/2—x	0,8359

Er_3Ru_2

Z. Kristallogr., 192, 249-254.

$P6_3/m$, 7.875, 3.931, Z = 2, R = 0.060. Er in 6(h): 0.3847,0.0915,1/4; Ru(1) in 2(c): 1/3,2/3,1/4; Ru(2) in 2(b): 0,0,0. Columns along c of $Ru(1)Er_6$ trigonal prisms and of $Ru(2)Er_6$ octahedra; a short Ru(2)-Ru(2) distance of 1.97 A suggests incommensurate Ru chains with Ru deficiency.

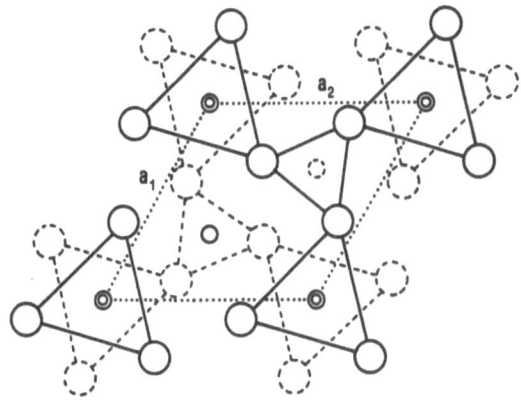

Er₅Sb₃

Acta Cryst., C46, 2456-2457.

Pnma, 11.662, 9.136, 8.007, Z = 4, R = 0.040. Isostructural with β-Yb₅S₃ (37A, 8). Sb atoms have 9- (tricapped trigonal prismatic) and 8-coordinations (dodecahedral) to Er.

		x	y	z
Er1	8(d)	0·0666	0·0575	0·1906
Er2	4(c)	0·0043	0·25	0·5255
Er3	4(c)	0·2280	0·25	0·8258
Er4	4(c)	0·2914	0·25	0·3444
Sb1	8(d)	0·3261	0·0094	0·0646
Sb2	4(c)	0·4767	0·25	0·5948

Eu₃Ga₂, EuGa, Eu₂In, EuIn, EuIn₄

Z. Kristallogr., 190, 295-304.

	Pearson code	Space group	a (Å)	b (Å)	c (Å)	
Eu₃Ga₂	mC20	C2/c	8.680	8.491	8.658	$\beta = 108.59$ °
EuGa	aP8	P1̄	5.878	6.293	6.546	$\alpha = 68.00$
						$\beta = 78.96$
						$\gamma = 84.51$
Eu₂In	oP12 (Co₂Si-type)	Pnma	7.445	5.573	10.306	
EuIn	tP4 (AuCu-type)	P4/mmm	5.351		4.288	
EuIn₄	mC20	C2/m	11.928	5.099	9.865	$\beta = 114.56$

	x	y	z
Eu$_3$Ga$_2$ (*C2/c*)			
Eu1 8(f)	0.2948	0.0398	0.0296
Eu2 4(e)	0	0.1745	1/4
Ga 8(f)	0.4067	0.2886	0.3403
EuGa (*P$\bar{1}$*). All atoms are in the equipoint 2(i)			
Eu1	0.2360	0.1484	0.0772
Eu2	0.6738	0.3014	0.3494
Ga1	0.1354	0.2440	0.5683
Ga2	0.2098	0.6086	0.1951
Eu$_2$In (*Pnma*). All atoms are in the equipoint 4(c)			
Eu1	0.0299	1/4	0.7021
Eu2	0.1788	1/4	0.0704
In	0.2239	1/4	0.3936
EuIn$_4$ (*C2/m*). All atoms are in the equipoint 4(i)			
Eu	0.1651	0	0.7068
In1	0.1536	0	0.3466
In2	0.1546	0	0.0513
In3	0.4227	0	0.5758
In4	0.4382	0	0.1034

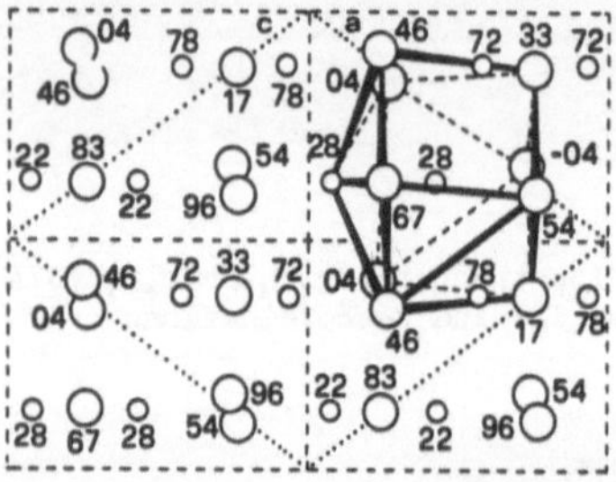

Eu$_3$Ga$_2$

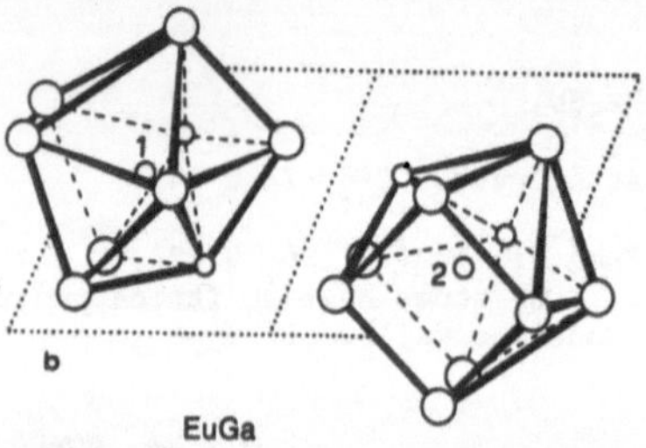

EuGa

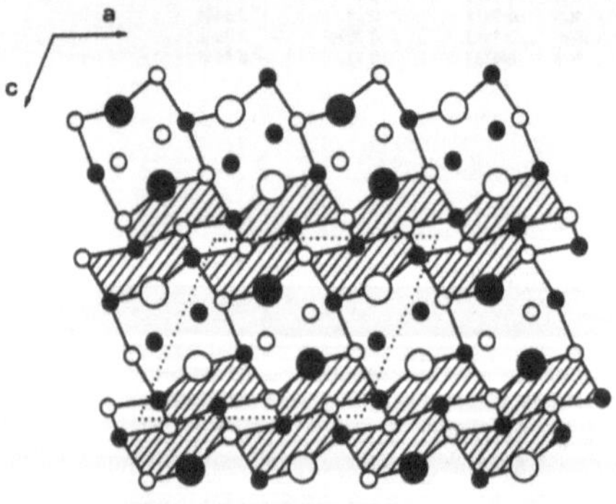

EuIn$_4$

Fe$_{17}$Nd$_5$

J. Less-Common Metals, 163, 245-251.

P6$_3$/mcm, 20.214, 12.329, Z = 12, R = 0.08. Two layers of triangles, pentagons, hexagons, and heptagons, the stacking of which forms antiprisms.

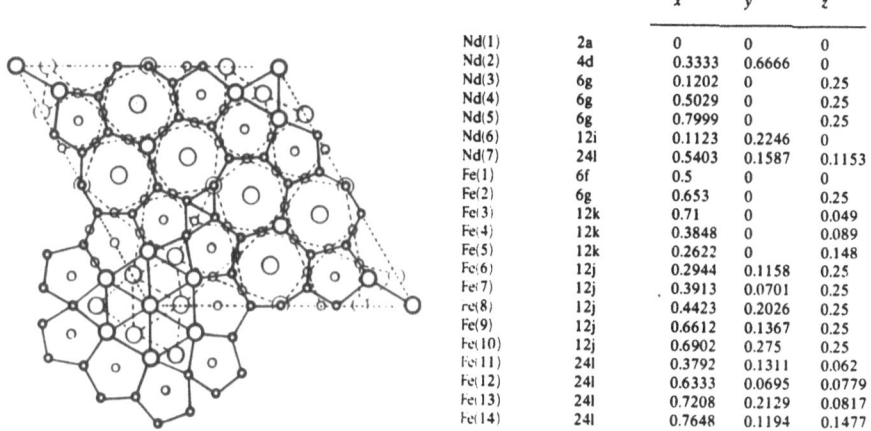

		x	y	z
Nd(1)	2a	0	0	0
Nd(2)	4d	0.3333	0.6666	0
Nd(3)	6g	0.1202	0	0.25
Nd(4)	6g	0.5029	0	0.25
Nd(5)	6g	0.7999	0	0.25
Nd(6)	12i	0.1123	0.2246	0
Nd(7)	24l	0.5403	0.1587	0.1153
Fe(1)	6f	0.5	0	0
Fe(2)	6g	0.653	0	0.25
Fe(3)	12k	0.71	0	0.049
Fe(4)	12k	0.3848	0	0.089
Fe(5)	12k	0.2622	0	0.148
Fe(6)	12j	0.2944	0.1158	0.25
Fe(7)	12j	0.3913	0.0701	0.25
Fe(8)	12j	0.4423	0.2026	0.25
Fe(9)	12j	0.6612	0.1367	0.25
Fe(10)	12j	0.6902	0.275	0.25
Fe(11)	24l	0.3792	0.1311	0.062
Fe(12)	24l	0.6333	0.0695	0.0779
Fe(13)	24l	0.7208	0.2129	0.0817
Fe(14)	24l	0.7648	0.1194	0.1477

Ga_8Yb_3 (I), $Ga_{2.64}Yb$ (II)

J. Less-Common Metals, <u>163</u>, 331-338.

I, Immm, 4.225, 4.340, 25.665, Z = 2, powder data, atomic positional parameters not determined. Eu_3Ga_8-type (<u>52</u>A, 43).

II, P6/mmm, 13.025, 8.360, Z = 15, R = 0.080. AlB_2-type superstructure (3a x 3b x 2c).

		x	y	z	*Occupation* (%)
Yb(1)	6(k)	0.3367	0.0	0.5	100
Yb(2)	6(j)	0.3480	0.0	0.0	100
Yb(3)	2(d)	1/3	2/3	0.5	100
Yb(4)	1(a)	0.0	0.0	0.0	86
Ga(1)	12(o)	0.1118	2x	0.201	100
Ga(2)	12(o)	0.223	2x	0.244	62
Ga(3)	12(o)	0.445	2x	0.265	100
Ga(4)	6(m)	0.067	2x	0.5	50
Ga(5)	6(l)	0.397	2x	0.0	84

Ge_4Hf_5

J. Less-Common Metals, <u>162</u>, L27-L29.

Pnma, 7.017, 13.434, 7.105, Z = 4, R = 0.050. Ge_4Sm_5-type (<u>32</u>A, 87).

		x	y	z
Hf(1)	8(d)	0.0141	0.0946	0.3229
Hf(2)	8(d)	0.1457	0.6239	0.1671
Ge(1)	8(d)	0.3039	0.5407	0.4648
Ge(2)	4(c)	0.0534	1/4	0.6110
Hf(3)	4(c)	0.1811	1/4	0.0021
Ge(3)	4(c)	0.3212	1/4	0.3540

GeLi

Angew. Chem., 99, 69-71 (1987) [Angew. Chem. Int. Edn. Engl., 26, 76-78 (1987)].

$I4_1/amd$, 4.0529, 23.282, Z = 12, powder data, high-pressure phase. Li(1) in 4(a); Li(2) in 8(e): z = 0.713; Ge(1) in 4(b); Ge(2) in 8(e): z = 0.5486. As in the normal-pressure phase (41A, 115) Ge atoms have 8 Li neighbours and vice versa. [See also LiSn (this volume, p.11).]

Ge_7Nb_{10} 2 x $Ge_3Ge_{0.5}Nb_5$

J. Solid State Chem., 84, 386-400.

$P\bar{3}m1$, 13.37, 5.37, Z = 3, electron microscopic and diffraction data. Superstructure ($\sqrt{3}a$ x c) of the Ge_3Nb_5 (D8$_8$) type (20, 109), with insertion of additional Ge at ordered octahedral sites. [Cf. 37A, 91.]

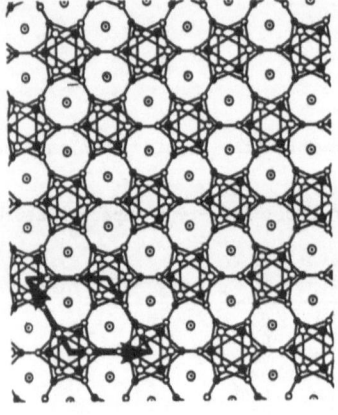

		x	y	z
Nb(1) in	6i	0.9167	0.0833	0.25
Nb(2)	6i	0.5833	0.4167	0.25
Nb(3)	6i	0.5	0.5	0.25
Nb(4)	6h	0.3333	0	¼
Nb(5)	6g	0.3333	0	0
Ge(1)	6i	0.7950	0.2050	0.25
Ge(2)	6i	0.4167	0.5833	0.75
Ge(3)	6i	0.1283	0.8717	0.75
Ge(4)	2d	⅓	⅔	0.75
Ge(5)	1a	0	0	0

Structure model of Nb₅Ge₃ with the D8₈-structure, projected onto the (00·1)-plane. The small and large circles denote Ge- and Nb-atoms, respectively. Black and white circles indicate atoms at z = 0 and ½, respectively. Circles with dots are Nb-atoms at z = ±¼ forming the Nb-chains along the [00·1]-direction. The parallelogram indicates the unit cell. The octahedral sites are at the corners of the parallelogram.

Hg_3Na_8

Z. anorg. Chem., 587, 103-109.

$R\bar{3}c$, a = 9.228, c = 52.638 A, Z = 12, R = 0.061. Isolated Hg atoms (Hg-Hg = 5.03-5.64 A) arranged as in close-packing, with Na in all the octahedral and 5/6 of the tetrahedral holes; Na-Hg = 3.15-4.76 A.

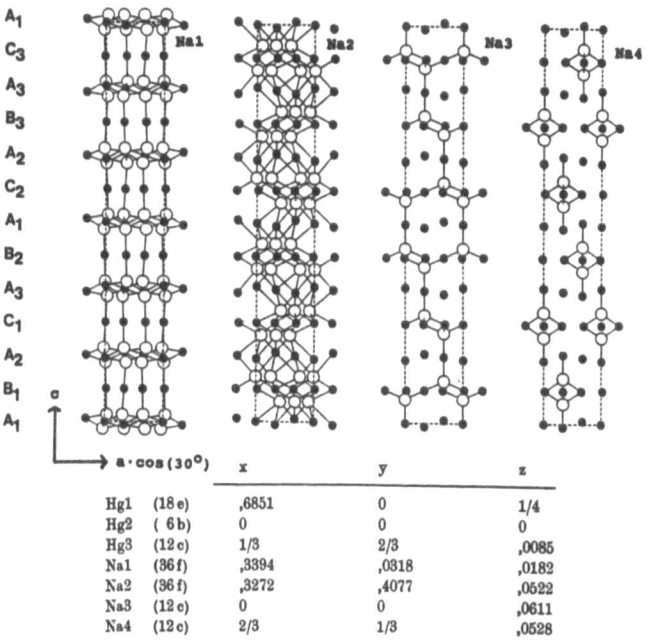

		x	y	z
Hg1	(18 e)	,6851	0	1/4
Hg2	(6 b)	0	0	0
Hg3	(12 c)	1/3	2/3	,0085
Na1	(36 f)	,3394	,0318	,0182
Na2	(36 f)	,3272	,4077	,0522
Na3	(12 c)	0	0	,0611
Na4	(12 c)	2/3	1/3	,0528

$Ir_{0.46}V_{0.54}$

J. Less-Common Metals, <u>159</u>, 343-347.

Above 779 K, P4/mmm, 3.651, 2.770, at 829K, Z=2, powder data. (0.9Ir + 0.1V) in 1(a): 0,0,0; (0.97V + 0.03Ir) in 1(b): 1/2,1/2,1/2. AuCu-type (<u>1</u>, 484).

Below 779K, Cmmm, 5.797, 6.762, 2.805 ($2a_T$ x c_T x a_T, where a_T, c_T correspond to tetragonal phase), at 293K, Z = 8, powder data. Ir in 4(j): 0,0.2178,1/2; V in 4(g): 0.2871,0,0. Lower-symmetry variant of the tetragonal phase, as previously described (<u>30</u>A, 64).

β-LiSn

Z. Kristallogr., <u>193</u>, 317-318.

$I4_1/amd$, 4.387, 25.511, Z = 12, R = 0.069. Sn(1) in 4(a): 1/2,3/4,3/8; Sn(2) in 8(e): 1/2,1/4,0.4533; Li(1) in 8(e): 1/2,1/4,0.293; Li(2) in 4(b): 0,1/4,3/8. Isostructural with high-pressure LiGe (this volume, p. 10).

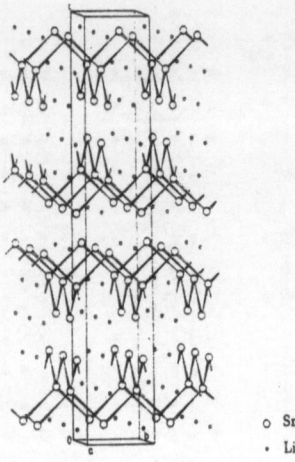

o Sn
· Li

Ni$_2$Y$_{0.95}$

J. Less-Common Metals, <u>161</u>, L27-L31.

F$\bar{4}$3m, 14.350, Z = 64, powder data.   Superstructure of C15-type, with doubled
cell parameter (compare <u>55A</u>, 66).

		x	y	z
Y1	4a	0	0	0
Y2	4b	1/2	1/2	1/2
Y3	16e	0.1000	0.1000	0.1000
Y4	16e	0.6280	0.6280	0.6280
Y5	24g	0.0099	1/4	1/4
Ni1	16e	0.3109	0.3109	0.3109
Ni2	16e	0.8141	0.8141	0.8141
Ni3	48h	0.0669	0.0669	0.8062
Ni4	48h	0.0634	0.0634	0.3133

Y(1) has 24% occupancy

Rh$_2$Sb

Z. Naturforsch., <u>45</u>B, 947-951.

Pnma, 5.721, 4.171, 7.928, Z = 4, R = 0.066. Rh(1), Rh(2), Sc each in 4(c):
x,1/4,z, x = 0.8384, 0.9591, 0.2861, z = 0.0677, 0.7032, 0.1079.
Anti-PbCl$_2$-type (<u>2</u>, 16), as for Rh$_2$Si (<u>23</u>, 57, 60; <u>28</u>, 40).

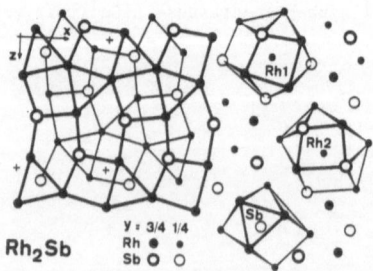

Rh$_2$Sb

y = 3/4 1/4
Rh ●   ·
Sb O   o

$Ag_{0.5}CaGa_{3.5}$ and related compounds.

J. Less-Common Metals, 166, 115-124.

Cr_2ThSi_2-type structures (43A, 99; 55A, 31), I4/mmm, cell parameters in Table below, Z = 2, R = 0.027-0.059. Ca/Sr/Ba in 2(a): 0,0.0; Ga in 4(e): 0,0,z, z = 0.3880, 0.3842, 0.3772 (for Ca, Sr, Ba compounds); Ga/Ag in 4(d): 0,1/2,1/4.

Al_2CaZn_2-type structures (51A, 11), I4/mmm, cell parameters in Table below, Z = 2, R = 0.012-0.041. Ca/Sr/Ba/Ln in 2(a): 0,0,0; Ag/Al or Cu/Al or Au/Ga in 4(e): 0,0,z, z = 0.3797-0.3910; Al or Ga in 4(d): 0,1/2,1/4.

	a (pm)	c (pm)		a (pm)	c (pm)
$CaAg_{0.5}Ga_{3.5}$*	427.8	1102.5	$SrCu_{0.6}Al_{3.4}$⁺	431.0	1132.2
$SrAg_{0.5}Ga_{3.5}$*	440.5	1091.4	$BaCu_{0.2}Al_{3.8}$⁺	451.1	1135.7
$BaAg_{0.4}Ga_{3.6}$*	461.7	1075.5	$CaAu_{0.8}Ga_{3.2}$⁺	418.3	1138.4
$CaAg_{0.7}Al_{3.3}$⁺	419.3	1179.4	$LaCu_{0.5}Al_{3.5}$⁺	430.9	1081.7
$SrAg_{0.8}Al_{3.2}$⁺	434.3	1161.0	$CeAg_{0.6}Al_{3.4}$⁺	432.2	1105.2
$BaAg_{0.3}Al_{3.7}$⁺	452.8	1145.3	$PrAg_{0.7}Al_{3.3}$⁺	430.3	1106.7

$ThCr_2Si_2$(*) and $CaZn_2Al_2$(+) structures

AgCaSb; CuSbYb, AuSbYb; BiCaCu, AgBiSr, BaCuSb

J. Less-Common Metals, 166, 319-327.

AgCaSb, Pnma, 7.708, 4.590, 8.393, Z = 4, R = 0.046. Ag, Ca, Sb in 4(c): x,1/4,z, x = 0.1440, 0.0173, 0.2592, z = 0.0661, 0.6967, 0.3887. NiTiSi-type (30A, 75).

CuSbYb, AuSbYb, $P6_3mc$, a = 4.452, 4.635, c = 7.995. 7.765 A, Z = 2, R = 0.044, 0.055. Cu or Au in 2(b): 1/3,2/3,z, z = 0.0089, 0.044; Sb in 2(b): z = 0.4921, 0.480; Yb in 2(a): 0,0,1/4. GaGeLi-type (35A, 61; 41A, 65).

BiCaCu, AgBiSr, BaCuSb, $P6_3/mmc$, a = 4.559, 4.876, 4.614, c = 8.053, 8.480, 9.412 A, Z = 2, R = 0.039, 0.029, 0.028. Ca/Sr/Ba in 2(a); 0,0,0; Cu/Ag/Cu in 2(c): 1/3,2/3,1/4; Bi/Bi/Sb in 2(d): 1/3,2/3,3/4. BeZrSi-type (17, 50).

Al_8Cr_4Er, Al_8LnMn_4 (Ln = Pr, Nd, Tb, Dy, Ho)

J. Less-Common Metals, 166, 329-334.

I4/mmm, a ∿ 8.9, c ∿ 5.1 A, Z = 2, neutron powder data. $Mn_{12}Th$-type (16, 113), with Cr or Mn mainly in the 8(f) site (42A, 5; 44A, 6).

Al_8Cr_4U, Al_8Mn_4U

Solid State Comm., 75, 929-933.

I4/mmm, a = 8.908-8.891, c = 5.107-5.097 A, for Cr compound at 300-4.2K; a = 8.845-8.826, c = 5.096 - 5.064 A, for Mn compound at 300-1.45K, Z = 2, neutron powder data. U in 2(a): 0,0,0; Cr/Mn in 8(f): 1/4,1/4,1/4; Al(1) in 8(i):

x,0,0, x = 0.3444-0.3435, 0.345-0.344; Aℓ(2) in 8(j): x,1/2,0, x = 0.2816-0.2840, 0.282-0.285. $Mn_{12}Th$-type (16, 113), as for related materials (51A, 11; 52A, 93; 56A, 12).

$AℓCs_6Sb_3$

Z. Kristallogr., 193, 283-284.

$P2_1/m$, 10.845, 6.507, 12.707, 100.95, Z = 2, R = 0.031. Trigonal planar $AℓSb_3{}^{6-}$ ions linked by Cs^+ ions; Aℓ-Sb = 2.651 A.

$Aℓ_2CuLi$

J. Solid State Chem., 85, 293-298.

$P6/mmm$, 4.954, 9.327, Z = 3.33, R = 0.087. Four layers along c̲: Aℓ, Aℓ/Cu, Li, Aℓ/Cu (Li sites are not well determined).

	x	y	z	s.o.f.
Cu(1) 6i	⅓	0	0.2363	0.556
Aℓ(1) 6i	⅓	0	0.2363	0.444
Aℓ(2) 2c	⅓	⅔	0	1.0
Aℓ(3) 2e	0	0	0.3568	1.0
Li(1) 2d	⅓	⅔	1/2	1.0
Li(2) 2e	0	0	0.051	0.666

Central atom	Coordinating atom	Distance (Å)
Aℓ/Cu(1)	2 × Aℓ/Cu(1)	2.477(2)
	2 × Aℓ(2)	2.627(2)
	2 × Aℓ(3)	2.720(3)
	2 × Li(1)	2.845(2)
	2 × Li(2)	3.02(2)
Aℓ(2)	6 × Aℓ/Cu(1)	2.627(2)
	3 × Aℓ(2)	2.860(2)
	3 × 1.3 × Li(2)	2.90(1)
Aℓ(3)	1 × Aℓ(3)	2.671(3)
	6 × Aℓ/Cu(1)	2.720(3)
	1 × Li(2)	2.85(8)
	6 × Li(1)	3.157(2)
Li(1)	6 × Aℓ/Cu(1)	2.845(2)
	3 × Li(1)	2.860(2)
	6 × Aℓ(3)	3.157(2)
Li(2)	1 × Li(2)	0.95(12)
	1 × Aℓ(3)	2.85(8)
	6 × Aℓ(2)	2.90(1)
	6 × Aℓ/Cu(1)	3.02(2)

$Aℓ_7Cu_{16}Lu_6$

Dopov. Akad. Nauk Ukr. SSR, Ser B: Geol., Khim. Biol. Nauki, No. 6, 60-62.

Fm3m, 12.198, Z = 4, powder data. 24Lu in 24(e): 0.207,0,0; 28Cu+4Aℓ(1) in 32(f): x,x,x, x = 0.163; 28Cu+Aℓ(2) in 32(f): x = 0.387; 4Cu+20Aℓ(3) in 24(d): 0,1/4,1/4; 4Cu in 4(b): 1/2,1/2,1/2. $Mn_{23}Th_6$-type (16, 113), with partial Cu/Aℓ ordering.

$Aℓ_{0.33}Ge_{1.67}Hf$

J. Less-Common Metals, 162, L39-L43.

Cmcm, 3.805, 14.780, 3.776, Z = 4, R = 0.055. 4 Hf, 4 Ge, (1.33Aℓ + 2.67Ge) each in 4(c): 0,y,¼, y = 0.39838, 0.0627, 0.7498. Ge_2Hf ($ZrSi_2$) type (21, 124), with Aℓ in only one Ge site.

AlGeY

Acta Cryst., C46, 2276-2279.

Cmcm, 4.0504, 10.440, 5.7646, Z = 4, R = 0.068. Al in 4(a); 0,0,0; Ge in 4(c):
0,y,1/4, y = 0.6058; Y in 4(c): y = 0.3099. Substitution variant of the
Pt$_2$U-type (23, 210; the structure conforms to Cmcm, 4.12, 9.68, 5,60, Z = 4;
Pt(1) in 4(a); Pt(2) in 4(c): y = 0.62; U in 4(c): y = 0.33). The structure
contains sheets of Ge-centred Al/Y trigonal prisms. AlGeLn compounds are
isostructural when Ln = small lanthanon; for large lanthanons, the tetragonal
α-ThSi$_2$-type (or an ordered variant) is found.

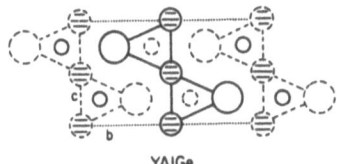

YAlGe

HEIGHT	Y	Al	Ge
0			
1/2			

The [100] projection of the YAlGe structure.

Interatomic distances up to 3·5 Å in YAlGe;
e.s.d.'s are given in parentheses

Y-2Ge	2·939 (3)	Ge-4Al	2·720 (1)
2Ge	3·014 (1)	2Y	2·939 (3)
Ge	3·090 (4)	2Y	3·014 (1)
4Al	3·181 (2)	Y	3·090 (4)
Al-4Ge	2·720 (1)		
2Al	2·8823 (5)		
4Y	3·181 (2)		

AlMg$_3$Pt$_2$

Acta Cryst., C46, 2454-2455.

Fd3m, 11.802, Z = 16, R = 0.017. Ordered ternary variant of the NiTi$_2$-type
(23, 195; 28, 20).

	x	y	z	Interatomic distances (Å)			
Al				Pt—Al	2·599 (1) (3×)	Al—Pt	2·599 (1) (6×)
Mg	0·9288			Pt—Mg	2·714 (1) (2×)	Al—Mg	2·967 (3) (6×)
Pt	0·71438	0·71438	0·71438	Pt—Mg	2·938 (3) (3×)	Mg—Pt	2·714 (1) (2×)
				Pt—Pt	2·984 (1) (3×)	Mg—Pt	2·938 (3) (2×)
						Mg—Al	2·967 (3) (2×)

ω-Al$_{2.25}$Nb$_{0.75}$Ti$_3$

Acta Cryst., C46, 374-377.

P$\bar{3}$m1, 4.5554, 5.5415, Z = 1, R = 0.032 (twinned crystal). Distortion of CsCl
(B2) type.

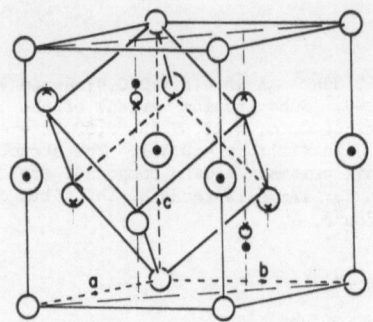

Trigonal cell for ω-Ti$_3$Al$_{2.25}$Nb$_{0.75}$. Large open circles: mainly Nb; intermediate open circles: mainly Ti; small open circles: mainly Al (see Table). The cube outlines the framework of the A atoms in the parent $B2$ structure; the filled circles correspond to the B atoms of that structure. The crosses indicate the positions of the atoms in the Ni$_2$In structure at $z = \frac{1}{4}$ and $\frac{3}{4}$.

Interatomic distances (Å) *in*
ω-Ti$_3$Al$_{2.25}$Nb$_{0.75}$

Distances in the parent $B2$ phase are $A\!-\!A$ (or $B\!-\!B$) = 3·203 Å and $A\!-\!B$ = 2·774 Å.

	N		Distance	Distance in $B2$
Ti(1)	2	Nb	2·771 (1)	8 $A\!-\!B$
	6	Al	2·909 (1)	
	6	Ti(2)	3·061 (1)	6 $A\!-\!A$
Nb	2	Ti(1)	2·771 (1)	8 $B\!-\!A$
	6	Ti(2)	2·893 (1)	
	6	Al	3·041 (2)	6 $B\!-\!B$
Al	3	Ti(2)	2·650 (1)	
	1	Ti(2)	2·732 (3)	
	1	Ti(2)	2·810 (3)	8 $B\!-\!A$
	3	Ti(1)	2·909 (1)	
	3	Nb	3·041 (2)	6 $B\!-\!B$
	3	Al	3·620 (3)	
Ti(2)	3	Al	2·650 (1)	
	1	Al	2·732 (3)	
	1	Al	2·810 (3)	8 $A\!-\!B$
	3	Nb	2·893 (1)	
	3	Ti(1)	3·061 (1)	6 $A\!-\!A$
	3	Ti(2)	3·567 (2)	

	Site	x	y	z	No. of atoms		
					Ti	Al	Nb
Ti(1)	1(a)	0	0	0	0·88		0·12
Nb	1(b)	0	0	$\frac{1}{2}$		0·32	0·68
Al	2(d)	$\frac{1}{3}$	$\frac{2}{3}$	0·22449	0·06	1·94	
Ti(2)	2(d)	$\frac{1}{3}$	$\frac{2}{3}$	0·71747	2·00		

Au$_3$LiNa$_2$

Z. Naturforsch., 45B, 1333-1334.

P6$_3$/mmc, 5.5302, 8.674, Z = 2, R = 0.045. Au in 6(h): x,2x,1/4, x = 0.16805; Li in 2(a): 0,0,0; Na in 4(f): 2/3,1/3,0.069. Ternary variant of the MgZn$_2$-type (1, 180). Au-Li = 2.701, Au-Au = 2.742, 2.788, Au-Na = 3.180, 3.188, Li-Na = 3.248, Na-Na = 3.14, 3.41 A.

CeCu$_{1.5}$In$_{1.5}$ (I), Ce$_{0.78}$Cu$_{0.64}$(Cu$_{0.68}$In$_{0.32}$)$_{12}$ (II)

Izv. Akad. Nauk SSSR, Neorg. Mater., 26, 2316-2318.

I, P4/nmm, 4.245. 10.550, Z = 2, R = 0.074. 2 Ce in 2(c): 1/4,1/4,z, z = 0.2259; 2 Cu(1) in 2(c): z = 0.6573; 1 Cu(2) in 2(c): z = 0.8824; 1 In(1) in 2(a): 3/4,1/4,0; 2 In(2) in 2(b): 3/4,1/4,1/2. CaBe$_2$Ge$_2$-type (43A, 28), with partial occupancy of some sites.

II, Fm3c, 12.482, Z = 8, powder data. 6 Ce in 8(a): 1/4,1/4,1/4; 5 Cu in 8(b): 0,0,0; 96 Cu/In in 96(i): 0,0.1805,0.1223. NaZn$_{13}$-type (6, 8, 157; 16, 139).

$Ce_3Ge_4Rh_4$, $Ce_3Ge_4IrRh_3$

J. Solid State Chem., $\underline{88}$, 429-434.

Immm, a = 4.0915, 4.0839, b = 4.2400, 4.2437, c = 25.0673, 25.0403 A, Z = 2, R = 0.085, 0.069. $U_3Ni_4Si_4$-type ($\underline{45A}$, 96).

		x	y	z	Occ.
					$Ce_3Rh_4Ge_4$
Ce1	2a	0	0	0	1.0
Ce2	4j	½	0	0.3545	1.0
Rh1	4j	½	0	0.0992	0.98
Rh2	4i	0	0	0.2506	1.0
Ge1	4j	½	0	0.1969	1.0
Ge2	4i	0	0	0.4501	1.0
					$Ce_3Rh_3IrGe_4$
Ce1	2a	0	0	0	1.0
Ce2	4j	½	0	0.3540	1.0
Rh1 } Ir1 }	4j	½	0	0.0984	{ 0.71 { 0.29
Rh2 } Ir2 }	4i	0	0	0.2504	{ 0.74 { 0.26
Ge1	4j	½	0	0.1967	1.0
Ge2	4i	0	0	0.4499	1.0

$Ce_{1.22}Ge_4Sc_3$ (I), $Ce_{3.66}Ge_4Y_{0.86}$ (II)

Izv. Akad. Nauk SSSR, Neorg. Mater., $\underline{26}$, 969-972.

Pnma, a = 7.188, 7.754, b = 13.99, 14.99, c = 7.416, 7.918 A, Z = 4, R = 0.064, 0.047. Defect derivative of the $Ce_2Sc_3Si_4$-type ($\underline{45A}$, 52).

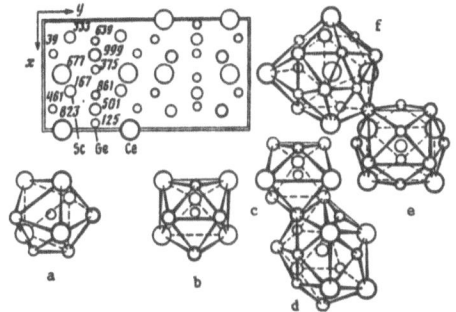

Projection of the $Ce_{1.22}Sc_3Ge_4$ structure on the XY plane and coordination polyhedra of [Ge3GeCe$_4$-Sc$_4$] (a), [Ge1GeCe$_4$Sc$_4$] (b), [Ge2GeCe$_2$Sc$_6$] (c), [Sc2-Ge$_6$Ce$_5$Sc$_5$] (d), [Sc1Ge$_6$Ce$_4$Sc$_4$] (e), and [CeGe$_7$Ce$_4$Sc$_6$] (f).

	a, %	x	y	z
Ce	61,3	0,0039	0,5949	0,1767
Sc (1)	100	0,183	3/4	0,499
Sc (2)	100	0,157	0,3753	0,333
Ge (1)	100	0,310	3/4	0,139
Ge (2)	100	0,4556	1/4	0,3754
Ge (3)	100	0,3153	0,4593	0,0395

	G, %	x	y	z
Ce (1)	100	0,0107	0,5986	0,1808
(Y+Ce)	86 (1)+14 (1)	0,1966	3/4	0,4997
Ce (2)	76,4 (1)	0,1384	0,3822	0,3361
Ge (1)	100	0,3117	3/4	0,1317
Ge (2)	100	0,4352	1/4	0,3815
Ge (3)	100	0,2980	0,4564	0,0377

G is the population factor.

$CeLiSn_2$

Izv. Akad. Nauk SSSR, Neorg. Mater., 25, 1145-1148 (1989).

Cmcm, 4.445, 18.068, 4.524, Z = 4, powder data. Ce, Li, Sn(1), Sn(2) in 4(c): 0,y,1/4, y = 0.099, 0.316, 0.459, 0.754. $CeNiSi_2$-type (34A, 63; 35A, 96).

$Ce_4Pt_{12}Sn_{25}$

Mater. Res. Bull., 25, 807-814.

Im3, 12.281, Z = 2, R = 0.036. Ce in 8(c): 1/4,1/4,1/4; Pt in 24(g): 0,y,z, y = 0.3183, z = 0.1671; Sn(1) in 2(a): 0,0,0; Sn(2) in 24(g): y = 0.3793, z = 0.3738; Sn(3) in 24(g): y = 0.1230, z = 0.2505. New structure type related to that of $Rh_4Sn_{13}Yb_3$ (46A, 115; 53A, 16), with Ce-centered Sn cuboctahedra, Sn-centered Sn octahedra, and Pt-centered Sn trigonal prisms. High thermal parameters for Sn(3) may indicate disorder of this site.

$Co_6Er_{6-x}Ge_4$ (x = 3.94), $Er_{6-x}Fe_6Ge_4$ (x = 4.17).

Dopov. Akad. Nauk Ukr.SSR, Ser B: Geol., Khim. Biol. Nauki, No. 2, 30-34.

$P6_3/mcm$, a = 5.091, 5.085, c = 7.861, 7.851 A, Z = 1, R = 0.036, 0.052. 6Co or Fe in 6(g): x,0,1/4, x = 0.5187, 0.519; 4Ge in 4(d): 2/3,1/3,0; 1Er(1) in 2(b): 0,0,0; 1Er(2) in 4(e): 0,0,z, z = 0.1533, 0.155.

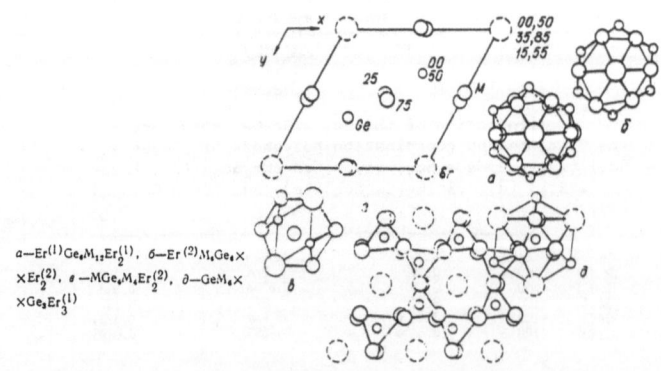

$a—Er^{(1)}Ge_6M_{12}Er_2^{(1)}$, $б—Er^{(2)}M_4Ge_6 \times$
$\times Er_2^{(2)}$, $в—MGe_6M_4Er_2^{(2)}$, $г—GeM_4 \times$
$\times Ge_2Er_3^{(1)}$

$Co_2Ge_4Sm_3$

Izv. Akad. Nauk SSSR, Neorg. Mater., 25, 2023-2026 (1989).

B2/m (c unique), 10.846, 8.1388, 4.1851, γ = 107.70°, Z = 2, powder data. $Co_2Ge_4Tb_3$-type (53A, 11).

	x	y	z
Co	0.702	0.377	0
Ge(1)	0.104	0.407	0
Ge(2)	0.716	0.088	0
Sm(1)	0	0	0
Sm(2)	0.3826	0.3098	0

$Co_{17.6}In_{14}Lu_6$

Kristallografija, 35, 493-494 [Soviet Physics - Crystallography, 35, 286].

Pm3, 8.652, Z = 1, R = 0.047. New structure type, with some partially-occupied Co sites. Coordination numbers are: Co = 10-14, In = 12,13, Lu = 15.

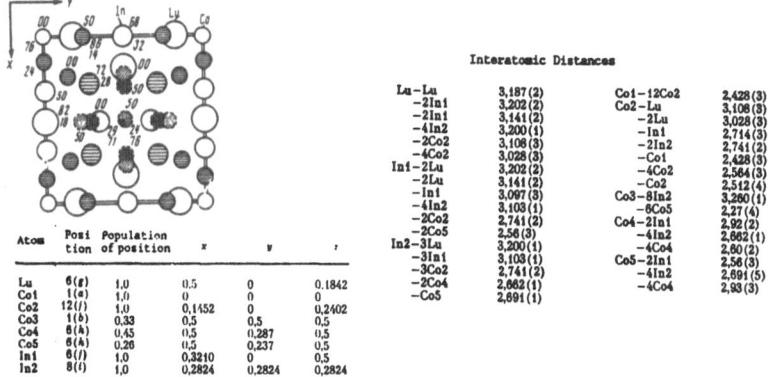

Interatomic Distances

Lu–Lu	3,187(2)	Co1–12Co2	2,428(3)
–2In1	3,202(2)	Co2–Lu	3,108(8)
–2In1	3,141(2)	–2Lu	3,028(3)
–4In2	3,200(1)	–In1	2,714(3)
–2Co2	3,108(3)	–2In2	2,741(2)
–4Co2	3,028(3)	–Co1	2,428(3)
In1–2Lu	3,202(2)	–4Co2	2,564(3)
–2Lu	3,141(2)	–Co2	2,512(4)
–In1	3,097(3)	Co3–8In2	3,260(1)
–4In2	3,103(1)	–6Co5	2,27(4)
–2Co2	2,741(2)	Co4–2In1	2,92(2)
–2Co5	2,56(3)	–4In2	2,662(1)
In2–3Lu	3,200(1)	–4Co4	2,60(2)
–3In1	3,103(1)	Co5–2In1	2,56(3)
–3Co2	2,741(2)	–4In2	2,691(5)
–2Co4	2,662(1)	–4Co4	2,93(3)
–Co5	2,691(1)		

Atom	Position of position	Population	x	y	z
Lu	6(g)	1,0	0,5	0	0,1842
Co1	1(a)	1,0	0	0	0
Co2	12(j)	1,0	0,1452	0	0,2402
Co3	1(b)	0,33	0,5	0,5	0,5
Co4	6(k)	0,45	0,5	0,287	0,5
Co5	6(k)	0,26	0,5	0,237	0,5
In1	6(f)	1,0	0,3210	0	0,5
In2	8(i)	1,0	0,2824	0,2824	0,2824

Cu_4InYb

J. Phys. Soc. Japan, 59, 792-795.

F$\bar{4}$3m, 7.1575, Z = 4, neutron powder data. Cu in 16(e): x,x,x, x = 0.6254; In in 4(a): 0,0,0; Yb in 4(c): 1/4,1/4,1/4. $AuBe_5$-type (3, 330; 22, 48), with Yb ordered in 4(c).

ErGeLi

Z. anorg. Chem., 580, 45-49.

P$\bar{6}$2m, 6.965, 4.022, Z = 3, R = 0.031. Er in 3(g): 0.5750,0,1/2; Ge(1) in 1(a): 0,0,0; Ge(2) in 2(d): 1/3,2/3,1/2; Li in 3(f): 0.234,0,0. Fe_2P-type (2, 15; 45A, 86).

$Fe_{10}LnV_2$ (Ln = Nd, Tb, Dy, Ho, Er, Y)

J. Less-Common Metals, <u>162</u>, 285-295.

I4/mmm, a ∿ 8.5, c ∿ 4.7, Z = 2, neutron powder data. Ln in 2(a): 0,0,0;
4Fe+4V in 8(i): x,0,0, x ∿ 0.36; Fe in 8(j): x,1/2,0, x ∿ 0.27; Fe in 8(f):
1/4,1/4,1/4. $Mn_{12}Th$-type (<u>16</u>, 113).

$Ga_6K_{20}Sb_{12.66}$, $Ga_6K_{20}As_{12.66}$

Z. Naturforsch., <u>45</u>B, 277-282.

P6$_3$/m, a = 17.800, 16.780, c = 5.438, 5.245 A, Z = 1 [not 4], R = 0.09, 0.10.
K(1) in 2(c): 1/3,2/3,1/4; 0.66 X(3) in 2(b): 0,0,0; K(2,3,4), Ga, X(1,2) in
6(h): x,y,1/4, x=0.5480, 0.2732, 0.8058, 0.5036, 0.6113, 0.3414, y = 0.9355,
0.8801, 0.9935, 0.2236, 0.1627, 0.1139 for X = Sb; x = 0.5470, 0.2774, 0.8045,
0.5054, 0.6116, 0.3455, y = 0.9368, 0.8850, 0.9907, 0.2248, 0.1656, 0.1179 for
X = As. Planar cyclic $Ga_3X_6{}^{9-}$ anions linked by K$^+$ ions, with an additional X^{3-}
ion (1/3 occupancy).

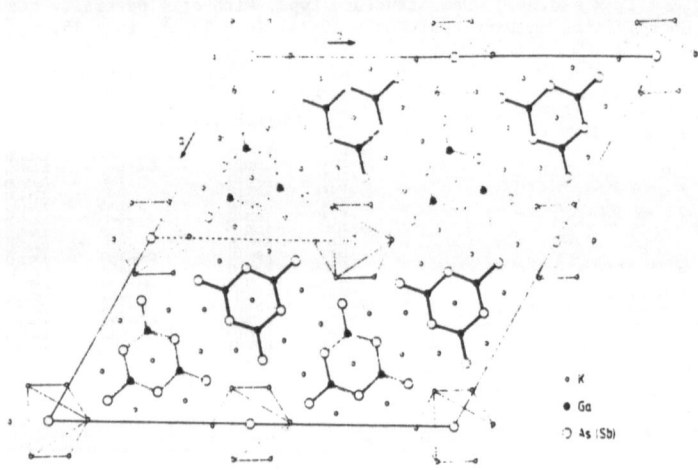

Ga_2LiM (M = Rh, Pt)

J. Less-Common Metals, <u>161</u>, 303-312.

Fm3m, a = 6.007, 6.055 A, Z = 4, R = 0.035 and neutron powder data. Ga in
8(c): 1/4,1/4,1/4; Li in 4(b): 1/2,1/2,1/2; Rh or Pt in 4(a): 0,0,0. Filled
CaF$_2$-type, with Li in M$_6$ octahedra.

Ga₃NiPr

Ga$_3$NiPr

J. Less-Common Metals, <u>162</u>, 361-369, 371-377.

Fmm2, 5.9522, 6.0053, 10.249, Z = 4, powder data. Pr in 4(a): 0,0,0; Ga(1) in 8(b): 1/4,1/4,0.255; Ga/Ni(2) in 4(a): 0,0,0.355; Ga/Ni(3) in 4(a): 0,0,0.591. (Ga/Ni occupancies = 50:50). CeGa$_3$Pt-type (<u>55</u>A, 14).

Ga$_3$RuTm

Kristallografija, <u>34</u>, 1571-1573 (1989) [Soviet Physics - Crystallography, <u>34</u>, 939-940 (1989)].

Pm3m, a = 6.352 A, Z = 3, R = 0.050. Ga(1) in 8(g): x,x,x, x = 0.2113; Ga(2) in 1(b): 1/2,1/2,1/2; Ru in 3(d): 1/2,0,0; Tm in 3(c): 0,1/2,1/2. New structure type.

Interatomic Distances (nm)

Ru—4Ru	0.3176(1)	Ga2—Ga2	0.3175(1)
—2Ga2	0.3176(1)	—3Tm	0.2920(1)
—8Ga1	0.2920(1)	—3Ga1	0.2684(3)
		—3Ru	0.2639(2)
Tm—4Tm	0.3176(1)		
—8Ga1	0.2639(2)	Ga1—6Tm	0.3176(1)
		—8Ga1	0.3175(1)

Ga$_2$SnTa$_5$

Acta Cryst., C<u>46</u>, 1193-1195.

I4/mcm, 10.354, 5.1795, Z = 4, R = 0.053. W$_5$Si$_3$-type structure (<u>19</u>, 277), with Ga$_2$Ta$_8$ columns along <u>c</u>, connected by Sn and Ga atoms; one site is statistically occupied by Ga and Sn.

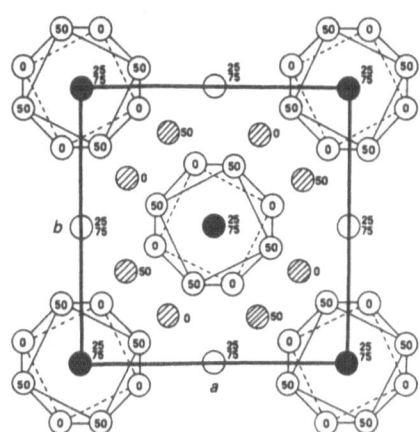

Projection of the crystal structure of Ta$_5$SnGa$_2$ along the c axis. Atom key: open circles Ta, filled circles Ga, shaded circles MX. Numbers correspond to z parameters.

		Occupancy	x	y	z
Ta(1)	16(k)	1·0	0·0734	0·2208	0
Ta(2)	4(b)	1·0	0	½	¼
Ga	4(a)	1·0	0	0	¼
MX	8(h)	0·5Sn, 0·5Ga	0·3322	0·1678	0

Interatomic distances (Å) and angles (°)

Ta(1)—Ta(1')	3·0027 (9)		Ta(1)—Ta(1ⁱᵛ)	3·013 (2)
Ta(1)—Ta(1ⁱⁱⁱ)	3·372 (1)		Ta(1)—Ta(1ⁱⁱⁱ)	3·407 (2)
Ta(1)—Ta(2)	3·257 (1)		Ta(1)—Ga	2·735 (1)
Ta(1)—MX	2·735 (5)		Ta(1)—MXⁱⁱⁱ	2·999 (3)
Ta(1)—MXⁱᵛ	2·751 (5)		Ta(2)—Ta(2ᵛ)	2·5898 (3)
Ta(2)—MXⁱᵛ	2·777 (4)		Ga—Ga'	2·5898 (3)
Ga—MX	4·065 (5)		MX-MXⁱⁱⁱ	3·536 (5)

Ta(1')—Ta(1)—Ta(1ⁱⁱⁱ)	64·32 (3)		Ta(1ⁱⁱⁱ)—Ta(1)—Ta(1ⁱⁱⁱ)	90
Ta(1)—Ta(2)—Ta(1')	54·89 (3)		Ta(1)—Ta(2)—Ta(1ᵛ)	55·10 (3)
MXⁱᵛ—Ta(2)—MXⁱᵛ	102·6 (1)		Ta(1)—Ga—Ta(1')	66·58 (3)
Ta(1)—Ga—Ta(1ⁱᵛ)	76·09 (3)		Ta(1)—Ga—Ta(1ⁱⁱⁱ)	77·05 (3)

Symmetry code: none x,y,z; (i) 1 − x,y,½ − z; (ii) ½ − y,½ − x,1 − z; (iii) y,x,½ − z; (iv) 1 − y, x,z; (v) − x, − y, − z; (vi) y,1 − x,z.

$GaSn_2V_2$

Acta Cryst., C$\underline{46}$, 1195-1197.

Acam, 6.7191, 18.798, 5.603, Z = 8, R = 0.044. New structure type, with two types of bands, one of linked GaV_5 pentagons and one of Sn-atom tetrahedra, alternating along $\underline{b}$.

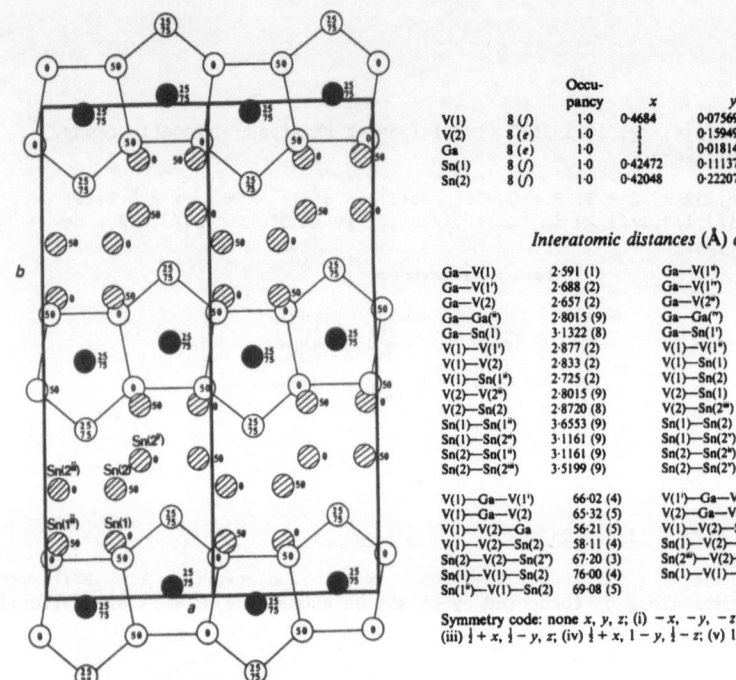

	Occupancy	x	y	z
V(1)	8 (f)	1·0	0·4684	0·07569	¼
V(2)	8 (e)	1·0	¼	0·15949	¼
Ga	8 (e)	1·0	¼	0·01814	¼
Sn(1)	8 (f)	1·0	0·42472	0·11137	0
Sn(2)	8 (f)	1·0	0·42048	0·22207	¼

Interatomic distances (Å) and angles (°)

Ga—V(1)	2·591 (1)	Ga—V(1^v)	2·591 (1)
Ga—V(1')	2·688 (2)	Ga—V(1^{iv})	2·688 (2)
Ga—V(2)	2·657 (2)	Ga—V(2^v)	3·861 (1)
Ga—Ga(^{ii})	2·8015 (9)	Ga—Ga(^{iii})	3·4281 (4)
Ga—Sn(1)	3·1322 (8)	Ga—Sn(1')	3·044 (1)
V(1)—V(1')	2·877 (2)	V(1)—V(1^{iv})	4·057 (2)
V(1)—V(2)	2·833 (2)	V(1)—Sn(1)	2·8955 (9)
V(1)—Sn(1^{iv})	2·725 (2)	V(1)—Sn(2)	2·771 (2)
V(2)—V(2')	2·8015 (9)	V(2)—Sn(1)	2·7490 (7)
V(2)—Sn(2)	2·8720 (8)	V(2)—Sn(2^{iii})	2·869 (1)
Sn(1)—Sn(1^{iv})	3·6553 (9)	Sn(1)—Sn(2)	3·4899 (9)
Sn(1)—Sn(2^{ii})	3·1161 (9)	Sn(1)—Sn(2')	3·2993 (9)
Sn(2)—Sn(1^{iv})	3·1161 (9)	Sn(2)—Sn(2')	3·6189 (9)
Sn(2)—Sn(2^{iii})	3·5199 (9)	Sn(2)—Sn(2^v)	3·1769 (9)

V(1)—Ga—V(1')	66·02 (4)	V(1')—Ga—V(1")	97·99 (5)
V(1)—Ga—V(2)	65·32 (5)	V(2)—Ga—V(2^{iv})	46·51 (4)
V(1)—V(2)—Ga	56·21 (5)	V(1)—V(2)—Sn(1)	62·48 (3)
V(1)—V(2)—Sn(2)	58·11 (4)	Sn(1)—V(2)—Sn(2)	76·73 (2)
Sn(2)—V(2)—Sn(2')	67·20 (3)	Sn(2^{iii})—V(2)—Sn(2')	78·21 (4)
Sn(1)—V(1)—Sn(2)	76·00 (4)	Sn(1)—V(1)—Sn(1^{iv})	81·07 (4)
Sn(1^{iv})—V(1)—Sn(2)	69·08 (5)		

Symmetry code: none x, y, z; (i) $-x, -y, -z$; (ii) $\frac{1}{2}-x, y, \frac{1}{2}+z$; (iii) $\frac{1}{2}+x, \frac{1}{2}-y, z$; (iv) $\frac{1}{2}+x, 1-y, \frac{1}{2}-z$; (v) $1-x, \frac{1}{2}-y, \frac{1}{4}-z$.

Projection of the crystal structure of V_2Sn_2Ga along the c axis (two unit cells). Atom key: open circles V, filled circles Ga, shaded circles Sn. Numbers correspond to z parameters.

$Gd_2InNi_{1.78}$, Ce_2InNi_2

Izv. Akad. Nauk SSSR, Neorg. Mater., $\underline{26}$, 94-96.

P4/mbm, a = 7.429, 7.499, c = 3.707, 3.751 A, Z = 2, R = 0.048, 0.059. 4 Gd or 4 Ce in 4(h): x,1/2+x,1/2, x = 0.17529, 0.1734; 3.56 Ni or 4 Ni in 4(g): x,1/2+x,0, x = 0.37990, 0.3768; 2In in 2(a): 0,0,0. U_3Si_2-type ($\underline{11}$, 285).

$Ge_3Nd_4Rh_4$

Kristallografija, $\underline{35}$, 495-497 [Soviet Physics - Crystallography, $\underline{35}$, 287-288].

B2/b ($\underline{c}$ unique), 21.021, 7.941, 5.652, γ = 110.08°, Z = 4, R = 0.034. New structure type; coordination numbers are: Ge = 10, 12, Nd = 17, 18, Rh = 10, 12.

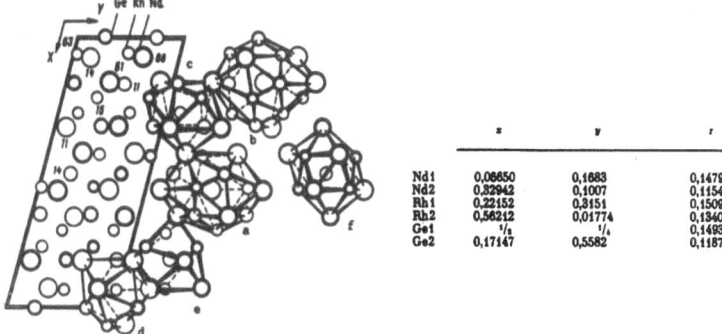

	x	y	z
Nd1	0,06650	0,1683	0,1479
Nd2	0,32942	0,1007	0,1154
Rh1	0,22152	0,3151	0,1509
Rh2	0,56212	0,01774	0,1340
Ge1	1/2	1/4	0,1493
Ge2	0,17147	0,5582	0,1187

Projection of unit cell of $Nd_4 Rh_4 Ge_3$ on XY plane and
coordination polyhedra of atoms of Nd (a, b), Rh (c, d), and
Ge (e, f).

Ge_3Ni_3Sm

Kristallografija, <u>35</u>, 202-204 [Soviet Physics - Crystallography, <u>35</u>, 122-123].

I4/mmm, 4.0659, 25.137, Z = 4, R = 0.038. Sm in 4(e): 0,0,z, z = 0.3465;
Ni(1) in 8(g): 0,1/2,z, z = 0.0553; Ni(2) in 4(e): z = 0.1999; Ge(1) in 4(e):
z = 0.1063; Ge(2) in 4(d): 0,1/2,1/4; Ge(3) in 2(a): 0,0,0; Ge(4) in 2(b):
0,0,1/2. New structure type, which is a filled derivative of the Ni_2ScSi_3-type
(<u>44</u>A, 83), containing fragments of Al_2Ga_2Ge and CsCl types.

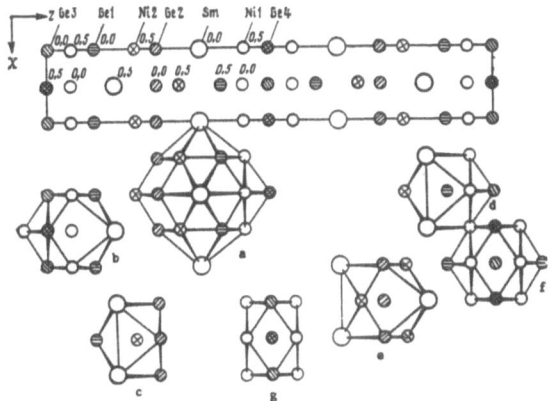

Projection of unit cell on XZ plane and coordination
polyhedra of atoms. a) Sm; b) Ni1; c) Ni2; d) Ge1; e) Ge2;
f) Ge3; g) Ge4.

$Ge_6Pt_4Y_3$

J. Less-Common Metals, <u>167</u>, 45-52.

$P2_1/m$, 8.6922, 4.3062, 13.1615, 99.45, Z = 2, R = 0.047. The structure is
built from an alternate stacking of $BaAl_4$- and $YIrGe_2$-type slabs, with $PtGe_5$
square pyramids, and Y with hexagonal- and pentagonal-prismatic coordinations.

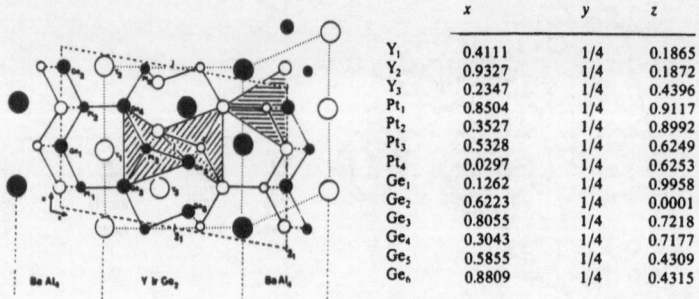

	x	y	z
Y_1	0.4111	1/4	0.1865
Y_2	0.9327	1/4	0.1872
Y_3	0.2347	1/4	0.4396
Pt_1	0.8504	1/4	0.9117
Pt_2	0.3527	1/4	0.8992
Pt_3	0.5328	1/4	0.6249
Pt_4	0.0297	1/4	0.6253
Ge_1	0.1262	1/4	0.9958
Ge_2	0.6223	1/4	0.0001
Ge_3	0.8055	1/4	0.7218
Ge_4	0.3043	1/4	0.7177
Ge_5	0.5855	1/4	0.4309
Ge_6	0.8809	1/4	0.4315

$LiPd_2Tl$

Z. Naturforsch., $\underline{45}$B, 1451-1452.

I4/mmm, 4.137, 7.318, Z = 2, R = 0.051. Li in 2(b): 0,0,1/2; Pd in 4(d): 0,1/2,1/4; Tl in 2(a): 0,0,0. Ternary variant of the Al_3Ti-type ($\underline{7}$, 13, 100; $\underline{53}$A, 3). Tl-4Tl = 4.137, Tl-8Pd = 2.763, Tl-6Li = 2.937 (x 4), 3.659 (x 2); Pd-6Pd = 2.937 (x 4), 3.659 (x 2), Pd-4Li = 2.763 A.

Na_5Sn_3Tl

Z. Kristallogr., $\underline{193}$, 319-320.

$P2_1/c$, 7.344, 13.443, 11.406, 105.07, Z = 4, R = 0.088. $TlSn_3{}^{5-}$ tetrahedra linked by Na^+ ions; Tl-Sn = 3.063-3.091(2), Sn-Sn = 2.923-3.031(3) A.

Ru_4Sn_6Y

Mater. Res. Bull., $\underline{25}$, 1541-1546.

$I\bar{4}2m$, 6.840, 9.786, Z = 2, R = 0.031. Ru in 8(i): x,x,z, x = 0.1726, z = 0.5772; Sn(1) in 8(i): x = 0.1763, z = 0.2933; Sn(2) in 4(c): 0,1/2,0; Y in 2(a): 0,0,0. New structure type related to $AuCu_3$ and $Rh_4Sn_{13}Yb_3$ types. Y is surrounded by a deformed cuboctahedron of 12Sn, capped by four Ru; Ru atoms form Ru_4 clusters.

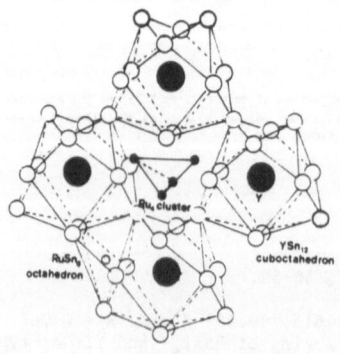

$Al_3Ge_2NiY_3$

Acta Cryst., C$\underline{46}$, 2273-2276.

P$\bar{6}$2m, 6.9481, 4.1565, Z = 1, R = 0.062. Al in 3(g): 0.2268,0,1/2; Ge in 2(d): 1/3,2/3,1/2; Ni in 1(a): 0,0,0; Y in 3(f): 0.5963,0,0. Quarternary substitution variant of the Fe$_2$P-type ($\underline{2}$, 15; $\underline{45A}$, 86).

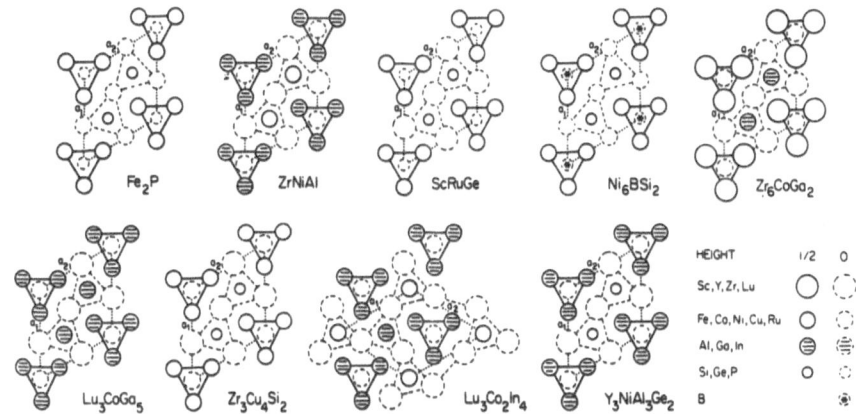

The [0001] projections of Y$_3$NiAl$_3$Ge$_2$, Fe$_2$P and seven of its ternary substitution variants.

Interatomic distances up to 3·5 Å in
Y$_3$NiAl$_3$Ge$_2$

Y···Ni	2·805 (2)	Al··2Ni	2·608 (3)
4Ge	2·965 (1)	2Al	2·729 (5)
4Al	3·202 (1)	2Ge	2·762 (4)
2Al	3·303 (3)	4Y	3·202 (1)
		2Y	3·303 (3)
Ni···6Al	2·608 (3)		
3Y	2·805 (2)	Ge···3Al	2·762 (4)
		6Y	2·965 (1)

$Al_7Cu_{16}Zr_6H_x$ (x = 3.0, 8.2)

J. Less-Common Metals, $\underline{159}$, 291-298.

Fm3m, a = 11.952, 12.058, Z = 4, powder data. Cu(1) in 32(f): x,x,x, x = 0.1723, 0.1741; Cu(2) in 32(f): x = 0.3796, 0.3791; Zr in 24(e): x,0,0, x = 0.2015, 0.2020; Al(1) in 24(d): 0,1/4,1/4; Al(2) in 4(b): 1/2,1/2,1/2; H(1) in 4(a): 0,0,0; H(2) in 32(f): x = 0.0901, 0.0912. Cu$_{16}$Mg$_6$Si$_7$-type ($\underline{16}$, 84; $\underline{17}$, 153; $\underline{20}$, 95), with H in 4(a) and 32(f) sites.

$CaPdH_2$

J. Less-Common Metals, $\underline{161}$, 299-302.

[Pm3m], 3.690, Z = 1, X-ray and neutron powder data. Perovskite structure, with 2/3 occupancy of H sites.

$Fe_{17}Ho_2$, $Fe_{17}Ho_2D_{3.6}$; $Fe_{17}Nd_2$, $Fe_{17}Nd_2D_{4.8}$; Ce_2Fe_{17}, $Ce_2Fe_{17}D_{4.8}$

J. Less-Common Metals, 162, 273-284.

Ho, $P6_3/mmc$, a = 8.415, 8.540, c = 8.290, 8.350 A, Z = 2, neutron powder data at 280 and 4.2K. Structure as previously described (46A, 58), with D in 6(h) and 12(i) sites.

Nd, Ce, R3m, a = 8.48-ε.69, c = 12.39-12.56 A, Z = 3, neutron powder data at 280 and 4.2K. Th_2Zn_{17}-type (20, 196; 27, 6) with D in 9(e) and 18(g) sites.

$GaKD_4$ (I), $GaKH_{2.56}D_{1.44}$ (II)

Koord. Khim., 16, 1210-1214.

Pnma, a = 8.987, 8.883, 8.995, b = 5.613, 5.559, 5.641, c = 7.262, 7.193, 7.263 A, for I at 295, 80K, II at 295K, Z = 4. Barite-type (1, 343).

K_3PdD_3

J. Less-Common Metals, 158, 163-167.

Room temperature, $P4_2/mnm$, 10.733, 10.524, Z = 8, neutron powder data. Linear PdH_2^{2-} ions, and additional H^- ions coordinated only to K.

High-temperature, Pm3m, a = 5.410 A, for the H compound at 503K, X-ray powder data. Metal atom sites correspond to $AuCu_3$-type (1, 486); H positions not determined.

Mg_2OsH_6

J. Less-Common Metals, 161, 337-340.

Fm3m, 6.6828, Z = 4, X-ray and neutron powder data. Mg in 8(c): 1/4,1/4,1/4, Os in 4(a): 0,0,0; D in 24(e): 0.252,0,0. K_2PtCl_6-type (1, 429).

SrD_2 (I), SrND (II)

J. Solid State Chem., 88, 571-576.

I, Pnma, 6.3706, 3.8717, 7.3021, Z = 4, neutron powder data. Atoms in 4(c): x,1/4,z, x = 0.2438, 0.3570, -0.0307, z = 0.1108, 0.4281, 0.6825, for Sr, D(1), D(2). $PbCl_2$-type (2, 16).

II, Fm3m, 5.4474, Z = 4, neutron powder data, for mixture with I. 4Sr in 4(b): 1/2,1/2,1/2; 4N in 192(ℓ): -0.0071,0.0072,0.0209; 4D in 192(ℓ): 0.050,-0.051,-0.146. NaCl-type, with rotational disorder of the imide group.

β-UD₃

J. Less-Common Metals, 158, 267-274.

Pm3n, a = 6.635, 6.627 A, at 307, 14K (minimum of 6.624 A at about 150K), Z = 8, neutron powder data. U(1) in 2(a): 0,0,0; U(2) in 6(c): 1/4,0,1/2; D in 24(k): 0,0.15595,0.30344. The U atoms occupy the Cr and Si sites of the Cr₃Si-type structure (3, 628), with the D atoms in a 24(k) site, as previously described (11, 128; 15, 81).

Aℓ₃HoB$_x$ (x = 0.4-0.5)

Izv. Akad. Nauk SSSR, Met., No. 2, 212-214.

R3̄m, a = 6.132, c = 20.986 A, Z = 9, R = 0.042. Ho(1) in 3(a): 0,0,0; Ho(2) in 6(c): 0,0,0.21754; Aℓ(1) in 9(e): 1/2,0,0; Aℓ(2) in 18(h): 0.1861, 0.3722, 0.1103; B positions not determined. BaPb₃ type (29, 30).

CeB₂C

J. Less-Common Metals, 157, 109-120.

R3̄m, a = 6.6138, c = 11.2594 A, Z = 9, powder data. Ce(1) in 3(a): 0,0,0; Ce(2) in 6(c): 0,0,0.3156; B in 18(g): 0.2762,0,1/2; C in 9(d): 1/2,0,1/2. ThB₂C-type (56A, 26), with (001) B/C sheets interleaved by sheets of Ce ions.

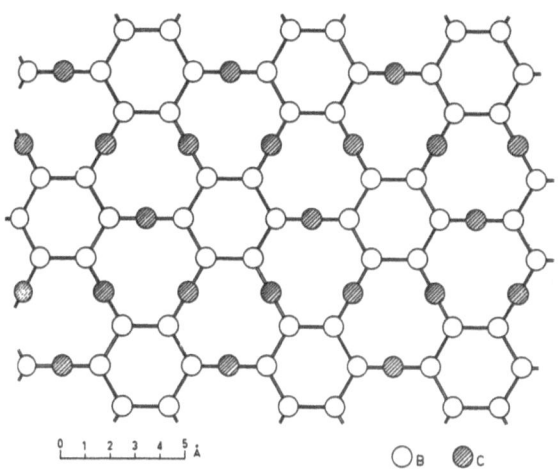

$CoFe_{13}Nd_2B$

J. Less-Common Metals, 162, 335-342.

$P4_2/mnm$, a = 8.7946, 8.7926, 8.7980, c = 12.1563, 12.1707, 12.1988 A, at 30, 293, 740K, Z = 4, neutron powder data. $Fe_{14}Nd_2B$-type (51A, 23), with Co statistically distributed in the Fe sites.

$Co_{40}Sm_{11}B_{40}$

Kristallografija, 35, 638-641 [Soviet Physics - Crystallography, 35, 372-373].

$P\bar{4}2_1c$, 7.051, 38.709, Z = 2, R = 0.051. The structure can be regarded as a combination of 11 Sm sublattices (I4/mmm, c = 3.519 A) and 10 Co_4B_4 sublattices ($P4_2/ncm$, c = 3.871 A).

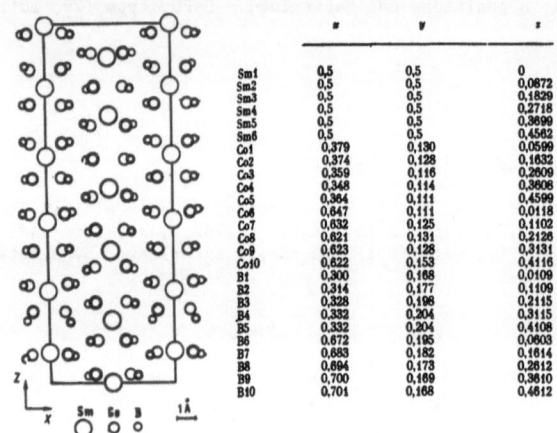

	x	y	z
Sm1	0,5	0,5	0
Sm2	0,5	0,5	0,0872
Sm3	0,5	0,5	0,1829
Sm4	0,5	0,5	0,2718
Sm5	0,5	0,5	0,3699
Sm6	0,5	0,5	0,4562
Co1	0,379	0,130	0,0599
Co2	0,374	0,128	0,1632
Co3	0,359	0,116	0,2609
Co4	0,348	0,114	0,3608
Co5	0,364	0,111	0,4599
Co6	0,647	0,111	0,0118
Co7	0,632	0,125	0,1102
Co8	0,621	0,131	0,2126
Co9	0,623	0,128	0,3131
Co10	0,622	0,153	0,4116
B1	0,300	0,168	0,0109
B2	0,314	0,177	0,1109
B3	0,328	0,198	0,2115
B4	0,332	0,204	0,3115
B5	0,332	0,204	0,4108
B6	0,672	0,195	0,0603
B7	0,683	0,182	0,1614
B8	0,694	0,173	0,2612
B9	0,700	0,169	0,3610
B10	0,701	0,168	0,4612

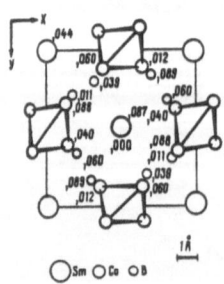

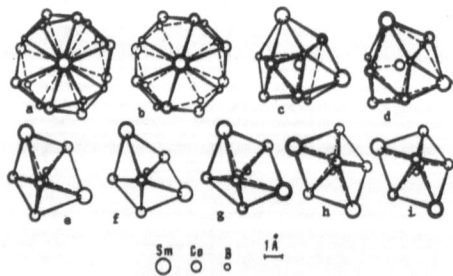

Characteristic coordination polyhedra in structure of $Sm_{11}(Co_4B_4)_{10}$. a) $[Sm1Sm2Co_8B_8]$, b) $[Co1Sm_2Co_{10}B_8]$, c) $[Co1Sm_2Co_6B_5]$, d) $[Co4Sm_3Co_6B_4]$, e) $[B1Sm_2Co_5]$, f) $[B2Sm_2Co_5]$, g) $[B5Sm_3Co_4]$, h) $[B9Sm_2Co_5]$, i) $[B10Sm_3Co_5]$.

ErMoB$_3$

Kristallografija, <u>34</u>, 1419-1421 (1989) [Soviet Physics - Crystallography, <u>34</u>, 852-853 (1989)].

P2$_1$/m, 6.790, 3.1513, 5.396, 101.60, Z = 2, R = 0.036. Coordination numbers are: Er = 19, Mo = 17, B = 9.

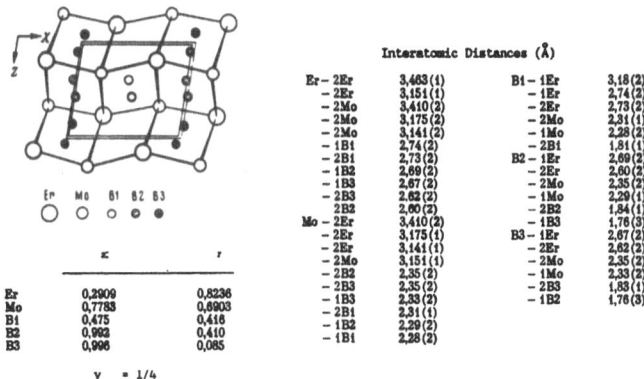

Interatomic Distances (Å)

Er – 2Er	3,463(1)	B1 – 1Er	3,18(2)
– 2Er	3,151(1)	– 1Er	2,74(2)
– 2Mo	3,410(2)	– 2Er	2,73(2)
– 2Mo	3,175(2)	– 2Mo	2,31(1)
– 2Mo	3,141(2)	– 1Mo	2,28(2)
– 1B1	2,74(2)	– 2B1	1,81(1)
– 2B1	2,73(2)	B2 – 1Er	2,69(2)
– 1B2	2,69(2)	– 2Er	2,60(2)
– 1B3	2,67(2)	– 2Mo	2,35(2)
– 2B3	2,62(2)	– 1Mo	2,29(1)
– 2B2	2,60(2)	– 2B2	1,84(1)
Mo – 2Er	3,410(2)	– 1B3	1,76(3)
– 2Er	3,175(1)	B3 – 1Er	2,67(2)
– 2Er	3,141(1)	– 2Er	2,62(2)
– 2Mo	3,151(1)	– 2Mo	2,35(2)
– 2B2	2,35(2)	– 1Mo	2,33(2)
– 2B3	2,35(2)	– 2B3	1,83(1)
– 1B3	2,33(2)	– 1B2	1,76(3)
– 2B1	2,31(1)		
– 1B2	2,29(2)		
– 1B1	2,28(2)		

	x	z
Er	0,2909	0,8236
Mo	0,7783	0,6903
B1	0,475	0,416
B2	0,993	0,410
B3	0,996	0,085

y = 1/4

Ho$_3$Ni$_{19}$B$_{10}$

Kristallografija, <u>35</u>, 998-999 [Soviet Physics - Crystallography, <u>35</u>, 587-588].

C2/m, 13.177, 8.719, 5.780, 91.33, Z = 2, R = 0.055 Corrugated (010) layers of metal ions separated by B atoms. Coordination numbers are: Ho = 21, 24, Ni = 12-14.

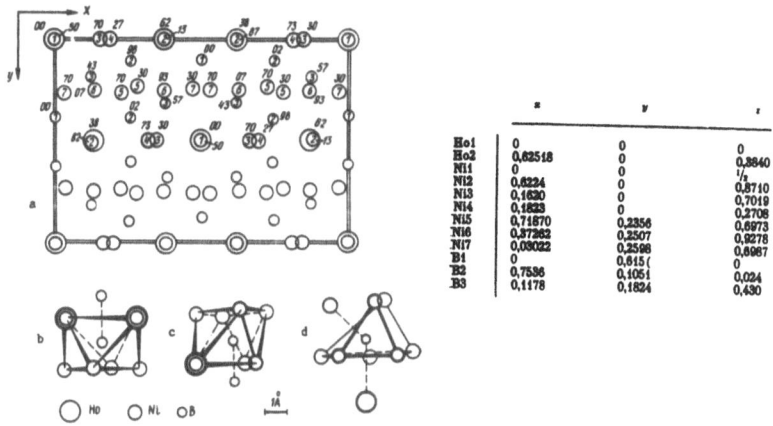

	x	y	z
Ho1	0	0	0
Ho2	0,62518	0	0,3640
Ni1	0	0	1/2
Ni2	0,6224	0	0,8710
Ni3	0,1620	0	0,7019
Ni4	0,1823	0	0,2708
Ni5	0,71870	0,2356	0,6973
Ni6	0,37262	0,2507	0,9278
Ni7	0,03022	0,2598	0,8967
B1	0	0,615(	0
B2	0,7536	0,1051	0,024
B3	0,1178	0,1824	0,430

a) Projection of structure on XY plane. Coordination polyhedra of boron atoms: b) [B1Ho$_2$ Ni$_6$B]; c) [B2HoNi$_7$B]; d) [B3Ho$_1$Ni$_8$].

$Ir_5Mg_2SiB_2$, $Ir_5Mg_2P_{0.7}B_2$

Z. anorg. Chem., <u>581</u>, 135-140.

P4/mbm, a = 9.317, 9.286, c = 2.894, 2.900 A, Z = 2, R = 0.049, 0.065. Ir(1)
in 8(j): x,y,1/2, x = 0.2104, 0.2084, y = 0.0708, 0.0715; Ir(2) in 2(c):
0,1/2,1/2; Mg in 4(g): x,1/2+x,0, x = 0.324, 0.3258; 2 Si or 1.44 P in 2(a):
0,0,0; B in 4(g): x = 0.126, 0.128. Substitution variant of the $Co_5Ti_3B_2$-type
(<u>38A</u>, 45).

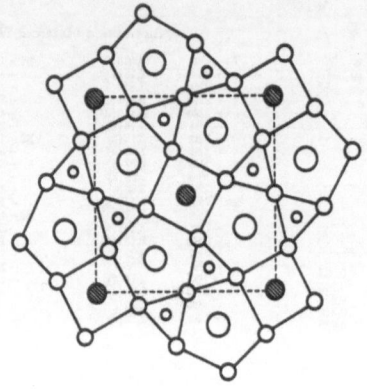

$LaNi_{12}B_6$

Izv. Akad. Nauk SSSR, Neorg. Mater., <u>26</u>, 2220-2222.

Cmc2₁, 9.598, 7.414, 11.075, Z = 4, R = 0.045. $CeNi_{12}B_6$-type (<u>52A</u>, 18).
Coordination numbers are: La = 24, Ni = 13, B = 6+2.

$La_2Rh_5B_4$, $Eu_2Rh_5B_4$, $Eu_3Rh_8B_6$

J. Less-Common Metals, <u>161</u>, 375-384.

Fmmm, a = 5.515, 5.492, 5.537, b = 9.964, 9.914, 9.847, c = 11.343, 11.237,
17.099 A, Z = 4, R = 0.023, 0.055, 0.050. Isostructural with the Ca compounds
(<u>51A</u>, 21). The structures are formed from elements of the $CeCo_3B_2$- and
$CaRh_2B_2$-types (<u>44A</u>, 35; <u>45A</u>, 41).

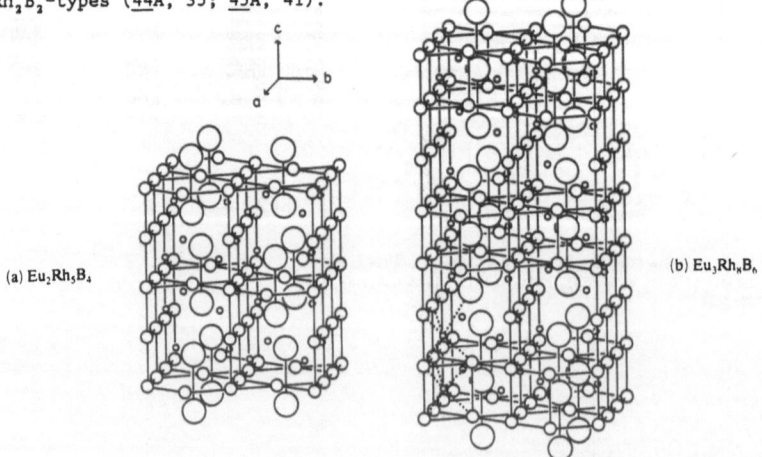

(a) $Eu_2Rh_5B_4$

(b) $Eu_3Rh_8B_6$

$Pr_8Re_{12.62}B_{12}$

Kristallografija, <u>35</u>, 621-624 [Soviet Physics - Crystallography, <u>35</u>, 363-364].

$R\bar{3}m$, a = 10.665, c = 14.830 A, Z = 3, R = 0.04.   Columns of deformed trigonal prisms, with channels containing Pr and Re atoms and B_6 rings with each B linked to one further B to form planar B_{12} groups.

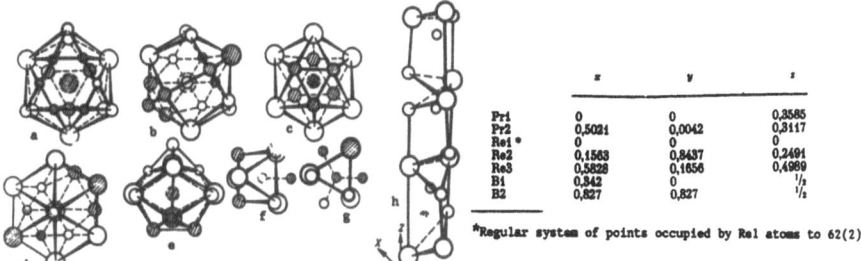

	x	y	z
Pr1	0	0	0,3585
Pr2	0,5021	0,0042	0,3117
Re1 *	0	0	0
Re2	0,1563	0,8437	0,2491
Re3	0,5828	0,1656	0,4989
B1	0,342	0	$^1/_2$
B2	0,827	0,827	$^1/_2$

*Regular system of points occupied by Re1 atoms to 62(2)%

Coordination polyhedra in structure of $Pr_8Re_{13-x}B_{12}$.
a) $[Pr1Pr_2Re_9B_6]$, b) $[Pr2Pr_3Re_{10}B_6]$, c) $[Re1Pr_8Re_6]$, d) $[Re2Pr_7Re_3B_4]$, e) $[Re3Pr_5 \cdot Re_6B_2]$, f) $[B1Pr_4Re_4B]$, g) $[B2Pr_4 \cdot Re_2B_3]$; h) column of trigonal prisms and empty tetrahedra $[EPr_2Re_2]$.

Sm_2B_5

Poroshk. Metall. (Kiev), No. 6, 61-63.

$P2_1/b$ (<u>c</u> unique), 7.179, 7.205, 7.180, γ = 102.02°, Z = 4, R = 0.089.
Gd_2B_5-type (<u>54A</u>, 26).

	x	y	z
Sm1	0,233	0,131	0,191
Sm2	0,240	0,756	—0,186
B1	—0,493	0,043	—0,326
B2	0,502	0,909	0,090
B3	0,502	0,671	0,041
B4	0,071	0,425	—0,009
B5	—0,313	0,537	0,002

Ta_5B_6

J. Less-Common Metals, <u>161</u>, 341-345.

Cmmm, 22.602, 3.1385, 3.2895, Z = 2, R = 0.023.   Capped BTa_6 trigonal prisms are packed to give single and double B chains.

		x	y	z
Ta(1)	2a	0	0	0
Ta(2)	4g	0.385543	0	0
Ta(3)	4h	0.194344	0	1/2
B(1)	4h	0.0832	0	1/2
B(2)	4g	0.2734	0	0
B(3)	4h	0.4603	0	1/2

Ta. z=0   O Ta. z=1/2
B. z=0   O B. z=1/2

$CrSc_2C_3$

Kristallografija, $\underline{35}$, 47-49 [Soviet Physics - Crystallography, $\underline{35}$, 25-26].

Pbam, 14.111, 5.809, 3.271, Z = 4, R = 0.057. Sc_6 trigonal prisms centred alternately by C_2 pairs and Cr atoms, and body-centred cubes of Sc with four C atoms on the edges.

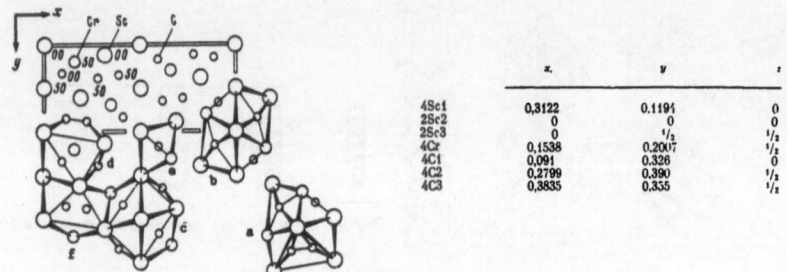

	x.	y	z
4Sc1	0,3122	0.1194	0
2Sc2	0	0	0
2Sc3	0	$^1/_2$	$^1/_2$
4Cr	0,1538	0.2007	$^1/_2$
4C1	0,091	0.326	0
4C2	0,2799	0.390	$^1/_2$
4C3	0,3835	0.355	$^1/_2$

Projection of structure of Sc_2CrC_3 on xy plane and co-ordination polyhedra of atoms. a) Sc1, b) Sc2, c) Sc3, d) Cr, e) C1, f) (C2—C3).

$Er_4Ni_{13}C_4$ (I), UW_4C_4 (II)

Kristallografija, $\underline{35}$, 337-341 [Soviet Physics - Crystallography, $\underline{35}$, 189-192].

I, Cmmm, 11.975, 11.694, 3.856, Z = 2, R = 0.066. New structure type, related to that of $La_2Ni_5C_3$ ($\underline{53}$A, 27; $\underline{56}$A, 26), with fragments of $CaTiO_3$, $CaCu_5$, and AlB_2 types.

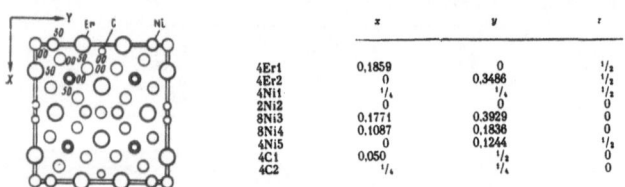

	x	y	z
4Er1	0,1859	0	$^1/_2$
4Er2	0,	0.3486	$^1/_2$
4Ni1	$^1/_4$	$^1/_4$	$^1/_2$
2Ni2	0	0	0
8Ni3	0,1771	0.3929	0
8Ni4	0,1087	0.1836	0
4Ni5	0	0.1244	$^1/_2$
4C1	0,050	$^1/_2$	0
4C2	$^1/_4$	$^1/_4$	0

II, P4/m, 8.328, 3.1345, Z = 2, R = 0.068. New structure type, related to that of UCr_4C_4 ($\underline{54}$A, 27; $\underline{56}$A, 27).

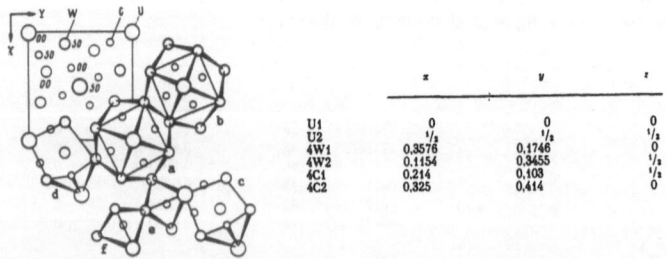

	x	y	z
U1	0	0	0
U2	$^1/_2$	$^1/_2$	$^1/_2$
4W1	0,3576	0.1746	0
4W2	0,1154	0.3455	$^1/_2$
4C1	0,214	0.103	$^1/_2$
4C2	0,325	0.414	0

$Fe_{17}Ln_2C$ (Ln = Y, Tb, Dy, Ho).

J. Less-Common Metals, 163, 353-359.

$R\bar{3}m$, a = 8.5927, 8.6068, 8.5947, 8.5493, at 300K, 8.5718, 8.5843, 8.5732, 8.5198, at 4.2K, c = 12.4616, 12.4619, 12.4504, 12.4585 at 300K, 12.4725, 12.4633, 12.4599, 12.4690 A, at 4.2 K, Z = 3, neutron powder data. Isostructural with $Fe_{17}Nd_2C_x$ (55A, 30; 56A, 28).

$Fe_{14}Lu_2C$

J. Less-Common Metals, 163, 361-368.

$P4_2/mnm$, a = 8.70, 8.7137, 8.71, c = 11.72, 11.722, 11.71 A, at 623, 295, 23K, Z = 4, neutron powder. Isostructural with $Fe_{14}Nd_2B$ (51A, 23).

295 K		x	y	z
Lu1	4f	0.2589	0.2589	0.000
Lu2	4g	0.1434	0.8566	0.000
Fe1	16k	0.2243	0.5632	0.1211
Fe2	16k	0.0351	0.3580	0.1738
Fe3	8j	0.0971	0.0971	0.2006
Fe4	8j	0.3155	0.3155	0.2450
Fe5	4e	0.000	0.000	0.6074
Fe6	4c	0.000	0.500	0.000
C	4g	0.3689	0.6311	0.000

Ln_2ReC_2 (Ln = Pr, Y, Er)

J. Solid State Chem., 89, 191-201.

Pnma, a = 6.6559, 6.5569, 6.4990, b = 5.3454, 5.0946, 5.0353, c = 10.1835, 9.8442, 9.7431 A, Z = 4, powder data for Ln = Pr, Y, single crystal data for Ln = Er, R = 0.032. The metal atom positions correspond to the $PbCl_2$ (Co_2Si)-type structure (2, 16; 3, 32), with C atoms in two octahedral holes (4Ln + 2Re, 5Ln + 1Re), giving ReC_2^{6-} chains. Other Ln compounds are isostructural.

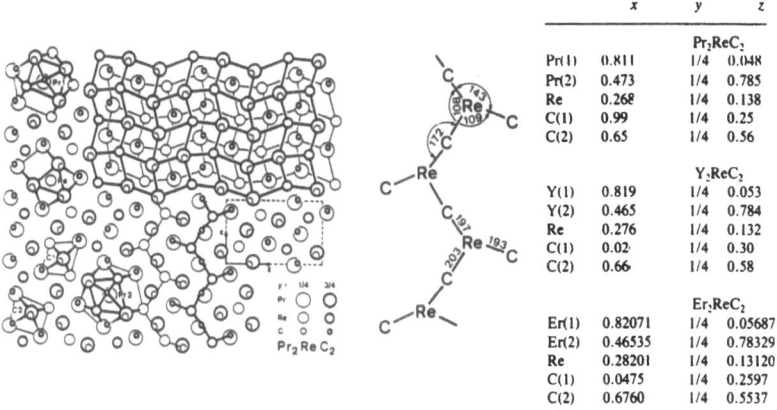

	x	y	z
	Pr_2ReC_2		
Pr(1)	0.811	1/4	0.048
Pr(2)	0.473	1/4	0.785
Re	0.268	1/4	0.138
C(1)	0.99	1/4	0.25
C(2)	0.65	1/4	0.56
	Y_2ReC_2		
Y(1)	0.819	1/4	0.053
Y(2)	0.465	1/4	0.784
Re	0.276	1/4	0.132
C(1)	0.02	1/4	0.30
C(2)	0.66	1/4	0.58
	Er_2ReC_2		
Er(1)	0.82071	1/4	0.05687
Er(2)	0.46535	1/4	0.78329
Re	0.28201	1/4	0.13120
C(1)	0.0475	1/4	0.2597
C(2)	0.6760	1/4	0.5537

δ'-Nb_6C_5

Kristallografija, 35, 1110-1115 [Soviet Physics - Crystallography, 35, 653-655].

C2/m, 5.447, 9.435, 5.447, 109.47, Z = 2, R = 0.06. NaCℓ-superstructure.

	x	y	z
Nb(1)	0.2554	0	0.7372
Nb(2)	0.2415	0.6718	0.7468
C(1)	0	1/2	1/2
C(2)	0	0.335	0
C(3)	0	0.163	1/2
Vacancy	0	0	0

$Ni_{60}Tm_{11}C_6$

Kristallografija, 35, 1378-1380 [Soviet Physics - Crystallography, 35, 812-813].

Im3m, 12.453, Z = 2, R = 0.050, New structure type; coordination numbers are: Ni = 11-14, Tm = 19, 22, C = 6 (octahedral).

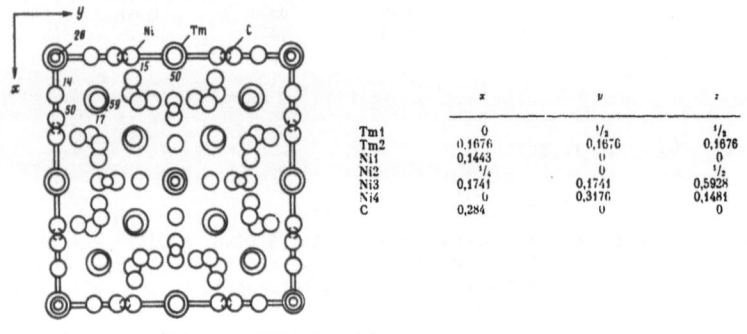

	x	y	z
Tm1	0	1/2	1/2
Tm2	0,1676	0,1676	0,1676
Ni1	0,1443	0	0
Ni2	1/4	0	1/2
Ni3	0,1741	0,1741	0,5928
Ni4	0	0,3176	0,1481
C	0,284	0	0

$Ru_6Th_2C_5$

Kristallografija, 35, 199-201 [Soviet Physics - Crystallography, 35, 120-121].

P4/mbm, 9.113, 4.186, Z = 2, R = 0.047. Ru_3ThC (deformed $CaTiO_3$) and Ru_3ThC_4 (deformed $CeCo_3B_2$) fragments.

	x	y	z
4Th	0,3263	0,8263	0
2Ru1	0	1/2	1/2
2Ru2	0	0	0
8Ru3	0,2113	0,0708	1/2
2C1	0	0	1/2
8C2	0,131	0,631	0,181

Projection of structure of compound $Th_2Ru_6C_5$ on xy plane and coordination polyhedra of atoms. a) Th; b) Ru1; c) Ru2; d) Ru3; e) C1; f) C2.

$Ru_{12}Th_{11}C_{18}$

Kristallografija, <u>35</u>, 487-490 [Soviet Physics - Crystallography, <u>35</u>, 282-284].

I$\bar{4}$3m, 10.7636, Z = 2, powder data. Th(1) in 2(a): 0,0,0; Th(2) in 12(d): 1/4,1/2,0; Th(3) in 8(c): x,x,x, x = 0.2030; Ru in 24(g): x,x,z, x = 0.4143, z = 0.2207; C(1) in 24(g): x = 0.235, z = 0.452; C(2) in 12(e): x,0,0, x = 0.560. New structure type, which is a derivative of the α-Mn type, and includes C_2 pairs.

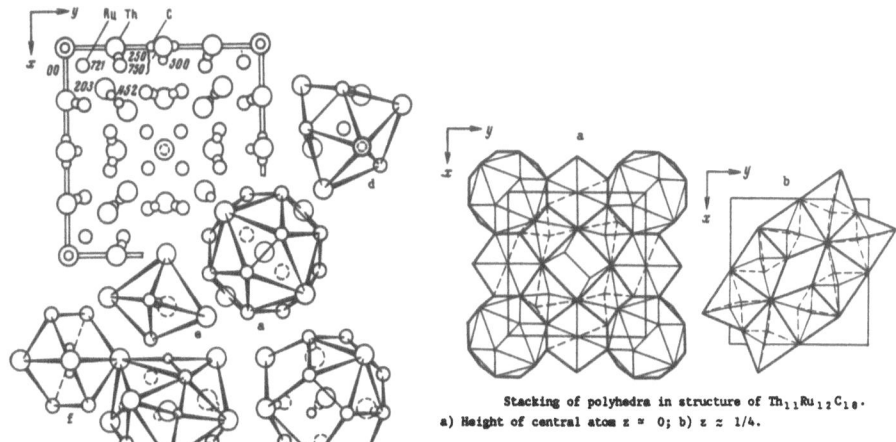

Projection of structure of $Th_{11}Ru_{12}C_{18}$ on xy plane and coordination polyhedra of atoms. a) Th1; b) Th2; c) Th3; d) Ru; e) C1; f) C_2.

Stacking of polyhedra in structure of $Th_{11}Ru_{12}C_{18}$. a) Height of central atom z ≈ 0; b) z ≈ 1/4.

$Ti_2C_{1.25}$

Kristallografija, <u>34</u>, 1513-1517 (1989) [Soviet Physics - Crystallography, <u>34</u>, 905-908 (1989)].

P3$_1$21, 3.060, 14,91, Z = 3. 6 Ti in 6(c): 2/3,0,0.0869; 3 C(1) in 3(b): 2/3,0,5/6; 0.75 C(2) in 3(a): 2/3,0,1/3. Ordered NaCℓ superstructure.

UW_4C_4

J. Less-Common Metals, <u>160</u>, 185-192.

P4/m, 8.3271, 3.1358, Z = 2, R = 0.046. U(1) in 1(a): 0,0,0; U(2) in 1(d): 1/2,1/2,1/2; W(1) in 4(j): 0.1539,0.3858,0; W(2) in 4(k): 0.6792,0.8563,1/2; C(1) in 4(j): 0.393,0.296,0; C(2) in 4(k): 0.909,0.763,1/2. Slight distortion of the UCr_4C_4 structure (<u>54</u>A, 27; <u>56</u>A, 27) (symmetry reduced from I4/m to P4/m). The structure contains CU_2W_4 octahedra. U-C = 2.47, 2.63, W-C = 2.07-2.19(4) A.

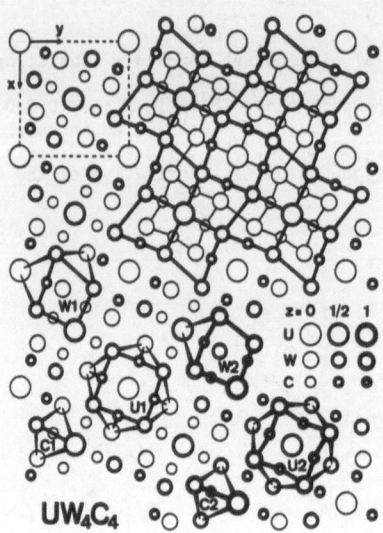

UW_4C_4

CoNdSi$_2$

Phys. Status Solidi, A, **116**, K29-K31 (1989).

Cmcm, a = 4.119, 4.126, b = 16.424, 16.440, c = 4.034, 4.036 A, at 1.36, 3.17K, Z = 4, neutron powder data. Atoms in 4(c): 0,y,1/4, y = 0.3237, 0.1018, 0.4667, 0.7564, at 1.36K; 0.3264, 0.1032, 0.4640, 0.7554, at 3.17K, for Co, Nd, Si(1), Si(2). CeNiSi$_2$-type (**34**A, 63; **35**A, 96).

Er$_2$Ru$_3$Si$_{4.6}$ (I), Er$_2$Ru$_3$Si$_5$ (II)

J. Less-Common Metals, **163**, L13-L17.

I, P4/mnc, 10.6556, 5.5781, Z = 4, R = 0.090. Sc$_2$Fe$_3$Si$_5$-type (**43**A, 56). Tb$_2$Os$_3$Si$_{4.6}$ is isostructural.

II, Ibam, 9.5418, 12.0801, 5.5502, Z = 4, R = 0.053. U$_2$Co$_3$Si$_5$-type (**43**A, 50). Tb$_2$Ru$_3$Si$_5$ is isostructural.

Er$_2$Ru$_3$Si$_{4.6}$		x	y	z
Er	8h	0.2628	0.4258	0
Ru1	8h	0.1445	0.1249	0
Ru2	4d	0	½	¼
Si1	8h	0.0247	0.3207	0
Si2	8g	0.1732	0.6732	¼
Si3	4e	0	0	0.2536

Er$_2$Ru$_3$Si$_5$		x	y	z
Er	8j	0.2681	0.3634	0
Ru1	8j	0.0902	0.1479	0
Ru2	4b	½	0	¼
Si1	8j	0.3298	0.0938	0
Si2	8g	0	0.2936	¼
Si3	4a	0	0	¼

$Fe_{0.5}LnSi_2$ (Ln = Ho, Dy, Tb)

J. Less-Common Metals, 161, 295-298.

Cmcm, a = 4.014, 4.050, 4.018, c = 15.709, 15.732, 16.289 A, Z = 4, R = 0.07, 0.067, 0.063. Ln, 0.5 Fe, Si(1), Si(2) in 4(c): 0,y,1/4, y = 0.3973, 0.750, 0.0459, 0.194 for Ho compound (similar values for others). CeNiSi$_2$-type (34A, 63; 35A, 96), with half-occupied Fe site.

$Fe_{13}Nd_6Si$

J. Less-Common Metals, 166, 73-79.

I4/mcm, 8.034, 22.78, Z = 4, R = 0.07. Ordered version of the Co$_{11}$Ga$_3$La$_6$-type (53A, 9), with three layers: A = Si-centred Nd antiprisms; B = Fe triangles and Nd-centred Fe octagons; C = Fe triangles, squares and pentagons.

		x	y	z
Nd(1)	8f	0	0	0.1104
Nd(2)	16l	0.1663	0.6663	0.1904
Fe(1)	4d	0	0.5	0
Fe(2)	16l	0.1788	0.6788	0.0607
Fe(3)	16l	0.3860	0.8860	0.0966
Fe(4)	16k	0.0666	0.2079	0
Si(1)	4a	0	0	0.25

GdSi

Acta Cryst., C46, 1197-1199.

Pnma, 7.973, 3.858, 5.753, Z = 4, R = 0.030. Gd, Si in 4(c): x,1/4,z, x = 0.1791, 0.5371, z = 0.3857, 0.6269. FeB-type structure (2, 7; 3, 12).

HfMoSi, HfRhSi

J. Less-Common Metals, 163, L7-L12.

Pnma, a = 6.932, 6.480, b = 3.5052, 3.9069, c = 8.196, 7.411 A, Z = 4, R = 0.047, 0.047. Atoms in 4(c): x,1/4,z, x = 0.0423, 0.1345, 0.2563, z = 0.6669, 0.0621, 0.3636, for Hf, Mo, Si; x = 0.02014, 0.1502, 0.2743, z = 0.68455, 0.0653, 0.3815, for Hf, Rh, Si. NiTiSi-type (30A, 75).

$Li_{13}Ni_{40}Si_{31}$ (I), $Ge_{31}Ni_{40.7}Sc_{12.3}$ (II)

Kristallografija, 35, 312-315 [Soviet Physics - Crystallography, 35, 173-175].

P6/mmm, a = 17.092, 17.865, c = 7.848, 8.220 A, Z = 2, R = 0.041, 0.043. New structure type with a four-layer close packing of atoms. Coordination numbers are: Li and Sc = 17-20, Ni = 9-14, 20, Si and Ge = 11-15.

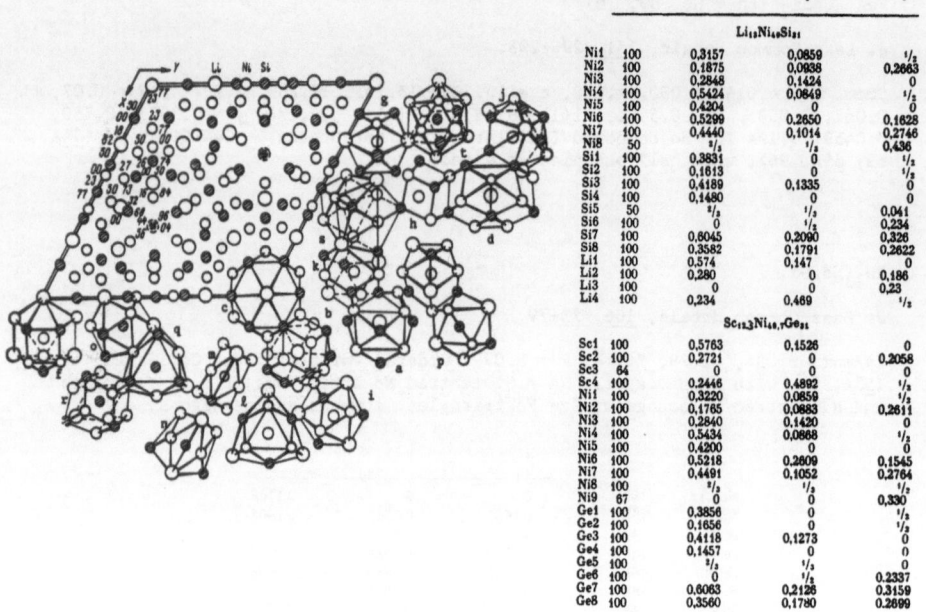

	G. %	x	y	z
		$Li_{13}Ni_{46}Si_{31}$		
Ni1	100	0,3157	0,0859	1/2
Ni2	100	0,1875	0,0938	0,2663
Ni3	100	0,2848	0,1424	0
Ni4	100	0,5424	0,0849	1/2
Ni5	100	0,4204	0	0
Ni6	100	0,5299	0,2650	0,1628
Ni7	100	0,4440	0,1014	0,2746
Ni8	50	2/3	1/3	0,436
Si1	100	0,3831	0	1/2
Si2	100	0,1613	0	1/2
Si3	100	0,4189	0,1335	0
Si4	100	0,1480	0	0
Si5	50	2/3	1/3	0,041
Si6	100	0	1/3	0,234
Si7	100	0,8045	0,2090	0,326
Si8	100	0,3582	0,1791	0,2622
Li1	100	0,574	0,147	0
Li2	100	0,280	0	0,186
Li3	100	0	0	0,23
Li4	100	0,234	0,469	1/2
		$Sc_{13}Ni_{46.7}Ge_{31}$		
Sc1	100	0,5763	0,1526	0
Sc2	100	0,2721	0	0,2058
Sc3	64	0	0	0
Sc4	100	0,2446	0,4892	1/2
Ni1	100	0,3220	0,0859	1/2
Ni2	100	0,1765	0,0883	0,2611
Ni3	100	0,2840	0,1420	0
Ni4	100	0,5434	0,0868	1/2
Ni5	100	0,4200	0	0
Ni6	100	0,5218	0,2609	0,1545
Ni7	100	0,4491	0,1052	0,2764
Ni8	100	2/3	1/3	1/2
Ni9	87	0	0	0,330
Ge1	100	0,3856	0	1/2
Ge2	100	0,1656	0	1/2
Ge3	100	0,4118	0,1273	0
Ge4	100	0,1457	0	0
Ge5	100	2/3	1/3	0
Ge6	100	0	1/3	0,2337
Ge7	100	0,6063	0,2126	0,3159
Ge8	100	0,3560	0,1780	0,2699

Mg_2PtSi

Acta Cryst., C46, 1092-1093.

$P6_3/mmc$, 4.254, 8.542, Z = 2, R = 0.035. Mg in 4(f): 1/3,2/3,0.5782; Pt in 2(c): 1/3,2/3,1/4; Si in 2(b): 0,0,1/4. Ordered ternary variant of the Na_3As-type (5, 6).

Interatomic distances (Å)

Pt—Si	2·456	(3×)	Mg—Mg	2·796 (4)	(3×)
—Mg	2·803 (6)	(2×)	—Pt	2·803 (6)	
—Mg	2·861 (3)	(6×)	—Pt	2·861 (3)	(3×)
Si—Pt	2·456	(3×)	—Si	2·861 (3)	(3×)
—Mg	2·861 (3)	(6×)	—Mg	2·935 (9)	
	3·727 (5)	(6×)			

$Ni_4Ti_4Si_7$

J. Mater. Res., 5, 1887-1893.

I4/mmm, 12.5229, 4.9343, Z = 4, neutron powder data. Isostructural with $Zr_4Co_4Ge_7$, (34A, 66).

Sb_3Zr_5Si, Sb_3ZnZr_5

Inorg. Chem., 29, 3274-3282.

$P6_3/mcm$, a = 8.5686, 8.6074, c = 5.7934, 5.8362 A, Z = 2, R = 0.021, 0.013. Sb in 6(g): x,0,1/4, x = 0.60630, 0.61504; Zr(1) in 4(d): 1/3,2/3,0; Zr(2) in 6(g): x = 0.2619, 0.27045; Si or Zn in 2(a): 0,0,0. Mn_5Si_3-type (4, 24), with Si or Zn in the centres of all Zr trigonal antiprisms.

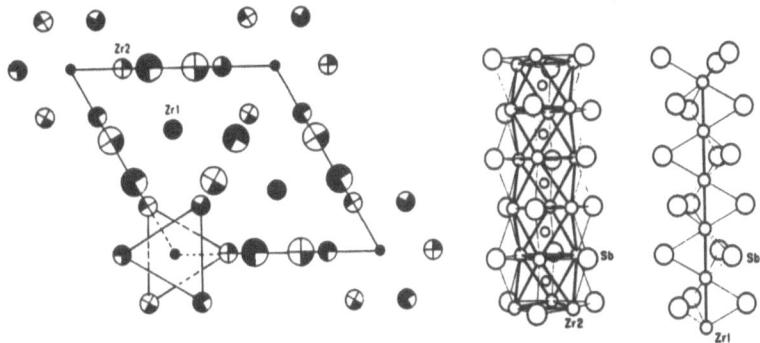

TbSi$_x$ (x = 1.75, 1.85)

J. Less-Common Metals, 162, 197-219.

Im2b, a = 4.050, 4.057, b = 3.958, 3.967, c = 13.389, 13.376 A, Z = 4, neutron powder data at 60K. Atoms in 4(b): 0,y,z (space group C_{2v}^{22}, No. 46, with b/c interchanged, and origin shift (-1/4,0,-1/4)), y = 0.254, 0.25, 0.192, z = 0.3735, 0.7973, 0.9678 for 4Tb, 4Si(1), 3 Si(2), respectively. Distortion of GdSi$_2$-type structure (23, 220), with vacancies in the Si(2) site. There is a 2a x 2a x c magnetic cell, possible C222$_1$.

BN

J. Phys. Chem. Solids, 51, 1011-1012.

F$\bar{4}$3m, a = 3.6160 A, Z = 4, R = 0.013-0.015 for 3 crystals. N in 4(a): 0,0,0; B in 4(c): 1/4,1/4,1/4. Zincblende-type, as previously described (21, 194; 28, 43; 40A, 103; 53A, 33).

BaNiN

J. Less-Common Metals, 159, L29-L31.

Pnma, 9.639, 13.674, 5.432, Z = 12, R = 0.062. Ni-N zigzag chains along b, linked by Ba ions so that N atoms have octahedral coordinations. Ni-2N = 1.78-1.83, Ba-4N = 2.79-2.97, Ni(1)...Ni(1) = 2.42 (short) A, N-Ni-N = 178, 180°.

		x	y	z
Ba1	8d	0.3281	0.0880	-0.0263
Ba2	4c	0.4974	1/4	0.5175
Ni1	8d	0.1479	0.1613	0.4460
Ni2	4a	0	0	0
N1	8d	0.0759	0.0773	0.2268
N2	4c	0.2241	1/4	0.6584

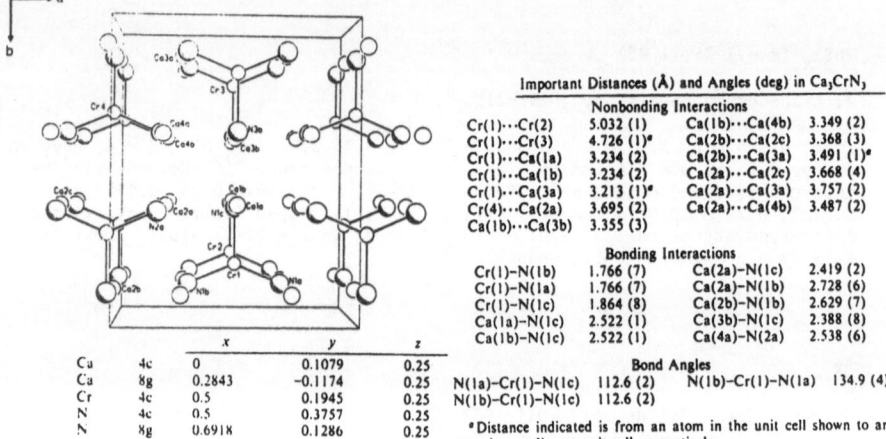

OBa   ON   ●Ni

Ca$_3$CrN$_3$

Inorg. Chem., 29, 4059-4062.

Cmcm, 8.503, 10.284, 5.032, R = 0.039. New structure type, with sheets of
planar triangular CrN$_3$$^{6-}$ units and Ca^{2+} ions.

Important Distances (Å) and Angles (deg) in Ca$_3$CrN$_3$

Nonbonding Interactions

Cr(1)···Cr(2)	5.032 (1)	Ca(1b)···Ca(4b)	3.349 (2)
Cr(1)···Cr(3)	4.726 (1)*	Ca(2b)···Ca(2c)	3.368 (3)
Cr(1)···Ca(1a)	3.234 (2)	Ca(2b)···Ca(3a)	3.491 (1)*
Cr(1)···Ca(1b)	3.234 (2)	Ca(2a)···Ca(2c)	3.668 (4)
Cr(1)···Ca(3a)	3.213 (1)*	Ca(2a)···Ca(3a)	3.757 (2)
Cr(4)···Ca(2a)	3.695 (2)	Ca(2a)···Ca(4b)	3.487 (2)
Ca(1b)···Ca(3b)	3.355 (3)		

Bonding Interactions

Cr(1)-N(1b)	1.766 (7)	Ca(2a)-N(1c)	2.419 (2)
Cr(1)-N(1a)	1.766 (7)	Ca(2a)-N(1b)	2.728 (6)
Cr(1)-N(1c)	1.864 (8)	Ca(2b)-N(1b)	2.629 (7)
Ca(1a)-N(1c)	2.522 (1)	Ca(3b)-N(1c)	2.388 (8)
Ca(1b)-N(1c)	2.522 (1)	Ca(4a)-N(2a)	2.538 (6)

		x	y	z
Ca	4c	0	0.1079	0.25
Ca	8g	0.2843	-0.1174	0.25
Cr	4c	0.5	0.1945	0.25
N	4c	0.5	0.3757	0.25
N	8g	0.6918	0.1286	0.25

Bond Angles

N(1a)-Cr(1)-N(1c)	112.6 (2)	N(1b)-Cr(1)-N(1a)	134.9 (4)
N(1b)-Cr(1)-N(1c)	112.6 (2)		

*Distance indicated is from an atom in the unit cell shown to an
atom in an adjacent unit cell, respectively.

Ca$_6$GaN$_5$, Ca$_6$FeN$_5$

Z. anorg. Chem., 591, 58-66.

P6$_3$/mcm, a = 6.277, 6.237, c = 12.198, 12.332 Å, Z = 2, R = 0.090, 0.087. Ca
in 12(k): x,0,z, x = 0.5935, 0.6049, z = 0.1226, 0.1190; Ga or Fe in 2(a):
0,0,1/4; N(1) in 4(d): 2/3,1/3,0; N(2) in 6(g): x,0,1/4, x = 0.3126, 0.2837.
Trigonal planar MN(2)$_3$$^{6-}$ anions (M = Ga, Fe) in sheets, separated by N(1)Ca$_3$$^{3+}$
layers of edge-sharing N(1)Ca$_6$ octahedra. M-3N = 1.95, 1.77, Ca-5N = 2.37-2.79
Å.

CaNiN

J. Solid State Chem., 88, 459-464.

P4$_2$/mmc, 3.5809, 7.0096, Z = 2, powder data. Ca in 2(e): 0,0,1/4; Ni in 2(b):
1/2,1/2,0; N in 2(c): 0,1/2,0. Ni-N-Ni-N- linear chains, with Ca in
tetrahedral holes.

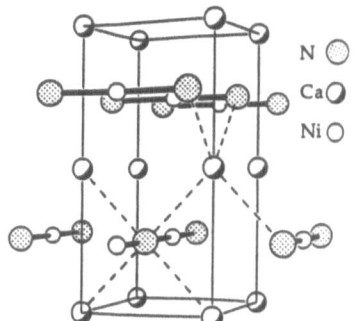

N	⊙
Ca	◯
Ni	◯

INTERATOMIC DISTANCES FOR CaNiN			
Ni–Ni[a]	3.5809 Å	Ni–Ni[b]	3.5048 Å
Ni–N	1.7904 Å	Ca–N	2.5053 Å
Ni–Ca	3.0352 Å		

[a] Distance between parallel chains.
[b] Distance between perpendicular chains.

Ca$_2$ZnN$_2$

J. Solid State Chem., **88**, 528-533.

I4/mmm, 3.5835, 12.6583, Z = 2, powder data. Ca in 4(e): 0,0,0.3360; Zn in 2(a): 0,0,0; N in 4(e): 0,0,0.1455. Double sheets of edge-sharing NCa$_5$Zn octahedra are linked by sharing Zn atoms. Ca–N = 2.413, 2.545, Zn–N = 1.842 A.

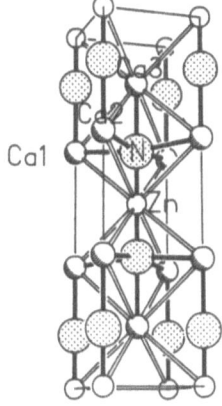

The crystal structure of Ca$_2$ZnN$_2$. The N is octahedrally coordinated to one Zn and five Ca atoms; the Ca has five nearest neighbors; the Zn is linearly coordinated to two nitrogen atoms.

Ce$_3$B$_2$N$_4$

J. Amer. Ceram. Soc., **73**, 2634-2639.

Immm, 3.5653, 6.3160, 10.7131, Z = 2, neutron powder data. Ce(1) in 2(a): 0,0,0; Ce(2) in 4(j): 1/2,0,0.2944; B in 4(h): 0,0.1416,1/2; N in 8(ℓ): 0,0.2552,0.3807. Planar N$_2$B–BN$_2$ units, with each B in the centre of a Ce$_6$ trigonal prism. B–B = 1.79, B–2N = 1.46, B–6Ce = 2.88, 2.97, N–5Ce = 2.43–2.68 A.

Cr_2N

J. Less-Common Metals, 158, L9-L10.

P$\bar{3}$1m, 4.752, 4.429, Z = 3, R = 0.037. Cr in 6(k): 0.333,0,0.2491; N(1) in
2(d): 1/3,2/3,1/2; N(2) in 1(a): 0,0,0. Isostructural with related
materials (3, 586; 21, 140; 45A, 100), with h.c.p. Cr, and N in holes.

Cu_3N

J. Less-Common Metals, 161, 175-184.

Pm3m, a = 3.819, 3.814 A, at 300, 130, Z = 1, R = 0.009-0.059. Cu in 3(d):
1/2,0,0; N in 1(a): 0,0,0. Anti-ReO$_3$-type (2, 31), but possibly with a
disordered tilting of the NCu$_6$ octahedra, so that N-Cu-N $\sim$ 176°.

$FeLi_3N_2$

J. Less-Common Metals, 161, 31-36.

Ibam, 4.872, 9.641, 4.792, Z = 4, R = 0.020. Fe in 4(a): 0,0,1/4; N in 8(j):
0.2237,0.8860,0; Li(1) in 8(g): 0,0.7411,1/4; Li(2) in 4(b): 1/2,0,1/4.
Fluorite superstructure, with chains along c of edge-sharing FeN$_4$ tetrahedra,
linked by LiN$_4$ tetrahedra. Fe-N = 1.957(2), Li-N = 2.111-2.179(4) A.

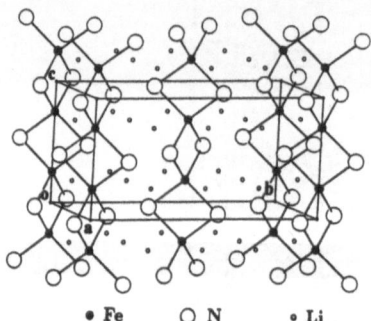

● Fe   ○ N   ● Li

Li_6MoN_4 (I), Li_6WN_4 (II), $Li_{15}Cr_2N_9$ (III), $Li_{14}Cr_2N_8O$ (IV)

Z. Naturforsch., 45B, 111-120.

I, II, P4$_2$/nmc, a = 6.673, 6.679, c = 4.925, 4.927 A, Z = 2, R = 0.016, 0.024.
III, P4/ncc, 10.233, 9.389, Z = 4, R = 0.056. IV, P3, 5.799, 8.263, Z = 1, R =
0.036. The compounds all have fluorite-type superstructures with Li and
transition metals in tetrahedral holes of a f.c.c. N. arrangement.

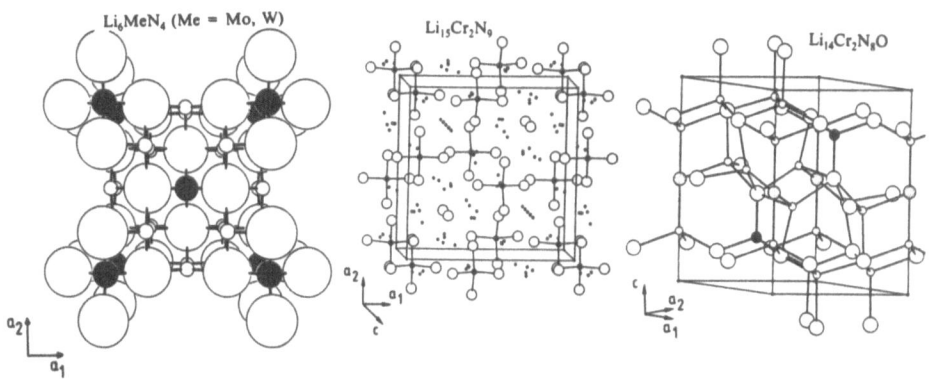

Li₆MeN₄ (Me = Mo, W)   Li₁₅Cr₂N₉   Li₁₄Cr₂N₈O

Li₆MoN₄	x	y	z
Mo (2a)	0	0	0
N (8g)	0	0,2405	−0,1983
Li1 (4d)	0	0,5	0,0858
Li2 (8f)	0,2868	0,2868(8)	0

Li₆WN₄	x	y	z
W (2a)	0	0	0
N (8g)	0	0,247	−0,197
Li1 (4d)	0	0,5	0,054
Li2 (8f)	0,289	0,289	0

	x	y	z
Cr (8f)	0,4195	0,5805	1/4
N1 (16g)	0,0617	0,4156	0,1447
N2 (16g)	0,0855	−0,2170	0,1326
N3 (4c)	1/4	1/4	0,3911
Li1 (16g)	−0,0850	−0,2699	0,2349
Li2 (16g)	0,2150	0,4123	0,0066
Li3 (16g)	0,4000	0,5655	−0,0221
Li4 (4a)	3/4	1/4	1/4
Li5 (4b)	3/4	1/4	0
Li6 (8f)*	−0,0969	0,0969	1/4

	x	y	z
Cr (2d)	1/3	2/3	0,2382
N1 (6g)	0,0608	0,6889	0,1676
N2 (2d)	1/3	2/3	0,4528
O (1b)	0	0	1/2
Li1 (6g)	−0,0717	0,3028	0,0923
Li2 (6g)	0,3555	0,3181	0,4101
Li3 (2c)	0	0	0,2501

LiPN₂

Z. anorg. Chem., **588**, 19-25

$I\overline{4}2d$, 4.575, 7.118, Z = 4, powder data. Li in 4(b): 0,0,1/2; P in 4(a): 0,0,0; N in 8(d): 0.1699,1/4,1/8. The structure in between a filled β-cristobalite type and chalcopyrite type, with a framework of corner-sharing PN_4 tetrahedra, and Li also with tetrahedral coordination. P-N = 1.65, Li-N = 2.09(1) A.

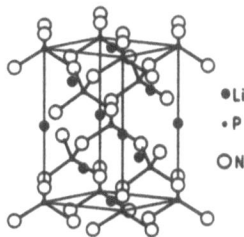

●Li
•P
◯N

Li₇PN₄

J. Solid State Chem., **87**, 101-106.

$P\overline{4}3n$, 9.3648, Z = 8, powder data. Antifluorite derivative structure, with isolated tetrahedral PN_4^{7-} anions linked by tetrahedrally-coordinated Li⁺ ions. P-N = 1.69, 1.73(1), Li-N = 1.95-2.24(4) A.

		x	y	z
Li (1)	6b	0	¼	¼
Li (2)	6d	¼	0	¼
Li (3)	8e(xxx)	0.2245	0.2245	¼
Li (4)	24i	0.2493	0.2386	0.2245
Li (5)	12f	0.2576	0	-0.0258
P (1)	6c	¼	0	0
P (2)	2a	0	0	¼
N (1)	24i	0.3532	0.3844	0.1012
N (2)	8e(xxx)	0.1040	0.1040	0.1040

Na_3BN_2

J. Less-Common Metals, 162, L17–L22

$P2_1/c$, 5.717, 7.931, 7.883, 111.32, Z = 4, powder data. β-Li_3BN_2-type (53A, 33), with linear BN_2^{3-} anions linked by Na^+ ions. B–N = 1.34, Na–N = 2.41–2.78 A.

	x	y	z
Na(1)	0.2511	0.4687	0.4921
Na(2)	0.2413	0.0218	0.3778
Na(3)	0.7424	0.2080	0.3131
B	0.2133	0.3226	0.1760
N(1)	0.4158	0.4237	0.2241
N(2)	0.0120	0.2206	0.1303

$Na_3P_6N_{11}$

Latv. PSR Zinat. Akad. Vestis, Kim. Ser., No.3, 299–301.

$P2_13$, 10.112, Z = 4, R = 0.087. PN_4 tetrahedra with 3-coordinate N.

NbN_{1-x} (x = 0.42)

J. Less-Common Metals, 160, 193–196.

I4/mmm, [4.39, 8.67], Z = 8, R = 0.033. Nb(1) in 4(e): 0,0,0.2447; Nb(2) in 4(c): 0,1/2,0; 1.5 N(1) in 2(b): 0,0,1/2; 3.1 N(2) in 4(d): 0,1/2,1/4. Structure as previously described (46A, 104; see also 56A, 37).

Sr_2N

J. Solid State Chem., 87, 134–140.

$R\bar{3}m$, a = 3.8566, c = 20.6958, Z = 3, neutron powder data. Sr in 6(c): 0,0,0.26737; N in 3(a): 0,0,0. Anti-$CdCl_2$ type, as previously described (38A, 135). Sr–N = 2.612(1) A.

AgSmZnP$_2$

J. Solid State Chem., <u>89</u>, 227-236.

P$\bar{3}$m1, 4.1247, 6.6920, Z = 1, R = 0.021. Sm in 1(a): 0,0,0; Ag/Zn in 2(d): 1/3,2/3,z, z = 0.3666; P in 2(d): z = 0.7541. Hexagonal close-packed P, with Sm in half of the octahedral holes and Ag/Zn disordered in half of the tetrahedral holes. Sm-6P = 2.895(1), Ag/Zn-4P = 2.515, 2.593(2) A.

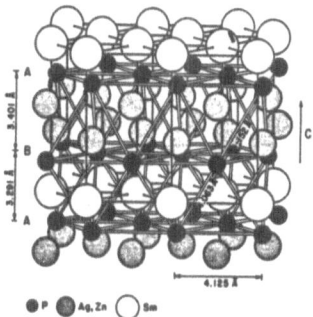

Al$_2$Cs$_6$P$_4$

Z. Kristallogr., <u>193</u>, 303-304.

P2$_1$/c, 11.233, 8.641, 18.986, 100.06, Z = 4, R = 0.050. Dimeric PAlP$_2$AlP^{6-} anions with 4-membered rings; Al-P = 2.34 (ring), 2.25 (terminal) A.

Al$_4$K$_{12}$P$_8$, Al$_4$K$_{12}$As$_8$

Z. Kristallogr., <u>193</u>, 299-300, 301-302.

P$\bar{1}$, a = 8.871, 9.062, b = 11.879, 12.164, c = 15.280, 15.570 A, α = 72.47, 72.40, β = 73.35, 73.05, γ = 71.62, 71.63°, Z = 2, R = 0.056, 0.098. Infinite AlX$_3^{3-}$ chains of edge-sharing AlX$_4$ tetrahedra, and isolated XAlX$_2$AlX^{6-} dimeric anions with 4-membered rings. Al-P = 2.43; 2.25, 2.34, Al-As = 2.53; 2.33, 2.44 A.

BaCu$_2$P$_4$

J. Less-Common Metals, <u>167</u>, 127-134.

Fddd, 5.435, 18.973, 10.244, Z = 8, R = 0.053. Ba in 8(a): 1/8,1/8,1/8; Cu in 16(f): 1/8,0.5041,1/8; P in 32(h): 0.1951,0.1789,0.8169. Chains of edge-sharing CuP$_4$ tetrahedra, with Ba in hexagonal tubes.

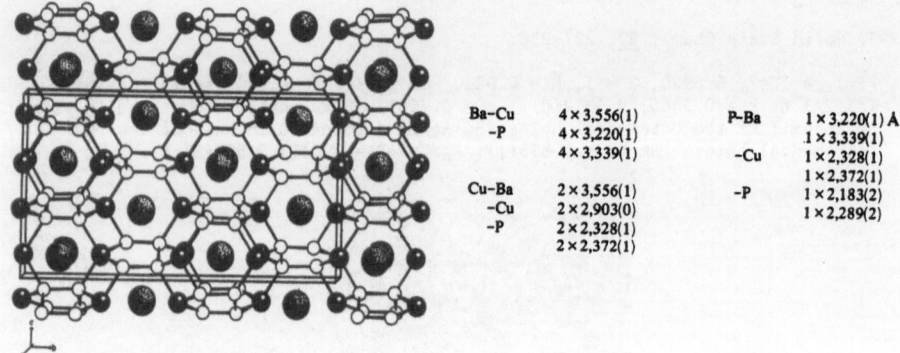

Ba–Cu	4 × 3,556(1)	P–Ba	1 × 3,220(1) Å
–P	4 × 3,220(1)		1 × 3,339(1)
	4 × 3,339(1)	–Cu	1 × 2,328(1)
			1 × 2,372(1)
Cu–Ba	2 × 3,556(1)	–P	1 × 2,183(2)
–Cu	2 × 2,903(0)		1 × 2,289(2)
–P	2 × 2,328(1)		
	2 × 2,372(1)		

$BaCu_8P_4$, $BaCu_8As_4$

Z. anorg. Chem., 582, 224; 588; 109–116.

I4/m, a = 10.192, 10.455, c = 3.878, 3.969 A, Z = 2, R = 0.045, 0.039. Ba in
2(a): 0,0,0; Cu(1), Cu(2), P (or As) in 8(h): x,y,0, for the phosphide, x =
0.2266, 0.0154, 0.2344, y = 0.6640, 0.3651, 0.4259; for the arsenide, x =
0.2150, 0.0179, 0.2420, y = 0.6720, 0.3712, 0.4289. Framework of edge-sharing
Cu(1)X_4 tetrahedra (X = P or As) with two channels along c; the larger channel
contains Ba, and the smaller one contains chains of edge-sharing Cu(2)$_4$
tetrahedra.

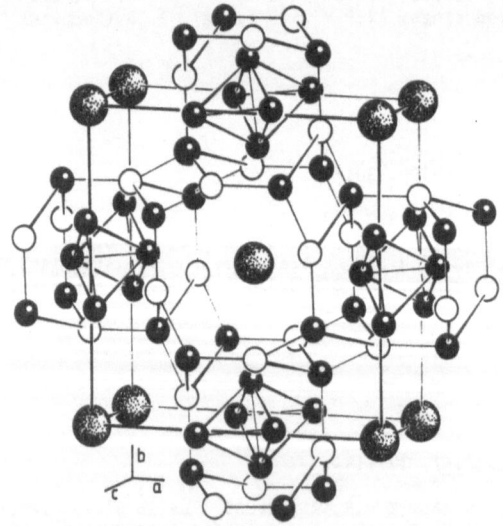

$BaNi_9P_5$

J. Solid State Chem., 87, 10–14.

P6$_3$/mmc, 6.5359, 10.854, Z = 2, R = 0.023. Hexagonal layers of Ba, with each
Ba surrounded by $Ni_{18}P_{12}$ cages. Ba–Ni = 3.49, 3.56, Ba–P = 3.35, Ni–Ni =
2.47–2.57, Ni–P = 2.16–2.53 A.

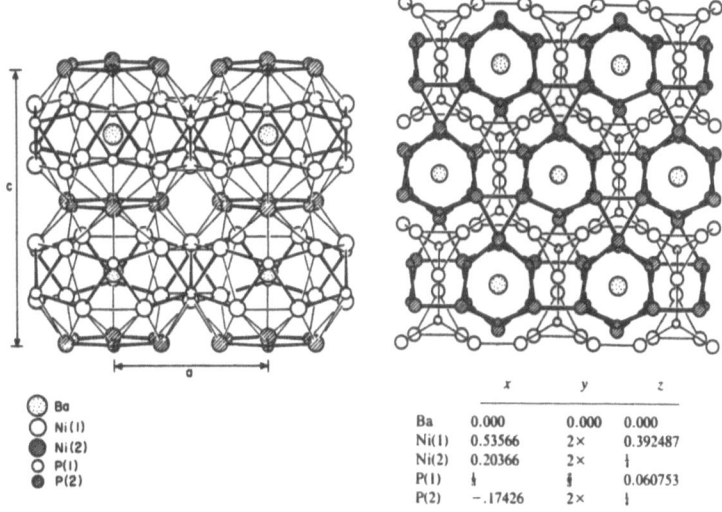

	x	y	z
Ba	0.000	0.000	0.000
Ni(1)	0.53566	2×	0.392487
Ni(2)	0.20366	2×	¼
P(1)	¼	¾	0.060753
P(2)	−.17426	2×	¼

- ◯ Ba
- ◯ Ni(1)
- ● Ni(2)
- ◯ P(1)
- ● P(2)

BeK_4P_2, BeK_4As_2

Z. Kristallogr., 192, 263-264, 265-266.

R$\bar{3}$m, a = 5.588, 5.681, c = 25.132, 25.80 A, Z = 3, R = 0.027, 0.060. Linear
BeX$_2^{4-}$ ions (X = P, As), linked by K$^+$ ions. Be-P = 1.98, Be-As = 2.06 A.

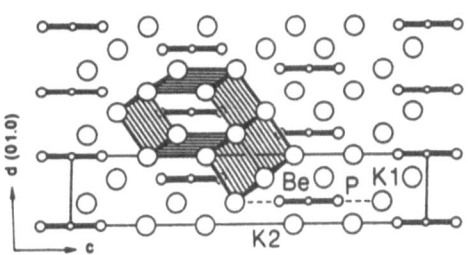

Cs_3BP_2, Cs_3BAs_2

Z. Kristallogr., 193, 295-296, 297-298.

C2/c, a = 9.834, 10.06, b = 9.674, 9.87, c = 9.859, 10.171 A, β = 109.77,
110.62, Z = 4, R = 0.031, 0.049. Linear BX$_2^{3-}$ anions; B-P = 1.771(2), B-As =
1.879(2) A.

$Cs_6Ga_2P_4$

Z. Kristallogr., 193, 287-288.

P2$_1$/c, 11.173, 8.661, 18.939, 99.64, Z = 4, R = 0.069. Isolated dimeric
PGaP$_2$GaP^{6-} ions with 4-membered rings; Ga-P = 2.358 (ring), 2.247 (terminal)
A.

CuSmP$_2$

Ž. Neorg. Khim., <u>35</u>, 1656-1658 [Russ. J. Inorg. Chem., <u>35</u>, 942-944].

I4/mmm, 3.839, 19.431, Z = 4, powder data. Cu in 4(d): 0,1/2,1/4; Sm in 4(e):
0,0,0.117; P(1) in 4(c): 0,1/2,0; P(2) in 4(e): 0,0,0.342. SrZnBi$_2$-type
(<u>42</u>A, 17), as for CuUP$_2$ (<u>54</u>A, 37).

Fe$_4$ScP$_2$, DyNi$_4$As$_2$

J. Less-Common Metals, <u>161</u>, 125-134.

P4$_2$/mnm, a = 6.962, 7.239, c = 3.622, 3.760 A, Z = 2, R = 0.014, 0.019. Sc or
Dy in 2(b): 0,0,1/2; Fe or Ni in 8(i): x,y,0, x = 0.08342, 0.08807, y =
0.34098, 0.34584; P or As in 4(g), x,x̄,0, x = 0.21754, 0.21764. Fe$_4$ZrSi$_2$-type
(<u>41</u>A, 82).

GaK$_2$NaP$_2$

Z. Kristallogr., <u>192</u>, 267-268.

Ibam, 6.613, 14.490, 6.401, Z = 4, R = 0.036. Ga in 4(b): 0,1/2,1/4; K in
8(j): 0.2421,0.8248,0; Na in 4(a): 0,0,1/4; P in 8(j): 0.1985,0.5920,0.
[GaP$_2^{3-}$]$_n$ chains of edge-sharing tetrahedra, as in SiS$_2$ (<u>49</u>A, 55); Ga-P =
2.462 A.

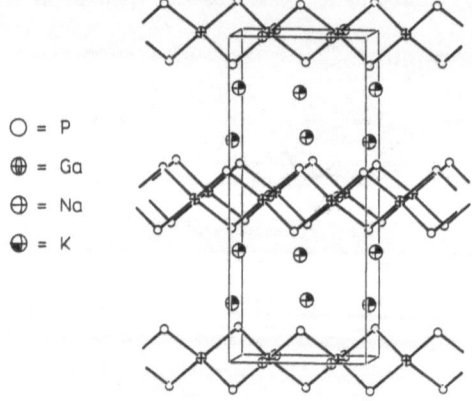

O = P

⊕ = Ga

⊖ = Na

◓ = K

GaRb$_3$P$_2$

Z. Kristallogr., <u>192</u>, 271-272.

Pbca, 14.634, 24.893, 9.163, Z = 16, R = 0.082. Isolated dimeric Ga$_2$P$_4^{6-}$ ions
with 4-membered rings; Ga-P = 2.376 (ring), 2.237 A (terminal).

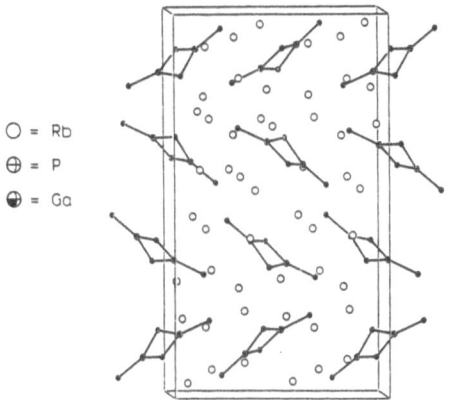

O = Rb
⊕ = P
◐ = Ga

GeNi$_2$P, Ni$_2$SiAs

Ž. Neorg. Khim., 35, 1938-1939 [Russ. J. Inorg. Chem., 35, 1104-1105].

Pbca, a = 6.0146, 6.274, b = 4.9864, 5.023, c = 13.889, 14.480 Å, Z = 8, powder data. Ni$_2$SiP-type (56A, 36).

In$_2$Ni$_{21}$P$_6$

J. Solid State Chem., 85, 315-317.

Fm3m, 11.1120, Z = 4, R = 0.086. Cr$_{23}$C$_6$-type (3, 59, 367; 27, 122; 38A, 60).

	x	y	z
Ni(1)	48h	0	0.1740	0.1740
Ni(2)	32f	0.3842	0.3842	0.3842
Ni(3)	4a	0	0	0
P(1)	24e	0.2623	0	0
In(1)	8c	¼	¼	¼

INTERATOMIC DISTANCES LESS THAN 4.0 (Å)

In(1)	–	4 Ni(2)	2.5830	Ni(2)	–	In(1)	2.5830
	–	12 Ni(1)	3.0239		–	6 Ni(1)	2.7442
	–	12 P(1)	3.9311		–	3 Ni(3)	3.6394
Ni(1)	–	2 P(1)	2.1680	P(1)	–	4 Ni(1)	2.1680
	–	Ni(1)	2.3890		–	4 Ni(2)	2.2687
	–	5 Ni(1)	2.7342		–	Ni (3)	2.9144
	–	4 Ni(2)	2.7442		–	4 Ni(1)	3.6912
	–	2 In(1)	3.0239		–	4 P(1)	3.7358
	–	2 P(1)	3.6912		–	4 In(1)	3.9311
	–	2 Ni(1)	3.8667		–	4 Ni(1)	3.9962
	–	2 P(1)	3.9962	Ni(3)	–	12 Ni(1)	2.7342
Ni(2)	–	3 P(1)	2.2687		–	6 P(1)	2.9144
	–	3 Ni(2)	2.5734				

All ESDs are smaller than 0.0002 Å.

K$_3$BP$_2$, K$_3$BAs$_2$

Z. Kristallogr., 191, 311-312, 313-314.

C2/c, a = 9.362, 9.609, b = 8.894, 9.109, c = 9.013, 9.194 Å, β = 110.99, 111.68°, Z = 4, R = 0.042, 0.042. BX$_2$ groups, B-P = 1.767, B-As = 1.868 Å.

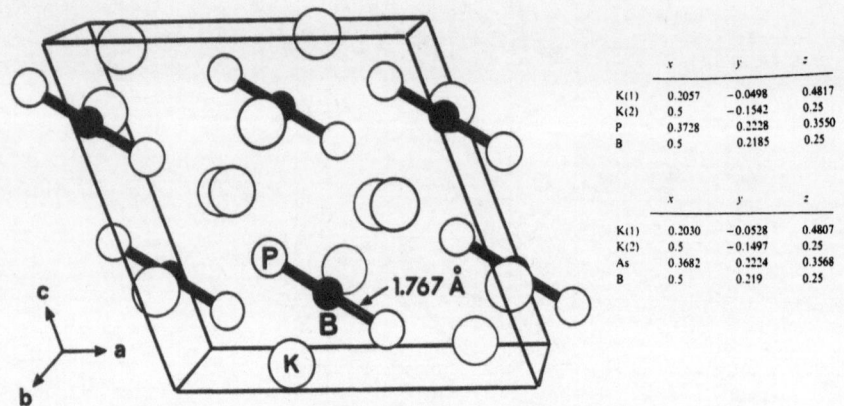

	x	y	z
K(1)	0.2057	−0.0498	0.4817
K(2)	0.5	−0.1542	0.25
P	0.3728	0.2228	0.3550
B	0.5	0.2185	0.25

	x	y	z
K(1)	0.2030	−0.0528	0.4807
K(2)	0.5	−0.1497	0.25
As	0.3682	0.2224	0.3568
B	0.5	0.219	0.25

K_2NiP_2, K_2NiAs_2

Z. Kristallogr., 193, 291-292, 293-294.

Cmcm, a = 6.430, 6.607, b = 13.65, 13.87, c = 5.644, 5.745 A, Z = 4, R = 0.024, 0.055. Chains of edge-sharing NiX_4 square planes; Ni-P = 2.265(2), Ni-As = 2.356, P-P (shared edge) = 2.187(4), As-As = 2.424(4) A.

LnPdP (Ln = La, Ce, Pr, Nd, Sm, Eu, Gd), ErPdP

Z. Naturforsch., 45B, 1262-1266.

Ln, $P6_3/mmc$, a = 4.269, 4.248, 4.232, 4.219, 4.198, 4.150, 4.179, c = 7.909, 7.799, 7.732, 7.690, 7.576, 8.112, 7.507 A, Z = 2, R = 0.097 for Ln = Ce. Ln in 2(a): 0,0,0; Pd in 2(c): 1/3,2/3,1/4; P in 2(d): 2/3,1/3,1/4. Ni_2In-type (9, 91).

ErPdP, Pnma, 6.826, 3.965, 7.689, Z = 4, R = 0.047. Er, Pd, P in 4(c): x,1/4,z, x = 0.5314, 0.6475, 0.2495, z = 0.6880, 0.0619, 0.1228. NiTiSi-type (30A, 75). The Tb, Dy, Ho, Tm, Yb, and Lu compounds are isostructural.

Na_3BP_2

Z. Kristallogr., 193, 281-282.

$P2_1/c$, 6.995, 9.279, 9.159, 111.03, Z = 4, R = 0.059. Linear BP_2^- ions linked by Na^+ ions; B-P = 1.770(3) A.

$Nd_2Ni_7P_4$

Dopov. Akad. Nauk. Ukr.RSR, Ser. B: Geol., Khim. Biol. Nauki, No. 9, 47-50.

$Pmn2_1$, 3.7588, 9.238, 10.413, Z = 2, R = 0.055. New structure type, with trigonal prismatic coordination for P.

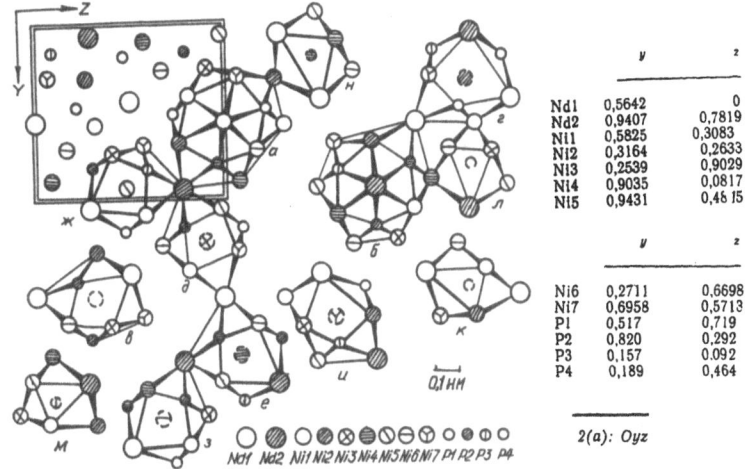

	y	z
Nd1	0,5642	0
Nd2	0,9407	0,7819
Ni1	0,5825	0,3083
Ni2	0,3164	0,2633
Ni3	0,2539	0,9029
Ni4	0,9035	0,0817
Ni5	0,9431	0,4615

	y	z
Ni6	0,2711	0,6698
Ni7	0,6958	0,5713
P1	0,517	0,719
P2	0,820	0,292
P3	0,157	0,092
P4	0,189	0,464

2(a): 0yz

PtP_2

Z. Kristallogr., 190, 143-146.

Pa3, 5.6968, Z = 4, R = 0.017, 0.040 (two data sets). Pt in 4(a): 0,0,0; P in 8(c): x,x,x, x = 0.3896. Pyrite-type as previously described (24, 47; 41A, 116). P-P = 2.180(1) A.

SnP

J. Solid State Chem., 87, 202-207.

$P\bar{3}m1$, 4.3922, 6.040, Z = 2, R = 0.068. 2 Sn in 2(d): 1/3,2/3,0.3025; 0.5 P(1) in 2(c): 0,0,0.185; 1.5 P(2) in 6(i): 0.863,0.137,0.054. CdI_2-related structure, with hexagonal packing of Sn, and P_2 pairs filling octahedra in every second layer; the P_2 pairs are disordered in four orientations. Sn-P = 2.62-2.67 A.

Zr_3P, Zr_3PD_{3-x}

J. Less Common Metals, 161, 269-278.

Zr_3P, $P4_2/n$, 10.7994, 5.3553, Z = 8, neutron powder data. Ti_3P-type structure as previously described (32A, 111).

Zr_3PD_{3-x}, $P4_2/nbc$, 11.1800, 5.4297, Z = 8, neutron powder data. β-V_3S-type (23, 242), with D in two tetrahedral interstices.

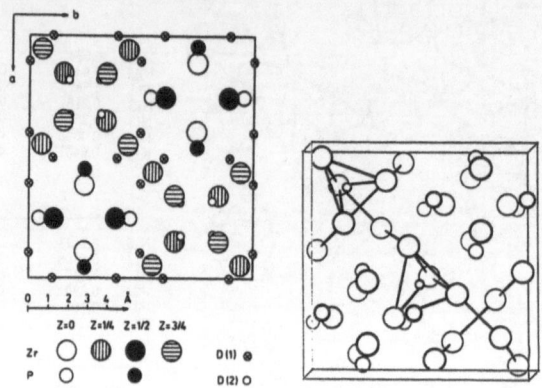

Zr_3PD_{3-x}

		x	y	z
Zr_3P. space group $P4_2/n$; origin at $\bar{1}$				
a = 10.7994 c = 5.3553 Å				
Zr(1)	8g	0.1653	0.6410	0.7122
Zr(2)	8g	0.1120	0.2786	0.5281
Zr(3)	8g	0.0719	0.5346	0.2358
P	8g	0.0424	0.2915	0.0347
Zr_3PD_{3-x}, space group $P4_2/nbc$; origin at $\bar{1}$				
a = 11.1800 c = 5.4297 Å				
Deuterium occupancies (%): D(1) 99.6 D(2) 44.7				
Zr(1)	8h	0.6108	1/4	0
Zr(2)	8j	0.1555	0.1555	1/4
Zr(3)	8j	0.5560	0.5560	1/4
P	8i	0.5451	1/4	1/2
D(1)	16k	0.9995	0.1050	0.0823
D(2)	16k	0.1813	0.1749	0.6338

Zr_3PO_{1-x}

J. Less-Common Metals, 167, 21-30.

Cmcm, 3.5409, 10.7248, 8.3634, Z = 4, R = 0.037. Zr(1) in 8(f): 0,0.13951,0.05325; Zr(2) in 4(c): 0,y,1/4, y = 0.43850; P in 4(c): y = 0.7460; 1.70 in 4(b): 0,1/2,0. Filled Re_3B-type (24, 73; 37A, 55).

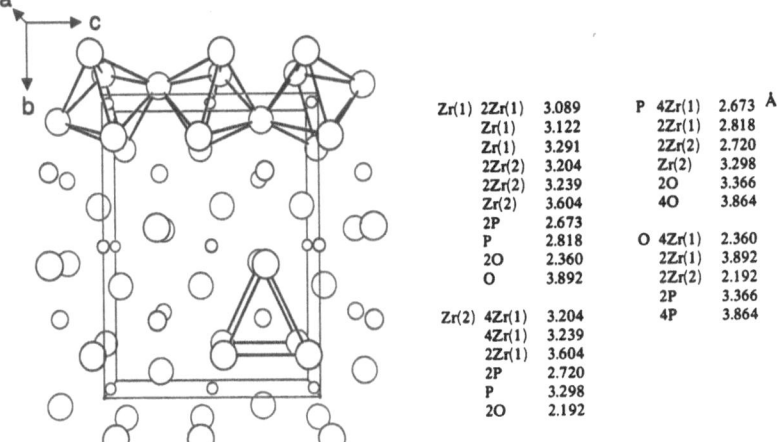

Zr(1)	2Zr(1)	3.089	P	4Zr(1)	2.673 Å
	Zr(1)	3.122		2Zr(1)	2.818
	Zr(1)	3.291		2Zr(2)	2.720
	2Zr(2)	3.204		Zr(2)	3.298
	2Zr(2)	3.239		2O	3.366
	Zr(2)	3.604		4O	3.864
	2P	2.673			
	P	2.818	O	4Zr(1)	2.360
	2O	2.360		2Zr(1)	3.892
	O	3.892		2Zr(2)	2.192
				2P	3.366
Zr(2)	4Zr(1)	3.204		4P	3.864
	4Zr(1)	3.239			
	2Zr(1)	3.604			
	2P	2.720			
	P	3.298			
	2O	2.192			

AℓCs₃As₂

Z. Kristallogr., 192, 269-270.

P2₁/c. 11.458, 8.831, 19.453, 99.68, Z = 8, R = 0.056. Isolated dimeric Aℓ₂As₄⁶⁻ ions with 4-membered rings; Aℓ-As = 2.438 (ring), 2.342 A (terminal).

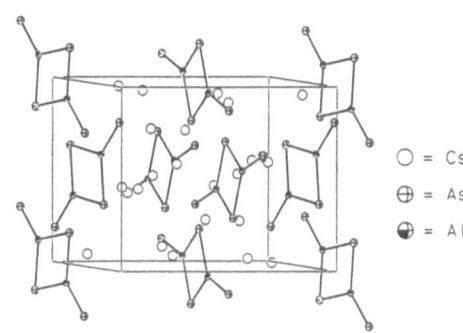

○ = Cs

⊕ = As

◐ = Aℓ

CaCu₂As₂ (I), Ca₂Cu₆P₅ (II), Ba₂Cu₃P₄ (III)

Z. anorg. Chem., 581, 173-182.

I, I4/mmm, 4.129, 10.251, Z = 2, R = 0.062; Ca in 2(a): 0,0,0; 3.5 Cu in 4(d): 0,1/2,/1/4; As in 4(e): 0,0,0.3799. II, I4/mmm, 4.015, 24.657, Z = 2, R = 0.046; Ca in 4(e): 0,0,z, z = 0.3511; Cu(1) in 8(g): 1/2,0,0.0537; Cu(2) in 4(d): 0,1/2,1/4; P(1) in 2(b): 0,0,1/2; P(2) in 4(e): z = 0.1067; P(3) in 4(e): z = 0.1974. III, Ibam, 9.347, 12.415, 6.188, Z = 4, R = 0.058; Ba in 8(j): x,y,0, x = 0.2959, y = 0.6097; Cu(1) in 8(g): 0,0.2787,1/4; Cu(2) in 4(b): 1/2,0,1/4; P(1) in 8(j): x = 0.3684, y = 0.3437; P(2) in 8(j): x = 0.1336, y = 0.3780. I has a Cr_2ThSi_2-type structure (43A, 99; 55A, 31); II contains a stacking of three unit cells of Cr_2ThSi_2-type, with two missing Ca-P layers; III is a variant of the Cr_2ThSi_2-type, with bands of edge-sharing CuP_4 tetrahedra and P_2 pairs.

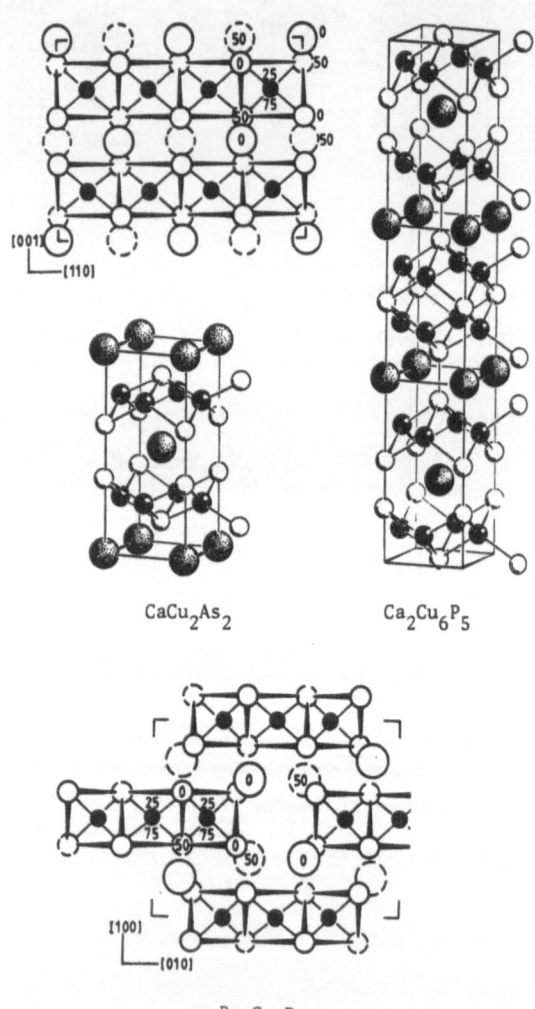

CaCu₂As₂ — $CaCu_2As_2$

$Ca_2Cu_6P_5$

$Ba_2Cu_3P_4$

Cs$_3$GaAs$_2$

Z. Kristallogr., <u>192</u>, 273-274.

P2$_1$/c, 11.371, 8.857, 19.460, 99.225, Z = 8, R = 0.055. Isolated dimeric
Ga$_2$As$_4$$^{6-}$ anions; Ga-As = 2.459 (ring), 2.343 A (terminal).

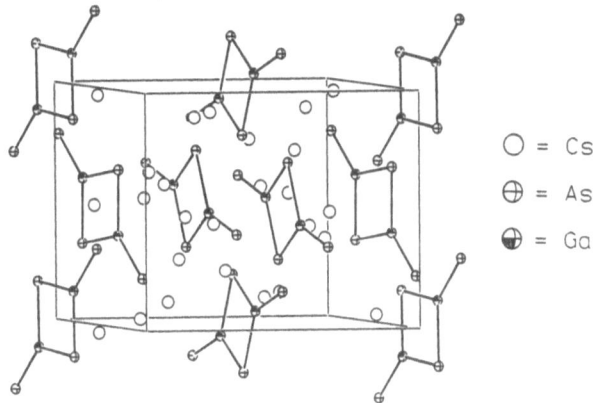

O = Cs

⊕ = As

◕ = Ga

Cs$_2$PdAs$_2$

Z. anorg. Chem., <u>584</u>, 138-142.

Cmcm, 7.068, 14.766, 6.366, Z = 4, R = 0.054. Cs(1) in 4(c): 0,0.4091,1/4;
Cs(2) in 4(c): 0,0.7911,1/4; Pd in 4(a): 0,0,0; As in 8(g):
0.1694,0.1016,1/4. K$_2$PdAs$_2$-type (<u>45</u>A, 26), with zigzag ribbons of edge-sharing
PdAs$_4$ square planes, linked by 6-coorindate Cs ions. Pd-As = 2.49, Cs-As =
3.65-3.95 A.

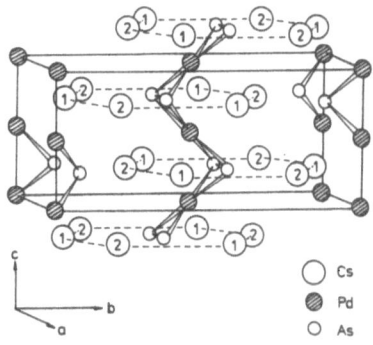

O Cs

◕ Pd

O As

Fe₃GaAs (I), Fe₃Ga₁.₇As₀.₃ (II)

J. Less-Common Metals, 157, 315-325.

P6₃/mmc, a = 4.02, 8.10, c = 5.03, 5.00 A, Z = 1, 4, powder data. I has the InNi₂ (B8₂)-type structure (9, 91) as previously described (56A, 58), with a fully occupied octahedral Fe(1) site, but partially-filled trigonal-bipyramidal Fe(2) site. II is a superstructure resulting from ordering of vacancies in the Fe(5) site.

									Fe₃Ga₁.₇As₃		
			Fe₃GaAs						x	y	z
			x	y	z						
1.95	Fe(1)	2a	0	0	0	2.0	Fe(3)	2a	0	0	0
1.28	Fe(2)	2d	1/3	2/3	3/4	5.5	Fe(4)	6g	1/2	0	0
2.0	Ga/As(1)	2c	1/3	2/3	1/4	0.0	Fe(5)	2c	1/3	2/3	1/4
						5.4	Fe(6)	6h	0.848	0.152	1/4
						0.6	Ga/As(2)	6h	0.848	0.152	1/4
						2.0	Ga/As(3)	2d	2/3	1/3	1/4
						6.0	Ga/As(4)	6h	0.194	0.806	1/4

Ga₃K₃As₄

Inorg. Chem., 29, 3892-3894.

Pnna, 6.597, 14.792, 10.589, at 100K, Z = 4, R = 0.071. Covalently bonded Ga₃As₄³⁻ (010) sheets, linked by K⁺ ions.

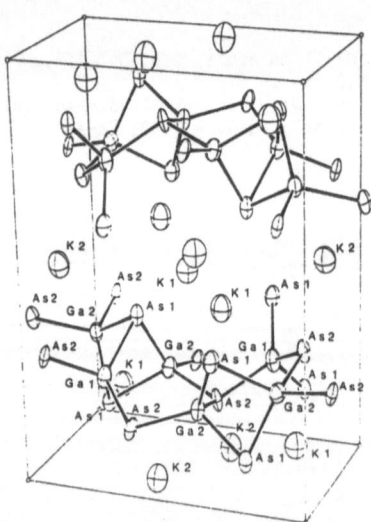

	x	y	z
As(1)	0.5411	0.1365	0.1932
As(2)	0.0407	0.1758	0.4047
Ga(1)	0.8083	0.2500	0.2500
Ga(2)	0.4025	0.2166	0.3889
K(1)	0.0593	0.0923	0.1202
K(2)	0.7500	0.0000	0.4219

Selected Distances (Å) and Angles (deg) in K₃Ga₃As₄

Bonding Distances

Ga(1)-As(1)	2.507 (2)	K(1)-As(1)	3.336 (3)
Ga(1)-As(2)	2.498 (1)	K(1)-As(1)	3.385 (3)
Ga(2)-As(1)	2.556 (2)	K(1)-As(2)	3.259 (3)
Ga(2)-As(1)	2.512 (2)	K(2)-As(1)	3.441 (3)
Ga(2)-As(2)	2.467 (2)	K(2)-As(2)	3.237 (1)
Ga(2)-As(2)	2.444 (2)	K(2)-As(2)	3.470 (2)

Nonbonding Distances

Ga(1)-Ga(2)	3.094 (2)	Ga(1)-K(1)	3.174 (3)
Ga(2)-Ga(2)	3.102 (2)		

Angles

As(1)-Ga(1)-As(1)	90.66 (7)	As(1)-Ga(2)-As(2)	106.75 (6)
As(1)-Ga(1)-As(2)	107.15 (4)	As(1)-Ga(2)-As(2)	118.49 (6)
As(1)-Ga(1)-As(2)	124.62 (4)	As(1)-Ga(2)-As(2)	125.96 (6)
As(2)-Ga(1)-As(2)	104.27 (7)	As(1)-Ga(2)-As(2)	112.76 (6)
As(1)-Ga(2)-As(1)	89.42 (6)	As(2)-Ga(2)-As(2)	103.86 (5)

GaK$_2$NaAs$_2$

Z. Kristallogr., <u>193</u>, 285-286.

Ibam, 6.7331, 14.8089, 6.574, Z = 4, R = 0.045. Infinite chains of
edge-sharing GaAs$_4$ tetrahedra, linked by K$^+$ and Na$^+$ ions; Ga-As = 2.550(1) A,
As-Ga-As = 99.8, Ga-As-Ga = 80.2°.

HfNa$_5$As$_3$

Z. Naturforsch., <u>45</u>B, 559-562.

P2$_1$/n, 13.844, 7.549, 8.433, 89.63, Z = 4, R = 0.057. Na$_5$SiP$_3$-type (<u>52</u>A, 76),
with Hf$_2$As$_6$$^{10-}$ anions of two edge-sharing tetrahedra, linked by Na$^+$ ions.
Hf-As = 2.566 (terminal), 2.613, 2.634(2) A (bridging).

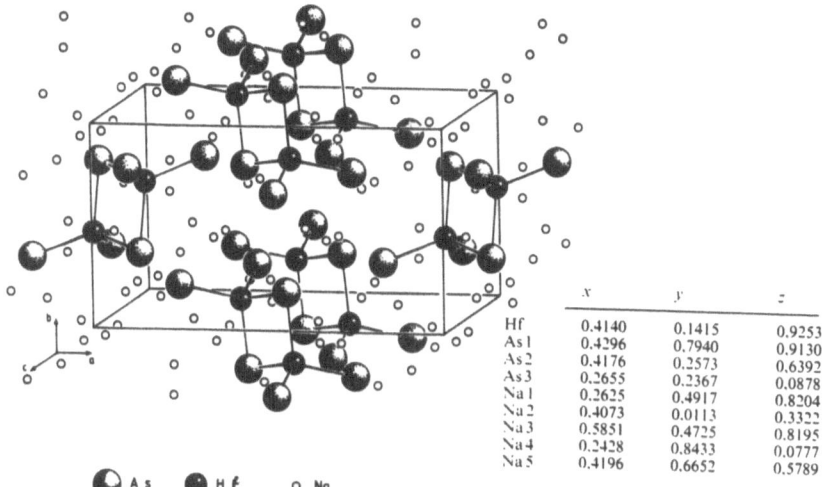

	x	y	z
Hf	0.4140	0.1415	0.9253
As1	0.4296	0.7940	0.9130
As2	0.4176	0.2573	0.6392
As3	0.2655	0.2367	0.0878
Na1	0.2625	0.4917	0.8204
Na2	0.4073	0.0113	0.3322
Na3	0.5851	0.4725	0.8195
Na4	0.2428	0.8433	0.0777
Na5	0.4196	0.6652	0.5789

◖ As   ● Hf   ○ Na

Mn$_3$As$_2$

Z. Kristallogr., <u>190</u>, 259-269.

C2/m, 13.856, 3.777, 13.622, 107.79, Z = 8, R = 0.034. Intermediate between
NiAs and Ni$_2$In (filled NiAs) types. The structure contains h.c.p. As layers
interrupted by some c.c.p. layers; each As has mono- or bicapped trigonal
prismatic coordination to Mn, and Mn atoms have 4, 5, and 6 coordinations to
As, plus weak bonds to Mn. Mn-As = 2.46-2.91, Mn-Mn = 2.79-3.29 A. The
material is slightly Mn-deficient.

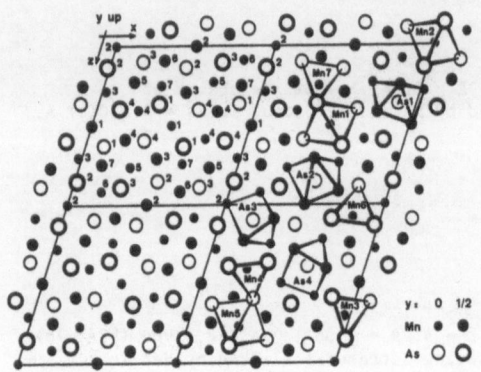

		x	y	z
Mn(1)	2d	0	1/2	1/2
Mn(2)	2a	0	0	0
Mn(3)	4i	0.02856	0	0.28431
Mn(4)	4i	0.26636	0	0.41035
Mn(5)	4i	0.67758	0	0.22468
Mn(6)	4i	0.82698	0	0.08903
Mn(7)[a]	4i	0.35736	0	0.24614
As(1)	4i	0.11743	0	0.63301
As(2)	4i	0.47489	0	0.13711
As(3)	4i	0.19746	0	0.10423
As(4)	4i	0.36435	0	0.60025

[a] The occupancy of the Mn(7) position was found to be 79.1 %

Na_3As

Z. Naturforsch., 45B, 1388-1392.

Below 3.6 GPa, α-form, $P6_3/mmc$, 4.874, 8.515, Z = 2, powder data. As in 2(c): 1/3,2/3,1/4; Na(1) in 2(b): 0,0,1/4; Na(2) in 4(f): 1/3,2/3,7/12. Na_3As-type (5, 6).

Above 3.6 GPa, β-form, Fm3m, 6.835, Z = 4, powder data. As in 4(a): 0,0,0; Na(1) in 4(b): 1/2,1/2,1/2; Na(2) in 8(c): 1/4,1/4,1/4. $BiLi_3$-type (3, 637).

$Ni_{1.6}UAs_2$

J. Less-Common Metals, 159, 121-125.

P4/nmm, 3.994, 9.281, Z = 2, R = 0.058. U in 2(c): 1/4,1/4,0.2509; Ni(1) in 2(b): 3/4,1/4,1/2; 1.2Ni(2) in 2(c): 1/4,1/4,0.8909; As(1) in 2(a): 3/4,1/4,0; As(2) in 2(c): 1/4,1/4,0.6423. $CaBe_2Ge_2$-type (43A, 28).

NiU_3As_4

J. Less-Common Metals, 157, L1-L3.

$I\bar{4}3d$, 8.672, Z = 4, R = 0.038. U in 12(a): 3/8,0,1/4; As in 16(c): x,x,x, x = 0.0795; 3.6 Ni in 12(b): 7/8,0,1/4. Filled U_3As_4 (Th_3P_4)-type (16, 24; 40A, 26). U-8As = 2.96, 3.04, U-4Ni = 2.65 A.

$Ag_{1.5}Bi_{5.5}S_9$ (I), $(Ag,Cu)Pb_3Bi_5Se_{11}$ (II), $Ag_{3.5}Bi_{7.5}S_{13}$ (III) (idealized compositions)

Neues Jb. Miner. Mh., 193-204.

C2/m, a = 13.37, 13.89, 13.41, b = 4.05, 4.22, 4.05, c = 14.71, 20.75, 21.51 A, β = 99.5, 115.5, 94.5°, Z = 2, film data. Homologues of pavonite, $AgBi_3S_5$ (⁴P, ⁶P, ⁸P) (45A, 39).

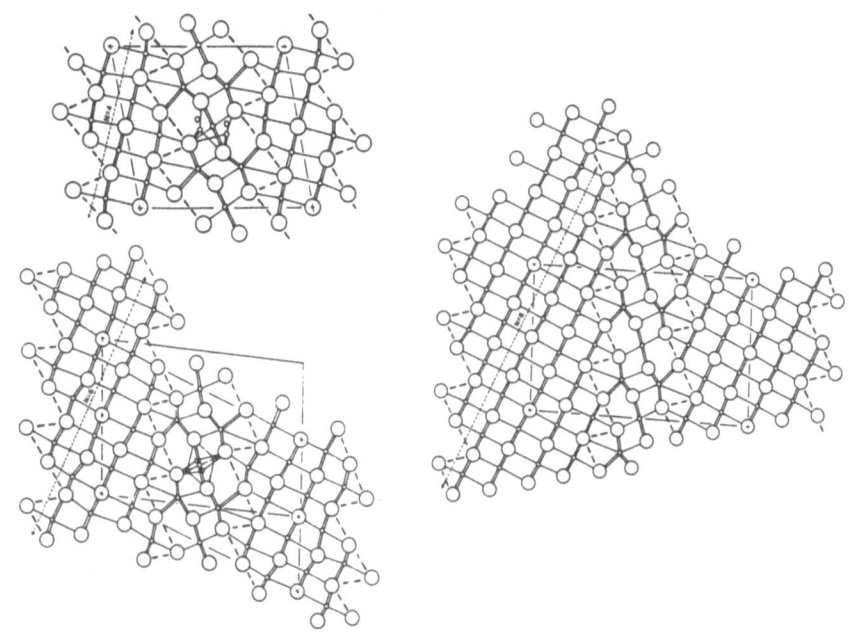

$AgMo_6S_8$

Kristallografija, 35, 349-354 [Soviet Physics-Crystallography, 35, 197-200.]

293K, R$\bar{3}$, a = 6.4820 A, α = 91.78°, Z = 1, R = 0.043. 140K, P$\bar{1}$, 6.4592, 6.4469, 6.4590, 91.77, 91.57, 91.72, Z = 1, R = 0.034. Structures as previously described for related materials (e.g. 43A, 53; 45A, 61); the Ag site in the low-temperature form may be split into two positions.

T, K		x ×10⁵	y	z
293	Mo	40845	21787	55290
	S1	73376	13070	38290
	S2	22427	22427	22427
	Ag	0	0	0
140	Mo1	55286	40807	21674
	Mo2	21671	55281	40813
	Mo3	41817	21672	55286
	S11	73472	13142	38312
	S12	38362	73450	13192
	S13	13140	38302	73440
	S2	22310	22300	22340
	Ag	0	0	0

$Ag_{0.6}NbS_2$

Acta Cryst., C<u>46</u>, 976–979.

$P6_3/mmc$, 3.354, 14.431, Z = 2, R = 0.037. 1.2 Ag in 4(f): 1/3,2/3,0.4801; 2Nb in 2(c): 1/3,2/3,1/4; 4S in 4(f): 2/3,1/3,0.14277. NbS_2 sandwiches, with Ag statistically distributed in tetrahedral holes.

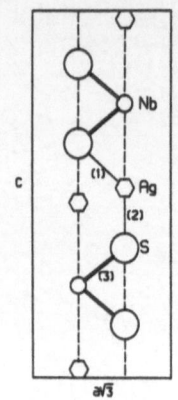

(1) 2.626(2)
(2) 2.347(2)
(3) 2.479(1)

$Ag_{0.167}TiS_2$

J. Solid State Chem., <u>84</u>, 355–364.

Stage II, P3̄m1, a = 3.4057, 3.4082, 3.4130, 3.4161, c = 12.033, 12.056, 12.092, 12.100 A, at 13, 150, 298, 305K, Z = 2, neutron powder data. 0.333 Ag in 1(a): 0,0,0; Ti in 2(c); 0,0,z, z = 0.2662, 0.2666, 0.2661, 0.2677; S(1) in 2(d): 1/3,2/3,z, z = 0.1452, 0.1449, 0.1452, 0.1452; S(2) in 2(d): 2/3,1/3,z, z = 0.3823, 0.3815, 0.3807, 0.3810.

Stage I, P3̄m1, a = 3.4676, 3.4812, c = 6.2247, 6.2564 A, at 1123, 1323K, Z = 1, neutron powder data. 0.167 Ag in 1(a): 0,0,0; Ti in 1(b): 0,0,1/2; S in 2(d): 1/3,2/3,z, z = 0.2738, 0.2766.

TiS_2 structure with intercalated Ag (<u>41</u>A, 109; <u>54</u>A, 44).

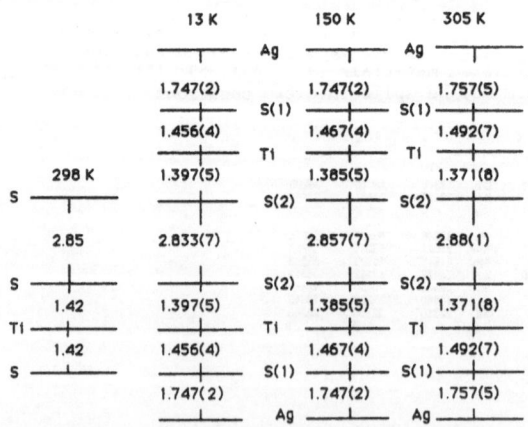

Comparison of layer separations for stage-II $Ag_{0.167}TiS_2$ and TiS_2

Al_2CdS_4, Ga_2HgS_4, Al_2HgS_4

Z. Kristallogr., <u>190</u>, 103-110.

Al_2CdS_4, Ga_2HgS_4, $I\bar{4}$, a = 5.5523, 5.5106, c = 10.1031, 10.2392 A, Z = 2, R = 0.032, 0.042. Al or Ga(1) in 2(b): 0,0,1/2; Al or Ga(2) in 2(c): 0,1/2,1/4; Cd or Hg in 2(a): 0,0,0; S in 8(g): (0.2660,0.2770,0.1352), (0.2718,0.2675,0.1374).

Al_2HgS_4, $I\bar{4}2m$, 5.5059, 10.1918, Z = 2, R = 0.046. Al(1) in 2(b): 0,0,1/2; 0.5 Al(2) in 4(d): 0,1/2,1/4; Hg in 2(a): 0,0,0; S in 8(i): x,x,z, x = 0.2272, z = 0.3633.

Isostructural with $CdGa_2S_4$ (<u>50A</u>, 18), but with a higher symmetry for Al_2HgS_4 and one partially-occupied Al site.

$BaCu_{5.65}S_{4.5}$

Mater. Res. Bull, <u>25</u>, 863-869.

$R\bar{3}$, a = 21.050, c = 9.258, Z = 18, R = 0.058. Framework of edge- and corner-sharing CuS_4 tetrahedra, and CuS_3 trigonal pyramids and trigonal planes; Ba has 10-coordination.

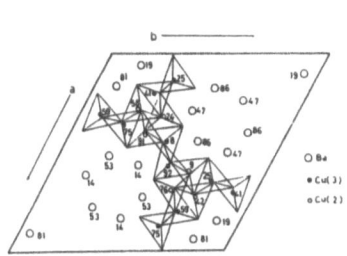

	x	y	z
Ba	0.05572	0.15938	0.19451
Cu1	-0.10049	0.18071	0.0872
Cu2	0.01321	0.27784	0.42446
Cu3	0.20119	0.30489	0.4120
Cu4	-0.18267	0.23882	-0.0281
Cu5	-0.08740	0.29957	0.2354
Cu6	0.1063	0.0686	0.4444
Cu6'	0.1217	0.0791	0.3876
S1	0.19477	0.12862	0.1422
S2	0.20201	0.31542	0.1543
S3	0.00655	0.28048	0.1692
S4	-0.03212	0.15538	0.4987
S5	0	0	0.3068
S6	0	0	0.0545

Occupancies of Cu6, Cu6' and S6 are 0.3781, 0.2736 and 0.50, respectively.

Ba_2TaS_5

Mater. Res. Bull., <u>25</u>, 723-730.

$P6_3/mmc$, 6.926, 49.43, Z = 10, powder data. Isostructural with the Nb compound (<u>52A</u>, 16).

$Ba_{16.5}Ta_9S_{39}$

Mater. Res. Bull., 24, 1491-1499 (1989).

$R\overline{3}m$, a = 6.883, c = 41.83 A, Z = 1, powder data. Isostructural with the Nb compound (50A, 14).

			x	y	z
	Ta(1)	in	6(c)	0	0	0.4220
	Ta(2)		3(b)	0	0	1/2
	Ba(1)		6(c)	0	0	0.2014
	Ba(2)		6(c)	0	0	0.0706
4.5	Ba(3)		18(h)	0.0372	-x	0.3315
	S(1)		18(h)	0.4982	-x	0.2010
	S(2)		18(h)	0.5004	-x	0.0634
3	S(3)		18(h)	-0.0561	-x	0.0003

$(CeS)_{1.16}NbS_2$

J. Solid State Chem., 89, 328-339.

Misfit layer structure; CeS lattice, Cm2a, 5.727, 5.765, 11.41, Z = 4, R = 0.048; NbS_2 lattice, Fm2m, 3.311, 5.765, 22.82, Z = 4, R = 0.069; common projection, R = 0.072. Alternate double layers of NaCℓ-type CeS and NbS_2 sandwiches.

$CoGaInS_4$

J. Solid State Chem., 87, 15-19.

$P\overline{3}m1$, 3.759, 12.184, Z = 1, R = 0.03. M(1) in 1(b): 0,0,1/2; M(2) in 2(d): 1/3,2/3,z, z = 0.1995; S(1) in 2(d): z = 0.3896; S(2) in 2(d): z = 0.8720; M(1) = octahedral site = 0.612 In + 0.388 Co; M(2) = tetrahedral site=0.5 Ga + 0.388/2 In + 0.612/2 Co. $FeGa_2S_4$-type structure (46A, 79). M(1)-6S = 2.55, M(2)-4S = 2.31, 2.34 A.

CoK_6S_4, $CoNa_6Se_4$, CoK_6Se_4

J. Less-Common Metals, 162, 309-314.

$P6_3mc$, a = 9.787, 9.306, 10.191, c = 7.614, 7.167, 7.886 A, Z = 2, R = 0.021, 0.025, 0.020. Na_6ZnO_4-type (40A, 214), as for $CoNa_6S_4$ (50A, 24), with isolated $CoX_4{}^{6-}$ tetrahedra, linked by alkali-metal cations which have tetrahedral and octahedral coordinations.

K_nCoS_4	x	y	z	Na_6CoSe_4	x	y	z	K_nCoSe_4	x	y	z
K1	0.1462	-0.1462	0.5462	Na1	0.1471	-0.1471	0.5423	K1	0.1466	-0.1466	0.5476
K2	0.5281	-0.5281	0.3655	Na2	0.5293	-0.5293	0.3684	K2	0.5270	-0.5270	0.3657
Co	1/3	2/3	0.25	Co	1/3	2/3	0.25	Co	1/3	2/3	0.25
S1	1/3	2/3	0.5651	Se1	1/3	2/3	0.5927	Se1	1/3	2/3	0.5680
S2	0.1986	-0.1986	0.1523	Se2	0.1880	-0.1880	0.1429	Se2	0.1978	-0.1978	0.1506

$CoNa_2S_2$, CoK_2S_2, $CoRb_2S_2$, $CoCs_2S_2$, $CoNa_2Se_2$, CoK_2Se_2, $CoRb_2Se_2$, $CoCs_2Se_2$

J. Less-Common Metals, 158, 169-176.

Ibam, a = 6.37-7.43, b = 11.21-14.22, c = 5.85-6.58 A, Z = 4, R = 0.014-0.037. Co in 4(a): 0,0,1/4, alkali metal and S/Se in 8(j): x,y,0; for $CoNa_2S_2$, x(Na) = 0.3494, y(Na) = 0.1448, x(S) = 0.2092, y(S) = 0.8906; similar values for the other compounds. K_2ZnO_2-type (33A, 325).

$CrInMnS_4$

Z. Kristallogr., 190, 277-285.

Fd3m, 10.4297, Z = 8, synchrotron powder data. Spinal structure (1, 350), with site occupancies $(Mn_{0.75}In_{0.25})(Cr_{0.5}Mn_{0.12}In_{0.38})_2S_4$, x(S) = 0.8881.

$Cs_6Re_6S_{15}$ (I), $Cs_4Re_6S_{13}$ (II), $Cs_4Re_6S_{13.5}$ (III)

Z. anorg. Chem., 587, 74-79, 91-102.

I, R$\bar{3}$c, a = 14.012, c = 27.779 A, Z = 6, R = 0.055. II, C2/c, 10.016, 17.214, 13.730, 100.65, Z = 4, R = 0.043. III, P2$_1$/n, 10.132, 17.639, 13.715, 100.71, Z = 4, R = 0.038. The structures contain Re_6S_8 clusters linked by sulphide, disulphide, and trisulphide bridges.

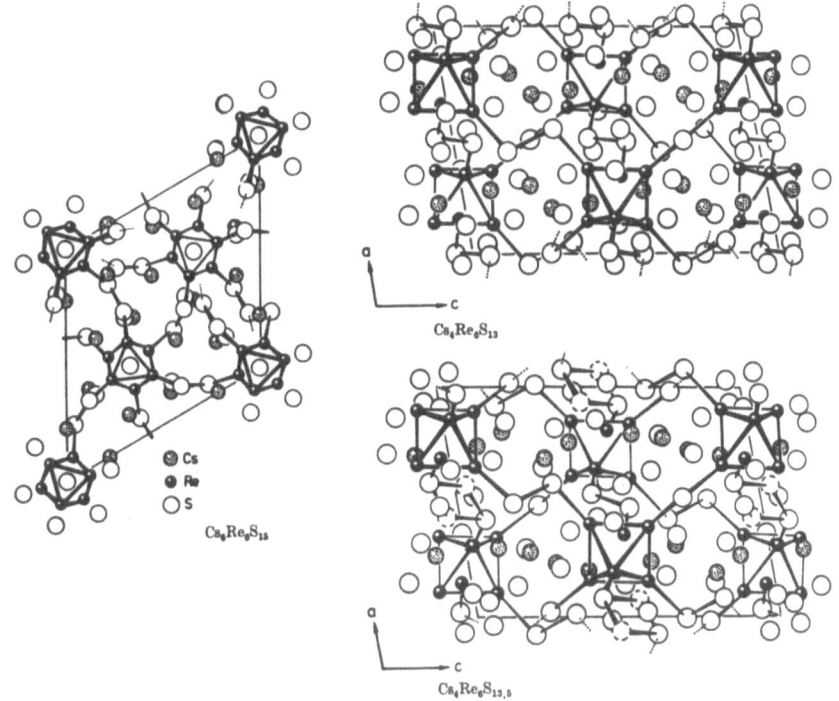

$Cs_6Re_6S_{15}$

$Cs_4Re_6S_{13}$

$Cs_4Re_6S_{13.5}$

$Cu_2Er_{0.67}S_2$

Phys. Status Solidi A, __121__, 21-28.

$P\bar{3}m1$, 3.879, 6.303, Z = 1, R = 0.066. 2 Cu in 2(d): 2/3,1/3,0.3717; 2 S in
2(d): 2/3,1/3,0.7543; 0.67 Er in 1(a): 0,0,0. anti-La_2O_3-type (__45A__, 220),
with stacking faults.

$Cu_{0.07}Li_xTi_{2.05}S_4$ (x = 0.92, 1.92)

Mater. Res. Bull., __25__, 533-538.

Fd3m, a = 9.948, 10.0573, Z = 8, neutron powder data. 16Ti(1) in 16(d):
1/2,1/2,1/2; 0.6 Cu in 8(a): 1/8,1/8,1/8; 8 or 14 Li + 0.4 Ti(2) in 16(c):
0,0,0; S in 32(e): x,x,x, x = 0.2538, 0.2542. Spinel structure (__1__, 350), with
Cu in the tetrahedral sites, Ti in the octahedral sites, and Li and excess
titanium in a second octahedral site.

$Cu_{0.46}Nb_6S_{2.54}$

J. Less-Common Metals, __161__, L37-L40.

Pnnm, 11.613, 14.434, 3.3954, Z = 4, R = 0.028. Isostructural with related
materials (__54A__, 69). The structure is of Ta_2P-type (__31A__, 58), with Cu
partially substituting in one non-metal site.

	x	y	z
Nb1	0.08696	0.91933	0
Nb2	0.3809	0.57822	0
Nb3	0.30741	0.79355	0
Nb4	0.3460	0.02259	0
Nb5	0.42021	0.25314	0
Nb6	0.02734	0.39180	0
M*	0.2556	0.4153	0
S1	0.0755	0.2054	0
S2	0.1869	0.6515	0

*M = 46 at.% Cu + 54 at.% S

$Cu_6T\ell S_4$, $Cu_{5.5}T\ell S_4$

J. Less Common Metals, __161__, 165-173.

I4/mmm, a = 3.9465, 3.9357, c = 24.230, 24.183 A, Z = 2, powder data for Cu_6,
single-crystal data, R = 0.061 for $Cu_{5.5}$. Layers of interconnected CuS_4
tetrahedra separated by layers of $T\ell S_8$ polyhedra. Cu-S = 2.29-2.52, $T\ell$-S =
3.40 A.

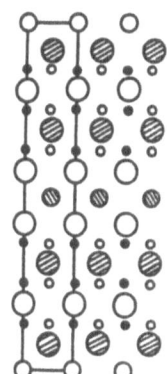

		Occupancy	y	z
Tl	2a	1	0	0
Cu1	8g	0.93	1/2	0.1294
Cu2	4d	0.90	1/2	1/4
S1	4e	1	0	0.1947
S2	4e	1	0	0.4193

All atoms are situated in x = 0

CuVP$_2$S$_6$

Inorg. Chem., 29, 4916-4920.

C2, 5.9462, 10.299, 6.6870, 107.247, Z = 2, R = 0.033. The structure resembles those of related materials (e.g. 52A, 80; 55A, 56), with an ordered cation arrangement but statistical distribution of Cu over three sites.

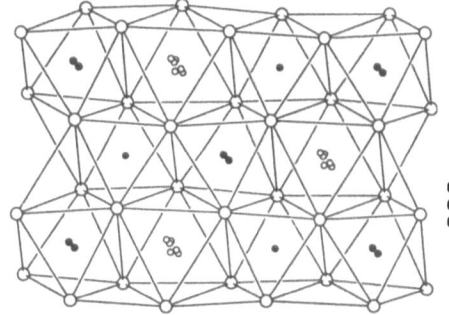

		x	y	z
	V	0	0.3308	0
	P	0.9442	0	0.8303
	S1	0.7270	0.6708	0.2539
	S2	0.2666	0.8418	0.2493
	S3	0.7476	0.4865	0.7510
0.139	Cu1	0.970	0.673	0.012
0.265	Cu2	0.0693	0.6667	0.215
0.066	Cu3	0.126	0.669	0.371

EuMo$_6$S$_8$, BaMo$_6$S$_8$

Acta Cryst., C46, 181-186.

112 and 177K, R$\bar{3}$, a = 6.5378, 6.6441 A, α = 88.809, 88.562°, for Eu, Ba, Z = 1, R = 0.038, 0.060. 40 and 173K, P$\bar{1}$, a = 6.4692, 6.5896, b = 6.5651, 6.6500, c = 6.5986, 6.6899 A, α = 89.179, 88.731, β = 89.184, 88.818, γ = 88.009, 88.059°, Z = 1, R = 0.058, 0.070. Structures as previously described (53A, 43, 46; 54A, 46; 56A, 40). The phase transformations involve discontinuous rearrangements of the octahedral Mo$_6$ clusters.

$Fe_{1-x}S$ (x = 0 - 0.125)

J. Solid State Chem., 84, 194-210, 211-225.

Low temperature, $P\bar{6}2c$, a = $\sqrt{3}a'$ = $\sqrt{3}$ x 3.444, 3.455, c = 2c' = 2 x 5.878, 5.871 A, at 294, 414K (where a', c' correspond to a NiAs-type substructure), Z = 12, R = 0.031, 0.036. Fe in 12(i): x,y,z,   x = 0.3786, 0.3761, y = 0.0553, 0.0527, z = 0.1230, 0.1233; S(1) in 2(a): 0,0,0; S(2) in 4(f): 1/3,2/3,z, z = 0.0200, 0.0189; S(3) in 6(h): x,y,1/4, x = 0.6653, 0.6651, y = -0.0030, -0.0027. NiAs-type superstructure, as previously described (20, 129; 24, 166; 49A, 40).

High-temperature, $P6_3mc$, a = 2a' = 2 x 3.479, 3.487, c = c' = 5.824, 5.821 A, at 429, 453K, Z = 8, R = 0.060, 0.068. Fe(1) in 6(c): x,2x,z, x = 0.5098, 0.511, z = 0; Fe in 2(a): 0,0,z, z = -0.006, -0.008; S(1) in 6(c): x = 0.1667, 0.1663, z = 0.240, 0.245; S(2) in 2(b): 1/3,2/3,z, z = 0.761, 0.7656. NiAs-type superstructure, rather than twinned MnP-type (49A, 40).

$(Fe,Mg)_2SiS_4$

Acta Cryst., C46, 1996-1998.

Pnma, a = 12.633, 12.677, b = 7.348, 7.405, c = 5.901, 5.913, for Fe/Mg = 0.6/1.4 and 0.14/1.86, Z = 4, R = 0.025, 0.029. Olivine-type (30A, 441; 42A, 88; 56A, 45), with Fe mainly in the M(1) site.

$Mg_{1.6}Fe_{0.4}SiS_4$		x	y	z	Occupancy	
M(1)	4(a)	0	0	0	0·283	Mg
					0·217	Fe
M(2)	4(c)	0·23039	¼	0·50848	0·414	Mg
					0·086	Fe
Si	4(c)	0·41096	¼	0·09344	0·5	
S(1)	4(c)	0·40830	¼	0·73625	0·5	
S(2)	4(c)	0·56613	¼	0·24055	0·5	
S(3)	8(d)	0·33430	0·02186	0·24750	1·0	

$Mg_{1.86}Fe_{0.14}SiS_4$		x	y	z	Occupancy	
M(1)	4(a)	0	0	0	0·444	Mg
					0·056	Fe
M(2)	4(c)	0·23044	¼	0·50820	0·487	Mg
					0·013	Fe
Si	4(c)	0·41038	¼	0·09352	0·5	
S(1)	4(c)	0·40767	¼	0·73832	0·5	
S(2)	4(c)	0·56518	¼	0·23935	0·5	
S(3)	8(d)	0·33372	0·02360	0·24687	1·0	

$GaInS_3$

Kristallografija, 35, 1298-1299 [Soviet Physics - Crystallography, 35, 766-767].

$P6_3mc$, 3.8134, 30.656, Z = 10/3, R = 0.074. Two-block polytype, with 10 S layers, octahedral In, tetrahedral Ga, and intercalated 4-aminopyridine.

$Ga_{0.5}In_{1.5}S_3$

Kristallografija, <u>35</u>, 332-336 [Soviet Physics - Crystallography, <u>35</u>, 187-189].

$R\bar{3}c$, a = 3.814, c = 100.04 A, Z = 11, R = 0.069. Alternating two- and three-layer packets of MS_4 tetrahedra and MS_6 octahedra.

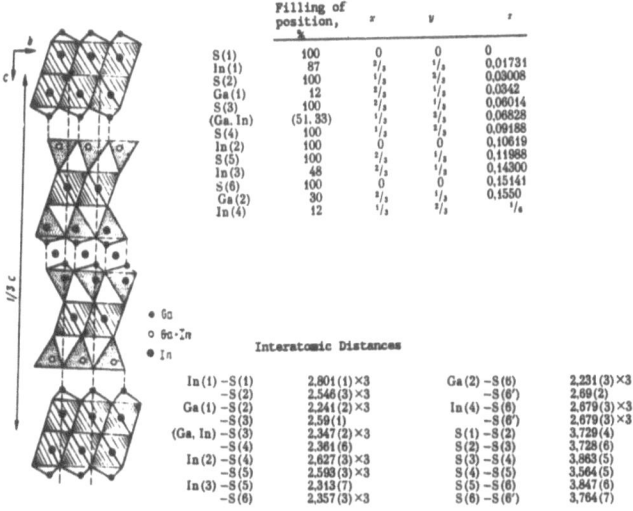

Filling of position, %		x	y	z
S(1)	100	0	0	0
In(1)	87	$^2/_3$	$^1/_3$	0,01731
S(2)	100	$^1/_3$	$^2/_3$	0,03008
Ga(1)	12	$^2/_3$	$^1/_3$	0,0342
S(3)	100	$^2/_3$	$^1/_3$	0,06014
(Ga, In)	(51,33)	$^1/_3$	$^2/_3$	0,06828
S(4)	100	$^1/_3$	$^2/_3$	0,09188
In(2)	100	0	0	0,10619
S(5)	100	$^2/_3$	$^1/_3$	0,11988
In(3)	48	$^2/_3$	$^1/_3$	0,14300
S(6)	100	0	0	0,15141
Ga(2)	30	$^2/_3$	$^1/_3$	0,1550
In(4)	12	$^1/_3$	$^2/_3$	$^1/_4$

Interatomic Distances

In(1) –S(1)	2,801 (1) ×3	Ga(2) –S(5)	2,231 (3) ×3
–S(2)	2,546 (3) ×3	–S(6')	2,69 (2)
Ga(1) –S(2)	2,244 (2) ×3	In(4) –S(6)	2,679 (3) ×3
–S(3)	2,59 (1)	–S(6')	2,679 (3) ×3
(Ga, In) –S(3)	2,347 (2) ×3	S(1) –S(2)	3,729 (4)
–S(4)	2,361 (6)	S(2) –S(3)	3,728 (6)
In(2) –S(4)	2,627 (3) ×3	S(3) –S(4)	3,863 (5)
–S(5)	2,593 (3) ×3	S(4) –S(5)	3,564 (5)
In(3) –S(5)	2,313 (7)	S(5) –S(6)	3,847 (6)
–S(6)	2,357 (3) ×3	S(6) –S(6')	3,764 (7)

GaLiS$_2$

Acta Cryst., <u>C46</u>, 2017-2019.

$Pna2_1$, 6.519, 7.872, 6.238, Z = 4, R = 0.049. Isostructural with β-NaFeO$_2$ and LiGaO$_2$ (<u>18</u>, 422; <u>28</u>, 133; <u>30A</u>, 314), with h.c.p. S and metal ions ordered in tetrahedral holes.

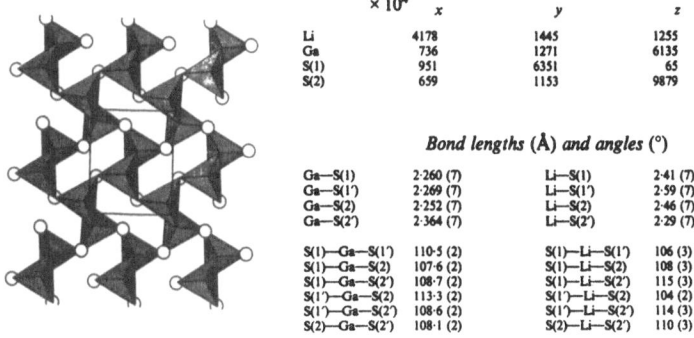

	× 10^4 x	y	z
Li	4178	1445	1255
Ga	736	1271	6135
S(1)	951	6351	65
S(2)	659	1153	9879

Bond lengths (Å) and angles (°)

Ga—S(1)	2·260 (7)	Li—S(1)	2·41 (7)
Ga—S(1')	2·269 (7)	Li—S(1')	2·59 (7)
Ga—S(2)	2·252 (7)	Li—S(2)	2·46 (7)
Ga—S(2')	2·364 (7)	Li—S(2')	2·29 (7)
S(1)—Ga—S(1')	110·5 (2)	S(1)—Li—S(1')	106 (3)
S(1)—Ga—S(2)	107·6 (2)	S(1)—Li—S(2)	108 (3)
S(1)—Ga—S(2')	108·7 (2)	S(1)—Li—S(2')	115 (3)
S(1')—Ga—S(2)	113·3 (2)	S(1')—Li—S(2)	104 (2)
S(1')—Ga—S(2')	108·6 (2)	S(1')—Li—S(2')	114 (3)
S(2)—Ga—S(2')	108·1 (2)	S(2)—Li—S(2')	110 (3)

GaMnScS$_4$ (I), Ga$_{1.6}$MnSc$_{0.4}$S$_4$ (II)

J. Less-Common Metals, <u>163</u>, 253-261.

I, Pnma, 12.796, 7.426, 6.086, Z = 4, R = 0.018. Olivine-type (<u>1</u>, 352), with Mn/Sc disordered in octahedral (O) sites.

II, P$\bar{3}$m1, 3.751, 12.231, Z = 1, powder data. FeGa$_2$S$_4$-type (<u>46</u>A, 79).

MnGaScS$_4$		x	y	z
O1	4a	0	0	0
O2	4c	0.2671	1/4	−0.0045
Ga	4c	0.0901	1/4	0.4145
S1	4c	0.0909	1/4	0.7814
S2	4c	0.4273	1/4	0.2473
S3	8d	0.1669	0.0099	0.2451

Ga$_2$Sr$_2$S$_5$, Ba$_2$In$_2$S$_5$, Ba$_2$In$_2$Se$_5$

Z. anorg. Chem., <u>580</u>, 151-159.

Pbca, a = 12.523, 13.167, 13.657, b = 12.034, 12.723, 13.182, c = 11.180, 11.784, 12.225 Å, Z = 8, R = 0.058, 0.066, 0.106. (100) Layers of corner-sharing MX$_4$ tetrahedra (M = Ga, In; X = S, Se), with rings of 4 and 8 tetrahedra; layers are linked by 7-coordinate Sr or Ba ions.

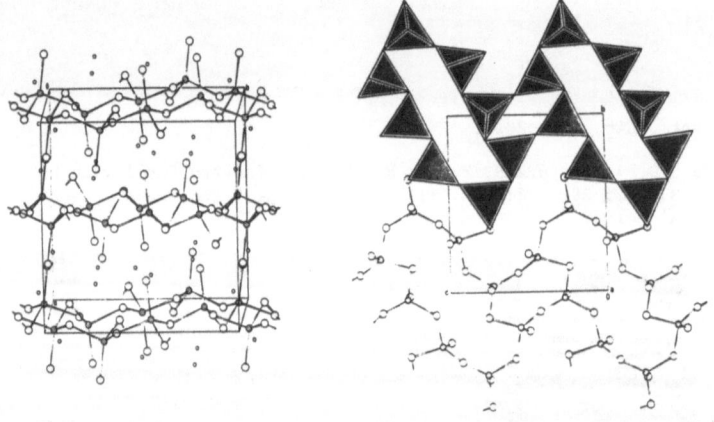

Gd$_2$S$_3$

Z. anorg. Chem., <u>590</u>, 111-119.

Pnma, 10.7447, 3.8985, 10.5462, Z = 4, R = 0.031. Gd(1,2), S(1,2,3) in 4(c): x,1/4,z, x = 0.99002, 0.30902, 0.0469, 0.8793, 0.2272, z = 0.31415, 0.50304, 0.8719, 0.5547, 0.1983. U$_2$S$_3$-type (<u>12</u>, 181; compare <u>33</u>A, 82). Gd(1)-7S = 2.77-2.83, Gd(2)-(7+1)S = 2.85-2.88, 3.33 Å.

$In_{10}Sn_6S_{21}$

J. Solid State Chem., 89, 275-281.

P2/m, 15.571, 3.842, 27.583, 95.44, Z = 2, R = 0.052. (010) Layers of corner-sharing InS_6 octahedra, linked by further corner-sharing along b, with cavities containing Sn ions which have bicapped trigonal prismatic coordination.

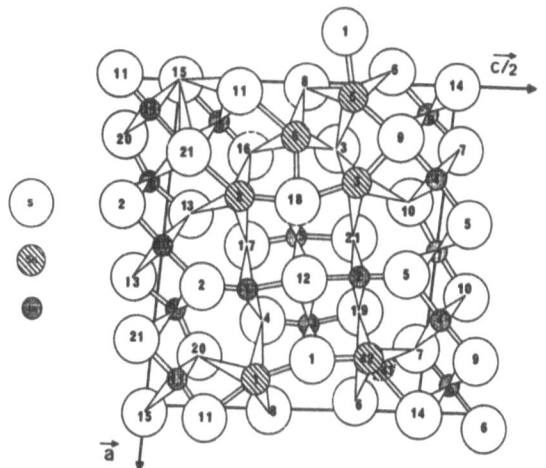

$In_2Zn_3S_6$

Izv. Akad. Nauk Mold.SSR, Ser. Fiz.-Tekh. Mat. Nauk, No.3, 68-70.

P3m1, 3.87, 18.9, Z = 1, R = 0.125. Structure as previously described (32A, 95).

IrS_2, $IrSe_2$

J. Solid State Chem., 89, 315-327.

Pnam, a = 19.791, 20.955, b = 5.6242, 5.9381, c = 3.5673, 3.7429 A, Z = 8, powder data. Marcasite-related structure as previously described (22, 145), with edge- and corner-sharing InX_6 octahedra and X_2^{2-} pairs.

$\alpha\text{-}K_2Pt_4S_6$

Z. anorg. Chem., 357, 264-272 (1968).

$R\bar{3}m$, a = 7.01, c = 19.14 A, Z = 3, film data. K in 6(c): 0,0,0.300; Pt(IV) in
3(b): 0,0,1/2; Pt(II) in 9(d): 1/2,0,1/2; S in 18(h): $x,\bar{x},z$, x not given, z =
0.1044 [see also the Se compound in this volume, p.81]. Layers of Pt(IV)S_6
octahedra and Pt(II)S_4 square planes. A second (ß) form has space group P6$_3$, a =
7.01, c = 12.77 A.

$Lu_{2+x}S_3$ (I), LuS_{1+x} (II)

J. Less-Common Metals, 157, 133-138.

I, Fddd, 10.7735, 7.7053, 22.873, Z = 16; 15.3Lu(1), 15.3Lu(2), 2.0Lu(3) each in
16(g): 1/8,1/8,z, z = 0.041, 0.376, 0.542; 16S(1) in 16(f): 1/8,0.375,1/8; 32S(2)
in 32(h): 0.125,0.375,0.458 (origin at $\bar{1}$). II, Fm3m, 5.3797, Z = 3; 3Lu in 4(a):
0,0,0; 4S in 4(b): 1/2,1/2,1/2. Powder data for a mixture of I and II. I is
Sc_2S_3-type (48A, 87), and II is a defect NaCl-type.

$(Mg,Mn)_2SiS_4$, $(Fe,Mn)_2SiS_4$, $(Fe,Mg)_2SiS_4$

Z. anorg. Chem., 586, 73-78.

Pnma, a = 12.676, 12.651, 12.586, b = 7.430, 7.361, 7.329, c = 5.927, 5.899,
5.870 A, Z = 4, R = 0.039, 0.040, 0.055. Olivine-type structures (30A, 441;
42A, 88; 56A, 45). Disordered cation distributions in the two octahedral sites,
except for the Fe/Mg compound, where Fe shows a preference for the M(1) site.

	x	y	z
$Mn_{1.4}Mg_{0.6}SiS_4$			
M(1)	0	0	0
M(2)	0.2297	1/4	0.5102
Si	0.4098	1/4	0.0934
S(1)	0.4072	1/4	0.7389
S(2)	0.5647	1/4	0.2375
S(3)	0.3331	0.0246	0.2461
$Mn_{1.4}Fe_{0.6}SiS_4$			
M(1)	0	0	0
M(2)	0.2299	1/4	0.5094
Si	0.4112	1/4	0.0936
S(1)	0.4088	1/4	0.7368
S(2)	0.5662	1/4	0.2413
S(3)	0.3347	0.0219	0.2475
$Mg_{1.14}Fe_{0.86}SiS_4$			
M(1)	0	0	0
M(2)	0.2303	1/4	0.5091
Si	0.4116	1/4	0.0935
S(1)	0.4088	1/4	0.7349
S(2)	0.5866	1/4	0.2411
S(3)	0.3345	0.0208	0.2482

Mo_6SrS_8

Mh. Chem., 121, 505-509.

Above 135K, $R\bar{3}$, a = 6.5630 A, α = 88.9982°, at 298K, Z = 1, powder data. Below
135K, $P\bar{1}$, 6.481, 6.572, 6.611, 89.246, 89.304, 88.169, at 20K, Z = 1. Powder
data. The room-temperature structure is as previously described (54A, 46), and
the low-temperature form involves a small triclinic distortion.

	R$\bar{3}$	x	y	z
Sr		0.000	0.000	0.000
Mo		0.2290	0.4192	0.5631
S1		0.379	0.1308	0.7357
S2		0.2410	0.2410	0.2410

	P$\bar{1}$	x	y	z
Sr		0.0000	0.0000	0.0000
Mo 1		0.231	0.564	0.417
Mo 2		0.418	0.232	0.568
Mo 3		0.567	0.418	0.226
S 1		0.126	0.385	0.765
S 2		0.232	0.255	0.225
S 3		0.373	0.718	0.123
S 4		0.748	0.144	0.359

NaRbS

Acta Cryst., C$\underline{46}$, 1596–1597.

P4/nmm, 4.696, 7.559, Z = 2, R = 0.022. Na in 2(a): 0,0,0; Rb in 2(c): 0,1/2,
0.6493; S in 2(c): 0,1/2,0.2056. PbFCl-type ($\underline{2}$, 45), with (001) layers of
corner-sharing NaS$_4$ tetrahedra, separated by double layers of 5-coordinate Rb
ions. Na–S = 2.816(1), Rb–S = 3.497 (x 4), 3.354(1) A.

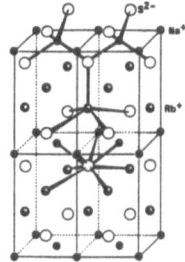

Nb$_{4.92}$Ta$_{6.08}$S$_4$

J. Solid State Chem., $\underline{86}$, 88–93.

Pnma, 31.210, 3.3507, 9.592, Z = 4, R = 0.051. Metal coordinations are capped
distorted cubic prisms, and S coordinations are capped trigonal prisms.

	Occupancy	x	y	z
M1	76% Ta + 24% Nb	0.2177	1/4	0.5355
M2	76% Ta + 24% Nb	0.3268	1/4	0.8482
M3	72% Ta + 28% Nb	0.4247	1/4	0.7579
M4	70% Ta + 30% Nb	0.0254	1/4	0.0853
M5	66% Ta + 34% Nb	0.2357	1/4	0.2166
M6	54% Ta + 46% Nb	0.3401	1/4	0.1834
M7	50% Ta + 50% Nb	0.4207	1/4	0.4183
M8	48% Ta + 52% Nb	0.2150	1/4	0.8917
M9	46% Ta + 54% Nb	0.1350	1/4	0.1206
M10	44% Ta + 56% Nb	0.1102	1/4	0.4761
M11	6% Ta + 94% Nb	0.0233	1/4	0.7255
S1		0.4200	1/4	0.153
S2		0.3044	1/4	0.580
S3		0.1367	1/4	0.865
S4		0.0296	1/4	0.380

$NbS_2(YS)_{1.23}$

J. Solid State Chem., **88**, 451-458.

Misfit layer compound, with alternating slices of YS (NaCl-type) and NbS_2 sandwiches. For the two sublattices: YS, Fm2m, 5.393, 5.658, 22.284, Z = 8, R = 0.060; NbS_2, C2, 3.322, 5.662, 11.13, 92.62, Z = 2, R = 0.075; for the common 0kℓ part, R = 0.063. Y has square-pyramidal coordination, Y-S = 2.75-3.00 A; Nb-6S = 2.47 A.

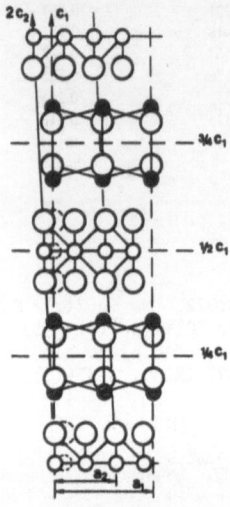

Structure of $(YS)_{1.23}NbS_2$ projected along [010]. Small hatched circles, small open circles, and large open circles for Y, Nb, and S, respectively.

P_4S_9

Z. anorg. Chem., **588**, 139-146.

$P2_1/n$, 8.555, 12.637, 12.453, 104.94, Z = 4, R = 0.032. Adamantane-like molecules with 3 terminal S; P-S = 2.07-2.11 (cage), 1.91-1.92 (terminal) A. A cubic form has been described previously (34A, 118).

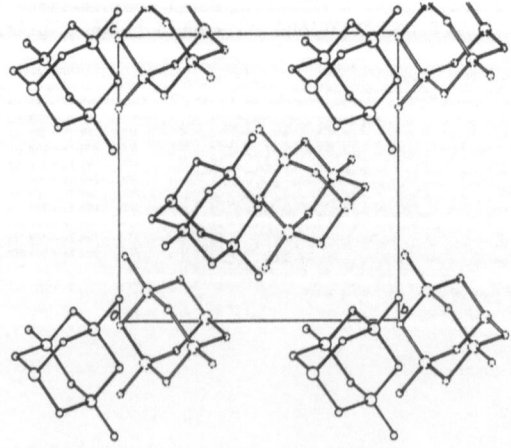

$(PbS)_{1.14}NbS_2$, $(LaS)_{1.14}NbS_2$

Acta Cryst., B46, 324-332.

Misfit layer structures, PbS and LaS parts, Cm2a, a = 5.834, 5.828, b = 5.801,
5.799, c = 11.902, 11.512 A, Z = 4, R = 0.062, 0.048; NbS_2 parts, Cm2m and
Fm2m, a = 3.313, 3.310, b = 5.801, 5.793, c = 23.807, 23.043 A, Z = 4, R =
0.085, 0.087; common **a**-axis projection, b = 2.901, 2.899, c = 23.807, 11.512 A,
Z = 2, 1, R = 0.138, 0.063. The two misfit layer structures are built from
alternate double layers of PbS (LaS) with distorted square-pyramidal coordination for Pb (La), and NbS_2 sandwiches with trigonal-prismatic coordination
for Nb.

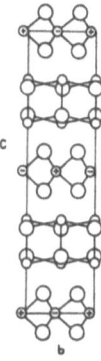

Mater. Res. Bull., 25, 855-861.

Supercell refinement of the La compound in Bb2b (cf. the centrosymmetric space
group in 56A, 43). The Nb sites are now fully occupied.

$Pb_2Sb_2S_5$

Acta Cryst., C46, 534-536.

Pbnm, 11.355, 19.783, 4.042, Z = 4, R = 0.071. Structure as previously proposed (50A, 7), with ribbons along **c** of $(Pb,Sb)S_5$ square pyramids; site occupancies are determined.

	x	y	z
$M(1)$	0·3757	0·3715	0·25
$M(2)$	0·7022	0·4758	0·25
$M(3)$	0·9983	0·4008	0·75
$M(4)$	0·6678	0·2914	0·75
$S(1)$	0·4950	0·2670	0·25
$S(2)$	0·8183	0·3685	0·25
$S(3)$	0·8690	0·5212	0·75
$S(4)$	0·5659	0·4221	0·75
$S(5)$	0·2495	0·3288	0·75

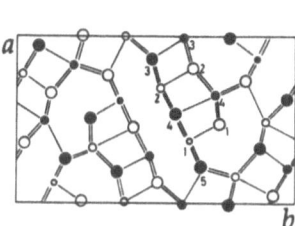

Interatomic distances (Å) *less than* 3·4 Å *in*
$Pb_2Sb_2S_5$

$M(1)-S(1)$	3·475 (7)	$M(3)-S(3)$	2·795 (7)
$S(5) \times 2$	2·621 (5)	$S(2) \times 2$	2·947 (5)
$S(4) \times 2$	3·124 (5)	$S(3) \times 2$	2·952 (5)
		$S(5)$	3·183 (8)
$M(2)$ $S(2)$	2·497 (7)		
$S(4) \times 2$	2·757 (5)	$M(4)-S(4)$	2·836 (7)
$S(3) \times 2$	2·915 (5)	$S(1) \times 2$	2·858 (5)
		$S(2) \times 2$	3·052 (5)
		$S(5) \times 2$	3·253 (6)

$(PbS)_{1.13}TaS_2$

J. Solid. State Chem., <u>84</u>, 118–129.

Misfit layer compound, with two Fm2m cells: PbS, 5.825, 5.779, 23.96, Z = 8; TaS_2, 3.304, 5.779, 23.96, Z = 4, a/a' irrational but close to 7/4, R = 0.074, 0.033 for the PbS and TaS_2 data, respectively; R = 0.056 for an a-axis projection of the complete structure (P11m, b = 2.889, c = 11.891 A). The structure contains alternate (001) double layers of PbS with distorted NaCl-type structure, and TaS_2 sandwiches with TaS_6 trigonal prisms.

	x	y	z	s.o.f.
(a) PbS				
Pb(1)	0.0	0.0	0.31847	0.896
Pb(2)	0.0	0.514	0.2167	0.102
S(1)	0.0	0.018	0.2007	1.0
(b) TaS_2				
Ta(1)	0.0	0.0	0.0	
S(1)	0.0	0.3438	0.0650	
Common part[b]				
Pb(1)	—	0.517	0.6366	0.52
S(1)	—	0.553	0.4011	0.52
Ta(1)	—	0.6877	0.0	1.0
S(2)	—	0.0	0.1300	1.0

[b] Coordinates in a unit cell with b = 2.889 Å, c = 11.981 Å.

$(PbS)_{1.12}VS_2$

Acta Cryst., B<u>46</u>, 487–492.

Incommensurate composite crystal: PbS part, monoclinic, F-lattice, a = 5.728, b = 5.789, c = 23.939 A, β = 98.947°; VS_2 part, monoclinic, C-lattice, a' = a, b' = 3.256 A, c'=c/2, β' = β. R = 0.083 for a model without modulation, space group C2 for each part; R = 0.059 for a modulated model. Alternate PbS layers with distorted NaCl-type structure, and VS_2 sandwiches with distorted CdI_2-type structure.

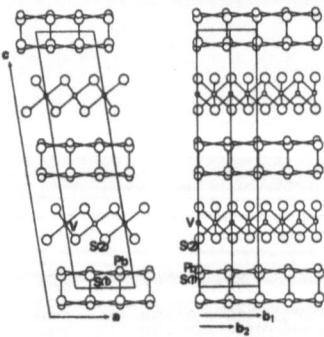

RuS_2, $RuSe_2$

Acta Cryst., C<u>46</u>, 2003–2005.

Pa3, a = 5.6106, 5.9336, Z = 4, R = 0.021, 0.021. Ru in 4(a): 0,0,0; S or Se in 8(c): x,x,x, x = 0.38831, 0.38065. Pyrite-type, as previously described (<u>1</u>, 153; <u>2</u>, 274; <u>32</u>A, 126; <u>33</u>A, 34).

Selected interatomic distances (Å) and angles
(°) for RuX$_2$ (X = S, Se) with e.s.d.'s in parentheses

	S	Se	
Ru—X	2·3520 (3)	2·4707 (2)	6×
	3·5445 (3)	3·8090 (3)	6×
	3·7736 (5)	3·9121 (1)	2×
Ru—Ru	3·9673 (2)	4·1957 (3)	12×
X—X	2·1707 (8)	2·4532 (2)	1×
	3·2060 (3)	3·3475 (2)	6×
	3·4423 (4)	3·6348 (3)	6×
X—Ru—X	85·93 (1)	85·29 (1)	6×
	94·07 (1)	94·71 (1)	6×
Ru—X—X	103·13 (1)	101·38 (1)	3×
Ru—X—Ru	115·00 (1)	116·29 (1)	3×
X—X—X	64·94 (1)	65·76 (1)	6×

SnS$_2$ (2H, 4H, and 18R POLYTYPES)

Acta Cryst., B46, 449–455.

P$\bar{3}$m1, P3m1, R$\bar{3}$m, a = 3.6470, c = 5.8990, 11.811, 53.118 A, Z = 1, 2, 9, R = 0.013, 0.026, 0.056, atomic positional parameters not given. The 2H-polytype has a CdI$_2$-type structure (1, 163; 44A, 100) and the other two polytypes are stacking variants. All three polytypes (especially 2H) have Sn and S vacancies.

SnTaS$_2$

Mater. Res. Bull., 25, 1011–1018.

P6$_3$/mmc, a = 3.3053, 3.3086, c = 17.434, 17.45 A, at 295, 425K, Z = 2, R = 0.023, 0.015. Sn in 2(a): 0,0,0; Ta in 2(c): 1/3,2/3,1/4; S in 4(e): 0,0,z, z = 0.1606, 0.1607. Structure as previously described (43A, 86), with 2H-TaS$_2$-type sandwiches in which Ta has trigonal prismatic coordination (Ta–S = 2.464(1) A), and Sn coordinated linearly to 2 S of neighbouring sandwiches (Sn–S = 2.800(2) A).

2H-TaS$_2$

Acta Cryst., C46, 1598–1599.

P6$_3$/mmc, 3.314, 12.097, Z = 2, R = 0.032. Ta in 2(b): 0,0,1/4; S in 4(f): 1/3,2/3,0.1212. Structure as previously described (18, 289), with sandwiches of TaS$_6$ trigonal prisms.

Tm$_2$S$_3$

J. Less-Common Metals, 158, L21–L25.

I$\bar{4}$3d, 8.223, Z = 5.33, R = 0.028. 10.67 Tm in 12(a): 3/8,0,1/4; 16 S in 16(c): x,x,x, x = 0.0674. Defect Th$_3$P$_4$-type (7, 15), with cation vacancies. Tm–S = 2.710, 2.993(4) (each x 4) A.

Acta Cryst., C46, 487–488.

R$\bar{3}$c, a = 6.768, c = 18.236 A, Z = 6, R = 0.041. Tm in 12(c): 0,0,0.35007; S in 18(e): 0.3026,0,1/4. α-Corundum-type (1, 240). Tm–S = 2.660, 2.743(1) A.

θ-Tm$_2$S$_3$

Ž. Neorg. Khim., 35, 869-873 [Russ. J. Inorg. Chem., 35, 488-490].

Ia3, 12.449, Z = 16, R = 0.018. Tm(1) in 8(b): 1/4,1/4,1/4; Tm(2) in 24(d):
1/4,0.4522,0; S in 48(e): 0.3834,0.3938,0.1609. Mn$_2$O$_3$-type, as for the Yb
compound (53A, 53). Tm-6S = 2.691(4) A.

Tm$_8$S$_{11}$

J. Less-Common Metals, 166, 135-140.

Cmcm, 3.7486, 12.6160, 34.932 A, Z = 4, R = 0.043. Tm atoms have distorted
octahedral, trigonal prismatic, and capped trigonal prismatic coordinations.
Tm-S = 2.64-2.83 A.

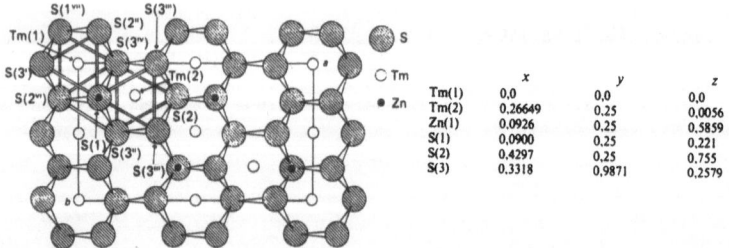

		x	y	z
Tm1	4a	0	0	0
Tm2	4c	0	0.5070	0.250
Tm3	8f	0	0.25787	0.05876
Tm4	8f	0	0.24166	0.68297
Tm5	8f	0	0.01927	0.13307
S1	4c	0	0.8541	0.250
S2	8f	0	0.0894	0.5682
S3	8f	0	0.1142	0.2967
S4	8f	0	0.3532	0.5159
S5	8f	0	0.3363	0.6113
S6	8f	0	0.3942	0.3377

Tm$_2$ZnS$_4$

Acta Cryst., C46, 365-368.

Pnma, 13.308, 7.769, 6.285, Z = 4, R = 0.039. Olivine-type structure (1, 352),
with Tm in octahedral and Zn in tetrahedral sites. Tm-S = 2.671-2.744, Zn-S =
2.294-2.388(9) A.

	x	y	z
Tm(1)	0.0	0.0	0.0
Tm(2)	0.26649	0.25	0.0056
Zn(1)	0.0926	0.25	0.5859
S(1)	0.0900	0.25	0.221
S(2)	0.4297	0.25	0.755
S(3)	0.3318	0.9871	0.2579

BOULANGERITE
Pb$_5$Sb$_4$S$_{11}$

Acta Cryst., C46, 531-534.

Pnam, 23.490, 21.245, 4.020, Z = 4, R = 0.077. Structure essentially as previously described (24, 373; 45A, 15), with ribbons along c of (Pb,Sb)S$_5$ square
pyramids; site occupancies are determined.

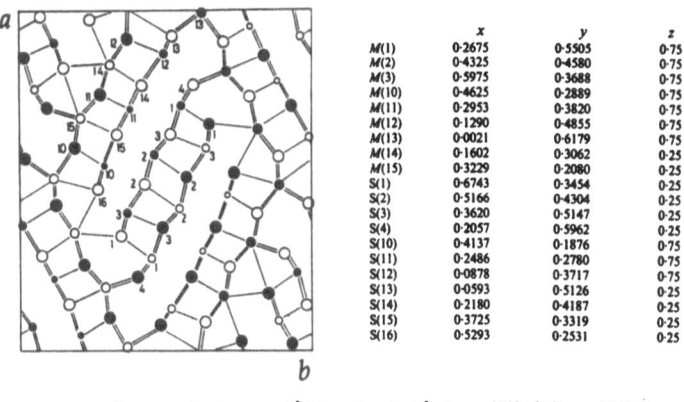

	x	y	z
M(1)	0·2675	0·5505	0·75
M(2)	0·4325	0·4580	0·75
M(3)	0·5975	0·3688	0·75
M(10)	0·4625	0·2889	0·75
M(11)	0·2953	0·3820	0·75
M(12)	0·1290	0·4855	0·75
M(13)	0·0021	0·6179	0·75
M(14)	0·1602	0·3062	0·25
M(15)	0·3229	0·2080	0·25
S(1)	0·6743	0·3454	0·25
S(2)	0·5166	0·4304	0·25
S(3)	0·3620	0·5147	0·25
S(4)	0·2057	0·5962	0·25
S(10)	0·4137	0·1876	0·75
S(11)	0·2486	0·2780	0·75
S(12)	0·0878	0·3717	0·75
S(13)	0·0593	0·5126	0·25
S(14)	0·2180	0·4187	0·25
S(15)	0·3725	0·3319	0·25
S(16)	0·5293	0·2531	0·25

Interatomic distances (Å) less than 3·4 Å in boulangerite

Narrow ribbon

M(1)—S(1)	2·602 (8)
S(4) × 2	2·662 (6)
S(3) × 2	3·090 (6)
M(2)—S(2)	2·659 (8)
S(3) × 2	2·869 (6)
S(2) × 2	2·880 (6)
M(3)—S(3)	2·651 (8)
S(1) × 2	2·744 (5)
S(2) × 2	3·059 (6)

Wide ribbon

M(10)—S(10)	2·432 (7)
S(16) × 2	2·661 (5)
S(15) × 2	3·058 (2)
M(11)—S(11)	2·468 (8)
S(14) × 2	2·817 (6)
S(15) × 2	2·907 (7)
M(12)—S(12)	2·609 (7)
S(13) × 2	2·656 (6)
S(14) × 2	3·230 (6)

M(13)—S(13)	3·130 (9)
S(12) × 2	2·920 (5)
S(16)	2·965 (8)
S(10) × 2	3·186 (6)
S(13) × 2	3·293 (7)
M(14)—S(14)	2·749 (8)
S(11) × 2	2·954 (5)
S(12) × 2	2·979 (5)
S(1)	3·236 (7)
S(16)	3·322 (8)
M(15)—S(15)	2·877 (9)
S(10) × 2	2·964 (5)
S(11) × 2	3·046 (5)
S(4) × 2	3·182 (7)

CHVILEVAITE
$Na(Cu,Fe,Zn)_2S_2$

Dokl. Akad. Nauk SSSR, **310**, 90–93.

P3m1, 3.873, 6.848, Z = 1, R = 0.045.   M–4S = 2.35–2.44, Na–6S = 2.87, 2.93 A.

	x	y	z
M(1)	2/3	1/3	0.1094
M(2)	0	0	0.3557
S(1)	0	0	0
S(2)	2/3	1/3	0.4629
Na	1/3	2/3	0.738

COBALTITE
CoAsS

Canad. Miner., **28**, 719–723.

$Pca2_1$, 5.5833, 5.5892, 5.5812, Z = 4, R = 0.016 (twinned crystal). Atoms in 4(a):
Co (0.99504,0.25909,0), As (0.61885,0.86935,0.61668), S (0.38266,0.63129,0.37996).
Derivative of the pyrite structure, with reinterpretation of the twinning (see
30A, 17; 44A, 21; 49A, 8).

GALENOBISMUTITE
Bi_2PbS_4

Acta Cryst., A<u>46</u>, 681-688.

Pnma, 11.79, 14.59, 4.10, Z = 4, Mo and synchrotron radiations, R = 0.057, 0.054.
Structure as previously described (<u>15</u>, 239; <u>27</u>, 81), with Bi/Pb position now more
clearly distinguished (results in agreement with <u>27</u>, 81).

MÜCKEITE
$(Bi,Sb)CuNiS_3$

Acta Cryst., C<u>46</u>, 127-128.

$P2_12_12_1$, 7.514, 12.557, 4.8880, Z = 4, R = 0.050. Bi,Sb is coordinated to 4 S and
2 M(2), M(1) (probably Cu) to 4 S, and M(2) (probably Ni) to 2 Bi,Sb and 4 S.
Bi,Sb-S = 2.60-2.92, Bi,Sb-M(2) = 3.18, 3.21, M(1)-S = 2.28-2.32, M(2)-S = 2.23-
2.25 A.

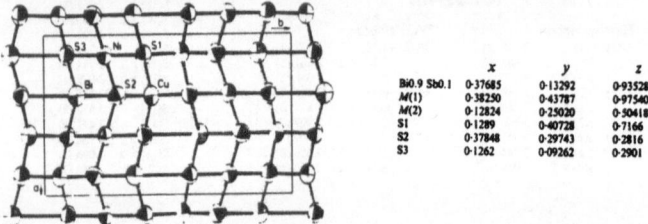

	x	y	z
Bi0.9 Sb0.1	0·37685	0·13292	0·93528
M(1)	0·38250	0·43787	0·97540
M(2)	0·12824	0·25020	0·50418
S1	0·1289	0·40728	0·7166
S2	0·37848	0·29743	0·2816
S3	0·1262	0·09262	0·2901

ROBINSONITE
$Pb_4Sb_6S_{13}$

Acta Cryst., C<u>46</u>, 527-531.

I2/m, 23.698, 3.980, 24.466, β = 93.9°, Z = 4, R = 0.071. Structure essentially
as previously described in P1 (<u>45A</u>, 15), with ribbons along **b** of face-sharing
$(Pb,Sb)S_5$ square pyramids; Pb/Sb occupancies (and Sn/Sb occupancies in $Sn_4Sb_6S_{13}$
(<u>46A</u>, 14)) were determined from bond valencies.

	x	y	z
M(1)	0·6901	0·5	0·7983
M(2)	0·5053	0·5	0·8117
M(3)	0·4365	0·0	0·9438
M(4)	0·6172	0·0	0·9228
M(10)	0·5068	0·5	0·3527
M(11)	0·4206	0·5	0·5114
M(12)	0·3431	0·5	0·6706
M(13)	0·1811	0·0	0·6627
M(14)	0·2698	0·0	0·5216
M(15)	0·3559	0·0	0·3694
S(1)	0·6921	0·5	0·9088
S(2)	0·5258	0·5	0·9251
S(3)	0·4199	0·0	0·8444
S(4)	0·5934	0·0	0·8235
S(5)	0·7627	0·0	0·8132
S(6)	0·3629	0·5	0·9507
S(10)	0·4090	0·5	0·3101
S(11)	0·3279	0·5	0·4579
S(12)	0·2509	0·5	0·6108
S(13)	0·3046	0·0	0·7281
S(14)	0·3782	0·0	0·5766
S(15)	0·4545	0·0	0·4347
S(16)	0·5335	0·0	0·2896

$Co_4Cs_7Se_8$

J. Less-Common Metals, <u>167</u>, 161-167.

C2/c, 19.473, 8.798, 17.265, 117.76, Z = 4, R = 0.040. Isostructural with
$Cs_7Fe_4Te_8$ (<u>50A</u>, 19), with cubane-like $Co_4Se_8^{7-}$ units. Co-Se = 2.30-2.45 A.

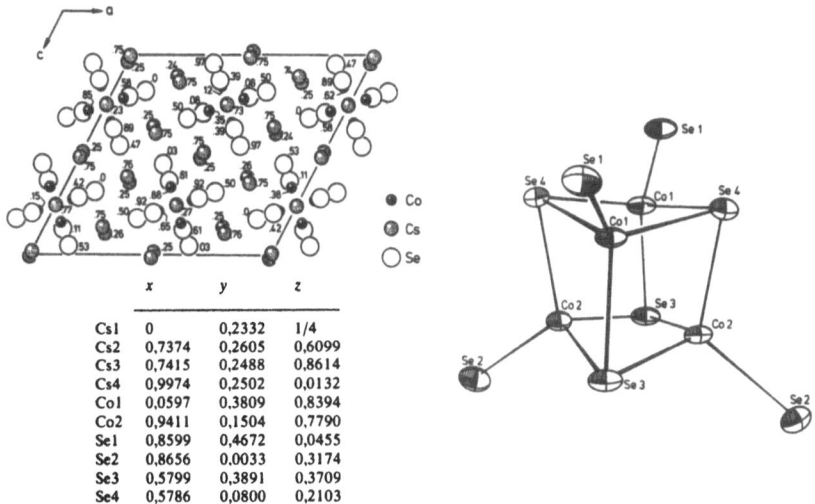

	x	y	z
Cs1	0	0,2332	1/4
Cs2	0,7374	0,2605	0,6099
Cs3	0,7415	0,2488	0,8614
Cs4	0,9974	0,2502	0,0132
Co1	0,0597	0,3809	0,8394
Co2	0,9411	0,1504	0,7790
Se1	0,8599	0,4672	0,0455
Se2	0,8656	0,0033	0,3174
Se3	0,5799	0,3891	0,3709
Se4	0,5786	0,0800	0,2103

$Cs_5Mo_{21}Se_{23}$

Acta Cryst., <u>C46</u>, 2284-2287.

$P6_3/m$, 9.6513, 29.939, Z = 2, R = 0.034. $Mo_{21}Se_{23}$ cluster with a central core of
six face-sharing Mo_6 octahedra. The clusters are linked via Mo-Se bonds to give
large voids and channels, which contain Cs ions with tri- and tetra-capped trigonal prismatic coordinations to Se.

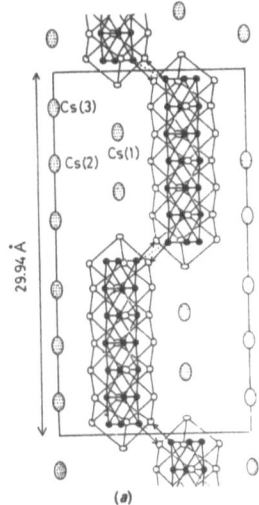

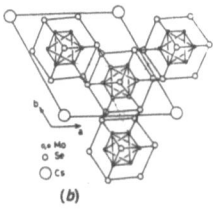

Projection of the structure of $Cs_5Mo_{21}Se_{23}$ on the hexagonal planes (a) (11$\bar{2}$0) and (b) (0001). Filled circles: Mo atoms;
empty circles: Se atoms. The thin lines define the Se polyhedra
surrounding the Mo_{21} cluster. Arrows show the Mo—Se inter-
unit bonds, the dotted lines the intercluster bonds.

Selected interatomic distances (Å)

Intratriangle distances		Intertriangle distances	
Mo(1)—Mo(1)	2 × 2·652 (2)	Mo(1)—Mo(2)	2·744 (1)
Mo(2)—Mo(2)	2 × 2·669 (2)	—Mo(2)	2·768 (2)
Mo(3)—Mo(3)	2 × 2·665 (2)	Mo(2)—Mo(3)	2·701 (2)
Mo(4)—Mo(4)	2 × 2·660 (3)	—Mo(3)	2·721 (1)
		Mo(3)—Mo(4)	2·721 (1)
		—Mo(4)	2·727 (2)

	x	y	z
Mo(1)	0·5195	0·1653	0·52395
Mo(2)	0·6847	0·1834	0·60037
Mo(3)	0·5184	0·1647	0·67488
Mo(4)	0·6878	0·1859	0·750
Se(1)	0·0440	0·3376	0·47042
Se(2)	0·6236	−0·0063	0·09918
Se(3)	0·7097	0·0439	0·67533
Se(4)	0·6223	0·0003	0·250
Se(5)	0·667	0·333	0·45672
Cs(1)	0·667	0·333	0·16787
Cs(2)	0·000	0·000	0·250
Cs(3)	0·000	0·000	0·09476

Intercluster distances			
Mo(1)—Mo(1)	3·342 (3)		

Mo—Se distances		Cs environment	
Mo(1)—Se(1)	2·601 (2)	Cs(1)—Se(2)	3 × 3·713 (2)
—Se(1)	2·611 (2)	—Se(3)	3 × 3·643 (1)
—Se(1)	2·646 (2)	—Se(4)	3 × 3·897 (2)
—Se(2)	2·686 (2)	—Se(5)	3·730 (3)
—Se(5)	2·529 (2)		
		Cs(2)—Se(3)	6 × 3·770 (1)
Mo(2)—Se(1)	2·575 (2)	—Se(4)	3 × 3·647 (2)
—Se(2)	2·586 (2)		
—Se(2)	2·602 (2)	Cs(3)—Se(1)	3 × 3·636 (2)
—Se(3)	2·690 (2)	—Se(2)	3 × 3·606 (1)
		—Se(3)	3 × 3·877 (2)
Mo(3)—Se(1)	2·694 (2)		
—Se(3)	2·606 (2)	Cs(1)···Cs(1)	4·918 (3)
—Se(3)	2·631 (2)	Cs(2)···Cs(3)	4·648 (2)
—Se(4)	2·697 (2)	Cs(3)···Cs(3)	5·674 (4)
Mo(4)—Se(3)	2 × 2·685 (2)		
—Se(4)	2·609 (3)		
—Se(4)	2·634 (3)		

$Cs_2Sn_3Se_7$

Z. Naturforsch., **45**B, 1643–1646.

C2/c, 23.214, 13.679, 14.657, 111.92, Z = 8, R = 0.066. $Sn_3Se_7^{2-}$ sheets of linked $SnSe_5$ trigonal bipyramids, with channels containing 7- and 8-coordinate Cs. Sn–Se = 2.49–2.80 A.

$CuInSe_2$

Fukuoka Daigaku Rigaku Shuho, **19**, 103–107.

$I\bar{4}2d$, 5.7810, 11.6103, Z = 4, R = 0.064. Chalcopyrite structure, as previously described (**17**, 19; **39**A, 51; **54**A, 62); x(Se) = 0.2258.

Cu_3TlSe_2

J. Less-Common Metals, **161**, 101–108.

C2/m, 15.2128, 14.0115, 8.3944, 111.700, Z = 4, R = 0.050. Ag_3CsS_2-type (**44**A, 43).

	x	z
Tl	0.1383	0.0477
Cu_1	0.5872	0.6303
Cu_2	0.0591	0.4026
Cu_3	0.6896	0.4460
Se_1	0.1938	0.6812
Se_2	0.5224	0.2282

All atoms in 4*i*, y = 0

Cu_5TlSe_3

J. Solid State Chem., <u>87</u>, 283-288.

$P4_2/mnm$, 12.900, 3.968, Z = 4, R = 0.033.   The structure contains $TlSe_8$ cubes, $CuSe_4$ tetrahedra, and $CuSe_3$ triangles; Tl-Se = 3.36, 3.48, Cu-Se = 2.39-2.63 A. The material is Cu-deficient.

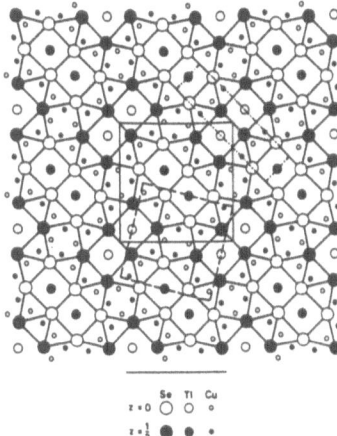

		Occupancy	x	y
Tl	4f	1	0.10749	x
Cu_1	4g	0.931	0.8196	$1 - x$
Cu_2	8i	0.940	0.2137	0.4183
Cu_3	8i	0.973	0.0159	0.3935
Se_1	4f	1	0.3979	x
Se_2	8i	1	0.6160	0.1714

All atoms are situated at $z = 0$

$K_{12}Mo_{12}Se_{56}$

J. Amer. Chem. Soc., <u>112</u>, 7400-7402.

$Cmc2_1$, 23.73, 17.70, 20.43, Z = 4, R = 0.080, atomic positional parameters not listed.   Discrete $Mo_{12}Se_{56}^{12-}$ cluster, which contains four trinuclear $Mo_3Se_{14}^{3-}$ subclusters.   Each Mo has 7-coordination (not counting the Mo-Mo bonds), including terminal and bridging Se^{2-}, Se_2^{2-}, and Se_3^{2-} ligands, and with one Se bridging three Se_2 to form a Se_7^{8-} unit.   Mean Mo-Mo = 2.75(2) A.

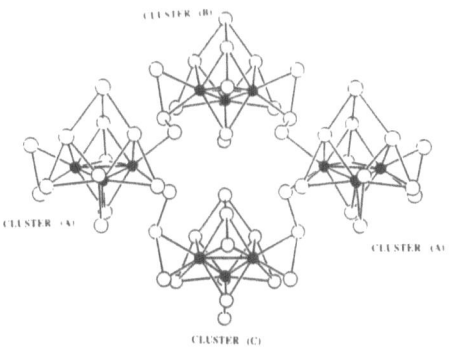

$M_2Pt_4Se_6$ (M = K, Rb, Cs; Na)

J. Less-Common Metals, <u>161</u>, 25-30.

M = K, Rb, Cs, $R\overline{3}m$, cell parameters not given [a ∿ 7, c ∿ 20 A], Z = 3, R = 0.046,

0.036, 0.056. M in 6(c): 0,0,z, z = 0.2001, 0.1994, 0.1989; Pt(1) in 3(a):
0,0,0; Pt(2) in 9(e): 1/2,0,0; Se in 18(h): x,$\bar{x}$,z, x = 0.1706, 0.1708, 0.1709,
z = 0.0654, 0.0626, 0.0589. Isostructural with $K_2Pt_4S_6$ (this volume , p. 70),
with Pt(II)Se$_4$ square planes and Pt(IV)Se$_6$ octahedra, and M ions in cuboctahedral
sites.

M = Na, P6$_3$mc, 7.346, 11.618, Z = 2, structure proposed.

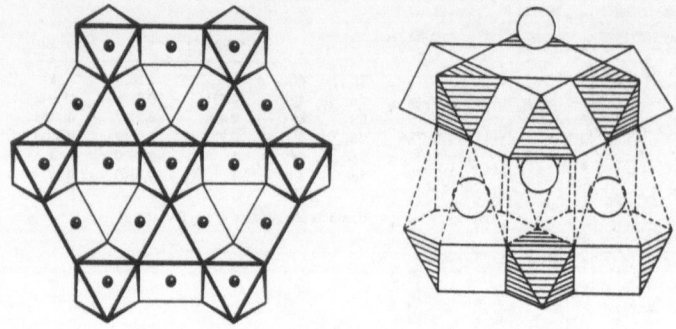

$M_2Pt_4Se_6$ (M = K, Rb, Cs)

α-$Mo_{15}Se_{19}$

J. Solid State Chem., 85, 332-336.

P6$_3$/m, 9.450, 19.600, Z = 2, R = 0.050. Mo_6Se_8 and Mo_9Se_{11} clusters, as in
related materials (39A, 85; 51A, 69).

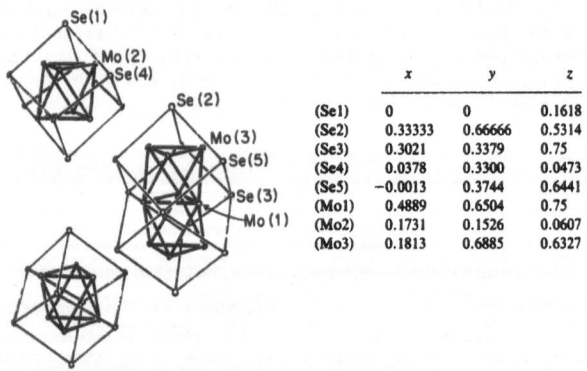

	x	y	z
(Se1)	0	0	0.1618
(Se2)	0.33333	0.66666	0.5314
(Se3)	0.3021	0.3379	0.75
(Se4)	0.0378	0.3300	0.0473
(Se5)	−0.0013	0.3744	0.6441
(Mo1)	0.4889	0.6504	0.75
(Mo2)	0.1731	0.1526	0.0607
(Mo3)	0.1813	0.6885	0.6327

$Pb_4Sb_4Se_{10}$

Acta Cryst., C46, 2287-2291.

Pnam, 24.591, 19.757, 4.166, Z = 4, R = 0.064. Isostructural with cosalite (24,
63; 40A, 37), the structure containing ribbons of square-pyramidal (Pb,Sb)Se$_5$
groups. Sb/Pb distributions are determined.

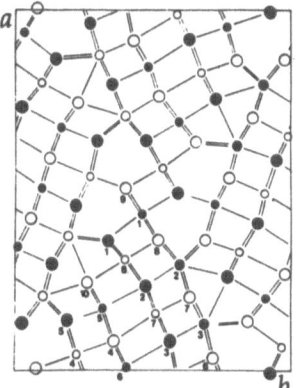

Percentage of Sb (*or* Bi) *on cation sites in*
$Pb_4Sb_4Se_{10}$ *and cosalite*

	$Pb_4Sb_4Se_{10}$		Cosalite
	X-ray refinement	Bond-valence method	
M(1)	82	80	54 Bi + 12 Ag
M(2)	0	0	0
M(3)	0	0	0
M(4)	32	33	23
M(5)	88	75	100
M(6)	60	46	23
M(7)	88	80	100
M(8)	92	86	100
M(9)	—	—	30 Cu
M(10)	—	—	24 Cu
Formula	$Pb_{3\cdot58}Sb_{4\cdot42}Se_{10}$	$Pb_4Sb_4Se_{10}$	$Ag_{0\cdot12}Cu_{0\cdot54}Pb_{3\cdot54}Bi_4S_{10}$

The unit cell of $Pb_4Sb_4Se_{10}$ projected down [001]. In order of decreasing size, the circles denote Se, Pb and mixed sites. Atoms at $z = 0\cdot25$ and $z = 0\cdot75$ are indicated by open and filled circles respectively.

Interatomic distances (Å) *less than* 3·6 Å *in*
$Pb_4Sb_4Se_{10}$

	x	y	z
M(1)	0·4359	0·5389	0·75
M(2)	0·2954	0·3910	0·75
M(3)	0·1273	0·2976	0·75
M(4)	0·0434	0·7710	0·25
M(5)	0·1741	0·6872	0·75
M(6)	0·0099	0·5912	0·75
M(7)	0·1560	0·4841	0·25
M(8)	0·3034	0·5929	0·25
Se(1)	0·3641	0·6548	0·75
Se(2)	0·2310	0·5296	0·75
Se(3)	0·0821	0·4380	0·75
Se(4)	0·0855	0·6275	0·25
Se(5)	0·1249	0·8069	0·75
Se(6)	0·0346	0·2833	0·25
Se(7)	0·2043	0·3628	0·25
Se(8)	0·3614	0·4774	0·25
Se(9)	0·5002	0·5886	0·25
Se(10)	0·2390	0·7289	0·25

M(1)—Se(9)	× 2	2·793 (3)
Se(1)		2·879 (4)
Se(9)		2·970 (4)
Se(8)	× 2	3·028 (3)
M(2)—Se(7)	× 2	3·109 (2)
Se(8)	× 2	3·146 (3)
Se(2)		3·163 (4)
Se(5)	× 2	3·308 (3)
Se(10)		3·313 (4)
M(3)—Se(3)		2·989 (4)
Se(6)	× 2	3·069 (2)
Se(7)	× 2	3·096 (2)
Se(1)	× 2	3·528 (3)
Se(10)		3·557 (3)
M(4)—Se(9)		2·969 (4)
Se(5)	× 2	2·976 (2)
Se(4)		3·019 (5)
Se(6)	× 2	3·055 (2)

M(5)—Se(5)		2·656 (4)
Se(10)	× 2	2·750 (4)
Se(4)	× 2	3·236 (3)
Se(2)		3·415 (4)
M(6)—Se(6)		2·728 (4)
Se(4)	× 2	2·883 (3)
Se(3)	× 2	3·131 (3)
Se(3)		3·509 (4)
M(7)—Se(7)		2·673 (4)
Se(3)	× 2	2·908 (3)
Se(2)	× 2	2·923 (3)
Se(4)		3·324 (4)
M(8)—Se(8)		2·692 (4)
Se(1)	× 2	2·830 (3)
Se(2)	× 2	3·011 (3)
Se(10)		3·118 (4)

Yb_5Se_7

J. Less-Common Metals, 157, L19–L22.

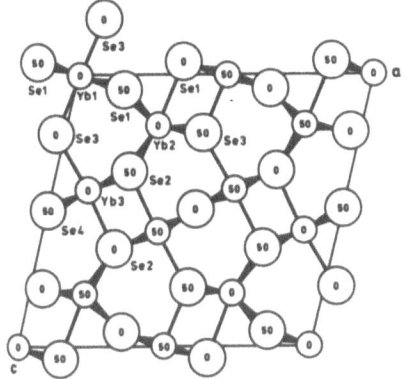

	x	y	z
Yb(1)	0	0	0
Yb(2)	0.69470	0	0.80996
Yb(3)	0.88492	0	0.57812
Se(1)	0.3425	0	0.9487
Se(2)	0.2597	0	0.6439
Se(3)	0.9646	0	0.2155
Se(4)[b]	0.4931	0	0.4859

[b] Split position

C2/m, 13.074, 3.9435, 12.046, 104.77, Z = 2, R = 0.036. Y_5S_7-type (55A, 54),
but with Se(4) split into two positions with half-occupancy (compare $\overline{Y}_5Se_7$
(55A, 63), which is described in Cm).

AuKTe

J. Less-Common Metals, 160, 181-184.

$P6_3/mmc$, 4.646, 9.744, Z = 2, R = 0.026. Au in 2(c): 1/3,2/3,1/4; K in 2(a):
0,0,0; Te in 2(d): 1/3,2/3,3/4. $InNi_2$ (filled NiAs) type (9, 91), with layers
of Au + Te hexagons, linked by K ions. Au-3Te = 2.682(1) A.

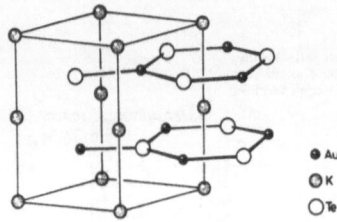

● Au
◉ K
○ Te

Ga_2MnTe_4

J. Less-Common Metals, 158, L27-L31.

Pnma, 27.448, 4.192, 6.993, Z = 4, R = 0.080, high-pressure phase. Isostructural
with the In compound (52A, 57), with $MnTe_6$ octahedra and $GaTe_4$ tetrahedra. Mn-Te
= 2.85-2.98, Ga-Te = 2.57-2.66(1) A. The normal-pressure form has a complex
tetragonal structure (not determined in detail).

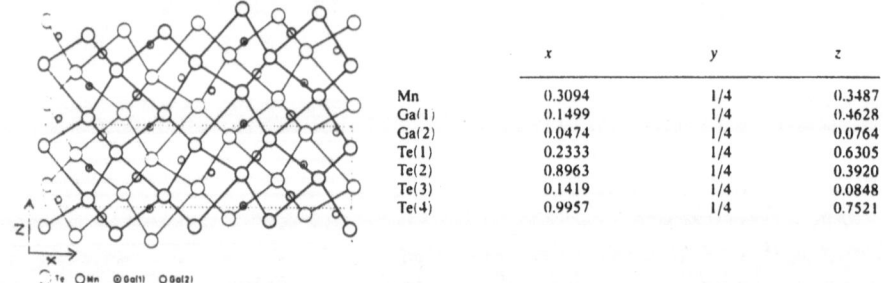

	x	y	z
Mn	0.3094	1/4	0.3487
Ga(1)	0.1499	1/4	0.4628
Ga(2)	0.0474	1/4	0.0764
Te(1)	0.2333	1/4	0.6305
Te(2)	0.8963	1/4	0.3920
Te(3)	0.1419	1/4	0.0848
Te(4)	0.9957	1/4	0.7521

○ Te ○ Mn ◉ Ga(1) ○ Ga(2)

$GeTl_2Te_5$

J. Solid State Chem., 84, 245-252; 87, 467-468.

P4/mbm, 8.243, 14.917, Z = 4, R = 0.077; the first paper concludes that the space
group is Cmmm, but the second suggests that the derived structure does not deviate
from P4/mbm symmetry. The structure contains (001) layers of $Ge_2Te_6^{4-}$ anions and
Te_4 clusters, linked by 8-coordinate Tl ions.

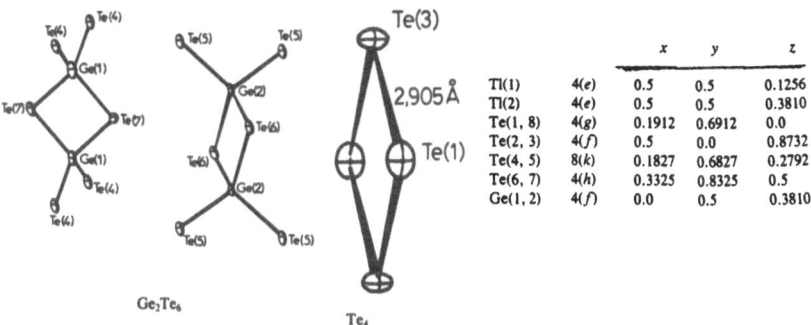

	x	y	z
Tl(1)	4(e)	0.5	0.5	0.1256
Tl(2)	4(e)	0.5	0.5	0.3810
Te(1, 8)	4(g)	0.1912	0.6912	0.0
Te(2, 3)	4(f)	0.5	0.0	0.8732
Te(4, 5)	8(k)	0.1827	0.6827	0.2792
Te(6, 7)	4(h)	0.3325	0.8325	0.5
Ge(1, 2)	4(f)	0.0	0.5	0.3810

InNa₅Te₄ (I), In₂Na₅Te₆ (II)

InNa$_5$Te$_4$ (I), In$_2$Na$_5$Te$_6$ (II)

Z. Naturforsch., 45B, 8–14.

I, Pbcn, 7.427, 20.154, 17.733, Z = 8, R = 0.071. II, Cmc2$_1$, 9.021, 15.361, 23.565, Z = 8, R = 0.046. I contains isolated InTe$_4$$^{5-}$ tetrahedra, linked by 6- and 4-coordinated Na$^+$ ions; In–Te = 2.76–2.81, Na–Te = 3.04–3.75 A. II contains chains of corner-sharing InTe$_4$ tetrahedra connected into ribbons by Te–Te bridges, with ribbons linked by 6-coordinate Na$^+$ ions; In–Te = 2.71–2.85, Te–Te = 2.86, Na–Te = 3.13–3.53 A.

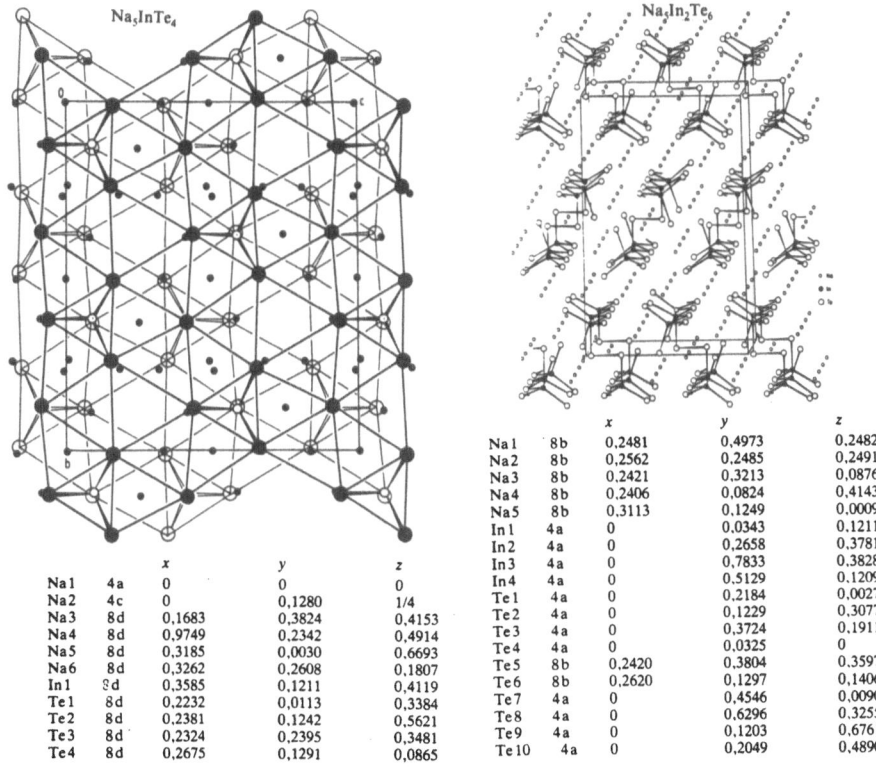

		x	y	z
Na1	4a	0	0	0
Na2	4c	0	0,1280	1/4
Na3	8d	0,1683	0,3824	0,4153
Na4	8d	0,9749	0,2342	0,4914
Na5	8d	0,3185	0,0030	0,6693
Na6	8d	0,3262	0,2608	0,1807
In1	8d	0,3585	0,1211	0,4119
Te1	8d	0,2232	0,0113	0,3384
Te2	8d	0,2381	0,1242	0,5621
Te3	8d	0,2324	0,2395	0,3481
Te4	8d	0,2675	0,1291	0,0865

		x	y	z
Na1	8b	0,2481	0,4973	0,2482
Na2	8b	0,2562	0,2485	0,2491
Na3	8b	0,2421	0,3213	0,0876
Na4	8b	0,2406	0,0824	0,4143
Na5	8b	0,3113	0,1249	0,0009
In1	4a	0	0,0343	0,1211
In2	4a	0	0,2658	0,3781
In3	4a	0	0,7833	0,3828
In4	4a	0	0,5129	0,1209
Te1	4a	0	0,2184	0,0027
Te2	4a	0	0,1229	0,3077
Te3	4a	0	0,3724	0,1911
Te4	4a	0	0,0325	0
Te5	8b	0,2420	0,3804	0,3597
Te6	8b	0,2620	0,1297	0,1406
Te7	4a	0	0,4546	0,0090
Te8	4a	0	0,6296	0,3255
Te9	4a	0	0,1203	0,6761
Te10	4a	0	0,2049	0,4890

K_5Te_3

Z. Naturforsch., 45B, 417-422.

I4/m, 13.742, 6.364, Z = 4, R = 0.057. The structure can be derived from W_5Si_3-type (19, 277) by replacing Si chains along c by chains of Te_2^{2-} ions; other Te atoms are involved in chains along c of edge-sharing KTe_4 tetrahedra, with an extremely short K...K distance, 3.18 A. Te-Te = 2.84 A.

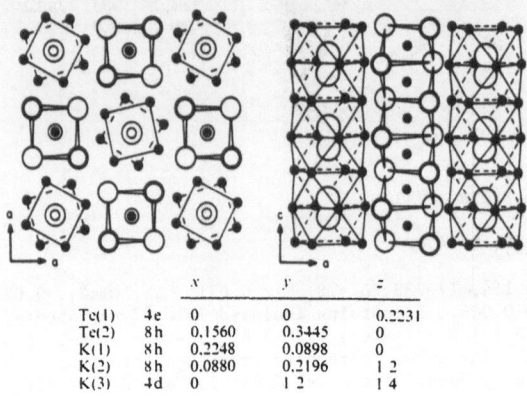

		x	y	z
Te(1)	4e	0	0	0.2231
Te(2)	8h	0.1560	0.3445	0
K(1)	8h	0.2248	0.0898	0
K(2)	8h	0.0880	0.2196	1 2
K(3)	4d	0	1 2	1 4

$(Nb_{0.28}Ta_{0.72})Te_4$

Acta Cryst., B46, 153-159.

Subcell, P4/mcc, a = 6.512, c = 6.818 A, Z = 2, R = 0.065. Nb/Ta in 2(a): 0,0,1/4; Te in 8(m): x,y,0, x = 0.1439, y = 0.3275. Superstructure, P4/ncc, 2a x 2a x c, modulation vector q = 2/3c*, R = 0.058. Structures similar to those of $NbTe_4$ and $TaTe_4$ (29, 77; 52A, 71; 53A, 59; 54A, 70).

Redetermined structures

Z. Kristallogr., 193, 217-242.

The following compounds, which were previously described as monoclinic, are in fact rhombohedral:

PtTe:	R3m. a = 3.963.	c = 19.98 Å. Z = 6
Pt_3Te_4:	R3m. a = 3.988.	c = 35.39 Å. Z = 3
Pt_2Te_3:	R3m. q = 4.003.	c = 50.89 Å. Z = 6
Li_8Pb_3:	R3m. a = 4.757.	c = 32.05 Å. Z = 3
$LiFe_6Ge_4$:	R3m. a = 5.045.	c = 19.66 Å. Z = 3
$LiFe_6Ge_5$:	R3m. a = 5.048.	c = 43.64 Å. Z = 6
$CaGa_6Te_{10}$:	R32. a = 14.42.	c = 17.65 Å. Z = 6
$La_{3.266}Mn_{1.1}S_6$:	R3m. a = 14.08.	c = 21.80 Å. Z = 15

Ta_4SiTe_4

Inorg. Chem., 29, 3952-3954.

Pbam, 10.536, 18.275, 4.799, Z = 4, R = 0.061. New structure type, with Si-centred Ta_8 antiprisms, with the square edges bridged by Te.

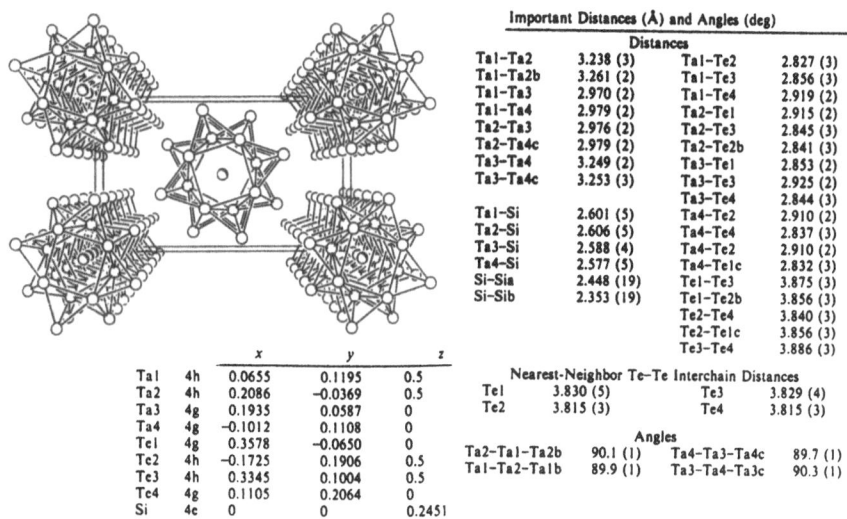

Important Distances (Å) and Angles (deg)			
Distances			
Ta1–Ta2	3.238 (3)	Ta1–Te2	2.827 (3)
Ta1–Ta2b	3.261 (2)	Ta1–Te3	2.856 (3)
Ta1–Ta3	2.970 (2)	Ta1–Te4	2.919 (2)
Ta1–Ta4	2.979 (2)	Ta2–Te1	2.915 (2)
Ta2–Ta3	2.976 (2)	Ta2–Te3	2.845 (3)
Ta2–Ta4c	2.979 (2)	Ta2–Te2b	2.841 (3)
Ta3–Ta4	3.249 (2)	Ta3–Te1	2.853 (2)
Ta3–Ta4c	3.253 (3)	Ta3–Te3	2.925 (2)
		Ta3–Te4	2.844 (3)
Ta1–Si	2.601 (5)	Ta4–Te2	2.910 (2)
Ta2–Si	2.606 (5)	Ta4–Te4	2.837 (3)
Ta3–Si	2.588 (4)	Ta4–Te2	2.910 (2)
Ta4–Si	2.577 (5)	Ta4–Te1c	2.832 (3)
Si–Sia	2.448 (19)	Te1–Te3	3.875 (3)
Si–Sib	2.353 (19)	Te1–Te2b	3.856 (3)
		Te2–Te4	3.840 (3)
		Te2–Te1c	3.856 (3)
		Te3–Te4	3.886 (3)

		x	*y*	*z*
Ta1	4h	0.0655	0.1195	0.5
Ta2	4h	0.2086	−0.0369	0.5
Ta3	4g	0.1935	0.0587	0
Ta4	4g	−0.1012	0.1108	0
Te1	4g	0.3578	−0.0650	0
Te2	4h	−0.1725	0.1906	0.5
Te3	4h	0.3345	0.1004	0.5
Te4	4g	0.1105	0.2064	0
Si	4c	0	0	0.2451

Nearest-Neighbor Te–Te Interchain Distances			
Te1	3.830 (5)	Te3	3.829 (4)
Te2	3.815 (3)	Te4	3.815 (3)

Angles			
Ta2–Ta1–Ta2b	90.1 (1)	Ta4–Ta3–Ta4c	89.7 (1)
Ta1–Ta2–Ta1b	89.9 (1)	Ta3–Ta4–Ta3c	90.3 (1)

TlTe,   Tl$_5$Te$_3$

J. Solid State Chem., 87, 229–236.

TlTe, I4/mcm, 12.961, 6.18, Z = 16, R = 0.059.   Tl in 16(k): 0.7704,0.0796,1/2;
Te(1) in 8(h): 0.1655,0.3345,0; Te(2) in 4(d): 1/2,0,0; Te(3) in 4(a): 0,0,1/4.
Structure as previously described (40A, 98).

Tl$_5$Te$_3$, I4/m, 8.917, 12.613, Z = 4, R = 0.054.   Tl(1) in 16(i): 0.3527,0.1475,
0.1590; Tl(2) in 2(b): 0,0,1/2; Te(3) in 2(a): 0,0,0; Te(1) in 4(e): 0,0,0.2493;
Te(2) in 8(h): 0.3410,0.1596,1/2.   Structure essentially as previously proposed
(35A, 108; 37A,137; 56A,57) [the present description differs significantly
from I4/mcm].

METALS

TABLE I

Some structural information has also been given for the following materials
(listed with abbreviated 1990 references).

Compound	Structure	Reference
LaSi NdSi GdSi HoSi ErSi	FeB	Acta Cryst., $\underline{20}$, 572 (1966); $\underline{22}$, 688 (1967)
HoSi ErSi	CrB	
$CrLaS_3$ $(Cr_{60}La_{72}S_{192})$	Superspace group description of the OD structure ($\underline{43A}$, 47)	Ibid., B$\underline{46}$, 39
$NbTe_4$	Incommensurate phases	Ibid., B$\underline{46}$, 587
$Al_{65}Co_{15}Cu_{20}$	5-Dimensional description	Ibid., B$\underline{46}$, 703
$ZrC_{0.95}$ $HfN_{0.98}$	NaCl	Acta Chem. Scand., $\underline{44}$, 851
La_5Sn_3	W_5Si_3 (and not Mn_5Si_3)	Chem. Mater., $\underline{2}$, 546
$CuTi_2$ CuTi Cu_4Ti_3 Cu_3Ti_2 Cu_4Ti	$MoSi_2$ γ-CuTi Cu_4Ti_3 Al_3Os_2 Au_4Zr	Dokl. Akad. Nauk SSSR, $\underline{306}$, 355 (1989)
$(Ti,Ta)C_{0.6}$	$TiC_{0.6}$	Izv. Akad. Nauk SSSR, Neorg. Mater., $\underline{26}$, 2103
MPt_5, M = Am, Cm, Bk, Cf	$CaCu_5$	J. Less-Common Metals, $\underline{157}$, 147
La_3Ru La_5Ru_2 La_7Ru_3 $LaRu_2$	Fe_3C Mn_5C_2 Sr_7Pt_3 Cu_2Mg	Ibid., $\underline{157}$, 307
Nd_3Ru Nd_5Ru_2 Nd_7Ru_3 $NdRu_2$	Fe_3C Mn_5C_3 Sr_7Pt_3 Cu_2Mg and $MgZn_2$	
IrTi	AuCu, NbRu, and CsCl	Ibid., $\underline{158}$, L11
$(Al,Ga)_2Tm$	Cu_2Mg [$\underline{56A}$, 12]	Ibid., $\underline{158}$, 319
Er_3Ru Er_5Ru_2 $Er_{44}Ru_{25}$	Fe_3C Mn_5C_2 $Y_{44}Ru_{25}$	Ibid., $\underline{159}$, L21
Nb_5B_6	Ta_5B_6 [this volume, p. 31]	Ibid., $\underline{159}$, L25

AX, A = Np, Pu, X = Sb, Te	NaCl, CsCl, and tetra-gonal phases	J. Less-Common Metals, $\underline{160}$, 35
Ge_2Lu Ge_2Tm	$ZrSi_2$	Ibid., $\underline{160}$, 197, 215

Ge_2LnM_x
LnM_xSn_2
  M = Mn, Fe etc.
  Ln = lanthanon
	$CeNiSi_2$	

$Ga_{4.5}Pd_{25}As_{4.5}$ $Ga_{4.5}Pd_{25}Sb_{4.5}$ $Pd_{25}Sn_{4.5}Si_{4.5}$	Ge_9Pd_{25}	Ibid., $\underline{161}$, 147
Fe_2LnSi_2 Fe_2Ge_2Ln	Cr_2ThSi_2	Ibid., $\underline{161}$, 185
$Fe_{14}Ho_2B$	$Fe_{14}Nd_2B$ ($\underline{51A}$, 23) at 300K; Cm at low temp-eratures	Ibid., $\underline{162}$, 237
$KMgH_3$	Perovskite, a = 4.025 A	Ibid., $\underline{163}$, 179
CoSb	NiAs	Ibid., $\underline{166}$, 103
$(Al,Fe)_{12}Y$	$Mn_{12}Th$, with Fe prefer-entially in 8(f) and Al in 8(i) sites	J. Phys.: Condens. Matter, $\underline{2}$, 1677
Al-Mn quasicrystal	Model for the structure	J. Non-Cryst. Solids, $\underline{117\text{-}118}$, 765
Na_2S	Antifluorite, cubic form II, and orthorhombic form III	J. Solid State Chem., $\underline{85}$, 283
$(Cr,Fe)_2Zr$	C14	Ibid., $\underline{87}$, 415
Ge_4Mn $CoGe_4$	β-Hg_4Ni superstructure	Ibid., $\underline{88}$, 384
$ErRh_3B_2$	$CeCo_3B_2$ superstructure	Kidorui, $\underline{16}$, 146
Fe_2ScSi_2	Fe_2HfSi_2	Kristallografija, $\underline{35}$, 223 [Soviet Physics –
Co_2ScSi_2 Ni_2ScSi_2	Ga_2CeAl_2	Crystallography, 35, 136]
$GaTlTe_2$ $InTlTe_2$ $InTlSe_2$	TlSe	Mater. Lett., $\underline{9}$, 269
$CoGa_{1.8}V_{0.2}S_4$ $CoGa_{1.8}Ti_{0.2}S_4$	$FeGa_2S_4$	Mater. Res. Bull., $\underline{25}$, 1371
Al-Co-Ni	Quasicrystal	Phys. Rev. Lett., $\underline{65}$ 1603

$(PbS)_{1.18}TiS_2$ $(PbS)_{1.13}VS_2$ $(SnS)_{1.20}TiS_2$	Misfit layer structures [see also this volume , pp. 72-74]	Solid State Comm., $\underline{75}$, 689
$CsRu_2As_2$ $CsRh_2As_2$	$ThCr_2Si_2$	Z. anorg. Chem., $\underline{584}$, 138
$GaAs_{0.065}P_{0.935}$	Zincblende	Z. Kristallogr., $\underline{190}$, 33

STRUCTURE REPORTS

SECTION II

INORGANIC COMPOUNDS

Edited by

J. Trotter

(University of British Columbia)

ARRANGEMENT

    To find particular inorganic compounds the subject index should be used.
The general arrangement in the text is: elements, boron hydrides, carbonyls,
phosphorus–nitrogen and sulphur–nitrogen compounds, halides, cyanides, oxides,
double oxides, hydroxides, sulphides, borates, carbonates, nitrates, phosphates,
arsenates, sulphates, perchlorates, iodates, silicates, silicate minerals. Only
complete structure analyses are described; incomplete structural data are given
in a Table, electron diffraction studies in the gas phase and compounds which
have been described only in preliminary communications are tabulated.

1,2-DIBISMUTHDECABORANE
$Bi_2B_{10}H_{10}$

Inorg. Chem., 29, 804-808.

$P\bar{1}$, 12.174, 14.070, 12.135, 91.07, 99.51, 106.70, at 115K, Z = 8 (4
molecules/asymmetric unit), R = 0.12; positional parameters listed for only 1
molecule. Distorted icosahedral molecule; Bi-Bi = 2.956(4), Bi-B = 2.39-2.63(8)
A.

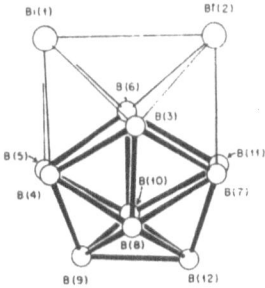

RHENIUM CARBONYL HYDRIDES
$[HRe(CO)_4]_2 \cdot [HRe(CO)_4]_3$ (I), $[HRe(CO)_4]_4$ (II)

J. Amer. Chem. Soc., 112, 9395-9397.

I, $P\bar{1}$, 8.722, 13.546, 14.179, 81.36, 74.35, 81.38, Z = 2, R = 0.024; II, C2/c,
14.429, 12.773, 13.483, 95.56, Z = 4, R = 0.020, atomic positional parameters not
listed. I is a 1:1 cocrystal of dimer and trimer, and II is a tetramer; Re-Re =
2.88, 3.24, 3.44 A, in the three molecules.

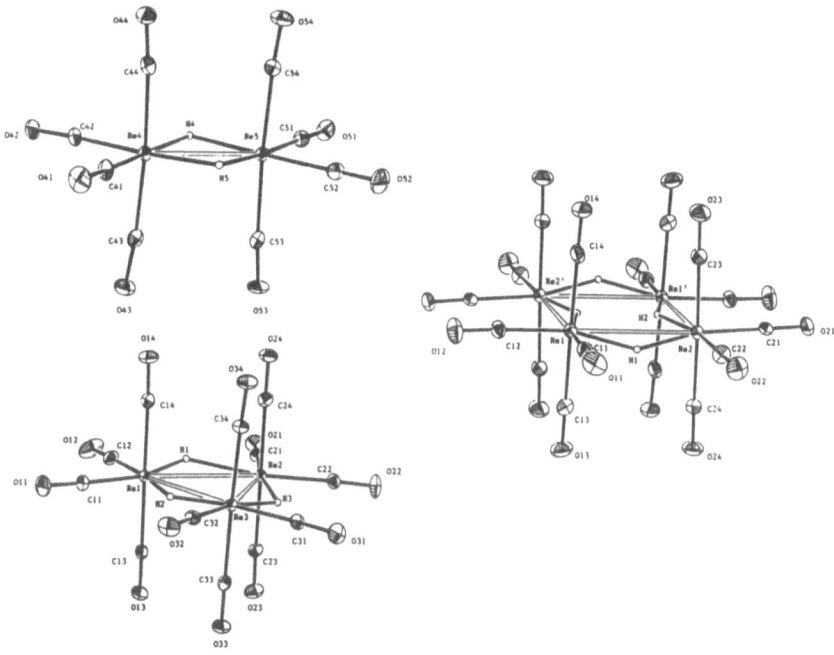

IRON PENTACARBONYL
$Fe(CO)_5$

Z. Kristallogr., 193, 289-290.

C2/c, 11.807, 6.821, 9.367, 107.72, at 200K, Z = 4, R = 0.030. Structure as
previously described (27, 650; 29, 241).

IRON CARBONYL SULPHIDE
$Fe_4(CO)_{11}S_2$

Polyhedron, 8, 1885-1890 (1989).

Pccn, 6.603, 15.429, 17.292, Z = 4, R = 0.043. Fe_4S_2 trans-octahedral cluster,
with one bridging, two semi-bridging, and eight terminal carbonyl groups. Fe-Fe
= 2.49-2.61, Fe-S = 2.28-2.35 A.

RUTHENIUM CARBONYL HYDRIDE (TRICLINIC)
$H_2Ru_4(CO)_{13}$

J. Organometal. Chem., 384, 209-216.

P$\bar{1}$, 9.062, 9.155, 26.649, 81.80, 88.03, 67.76, Z = 4 (2 molecules/asymmetric
unit), R = 0.023. The molecular structure is similar to that in the monoclinic
form, which also has two molecules per asymmetric unit (38A, 185). The
parameters associated with the two bent semi-bridging CO ligands vary in the four
molecules of the two forms (attributed to crystal packing forces). The H
ligands bridge the longest Ru-Ru edges.

RUTHENIUM CARBONYL HYDRIDE BOROHYDRIDE
$HRu_4(CO)_{12}BH_2$

Inorg. Chem., 29, 2874-2876.

Cm, 10.432, 15.709, 6.472, 112.86, at 223K, Z = 2, R = 0.020. Molecular
structure as shown below. Ru-Ru = 2.822-2.904(1), Ru-B = 2.11-2.20(6) A.

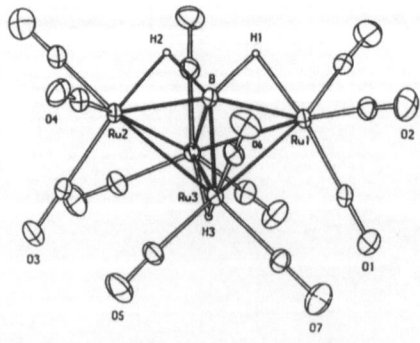

RUTHENIUM CARBONYL HYDRIDE
$H_2Ru_6(CO)_{17}$

J. Amer. Chem. Soc., 112, 8587-8589.

P$\bar{1}$, 8.131, 10.996, 15.350, 93.58, 97.89, 109.33, at 233K, Z = 2, R = 0.028.   The
molecular structure is shown below.   Ru-Ru = 2.64-3.01 A.

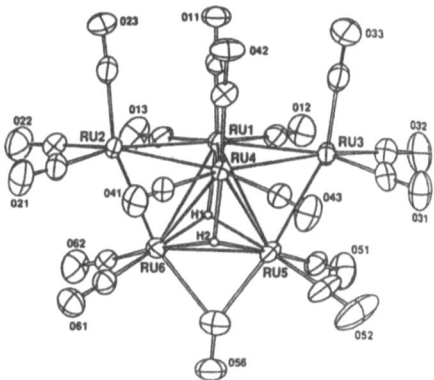

RUTHENIUM CARBONYL TELLURIDE
$Ru_4(CO)_{11}Te_2$

Inorg. Chem., 29, 4658-4665

Pccn, 6.924, 16.389, 18.054, Z = 4, R = 0.042.   Re$_4$Te$_2$ octahedral cluster, with
one bridging, two semi-bridging, and eight terminal carbonyl groups (see also
following report).   Ru-Ru = 2.806-2.945(1), Ru-Te = 2.691-2.758(1) A.

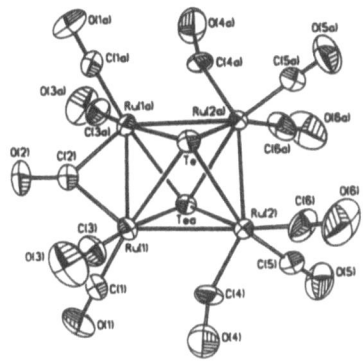

IRON RUTHENIUM CARBONYL TELLURIDE
$Fe_2Ru_2(CO)_{11}Te_2$

Inorg. Chem., 29, 4838-4840.

Pccn, 6.863, 16.064, 17.799, Z = 4, R = 0.052. Isostructural with the Fe/S
compound (this volume, p. 94 and preceding report), with an octahedral cluster
and one bridging, two semi-bridging, and eight terminal carbonyl groups.

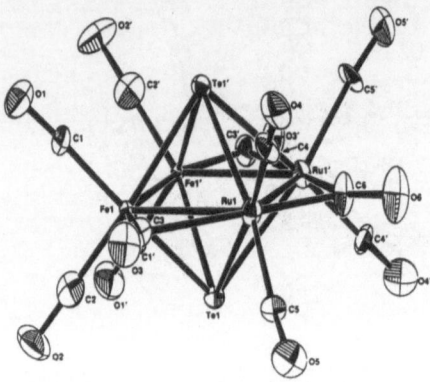

DIIRON OSMIUM DODECACARBONYL
$Fe_2Os(CO)_{12}$

Organometallics, 9, 446-452.

Pn (close to $P2_1/n$), 8.377, 22.715, 8.953, 96.510, Z = 4, R = 0.049. The
structure is similar to that of $Fe_3(CO)_{12}$ (34A, 160; 41A, 142), but with a
doubled b axis, lower symmetry, and two molecules per asymmetric unit, both with
a 12:1 disorder of metal atoms.

PLATINUM OSMIUM CARBONYL HYDRIDES
$Pt_2Os_5(CO)_{17}H_6$ (I), $PtOs_5(CO)_{16}H_6$ (II).

Inorg. Chem., 29, 3269-3270.

I, $P2_1/n$, 9.603, 16.246, 19.272, 91.49, Z = 4, R = 0.042. II, $P2_1/n$, 11.810,
15.656, 16.286, 109.15, Z = 4, R = 0.032. Molecular structures are shown below.

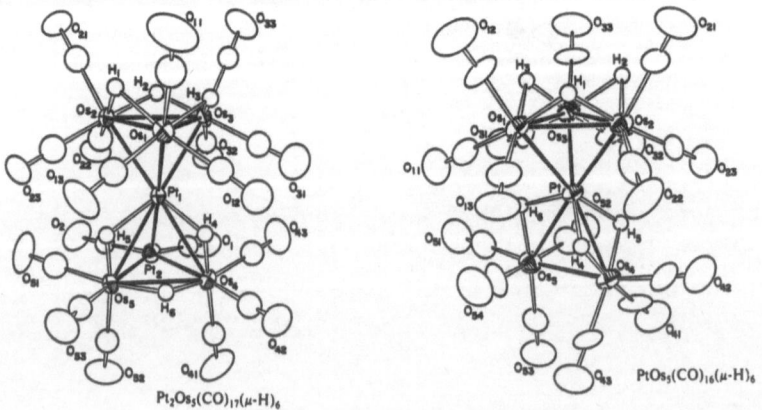

$Pt_2Os_5(CO)_{17}(\mu\text{-}H)_6$ $PtOs_5(CO)_{16}(\mu\text{-}H)_6$

DICHLOROBIS(TETRACARBONYLCOBALTIO)TIN(IV)(2Co-Sn)
SnCl$_2$[Co(CO)$_4$]$_2$

Acta Cryst., C46, 1759-1761.

P2$_1$/c, 11.716, 11.486, 12.765, 108.42, Z = 4, R = 0.045. Isolated molecules with
distorted tetrahedral geometry at Sn and trigonal bipyramidal at Co.

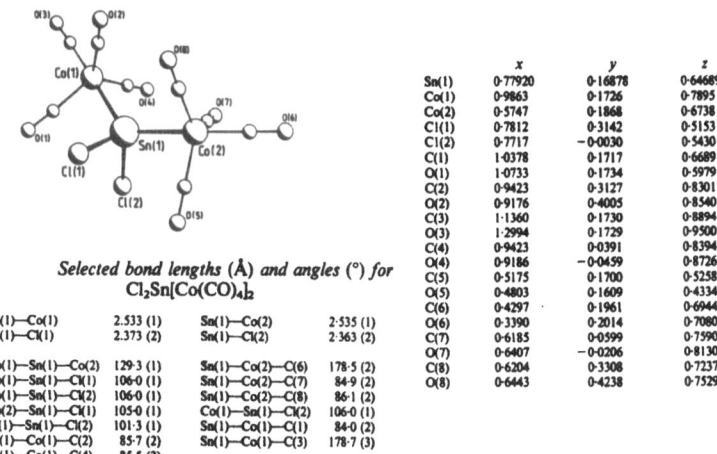

	x	y	z
Sn(1)	0·77920	0·16878	0·64689
Co(1)	0·9863	0·1726	0·7895
Co(2)	0·5747	0·1868	0·6738
Cl(1)	0·7812	0·3142	0·5153
Cl(2)	0·7717	−0·0030	0·5430
C(1)	1·0378	0·1717	0·6689
O(1)	1·0733	0·1734	0·5979
C(2)	0·9423	0·3127	0·8301
O(2)	0·9176	0·4005	0·8540
C(3)	1·1360	0·1730	0·8894
O(3)	1·2994	0·1729	0·9500
C(4)	0·9423	0·0391	0·8394
O(4)	0·9186	−0·0459	0·8726
C(5)	0·5175	0·1700	0·5258
O(5)	0·4803	0·1609	0·4334
C(6)	0·4297	0·1961	0·6944
O(6)	0·3390	0·2014	0·7080
C(7)	0·6185	0·0599	0·7590
O(7)	0·6407	−0·0206	0·8130
C(8)	0·6204	0·3308	0·7237
O(8)	0·6443	0·4238	0·7529

Selected bond lengths (Å) and angles (°) for
Cl$_2$Sn[Co(CO)$_4$]$_2$

Sn(1)—Co(1)	2.533 (1)	Sn(1)—Co(2)	2·535 (1)
Sn(1)—Cl(1)	2.373 (2)	Sn(1)—Cl(2)	2·363 (2)
Co(1)—Sn(1)—Co(2)	129·3 (1)	Sn(1)—Co(2)—C(6)	178·5 (2)
Co(1)—Sn(1)—Cl(1)	106·0 (1)	Sn(1)—Co(2)—C(7)	84·9 (2)
Co(1)—Sn(1)—Cl(2)	106·0 (1)	Sn(1)—Co(2)—C(8)	86·1 (2)
Co(2)—Sn(1)—Cl(1)	105·0 (1)	Co(1)—Sn(1)—Cl(2)	106·0 (1)
Cl(1)—Sn(1)—Cl(2)	101·3 (1)	Sn(1)—Co(1)—C(1)	84·0 (2)
Sn(1)—Co(1)—C(2)	85·7 (2)	Sn(1)—Co(1)—C(3)	178·7 (3)
Sn(1)—Co(1)—C(4)	85·5 (2)		

TETRAKIS(TETRACARBONYLCOBALTIO)TIN and -LEAD
M[Co(CO)$_4$]$_4$ (M = Sn, Pb)

Acta Cryst., C46, 732-736.

Sn, F$\bar{4}$3c, 17.255, Z = 8, R = 0.017. Pb, Pccn, 12.184, 12.267, 17.220, Z = 4, R =
0.077. Both structures contain molecules with a central M atom coordinated
tetrahedrally to four Co(CO)$_4$ groups, Co having trigonal bipyramidal
coordination. Sn-Co = 2.669(1), Pb-Co = 2.738, 2.761(5) A.

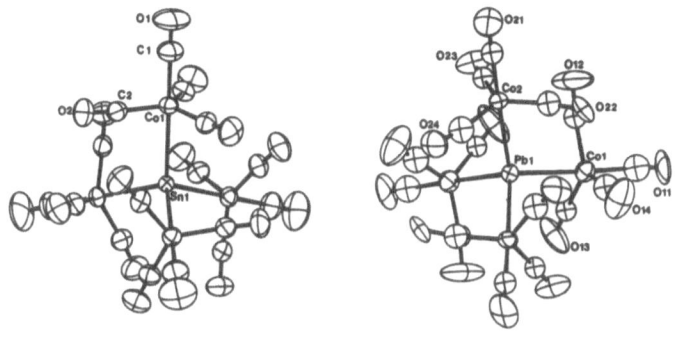

COPPER COBALT CARBONYLS
CuCo(CO)$_4$.xNH$_3$ (x = 0.5, 1, 2)

Angew. Chem., <u>102</u>, 825-826 [Angew. Chem. Int. Edn. Engl., <u>29</u>, 783-785].

x = 0.5, P$\bar{1}$, 8.598, 8.925, 10.373, 95.55, 98.50, 117.42, Z = 4, R = 0.026.
Tetrameric molecule.

x = 1, P2$_1$/n, 8.112, 11.263, 8.528, 99.0, Z = 4, R = 0.055. Dimeric molecule.

x = 2, P2$_1$/c, 8.682, 13.35, 8.184, 113.84, Z = 4, R = 0.049. Monomer.

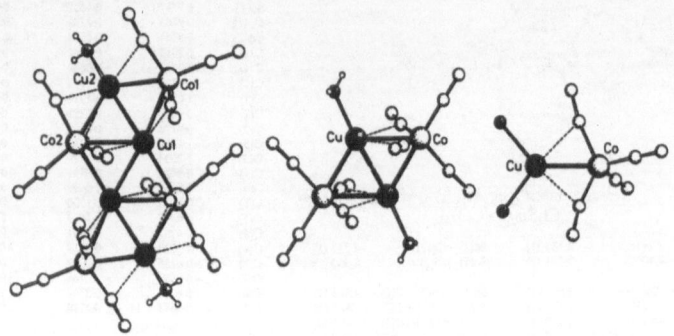

COPPER CARBONYL CHLORIDE
Cu(CO)Cl

Inorg. Chem., <u>29</u>, 5241-5244.

Pmn2$_1$, 3.672, 8.197, 4.947, at 129K, Z = 2, R = 0.054. Atoms in 2(a), Cu:
0,0.1396,0; Cl: 1/2, 0.1213, 0.3016; C: 0,0.325,-0.216; O: 0,0.438,-0.341.
Double layers of Cu(CO)Cl, tetrahedra sharing Cl corners.

CAESIUM TETRAAZIDOZINCATE
$Cs_2Zn(N_3)_4$

Mh. Chem., **121**, 91-97.

$Pca2_1$, 21.880, 6.762, 7.426, Z = 4, R = 0.053. Isolated $Zn(N_3)_4{}^{2-}$ tetrahedra,
linked by 9-coordinate Cs^+ ions; the azide groups are asymmetric, with some
groups having all three N coordinated to Cs. Zn-N = 1.95-2.04(2), N-N =
1.17-1.23, 1.13-1.17(3), Cs-N = 3.16-3.58(2) A, Zn-N-N = 117-131, N-N-N =
176-177°.

CAESIUM OCTAAZIDOEUROPATE(III)
$Cs_5Eu(N_3)_8$

Mh. Chem., **121**, 781-786.

Pbca, 16.811, 16.860, 16.964, Z = 8, R = 0.048. Islated $Eu(N_3)_8{}^{5-}$ anions, linked
by 8- and 7-coordinated Cs^+ ions. Eu-N = 2.44-2.57(2), N-N = 1.16-1.19(4), Cs-N
= 3.10-3.62(3) A, Eu-N-N = 117-132, N-N-N = 171-179°.

BARIUM IMIDE
BaND

J. Less-Common Metals, **167**, 81-90.

Above 192K, Fm3m, a = 5.861 A, at 294K, Z = 4, neutron powder data. Ba in 4(a):
0,0,0; N in 4(b): 1/2,1/2,1/2; 3.2 D in 32(g): x,x,x, x = 0.423. NaCl-type,
with orientational disorder of the ND^{2-} anion.

Below 192K, I4/mmm (or I$\bar{4}$m2), a = 4.062, c = 6.072 A, at 8K, Z = 2, neutron
powder data. Ba in 2(a): 0,0,0; 2 N in 4(e): 0,0,0.540; 1.6 D in 16(n): 0.09,
0,0.630. Statistical orientation of the ND^{2-} anion mainly along $\underline{c}$.

RUBIDIUM and CAESIUM TETRAAMIDOALUMINATE
$RbA\ell(NH_2)_4$, $CsA\ell(NH_2)_4$

J. Less-Common Metals, **159**, 315-325.

P4/n, a = 7.406, 7.563, c = 5.386, 5.354 A, Z = 2, R = 0.053, 0.046. $A\ell(NH_2)_4{}^-$
tetrahedra linked by alkali-metal cations. $A\ell$-N = 1.83(2), 1.84(1) A.

		x	y	z
Rb	2c	1/4	1/4	0.710
Cs	2c	1/4	1/4	0.723
Al*	2a	1/4	3/4	0
Al	2a	1/4	3/4	0
N*	8g	0.642	0.411	0.791
N	8g	0.665	0.418	0.782
H1*	8g	0.68	0.42	0.61
H1	8g	0.69	0.43	0.57
H2*	8g	0.61	0.50	0.83
H2	8g	0.63	0.50	0.87

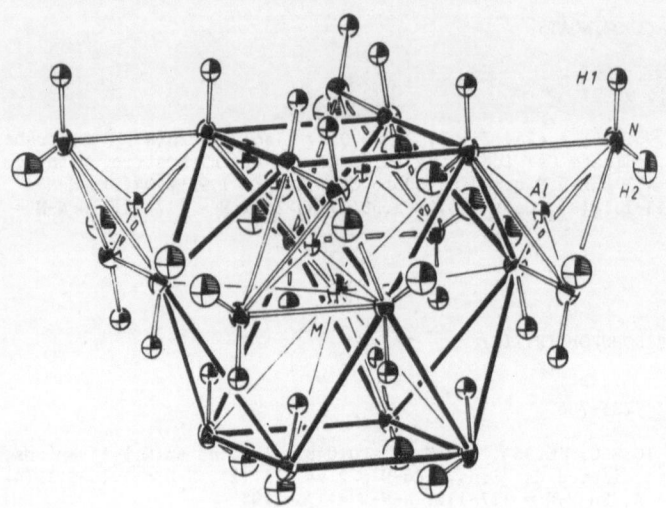

POTASSIUM IMIDONITRIDOSILICATE
$K_3Si_6N_5(NH)_6$

Z. anorg. Chem., <u>584</u>, 129-137.

$P4_332$, 10.789, Z = 4, R = 0.019.   K(1) in 4(a):   1/8,1/8,1/8; K(2) in 8(c):
x,x,x, x = 0.54146; Si in 24(e):   0.33827,0.43583,0.01469; N(1) in 24(e):
0.2753,0.4715,0.8708; N(2) in 12(d):   1/8,x,1/4-x, x = 0.2853; N(3) in 8(c): x =
0.9670; H in 24(e): 0.1938,0.4562,0.8602. Framework of corner-sharing SiN_4
tetrahedra, with K ions in holes. Si-N = 1.69-1.75 A.

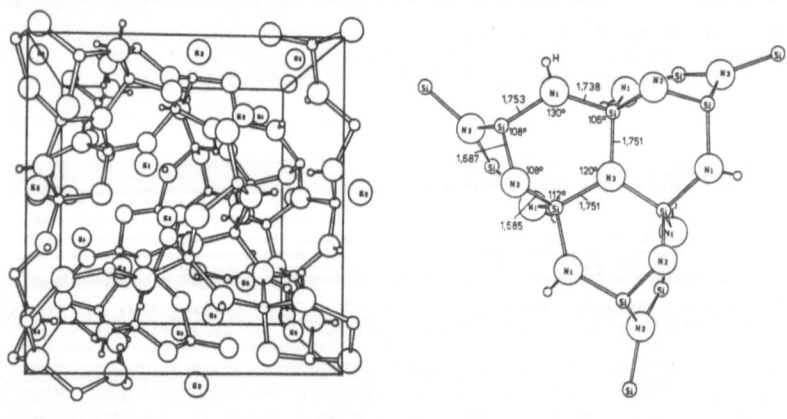

HEXAAMINOCYCLOTRIPHOSPHAZENE HEMIAMMONIATE
$P_3N_3(NH_2)_6 \cdot 0.5NH_3$

Z. anorg. Chem., <u>581</u>, 125-134.

Pbca, 11.395, 12.935, 12.834, Z = 8, R = 0.035. Cyclic molecules linked by
N-H...N hydrogen bonds.

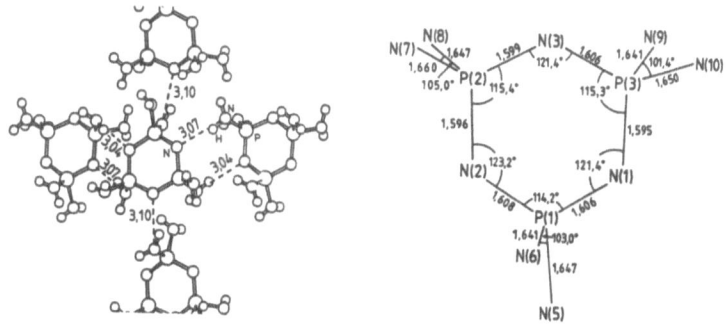

BIS(TRICHLOROPHOSPHORUS)IMINIUM TETRACHLOROOXOVANADATE
$[C\ell_3PNPC\ell_3][VOC\ell_4]$

Z. anorg. Chem., <u>591</u>, 137-142.

Pbca, 9.810, 15.789, 19.74, Z = 8, R = 0.10. Bent cations and square-pyramidal
anions. P-N = 1.52, 1.57, P-Cℓ = 1.93-1.96 A, P-N-P = 138°, V=O = 1.61, V-Cℓ =
2.17-2.26(1) A.

SULPHUR-NITROGEN COMPOUNDS
$SeS_2N_2C\ell_2$ (I), $(S_5N_5)(SeC\ell_5)$ (II), $(Se_xS_{3-x}N_2C\ell)(SbC\ell_6)$ (III)

Inorg. Chem., <u>29</u>, 1251-1259.

I, $P2_1/c$, 10.447, 9.259, 13.529, 105.93, Z = 8, R = 0.061. II, Pnma, 11.252,
12.075, 9.987, Z = 4, R = 0.046. III, $P2_1/c$, a = 7.000, 7.045, b = 14.294,
14.392, c = 13.486, 13.489 A, β = 97.52, 97.27°, for two samples, Z = 4, R =
0.046, 0.063. I contains (surprisingly) an $SeS_2N_2C\ell_2$ molecule, an $SeS_2N_2C\ell^+$
cation, and a $C\ell^-$ anion, with strong cation-anion interaction. II contains an
$S_5N_5^+$ cyclic cation and a square-pyramidal $SeC\ell_5^-$ anion, linked by a strong
Se...N interaction (2.794(5) A). III contains disordered $S_3N_2C\ell^+$-type cations
(<u>50A</u>, 94) and octahedral $SbC\ell_6^-$ anions.

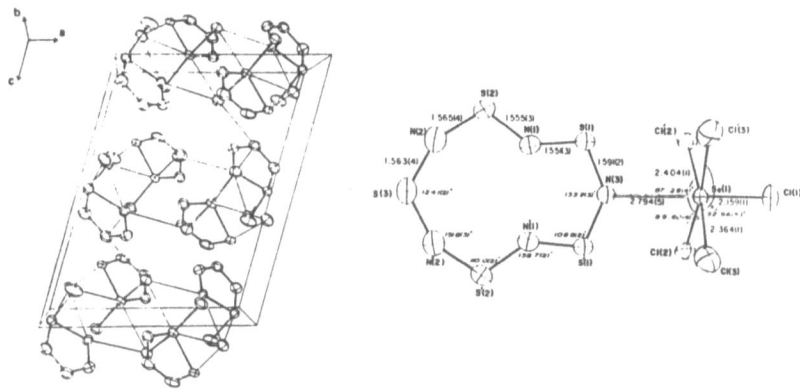

CHLOROSELENADITHIADIAZYL TETRACHLOROALUMINATE
[N_2S_2SeCl] [$AlCl_4$]

Inorg. Chem., <u>29</u>, 1643-1648.

$P2_1/n$, 9.237, 10.685, 11.332, 92.65, Z = 4, R = 0.049. Cyclic cation and
tetrahedral anion, linked by S/Se...Cl interactions. Se-S = 2.293(1), Se-Cl =
2.191(1), Se...Cl = 3.124(1), S...Cl = 3.190(2) A.

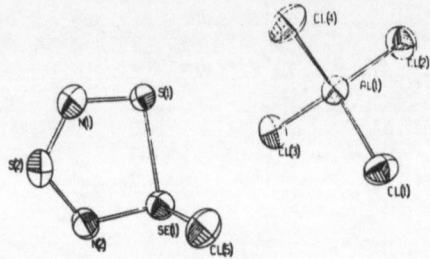

TIN(II,IV) FLUORIDE
γ-Sn_2F_6

Z. anorg. Chem., <u>590</u>, 173-180.

Fm3m, a = 8.321 A, at 497K, Z = 4, neutron powder data. Sn(II) in 4(a): 0,0,0;
Sn(IV) in 4(b): 1/2,1/2,1/2; F in 24(e): 0,276,0,0. Ordered ReO$_3$-type;
Sn(II)-6F = 2.29, Sn(IV)-6F = 1.86 A.

HAFNIUM(IV) FLUORIDE
THORIUM(IV) FLUORIDE
HfF_4, ThF_4

Z. anorg. Chem., <u>588</u>, 33-42.

C2/c, a = 11.725, 13.049, b = 9.869, 11.120, c = 7.636, 8.538 A, β = 126.15,
126.30°, Z = 12, R = 0.067, 0.045. β-ZrF_4-type (<u>29</u>, 262; <u>49A</u>, 83). Hf-8F =
2.05-2.15, Th-8F = 2.29-2.36 A.

VANADIUM(III) FLUORIDE
VF_3

Mater. Res. Bull., <u>25</u>, 413-420.

R$\bar{3}$c, a = 5.168, c = 13.438 A, Z = 6 (rhombohedral cell, a = 5.383 A, α = 57.384,
Z = 2), R = 0.026. V in 6(b): 0,0,0; F in 18(e): 0.4001,0,1/4. Structure as
previously described (<u>15</u>, 145). V-6F = 1.935(1) A.

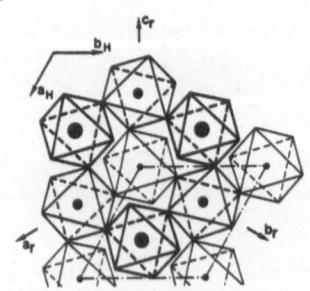

MOLYBDENUM(III) FLUORIDE
MoF_3

J. Less-Common Metals, 161, 135-140.

$R\bar{3}c$, a = 5.2118, c = 14.4069 A, Z = 6, R = 0.052. Mo in 6(b): 0,0,0; F in
18(e): 0.6307,0,1/4. Structure as previously described (24, 274).

MANGANESE(II) FLUORIDE
MnF_2

Acta Cryst., B46, 739-742.

$P4_2/mnm$, a = 4.8736,4.8736,4.8736, c = 3.3102,3.3020,3.3000 A, at 295, 60, 15K, Z
= 2, neutron radiation, R = 0.024-0.051. Mn in 2(a): 0,0,0; F in 4(f): x,x,0,
x = 0.3046,0.3048,0.3049. Rutile-type structure, as previously described (1,
158).

SILVER DIFLUORIDE
AgF_2

Z. anorg. Chem., 588, 77-83.

Pbca, 5.568, 5.831, 5.101, Z = 4, R = 0.059. Ag in 4(a): 0,0,0; F(1) in 8(c):
0.3050,-0.1309,0.1846. Structure as previously described (31A, 84; 37A, 178;
40A, 297), with puckered layers of corner-sharing AgF_4 square-planes (Ag-F = 2.09
A), linked by two longer Ag...F contacts (2.59 A).

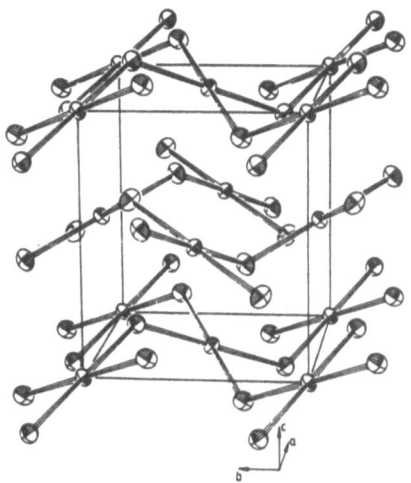

SODIUM LEAD(II) TETRAFLUOROBERYLLATE FLUORIDE
$Na_3Pb_2(BeF_4)_3F$

J. Less-Common Metals, 158, 123-130.

$P6_3/m$, 9.531, 7.028, Z = 2, R = 0.061. Apatite structure (2, 99), with F(4) in
two half-occupied sites displaced 0.67 A from 0,0,1/4. Be-F = 1.52-1.54(1) A.

		$10^4\ x$	y	z
Na	4(f)	1/3	2/3	62
M=Na+Pb	6(h)	2474.1	150.0	1/4
Bc	6(h)	3758	4055	1/4
F1	6(h)	4919	3391	1/4
F2	6(h)	4639	5891	1/4
F3	12(i)	2602	3413	779
F4	4(e)	0	0	3450

POTASSIUM DIHYDROGEN HEXAFLUOROALUMINATE DIHYDRATE
$KH_2AlF_6 \cdot 2H_2O$

Acta Cryst., C$\underline{46}$, 190-192.

Pa3, 8.6464. Z = 4, R = 0.026. Structure related to that of elpasolite, with trigonal pyramidal H_3O^+ ions and octahedral AlF_6^- ions, linked by very strong O-H...F hydrogen bonds.

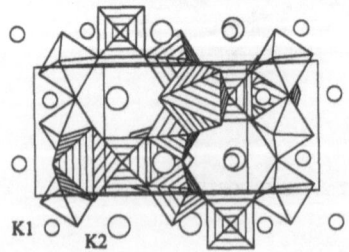

	x	y	z
K	0	0	0
Al	1/2	0	0
F	0,4399	0,1946	0,0488
O	0,1971	x	x
H	0,292	0,205	0,151

Al—F	6 ×	1,811 (1) Å	F—Al—F	89,06 (5) °
K—F	6 ×	2,724 (1)		90,94 (5)
K—O	2 ×	3,952 (2)		
O—H		0,91 (3)	H—O—H	111 (3)
H···F		1,56 (3)	O—H···F	168 (2)
O···F		2,460 (2)		

POTASSIUM SODIUM ALUMINUM FLUORIDE
RUBIDIUM SODIUM ALUMINUM FLUORIDE
$K_2NaAl_3F_{12}$, $Rb_2NaAl_3F_{12}$

Mater. Res. Bull., $\underline{25}$, 831-839.

$P2_1/m$, a = 11.882, 12.046, b = 6.983, 6.984, c = 6.942, 7.093 A, β = 125.59, 125.04°, Z = 2, R = 0.025 and 0.038 (twinned crystal for the Rb compound). Monoclinic distortion of the rhombohedral Cs structure ($\underline{40}$A, 137; $\underline{42}$A, 169), with layers of corner-sharing AlF_6 octahedra, linked by NaF_6 octahedra, and by 9- and 10-coordinate K or Rb. Al-F = 1.70-1.85, Na-F = 2.26-2.36 A.

SODIUM CALCIUM ALUMINUM FLUORIDES
β-NaCaAlF$_6$ (I), Na$_4$Ca$_4$Al$_7$F$_{33}$ (II)

J. Solid State. Chem., <u>84</u>, 153-164.

I, P321, 8.9295, 5.0642, Z = 3, powder data. Na_2SiF_6-type (<u>29</u>, 264; <u>53A</u>, 84; see also <u>43A</u>, 127; <u>49A</u>, 95).

II, Im3m, 10.781, Z = 2, R = 0.033. Framework of corner-sharing $A\ell F_6$ and CaF_6 octahedra, with 10-coordinate Na ions in holes.

I   II

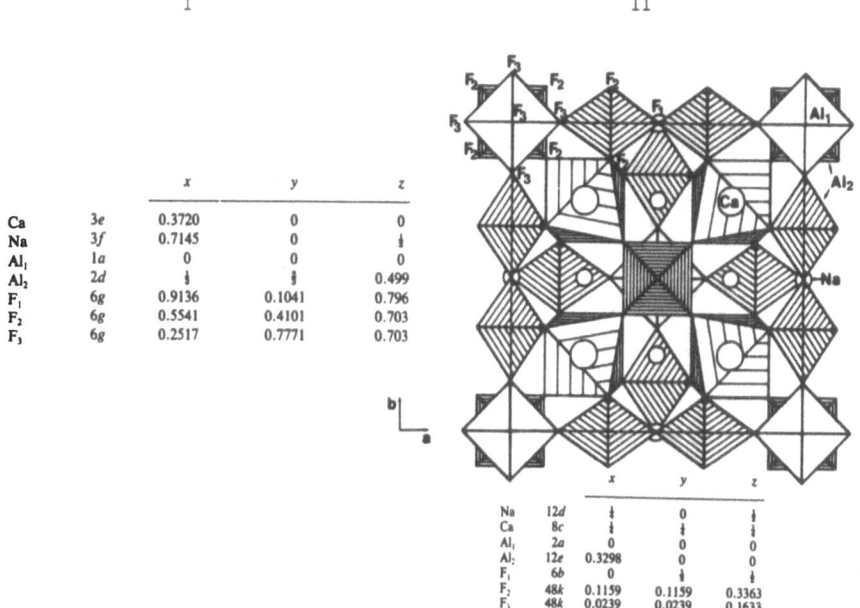

		x	y	z
Ca	3e	0.3720	0	0
Na	3f	0.7145	0	½
Al₁	1a	0	0	0
Al₂	2d	⅓	⅔	0.499
F₁	6g	0.9136	0.1041	0.796
F₂	6g	0.5541	0.4101	0.703
F₃	6g	0.2517	0.7771	0.703

		x	y	z
Na	12d	¼	0	½
Ca	8c	¼	¼	¼
Al₁	2a	0	0	0
Al₂	12e	0.3298	0	0
F₁	6b	0	½	½
F₂	48k	0.1159	0.1159	0.3363
F₃	48k	0.0239	0.0239	0.1633

BARIUM PENTAFLUOROALUMINATE
$BaA\ell F_5$

J. Solid State Chem., <u>89</u>, 282-291.

α-Form, P2₁2₁2₁, 13.7168, 5.6054, 4.9329, Z = 4; β-form, P2₁/n, 5.1517, 19.5666, 7.5567, 92.426, Z = 8; γ-form, P2₁, 5.2584, 9.7298, 7.3701, 90.875, Z = 4; X-ray and neutron powder data. The α-form is isostructural with the Ga compound, as previously described (<u>49A</u>, 369). All three forms contain infinite chains of cis-corner-sharing $A\ell F_6$ octahedra.

BARIUM ALUMINUM FLUORIDE (FORM I)
$Ba_3A\ell F_9$

Eur. J. Solid State Inorg. Chem., <u>27</u>, 571-580.

Pnma, 19.706, 5.599, 15.173, Z = 8, R = 0.048. Isolated $A\ell F_6$ octahedra, linked by Ba cations, with additional F⁻ ions, FBa_4 tetrahedra forming bands along <u>b</u>.

INORGANIC COMPOUNDS

SODIUM CADMIUM ALUMINUM FLUORIDE
NaCdAℓF$_6$

J. Solid State Chem., **86**, 249-254.

Pnma, 12.506, 3.6406, 9.902, Z = 4, R = 0.019. Chains along **b** of trans-corner-sharing AℓF$_6$ octahedra, and an additional F$^-$ ion, linked by 6-coordinate Na and 7-coordinate Cd ions. Aℓ-F = 1.76-1.82, Na-F = 2.28-2.51, Cd-F = 2.25-2.31 A.

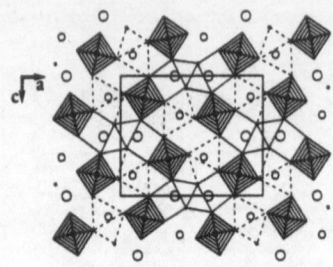

		x	y	z
Cd^{2+}	4c	0.1197	1/4	0.4998
Al^{3+}	4c	0.3354	1/4	0.6802
Na$^+$	4c	0.4142	1/4	0.3255
F1	4c	0.3000	1/4	0.5046
F2	4c	0.4728	1/4	0.6443
F3	4c	0.1663	1/4	0.1868
F4	4c	0.3567	1/4	0.8611
F5^a	4c	0.4574	1/4	0.0977
F6	4c	0.1925	1/4	0.7138

^{a} Independent fluorine ion

NaCdAlF$_6$ structure: (001) projection (Cd, Na, F: large, medium, and small circles, respectively; AlF$_6$ octahedra: shaded; NaF$_6$ octahedra: dotted lines; and CdF$_7$ polyhedra: heavy lines).

BARIUM COPPER(II) GALLIUM FLUORIDE
Ba$_3$CuGa$_2$F$_{14}$

Z. anorg. Chem., **590**, 200-212.

P2$_1$/n, 7.402, 27.88, 5.521, 90.12, Z = 4, R = 0.047. Chains of cis-corner sharing GaF$_6$ octahedra linked by chains of edge-sharing CuF$_7$ monocapped trigonal prisms; Ba ions have 11- and 12-coordinations. Ga-F = 1.84-1.88, 1.95-1.98, Cu-F = 1.86-2.75 A.

COBALT(II) HEXAFLUOROSILICATE HEXAHYDRATE (DEUTERATE)
CoSiF$_6$·6D$_2$O

Acta Cryst., **C46**, 186-189.

R$\bar{3}$, a = 9.369, c = 9.731 A, Z = 3, neutron radiation, R = 0.070. Structure as previously described (**39A**, 143), with disordered F sites (50:50). Co-O = 2.084(2), Si-F = 1.665, 1.680(3) A.

	x	y	z
Co	0	0	0
Si	0	0	0·5
F(1)	0·1553	0·1314	0·4009
F(2)	0·1672	0·0650	0·3997
O	0·1747	0·1861	0·1250
D(1)	0·1785	0·1654	0·2194
D(2)	0·2053	0·2981	0·1129

LITHIUM and SODIUM HEXAFLUOROGERMANATES
M_2GeF_6 (M = Li, Na)

Z. anorg. Chem., **582**, 111-120.

P321, a = 8.404, 9.058, c = 4.616, 5.107 A, Z = 3, R = 0.035, 0.092.
Isostructural with Na_2SiF_6 (**19**, 325; **29**, 264; **53A**, 84). Ge-6F = 1.78-1.80, Li-6F
= 1.96-2.19, Na-6F = 2.22-2.43 A.

		x	y	z			x	y	z
Ge1	1a	,0000	,0000	,0000	Ge1	1a	,0000	,0000	,0000
Ge2	2d	,3333	,6667	,4869	Ge2	2d	,3333	,6667	,4954
Li1	3f	,7060	,0000	,5000	Na1	3f	,7134	,0000	,5000
Li2	3e	,3610	,0000	,0000	Na2	3e	,3779	,0000	,0000
F1	6g	,4624	,2212	,2927	F1	6g	,4798	,2218	,3012
F2	6g	,4246	−,4446	,2668	F2	6g	,4089	−,4428	,2945
F3	6g	,2038	,0925	−,2187	F3	6g	,1864	,0820	−,1988

SODIUM HEXAFLUOROSTANNATE(IV)
Na_2SnF_6

J. Fluorine Chem., **48**, 219-227.

$P4_2/mnm$, 5.0532, 10.122, Z = 2, R = 0.045. Na in 4(e): 0,0,0.3367; Sn in 2(a):
0,0,0; F(1) in 8(j): x,x,z, x = 0.3149, z = 0.6428; F(2) in 4(f): x,x,0, x =
0.2741. Trirutile structure (**8**, 153; **53A**, 86), with SnF_6^{2-} octahedra linked by
octahedrally-coordinated Na^+ ions. Sn-F = 1.96, Na-F = 2.26-2.37 A.

BARIUM HEXAFLUOROSTANNATE(IV)   BARIUM HEXAFLUOROTITANATE(IV)
$BaSnF_6$ $BaTiF_6$

Z. anorg. Chem., **591**, 7-16.

Sn compounds, R$\overline{3}$, a = 7.428, c = 7.418 A, Z = 3, R = 0.036. Ba in 3(a): 0,0,0;
Sn in 3(b): 0,0,1/2; F in 18(f): 0.2586,0.8262,0.0047. $KOsF_6$-type (**20**, 232);
Sn-6F = 1.95 A.

Ti compound, R$\overline{3}$m, a = 7.368, c = 7.252 A, Z = 3, R = 0.023. Ba in 3(a): 0,0,0;
Ti in 3(b): 0,0,1/2; F in 18(h): 0.2302,0.1151,0.3452. $BaSiF_6$-type (**9**, 196;
51A, 126; **53A**, 81); Ti-6F = 1.85 A.

IRON(II) HEXAFLUOROSTANNATE(IV) HEXAHYDRATE
$FeSnF_6 \cdot 6H_2O$

Acta Cryst., **C46**, 2453-2454.

R$\overline{3}$, a = 9.826, c = 10.106 A, Z = 3, R = 0.029. Isostructural with related
materials (**2**, 102, 504; **39A**, 143), with $Fe(H_2O)_6^{2+}$ and SnF_6^{2-} octahedra, linked
by O-H...F hydrogen bonds. Fe-O = 2.116, Sn-F = 1.945(2) A.

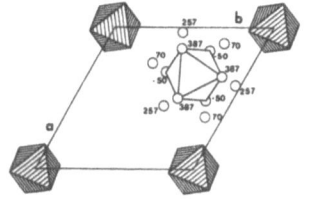

	x	y	z
Fe	0	0	$\frac{1}{2}$
Sn	0	0	0
F	0·1460	0·1718	0·1126
O	0·1840	0·1701	0·3813
H(1)	0·310	0·223	0·411
H(2)	0·161	0·151	0·271

STRONTIUM PENTAFLUOROANTIMONATE(III)
BARIUM PENTAFLUOROANTIMONATE(III)
SrSbF$_5$, BaSbF$_5$

Acta Cryst., C$\underline{46}$, 2294-2297.

Pbcm, a = 4.378, 4.676, b = 8.853, 9.313, c = 11.233, 11.213 A, Z = 4, R = 0.028, 0.015. Isolated SbF$_5$$^{2-}$ anions linked by 10-coordinate Sr or Ba ions. The anion is a square pyramid, with Sb below the basal plane and the lone-pair completing an octahedron.

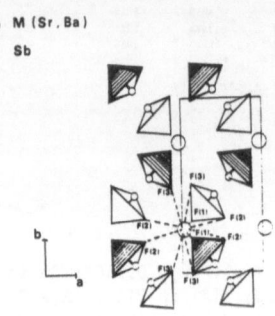

○ M (Sr, Ba)

○ Sb

Selected bond lengths (Å) and angles (°) in
MSbF$_5$ (M = Sr, Ba) (e.s.d.'s in parentheses)

	M = Sr	M = Ba
Sb—F(1)	1 × 1·963 (4)	1 × 1·970 (3)
Sb—F(2')	2 × 2·108 (4)	2 × 2·087 (2)
Sb—F(3")	2 × 2·103 (4)	2 × 2·090 (2)
F(1)—F(2')	2 × 2·493 (5)	2 × 2·502 (3)
F(1)—F(3")	2 × 2·514 (5)	2 × 2·520 (3)
F(2')—F(2"')	1 × 3·106 (8)	1 × 3·032 (4)
F(3")—F(3"')	1 × 3·055 (8)	1 × 2·974 (4)
F(2')—F(3")	2 × 2·680 (8)	2 × 2·731 (4)
F(1)—Sb—F(2')	75·4 (1)	76·1 (1)
F(1)—Sb—F(3")	76·3 (1)	76·7 (1)
F(2')—Sb—F(2"')	94·8 (2)	93·2 (1)
F(2')—Sb—F(3")	79·1 (1)	81·7 (1)
F(3")—Sb—F(3"')	93·2 (2)	90·7 (1)
M—F(1)	2 × 2·843 (1)	2 × 2·867 (1)
M—F(2')	2 × 2·515 (3)	2 × 2·672 (2)
M—F(2")	2 × 2·606 (3)	2 × 2·793 (2)
M—F(3")	2 × 2·494 (4)	2 × 2·649 (2)
M—F(3"')	2 × 2·680 (4)	2 × 2·825 (2)
(M—F)	2·628	2·761

Symmetry code: (i) $1-x$, $y-\frac{1}{2}$, z; (ii) x, $y-1$, z; (iii) $1-x$, $y-\frac{1}{2}$, $\frac{1}{2}-z$; (iv) x, $y-1$, $\frac{1}{2}-z$; (v) $-x$, $y-\frac{1}{2}$, z; (vi) $-x$, $1-y$, $-z$.

		Site symmetry	x	y	z
SrSbF$_5$					
Sr	4(c)	2..	0·0084	¼	0
Sb	4(d)	..m	0·3890	0·0537	¼
F(1)	4(d)	..m	0·0758	0·2124	¼
F(2)	8(e)	1	0·4793	0·7011	0·1118
F(3)	8(e)	1	0·0878	0·9870	0·1140
BaSbF$_5$					
Ba	4(c)	2..	0·02762	¼	0
Sb	4(d)	..m	0·40594	0·04943	¼
F(1)	4(d)	..m	0·1184	0·2040	¼
F(2)	8(e)	1	0·4573	0·6873	0·1148
F(3)	8(e)	1	0·1189	0·9856	0·1174

TETRAIODODISELENIUM UNDECAFLUORODIANTIMONATE
Se$_2$I$_4$(Sb$_2$F$_{11}$)$_2$ (I)

TETRAIODODISELENIUM HEXAFLUOROARSENATE SULPHUR DIOXIDE
Se$_2$I$_4$(AsF$_6$)$_2$·SO$_2$ (II)

Inorg. Chem., $\underline{29}$, 3529-3538.

I, P$\bar{1}$, 17.915, 9.276, 8.001, 96.04, 95.22, 91.83, Z = 2, R = 0.046. II, P2$_1$/c, 13.454, 9.930, 14.576, 93.20, Z = 4, R = 0.043. Both structures contain Se$_2$I$_4$$^{2+}$ ions with an eclipsed configuration and long Se-Se bonds (mean 2.84(1) A).

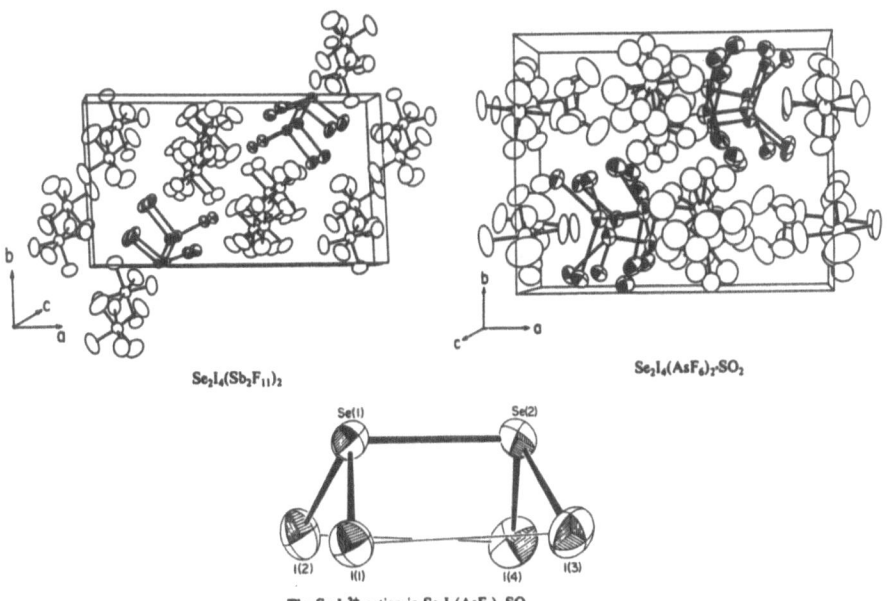

$Se_2I_4(Sb_2F_{11})_2$ $Se_2I_4(AsF_6)_7SO_2$

The $Se_2I_4^{2+}$ cation in $Se_2I_4(AsF_6)_7SO_2$

HYDRAZINIUM(2+) TRI-μ-FLUOROBIS[PENTAFLUOROZIRCONATE(IV)] FLUORIDE
$(N_2H_6)_3 [Zr_2F_{13}]F$

J. Cryst. Spect. Res., 20, 9-15.

$P2_1$, 5.670, 10.984, 10.601, 93.88, Z = 2, R = 0.021. The structure contains
complex anions, F^- anions, and hydrazinium cations. The complex anion consists
of two ZrF_8 bicapped trigonal prisms sharing a triangular face. One cation forms
strong hydrogen bonds to F^- to give chains along a; a second cation forms weaker
bi- and trifurcated hydrogen bonds to F ligands of the complex anion. Zr-F =
2.015-2.112 (terminal), 2.133-2.212(2) (bridging), shortest N-H...F = 2.437(5)
A.

COBALT(II) HEXAFLUOROZIRCONATE(IV)
$CoZrF_6$

J. Phys.: Condens. Matter, 2, 7373-7386, 7387-7394, 7395-7406.

300K, Fm3m, a = 7.789 A, Z = 4, neutron powder data. Co in 4(a): 0,0,0; Zr in
4(b): 1/2,1/2,1/2; F in 24(e): 0.252,0,0. $NaSbF_6$-type (6, 118; 20, 231; 46A,
162; 50A, 107). Zr-6F = 1.97, Co-6F = 2.04 A.

50K, R$\bar{3}$, a = 5.466, c = 13.982 A, Z = 3, neutron powder data. Co in 3(a):
0,0,0; Zr in 3(b): 0,0,1/2; F in 18(f): 0.084,0.333,0.084. $LiSbF_6$-type (27,
460; 46A, 162; 50A, 107). Zr-6F = 2.01, Co-6F = 2.02 A, Co-F-Zr = 154°.

POTASSIUM CHROMIUM(II,III) FLUORIDE
$K_5Cr_{10}F_{31}$ $K_5Cr(II)_4Cr(III)_6F_{31}$

J. Solid State Chem., 85, 151-158.

C2/m, 21.576, 7.6081, 32.865, 109.24, Z = 8, R = 0.040. Complex structure with
$Cr(III)F_6$ octahedra, $Cr(II)F_6$ distorted octahedra, $Cr(II)F_7$ pentagonal
bipyramids, and 7- to 10-coordinate K ions.

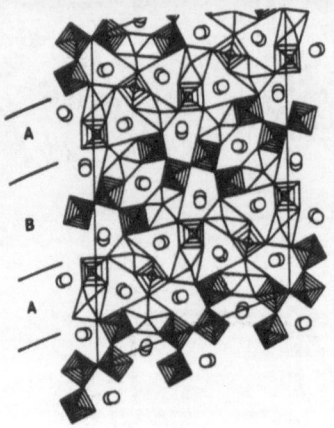

SODIUM STRONTIUM CHROMIUM(III) FLUORIDE
$NaSr_2CrF_8$

J. Solid State Chem., 87, 344-349.

$P2_1/c$, 7.7388, 6.2756, 14.827, 112.03, Z = 4, R = 0.040. Isolated $CrF_6{}^{3-}$
octahedra linked by Sr (8- and 9-coordinations) and Na ions (6-coordination);
additional F^- ions are coordinated to tetrahedra (3Sr + 1Na) which form double
chains along b. Cr-F = 1.88-1.95 A.

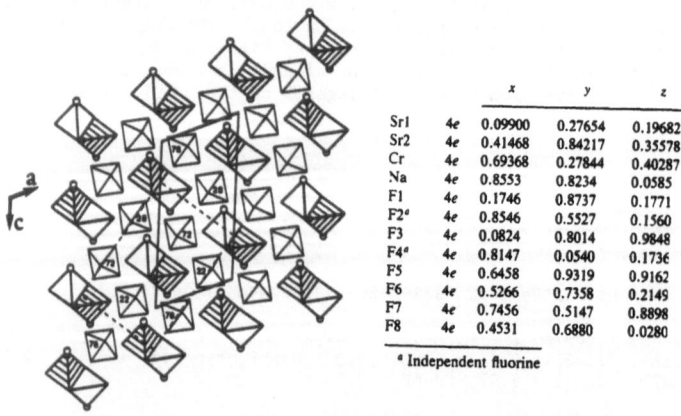

		x	y	z
Sr1	4e	0.09900	0.27654	0.19682
Sr2	4e	0.41468	0.84217	0.35578
Cr	4e	0.69368	0.27844	0.40287
Na	4e	0.8553	0.8234	0.0585
F1	4e	0.1746	0.8737	0.1771
F2ᵃ	4e	0.8546	0.5527	0.1560
F3	4e	0.0824	0.8014	0.9848
F4ᵃ	4e	0.8147	0.0540	0.1736
F5	4e	0.6458	0.9319	0.9162
F6	4e	0.5266	0.7358	0.2149
F7	4e	0.7456	0.5147	0.8898
F8	4e	0.4531	0.6880	0.0280

ᵃ Independent fluorine

LEAD CHROMIUM FLUORIDE
$Pb_5Cr_3F_{19}$

Acta Cryst., B46, 497-502.

I4cm, 14.384, 7.408, Z = 4, R = 0.044. The structure contains corner-, edge-,
and face-sharing CrF_6 octahedra and PbF_9 and PbF_{10} polyhedra; Cr-F = 1.83-1.95,
Pb-F = 2.27-3.03 A. There is a phase transition at 555K.

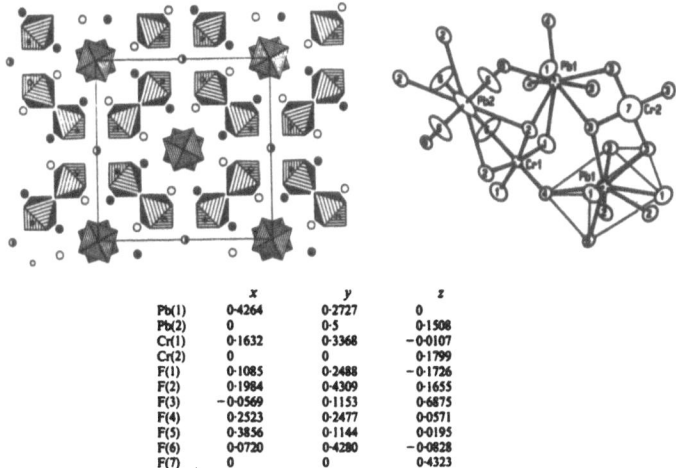

	x	*y*	*z*
Pb(1)	0·4264	0·2727	0
Pb(2)	0	0·5	0·1508
Cr(1)	0·1632	0·3368	−0·0107
Cr(2)	0	0	0·1799
F(1)	0·1085	0·2488	−0·1726
F(2)	0·1984	0·4309	0·1655
F(3)	−0·0569	0·1153	0·6875
F(4)	0·2523	0·2477	0·0571
F(5)	0·3856	0·1144	0·0195
F(6)	0·0720	0·4280	−0·0828
F(7)	0	0	0·4323

SODIUM TETRAFLUOROMANGANATE(III) DIHYDRATE and TRIHYDRATE
NaMnF$_4$.2H$_2$O   Na[MnF$_4$(H$_2$O)$_2$]
NaMnF$_4$.3H$_2$O   Na[MnF$_4$(H$_2$O)$_2$].H$_2$O

Z. Naturforsch., 45B, 593-597.

Dihydrate, C2/m, C2, or Cm, 8.166, 6.771, 4.968, 114.45, Z = 2, no structure
analysis; twinning and disorder.

Trihydrate, C2/c, 16.381, 6.676, 11.303, 103.78, Z = 8, R = 0.038. Distorted
trans-octahedral anions, linked by 7-coordinate Na ions and hydrogen bonds. Mn-F
= 1.827-1.855, Mn-O = 2.183, 2.246(2) A.

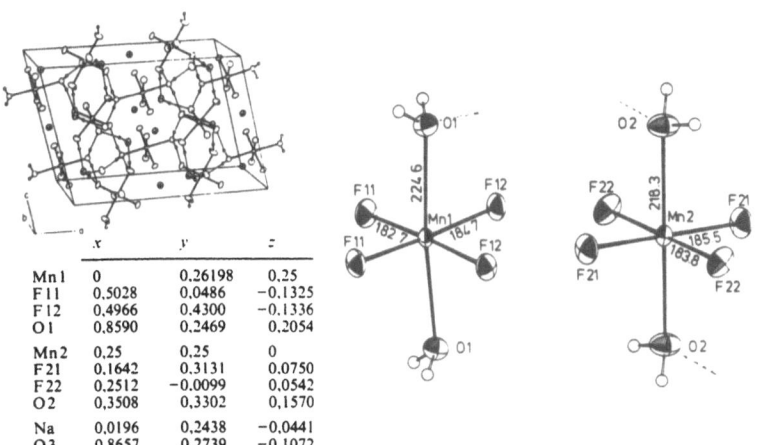

	x	*y*	*z*
Mn1	0	0.26198	0.25
F11	0.5028	0.0486	−0.1325
F12	0.4966	0.4300	−0.1336
O1	0.8590	0.2469	0.2054
Mn2	0.25	0.25	0
F21	0.1642	0.3131	0.0750
F22	0.2512	−0.0099	0.0542
O2	0.3508	0.3302	0.1570
Na	0.0196	0.2438	−0.0441
O3	0.8657	0.2739	−0.1072

CAESIUM PENTAFLUOROMANGANATE(III)
CAESIUM PENTAFLUOROMANGANATE(III) MONOHYDRATE
Cs$_2$MnF$_5$(I), Cs$_2$MnF$_5$.H$_2$O (II)

Z. Naturforsch., 45B, 1341-1348.

I, P4/mmm, 6.420, 4.229, Z = 1, R = 0.034. Cs in 2(e): 1/2,0,1/2; Mn in 1(a): 0,0,0; F(1) in 1(b): 0,0,1/2; F(2) in 4(j): 0.2045,0.2045,0. Chains of trans-corner-sharing MnF_6 octahedra, linked by Cs ions. Mn-F = 1.856(4) (terminal), 2.114(1) (bridging), Cs-10F = 3.13, 3.21 A.

II, Cmm2, 9.740, 8.674, 4.260, Z = 2, R = 0.026. Cs in 4(c): 1/4,3/4,0.4794; Mn in 2(a): 0,0,0; F(1) in 2(a): 0,0,0.499; F(2) in 4(d): 0.1886,0,-0.006; F(3) in 4(e): 0,0.2132,-0.011; O in 2(b): 1/2,0,0.712. Structure essentially as previously described in Cmmm (44A, 147), but with ordered water positions (indicating ferroelectric properties). Mn-F = 1.84 (terminal), 2.13 A (bridging). Dehydration gives an untwinned single crystal of Cs_2MnF_5 via a topotactic reaction.

CALCIUM PENTAFLUOROMANGANATE(III)
CADMIUM PENTAFLUOROMANGANATE(III)
$CaMnF_5$, $CdMnF_5$

Z. anorg. Chem., 583, 205-208.

C2/c, a = 8.938, 8.848, b = 6.369, 6.293, c = 7.830, 7.802 A, β = 116.23, 116.64°, Z = 4, R = 0.047, 0.047; previous descriptions in P2/c and $P2_1/n$ are incorrect (55A, 95; Rev. Chim. Minér., 23, 520), observed violations of the C-centering conditions being due to twinning. Isostructural with $CaCrF_5$ (37A, 186; 39A, 157).

	x	y	z
$CaMnF_5$			
Ca	0,0	0,4556	0,25
Mn	0,0	0,0	0,0
F(1)	0,0102	0,2943	-0,0327
F(2)	0,2244	-0,0173	0,1155
F(3)	0,0	0,0930	0,25
$CdMnF_5$			
Cd	0,0	0,45804	0,25
Mn	0,0	0,0	0,0
F(1)	0,0146	0,2948	-0,0354
F(2)	0,2278	-0,0211	0,1205
F(3)	0,0	0,1010	0,25

POTASSIUM IRON(III) PENTAFLUORIDE
K_2FeF_5

J. Solid State Chem., 84, 408-412.

Pbcn, 7.4059, 12.8771, 20.4282, Z = 16, R = 0.036. The structure is approximately the same as a previous (incorrect) $Pn2_1a$ description (43A, 129), but with some anomalously short interatomic distances removed. It contains chains of cis-corner-sharing FeF_6 octahedra, linked by 9- to 11-coordinate K ions. Fe-F = 1.883-2.025(3) A.

10^4	x	y	z
K1	0	729	2500
K2	0	5374	2500
K3	1593	216	4163
K4	1472	7881	1634
K5	4886	2321	-65
Fe1	6684	20	4219
Fe2	3372	8030	3286
F1	-59	1986	1406
F2	4957	9046	3767
F3	0	5000	0
F4	0	3218	2500
F5	2978	2057	2790
F6	2223	8801	402
F7	4724	1098	1119
F8	1817	9925	1518
F9	1791	9098	5299
F10	2070	9199	2966
F11	3098	2986	4038

RUBIDIUM TETRAFLUOROFERRATE(III) (PHASE IV)
$RbFeF_4$

J. Phys.: Condens. Matter, 2, 8269-8275.

Room temperature, Pmab, 7.6651, 7.6316, 6.2789, Z = 4, powder data. Fe in 4(a):
0,0,0; Rb in 4(d): 1/4,0.2824,0.5136; F(1) in 4(c): 0,1/4,-0.0711; F(2) in
4(d): 1/4,-0.0093,-0.0630; F(3) in 8(e): 0.0410,0.0497,0.2889. Structure as
previously described (38A, 206; 40A, 143); compare the $P2_12_12$ description in 53A,
88. Layers of corner-sharing FeF_6 octahedra, linked by Rb ions. Fe-F =
1.88-1.96 A.

RUBIDIUM PENTAFLUOROFERRATE(III)
Rb_2FeF_5

Z. Natur forsch., 45B, 161-169.

Pnma, 7.565, 5.810, 12.002, Z = 4, R = 0.039. Isostructural with the Cr compound
(40A, 142), with chains of FeF_6 octahedra sharing cis-corners, and 10-coordinate
Rb ions. Fe-F = 1.88-1.92, Rb-F = 2.81-3.17 A.

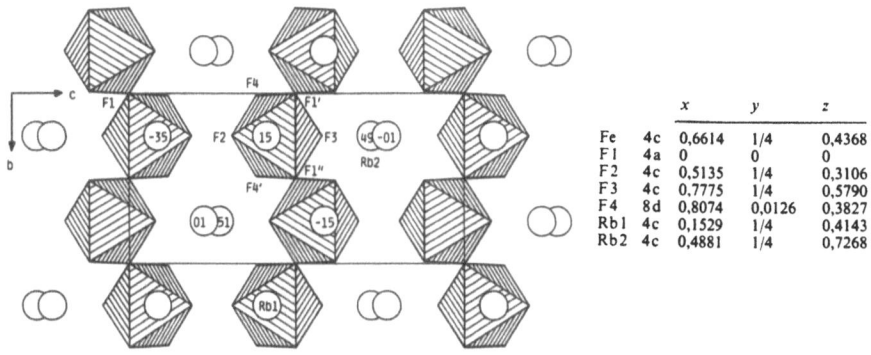

		x	y	z
Fe	4c	0,6614	1/4	0,4368
F1	4a	0	0	0
F2	4c	0,5135	1/4	0,3106
F3	4c	0,7775	1/4	0,5790
F4	8d	0,8074	0,0126	0,3827
Rb1	4c	0,1529	1/4	0,4143
Rb2	4c	0,4881	1/4	0,7268

RUBIDIUM IRON(III) FLUORIDE
$Rb_2Fe_5F_{17}$

Mater. Res. Bull., 25, 321-330.

Fmmm, 14.887, 25.814, 7.560, Z = 8, R = 0.068. Layers of corner-sharing FeF_6
octahedra with hexagonal and triangular tunnels. Rb ions are between the layers
and in the hexagonal tunnels; the latter sites have only 56% occupancy.

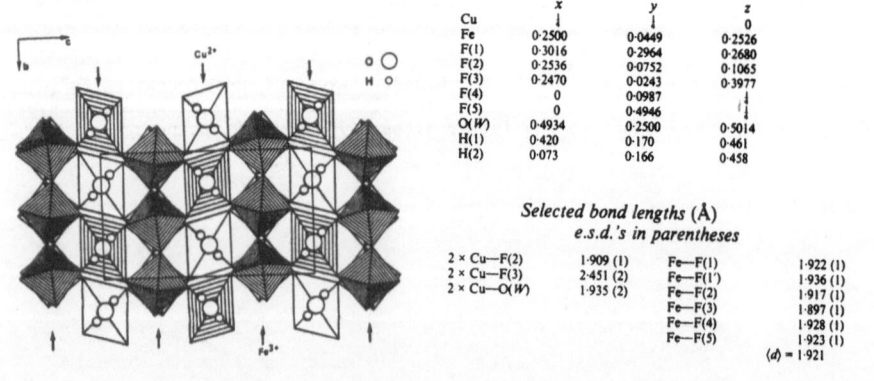

			x	y	z	Site occupancy factors
Rb(1)	(4a)	mmm	0	0	0	1
Rb(2)	(8g)	2mm	0.2199	0	1/2	0.94
Rb(3)	(8e)	..2/m	1/4	1/4	0	0.56
Fe(1)	(8c)	2/m..	0	1/4	1/4	1
Fe(2)	(32p)	1	0.1264	0.1260	0.2481	1
F(1)	(32p)	1	0.1525	0.0572	0.2107	1
F(2)	(32p)	1	0.0940	0.1996	0.2841	1
F(3)	(16o)	..m	0.1197	0.1441	0	1
F(4)	(8h)	m2m	0	0.2342	0	1
F(5)	(16k)	.2.	1/4	0.1494	1/4	1
F(6)	(16m)	m..	0	0.1105	0.2480	1
F(7)	(16o)	..m	0.1237	0.1155	1/2	1

CAESIUM TRIFLUOROFERRATE(II)
$CsFeF_3$

Z. anorg. Chem., **580**, 50-56.

$P6_3/mmc$, 6.1639, 14.870, Z = 6, R = 0.043. Hexagonal $BaTiO_3$-type (**11**, 448).
Fe-6F = 2.09-2.13, Cs-12F = 3.05-3.16 A.

		x	y	z
Cs(1)	(2b)	0	0	0,2500
Cs(2)	(4f)	0,3333	0,6667	0,0975
Fe(1)	(4f)	0,3333	0,6667	0,8483
Fe(2)	(2a)	0	0	0
F(1)	(6g)	0,5241	−,5241	0,2500
F(2)	(12k)	0,8332	−,8332	0,0780

COPPER(II) IRON(III) FLUORIDE HYDRATE
$CuFe_2F_8.2H_2O$

Acta Cryst., **C46**, 13-15.

C2/c, 7.541, 7.501, 13.027, 90.52, Z = 4, R = 0.036.   (001) Layers of
corner-sharing FeF_6 octahedra, linked by $CuF_4(H_2O)_2$ distorted octahedra.

	x	y	z
Cu		0·0449	0
Fe	0·2500	0·0449	0·2526
F(1)	0·3016	0·2964	0·2680
F(2)	0·2536	0·0752	0·1065
F(3)	0·2470	0·0243	0·3977
F(4)	0	0·0987	1/4
F(5)	0	0·4946	1/4
O(W)	0·4934	0·2500	0·5014
H(1)	0·420	0·170	0·461
H(2)	0·073	0·166	0·458

Selected bond lengths (Å)
e.s.d.'s in parentheses

2 × Cu—F(2)	1·909 (1)	Fe—F(1)	1·922 (1)
2 × Cu—F(3)	2·451 (2)	Fe—F(1')	1·936 (1)
2 × Cu—O(W)	1·935 (2)	Fe—F(2)	1·917 (1)
		Fe—F(3)	1·897 (1)
		Fe—F(4)	1·928 (1)
		Fe—F(5)	1·923 (1)
		⟨d⟩ =	1·921

SODIUM COPPER(II) IRON(III) FLUORIDE (MONOCLINIC)
Na_2CuFeF_7

Eur. J. Solid State Inorg. Chem., 27, 467-475.

C2/c, 12.46, 7.363, 12.93, 109.36, Z = 8, R = 0.038. New variant of the
weberite-type (12, 196; 44A, 142; 55A, 100), with FeF_6 octahedra, CuF_6 distorted
octahedra, and 8-coordinate Na.

HEXAFLUORIDES
$(Li,Ti)_{0.9}Ni_{1.2}F_6$ (I), $LiTiMnF_6$ (II), $(Li,Ti)_{1.2}Co_{0.6}F_6$ (III)

J. Solid State Chem., 88, 505-512.

I, $P4_2/mnm$, 4.664, 9.210, Z = 2, powder data. Trirutile-type structure (e.g. 8,
153).

II, P321, 8.741, 4.700, Z = 3, powder data. Na_2SiF_6-type structure (29, 264;
53A, 84).

III, P312 [not P321], [4.981, 4.589], Z = 1, powder data. Li/Ti/Co(1) in 1(a):
0,0,0; Li/Ti/Co(2) in 1(e): 2/3,1/3,0; Ti in 1(d): 1/3,2/3,1/2; F in 6(ℓ):
parameters not given. Structure similar to those of $PbSb_2O_6$ (8, 156; 54A, 152)
and Li_2NbF_6 (53A, 86). M-6F = 1.97-2.08 A.

$LiTiMnF_6$		x	y	z
1 Li	1a	0	0	0
2 Li	2d	1/3	2/3	0.4114
3 Mn	3e	0.6464	0	0
3 Ti	3f	0.3111	0	1/2
6 F(1)	6g	0.8912	0.7526	0.3387
6 F(2)	6g	0.4672	0.5825	0.6873
6 F(3)	6g	0.2298	0.7566	0.2827

$(Li,Ti)_{0.9}Ni_{1.2}F_6$		x	y	z
2 Ni	2a	0	0	0
4 $Li_9Ti_9Ni_2$	4e	0	0	0.3316
4 F(1)	4f	0.2990	0.2990	0
8 F(2)	8j	0.3005	0.3005	0.3393

SODIUM TRIFLUOROCUPRATE(II)   RUBIDIUM TRIFLUOROCUPRATE(II)
$NaCuF_3$ $RbCuF_3$

Z. anorg. Chem., 585, 93-104.

Na, $P\bar{1}$, 5.391, 5.552, 7.928, 90.66, 92.05, 86.95, Z = 4, R = 0.040. Rb, I4/mcm,
6.023, 7.912, Z = 4, R = 0.053. The Rb compound is isostructural with $KCuF_3$
(following report), and the Na compound is a distorted variant. Both compounds
have Jahn-Teller-distorted CuF_6 octahedra.

POTASSIUM TRIFLUOROCUPRATE(II)
$KCuF_3$

Acta Cryst., B46, 131-138.

I4/mcm, 5.8569, 7.8487, Z = 4, R = 0.020. Cu in 4(d): 0,1/2,0; K in 4(a):
0,0,1/4; F(1) in 4(b): 0,1/2,1/4; F(2) in 8(h): x,1/2+x,0, x = 0.22803.
Structure as previously described (26, 305; 39A, 158; 45A, 159; 46A, 398; 49A,
351). Cu-6F = 1.889, 1.962, 2.253 A.

SODIUM BARIUM COPPER(II) FLUORIDE
Na$_4$BaCu$_3$F$_{12}$

Eur. J. Solid State Inorg. Chem., 27, 771-782.

Ia3, 16.135, Z = 16, R = 0.058. Perovskite superstructure, with an octahedral
subnetwork of Cu(II) and one-quarter of the Na ions, in which Ba
(12-coordination) and the remainder of the Na (6-coordination) are inserted.

ALKALINE-EARTH HEXAFLUOROAURATES(V)
M(AuF$_6$)$_2$ (M = Mg, Ca, Sr, Ba)

Ž. Neorg. Khim., 35, 1970-1977 [Russ. J. Inorg. Chem., 35, 1122-1127].

Mg, Ca, P$\overline{4}$, a= 4.938, 5.019, c = 7.714, 8.428 A, Z = 1, powder data. AuF$_6^-$
octahedra linked by 12-coordinate Ca; Au-F = 1.86 A.

Sr, Ba, P$\overline{4}$3m or P23, a = 9.535, 9.901 A, Z = 4, powder data. AuF$_6^-$ octahedra
linked by Ba ions; Au-F = 1.86 A.

LANTHANON FLUORIDE TETRAFLUOROAURATES(III)
Ln$_2$F(AuF$_4$)$_5$ (Ln = La, Pr, Nd, Sm, Gd)

Z. anorg. Chem., 589, 51-61.

P4$_1$2$_1$2, a = 8.3707, 8.3011, 8.2700, 8.2173, 8.1726, c = 26.049, 25.885, 25.792,
25.716, 25.590 A, Z = 4, R = 0.044 for Sm compound (not given for the others).
Square-planar AuF$_4^-$ anions linked by 9-coordinate Ln ions (tricapped trigonal
prisms).

AMMONIUM TRIFLUOROCADMATE (ORTHORHOMBIC)
NH$_4$CdF$_3$

Physica B, 162, 231-236.

Pnma, 6.1791, 8.8786, 6.1655, Z = 4, powder data. N in 4(c): 0.5,1/4,0.0; Cd in
4(a): 0,0,0; F(1) in 8(d): 0.2848,0.0216,0.7851; F(2) in 4(c): -0.013,1/4,
0.0388. Orthorhombic perovskite (20, 273), with regular CdF$_6$ octahedra tilted
from the ideal cubic perovskite positions. Cd-F = 2.21-2.23(3) A.

CALCIUM LUTETIUM FLUORIDE
Ca$_2$LuF$_7$

J. Solid State Chem., 85, 133-143.

I4/m, 8.6633, 16.5252, powder data. Isostructural with the Yb compound (56A, 85).

			x	y	z
0.994	Ca(1)	16i	0.3849	0.1944	0.1633
0.87	Ca(2)	2b	0	0	1/2
0.78	Lu(1)	8h	0.1076	0.3106	0
0.78	Lu(2)	4e	0	0	0.1804
	F(1)	16i	0.292	0.421	0.0760
	F(2)	16i	0.082	0.228	0.1278
	F(3)	8h	0.304	0.155	0
	F(4)	8g	0	1/2	0.0811
	F(5)	2a	0	0	0
	F(6)	16i	0.296	0.400	0.2383
	F(7)	4d	0	1/2	1/4

GALLIUM(I,III) CHLORIDE
Ga_3Cl_7

Z. Naturforsch., 45B, 1-7.

Pna2_1, 11.818, 8.897, 10.575, Z = 4, R = 0.047. Isostructural with KGa_2Cl_7, (42A, 189), with $Cl_3Ga(III)-Cl-Ga(III)Cl_3^-$ anions (gauche conformation, approximate C_2 symmetry), linked by 10-coordinated Ga^+ ions. Ga(III)-Cl = 2.29, 2.31 (bridging), 2.13-2.16 (terminal), Ga(I)-Cl = 3.00-3.86(1) A.

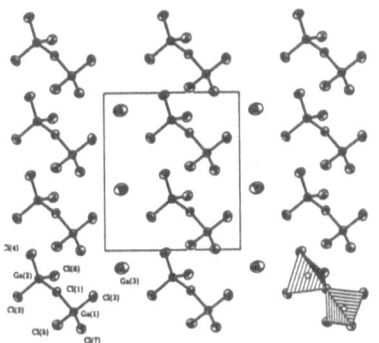

	x	y	z
Ga(1)	0.6138	0.6184	0.7601
Ga(2)	0.8262	0.5701	1/2
Ga(3)	0.8894	0.6384	0.1229
Cl(1)	0.7428	0.7415	0.6324
Cl(2)	0.7140	0.5460	0.3423
Cl(3)	0.7109	0.5194	0.9079
Cl(4)	0.9848	0.6661	0.4457
Cl(5)	0.5314	0.4588	0.6369
Cl(6)	0.8439	0.3719	0.6158
Cl(7)	0.5050	0.7980	0.8151

NIOBIUM(V) CHLORIDE
$NbCl_5$

Z. Kristallogr., 191, 139-140.

C2/m, 18.338, 17.861, 5.890, 90.69, Z = 12, R = 0.041. [The structure contains dimeric molecules, as previously described (22, 237).]

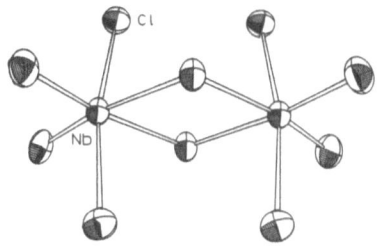

	x	y	z
Nb(1)	0	0.11091	0
Nb(2)	0.33334	0.11108	0.5246
Cl(1)	0.0561	0.1910	0.2406
Cl(2)	0.1005	0.0973	0.7738
Cl(3)	0.0525	0	0.2233
Cl(4)	0.2812	0	0.7449
Cl(5)	0.2772	0.1910	0.7629
Cl(6)	0.2328	0.0974	0.2920
Cl(7)	0.3856	0	0.3052
Cl(8)	0.3896	0.1908	0.2869
Cl(9)	0.4337	0.0976	0.7585

NIOBIUM CHLORIDE IODIDES
$Nb_6Cl_{12}I_2$, $Nb_6Cl_{10.8}I_{3.2}$

Z. anorg. Chem., <u>587</u>, 119-128.

Pa3, a = 12.578, 12.720 A, Z = 4, R = 0.032, 0.031. Nb_6X_{12} clusters connected by
I atoms.

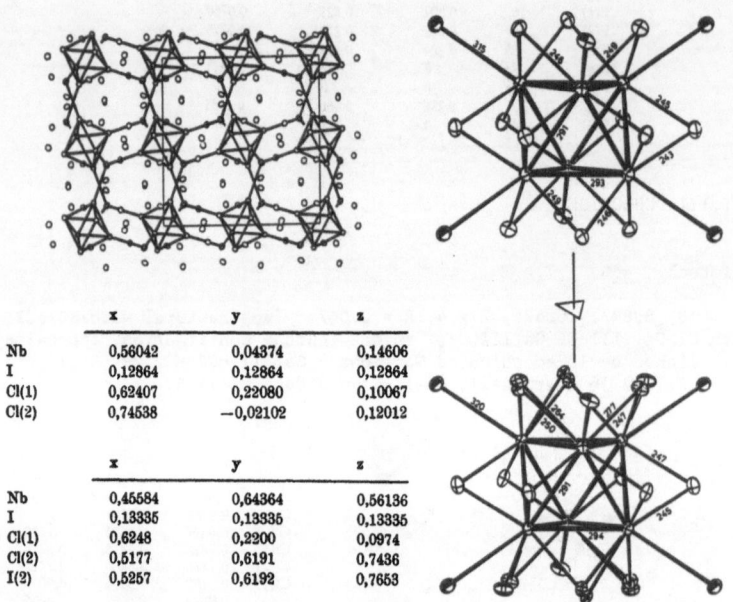

	x	y	z
Nb	0,56042	0,04374	0,14606
I	0,12864	0,12864	0,12864
Cl(1)	0,62407	0,22080	0,10067
Cl(2)	0,74538	−0,02102	0,12012

	x	y	z
Nb	0,45584	0,64364	0,56136
I	0,13335	0,13335	0,13335
Cl(1)	0,6248	0,2200	0,0974
Cl(2)	0,5177	0,6191	0,7436
I(2)	0,5257	0,6192	0,7653

MANGANESE(II) CHLORIDE
$MnCl_2$

Z. Kristallogr., <u>192</u>, 147-148.

R$\overline{3}$m, a = 3.711, c = 17.59 A, Z = 3, R = 0.105. Mn in 3(a): 0,0,0; Cl in 6(c):
0,0,0.2545. $CdCl_2$-type (<u>1</u>, 742), as previously described (<u>1</u>, 189, 192, 743, 773;
<u>2</u>, 245). Mn-Cl = 2.548(2) A.

RHENIUM TRICHLORIDE HYDRATE
$[Re_3Cl_9(H_2O)_3]\cdot 10H_2O$ 3 x $[ReCl_3\cdot(13/3)H_2O]$

Z. anorg. Chem., <u>581</u>, 104-110.

Pnma, 11.2516, 16.3030, 13.7884, Z = 4, R = 0.091. Trimeric molecules (<u>54A</u>, 104)
form columns along <u>b</u>, linked by hydrogen bonding via the crystal water molecules.
Re-Re = 2.440-2.448(1), Re-Cl = 2.300-2.422(3), Re-O = 2.223-2.260(9), O-H...O =
2.67-2.83(1), O-H...Cl = 3.20-3.63(1) A.

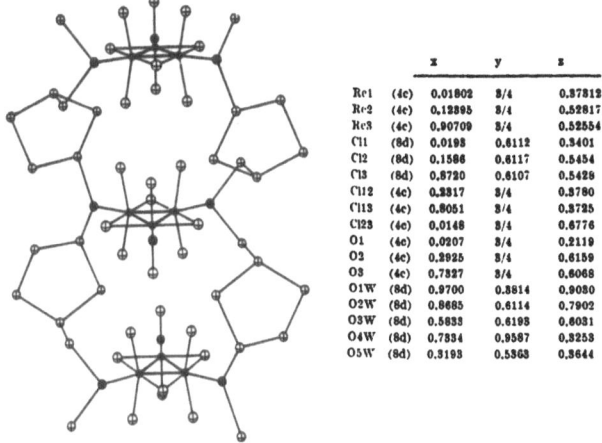

		x	y	z
Re1	(4c)	0.01802	3/4	0.37812
Re2	(4c)	0.12895	3/4	0.52817
Re3	(4c)	0.90709	3/4	0.52554
Cl1	(8d)	0.0193	0.6112	0.3401
Cl2	(8d)	0.1586	0.6117	0.5454
Cl3	(8d)	0.8720	0.6107	0.5428
Cl12	(4c)	0.2317	3/4	0.3780
Cl13	(4c)	0.8051	3/4	0.3795
Cl123	(4c)	0.0148	3/4	0.6776
O1	(4c)	0.0207	3/4	0.2119
O2	(4c)	0.2925	3/4	0.6159
O3	(4c)	0.7327	3/4	0.6068
O1W	(8d)	0.9700	0.3814	0.9030
O2W	(8d)	0.8685	0.6114	0.7902
O3W	(8d)	0.5633	0.6193	0.6031
O4W	(8d)	0.7334	0.9587	0.3253
O5W	(8d)	0.3193	0.5363	0.3644

TETRAAMMINENITROSYLOSMIUM COMPLEXES
[Os(NO)(NH$_3$)$_4$X]Y$_2$ (X/Y = Cℓ/Cℓ, Br/Cℓ, Br/Br, I/Cℓ) (I)
[Os(NO)(NH$_3$)$_4$NO$_3$]Cℓ$_2$.0.5H$_2$O (II)

Ž. Neorg. Khim., 35, 1760-1766 [Russ. J. Inorg. Chem., 35, 1003-1006].

I, I4mm, a = 7.372, 7.423, 7.625, 7.530, c = 8.475, 8.565, 8.718, 8.759 A, Z = 2, R = 0.020, 0.044, 0.052, 0.053. II, C2/c, 12.403, 6.720, 25.773, 97.01, Z = 8, R = 0.029. trans-Octahedral cations and halide anions linked by N-H...Y hydrogen bonds.

YTTRIUM CHLORIDE HEXAHYDRATE
YCℓ$_3$.6H$_2$O   [YCℓ$_2$(H$_2$O)$_6$]Cℓ

Acta Cryst., C46, 960-962.

P2/n, 7.8346, 6.4729, 9.5817, 93.768, Z = 2, R = 0.045. Isostructural with related compounds (26, 322; 30A, 275; 46A, 173; 50A, 118; 54A, 106), with YCℓ$_2$(H$_2$O)$_6$$^+$ square antiprisms.

	$\times 10^4$ x	y	z
Y(1)	2500	1556	2500
Cℓ(1)	2605	−1602	603
Cℓ(2)	7500	−3748	2500
O(1)	1071	2999	4362
O(2)	894	4247	1460
O(3)	−391	530	2181

Bond lengths (Å)

Y(1)—Cℓ(1)	2·740 (1)	Y(1)—O(1)	2·360 (3)
Y(1)—O(2)	2·333 (3)	Y(1)—O(3)	2·361 (3)
Y(1)—Cℓ(1a)	2·740 (1)	Y(1)—O(1a)	2·360 (3)
Y(1)—O(2a)	2·333 (3)	Y(1)—O(3a)	2·361 (3)

Hydrogen bonds (Å)

	H⋯X	O⋯X		H⋯X	O⋯X
O(1)—H(1)⋯Cℓ(2)	2·43	3·18	O(3)—H(5)⋯Cℓ(2)	2·51	3·25
O(2)—H(3)⋯Cℓ(1)	2·37	3·14	O(3)—H(6)⋯Cℓ(1)	2·40	3·16
O(2)—H(4)⋯Cℓ(2)	2·42	3·18			

H(2) does not appear to be involved in any hydrogen bonding.

GADOLINIUM CARBIDE HALIDES
Gd_2XC (X = Cl, Br, I)

J. Less-Common Metals, <u>167</u>, 65-79.

X = Cl, R$\bar{3}$m, a = 3.6902, c = 20.308 A, Z = 3, powder data. Gd in 6(c):
0,0,0.235; Cl in 3(a): 0,0,0; C in 3(b): 0,0,1/2. X = Br, I, P6$_3$/mmc, a =
3.7858, 3.8010, c = 14.209, 14.792 A, Z = 2, powder data for Br, single-crystal
data for I, R = 0.054. Gd in 4(f): 1/3,2/3,0.0871 (for I compound); Br or I in
2(d): 1/3,2/3,3/4; C in 2(a): 0,0,0. Both structure types contain Gd-C-Gd
layers connected by layers of X atoms, with octahedral coordination for Cl,
trigonal prismatic for Br and I.

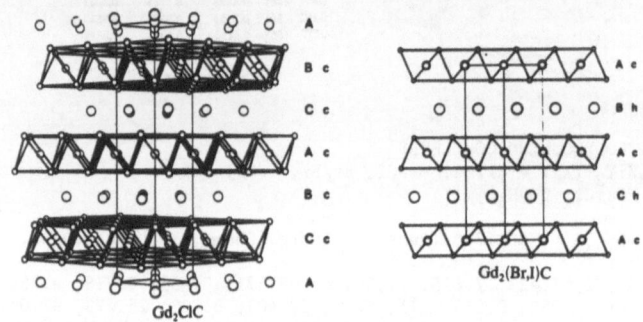

Gd$_2$ClC   Gd$_2$(Br,I)C

(pm)

	Gd₂ClC	*Gd₂BrC*	*Gd₂IC*	
Gd-Gd	349,89(1)	330,2(2)	338,5(2)	3×
	369,02(1)	378,58(5)	380,10(3)	6×
-C	254,26(1)	251,18(7)	254,50(8)	3×
-X	292,01(1)	318,4(1)	325,9(1)	3×
X-X	369,02(1)	378,58(5)	380,10(3)	6×
-C	399,94(2)	417,08(5)	430,06(4)	6×
C-C	369,02(1)	378,58(5)	380,10(3)	6×

DYSPROSIUM CHLORIDE
$DyCl_3$

Z. anorg. Chem., <u>586</u>, 99-105.

Cmcm, 3.816, 11.815, 8.507, Z = 4, R = 0.042. Dy in 4(c): 0,0.2432,1/4; Cl(1)
in 4(c): 0,0.5864,1/4; Cl(2) in 8(f): 0,0.1479,0.5665. PuBr$_3$-type (<u>11</u>, 282).

COPPER(I) TETRACHLOROGALLATE(III)
$CuGaCl_4$

Z. Kristallogr., <u>191</u>, 141-142.

P$\bar{4}$2c, 5.415, 10.197, Z = 2, R = 0.054. Cu in 2(e): 0,0,1/2; Ga in 2(b):
0,1/2,3/4; Cl in 8(n): -0.2392,-0.2711,-0.3715. [Isostructural with the Al
compound (<u>49A</u>, 111), with tetrahedral coordination for Cu and Ga.]

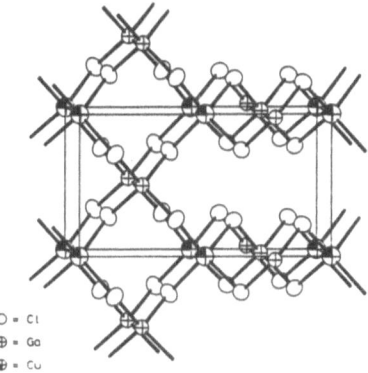

O = Cl
⊕ = Ga
⊛ = Cu

CAESIUM HEXACHLOROPLUMBATE(II)
Cs_4PbCl_6

Z. Kristallogr., 193, 217-242.

$R\bar{3}c$, a = 13.182, c = 16.641 A, Z = 6. Cs(1) in 18(e): 0.3718,0,1/4; Cs(2) in
6(a): 0,0,1/4; Pb in 6(b): 0,0,0; Cl in 36(f): 0.1900,0.0274,0.1009. A
previous monoclinic description is incorrect (54A, 109; compare 50A, 123).

CAESIUM HEXACHLOROANTIMONATE(V)
$CsSbCl_6$

Ž. Strukt. Khim., 31, No. 1, 104-109 [J. Struct. Chem., 31, 92-97].

Cc, 12.284, 6.433, 12.366, 102.38, Z = 4, R = 0.024. $SbCl_6^-$ octahedra linked by
6-coordinate Cs^+ ions. Sb-Cl = 2.344-2.367(4), Cs-Cl = 3.618-3.711 A.

CAESIUM HEPTACHLORODITITANATE(III)
$CsTi_2Cl_7$

Z. anorg. Chem., 580, 36-44.

P2/c, 7.0076, 6.2256, 12.000, 92.175, Z = 2, R = 0.026. (100) Layers of edge-
and corner-sharing $TiCl_6$ octahedra, linked by 12-coordinate Cs ions. Ti-Cl =
2.26 (not shared), 2.41-2.50, Cs-Cl = 3.60-3.74 [presumably not 2.74, as given in
the paper] A.

	x	y	z
Cs	1/2	0,37114	1/4
Ti	0,07168	0,75984	0,43195
Cl1	0,35911	0,8743	0,37706
Cl2	0	0,4176	3/4
Cl3	0,11367	0,07501	0,11700
Cl4	0.22037	0,58107	0,00303

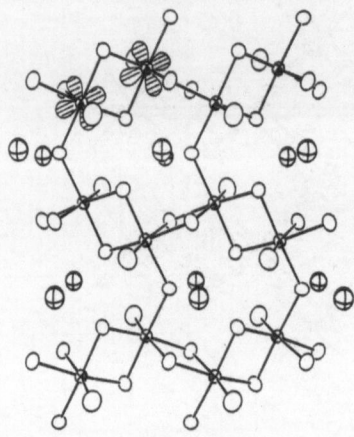

TRICHLOROSULPHONIUM ENNEACHLORODITITANATE(IV)
(SCℓ₃)(Ti₂Cℓ₉)

Ž. Neorg. Khim., 35, 1683-1685 [Russ. J. Inorg. Chem., 35, 957-959].

Rhombohedral, space group not given, a = 21.513, c = 17.318 A, Z = 18, R = 0.028.
Trigonal pyramidal SCℓ₃⁺ cations and dimeric anions (two octahedra sharing a
face), linked by S...Cℓ interactions which complete distorted octahedral
coordination at S. S-Cℓ = 1.974 A, Cℓ-S-Cℓ = 102°, Ti-Cℓ = 2.50 (bridging), 2.21
(terminal), S...Cℓ = 3.08-3.14 A.

HEXAAMMINECHROMIUM(III) HEXAAQUONICKEL PENTACHLORIDE - AMMONIUM CHLORIDE
[Cr(NH₃)₆][Ni(H₂O)₆]Cℓ₅.0.5(NH₄Cℓ)

J. Solid State Chem., 88, 498-504.

Fd3, 20.440, Z = 16, R = 0.037. Discrete octahedral complex cations, ammonium
cations, and chloride anions, linked by hydrogen bonds. Cr-N = 2.071(2), Ni-O =
2.040(3) A.

		10^4 x	y	z
Cr	16d	1/2	1/2	1/2
Ni	16c	0	0	0
Cl(1)	32e	2277	2277	2277
Cl(2)	48f	3983	1/8	1/8
Cl(3)	8b	5/8	5/8	5/8
N(1)	96g	3987	2495	2493
N(2)	8a	1/8	1/8	1/8
O	96g	867	2769	2085
H(1)	96g	3803	2947	2549
H(2)	96g	3801	2223	2862
H(3)	96g	3786	2317	2083
H(4)	96g	900	3183	1836
H(5)	96g	1324	2594	2150
H(6)	32e	1534	1534	1534

Projection of the structure of [Cr(NH₃)₆][Ni
(H₂O)₆]Cl₅ · 1/2(NH₄Cl) on the (100) plane. Octahedra
stand for [Cr(NH₃)₆]³⁺ (unhatched) and [Ni(H₂O)₆]²⁺
(hatched). Large circles represent the Cl⁻ anions while
small ones stand for the NH₄⁺ cations.

TETRATELLURIUM(2+) HEXACHLOROTUNGSTATE(V)
$Te_4^{2+}(WCl_6^-)_2$

Z. Naturforsch., 45B, 413-416.

P$\bar{1}$, 6.389, 7.532, 11.020, 101.19, 102.85, 90.21, Z = 1, R = 0.026. Nearly-ideal
square Te_4^{2+} cations and distorted octahedral anions linked by Te...Cl
interactions. Te-Te = 2.688, W-Cl = 2.275-2.373, Te...Cl = 3.192-3.657(1) A.

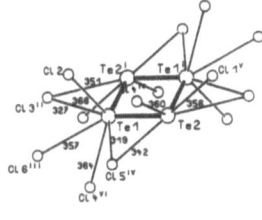

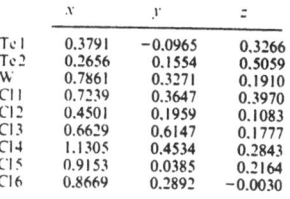

	x	y	z
Te1	0.3791	−0.0965	0.3266
Te2	0.2656	0.1554	0.5059
W	0.7861	0.3271	0.1910
Cl1	0.7239	0.3647	0.3970
Cl2	0.4501	0.1959	0.1083
Cl3	0.6629	0.6147	0.1777
Cl4	1.1305	0.4534	0.2843
Cl5	0.9153	0.0385	0.2164
Cl6	0.8669	0.2892	−0.0030

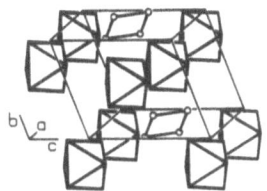

LITHIUM TETRACHLOROMANGANATE(II)
Li_2MnCl_4

Z. Naturforsch., 45B, 1543-1547.

Fd3m, 10.5027, Z = 8, R = 0.029. 8Li in 8(a): 1/8,1/8,1/8; 8Li + 8Mn in 16(d):
1/2,1/2,1/2; 32Cl in 32(e): x,x,x, x = 0.25623. Inverse spinel, as previously
described (41A, 172). Li-4Cl = 2.387, Li, Mn-6Cl = 2.562 A.

POTASSIUM HEXACHLORORHENATE(IV), -OSMATE(IV), AND -PLATINATE(IV)
K_2MCl_6 (M = Re, Os, Pt)

POTASSIUM TETRACHLOROPLATINATE(II)
K_2PtCl_4

Acta Cryst., B46, 166-174.

K_2MCl_6, Fm3m, a = 9.7953,9.7195,9.6911 A, at 120K, Z = 4, R = 0.022,0.019,
0.016/0.011 (Ag/Mo radiations). x(Cl) = 0.24037,0.24011,0.2390. K_2PtCl_6-type
(1, 429). Electron-density studies.

K_2PtCl_4, P4/mmm, 6.9961, at 120K, Z = 1, R = 0.019/0.015 (Ag/Mo radiations).
x(Cl) = 0.2338. Structure as previously described (1, 358). Electron-density
studies.

TETRAHALOGENOAQUONITROSYLRUTHENATES
$M_2[RuX_4(H_2O)(NO)]X.H_2O$ (M/X = Rb/Cl, Cs/Cl, Cs/Br)

Ž. Neorg. Khim., <u>35</u>, 1159-1166 [Russ. J. Inorg. Chem., <u>35</u>, 653-657].

$P2_1/c$, a = 11.476,11.748,12.181, b = 6.751,6.973,7.256, c = 15.914,16.204,16.878
A, β = 103.12,103.90,104.09, Z = 4, R = 0.056,0.050,0.056. Octahedral anions
with trans-NO/H_2O, linked via O-H...X⁻ hydrogen bonds and by 12-coordinate M⁺
ions. Ru-N = 1.704-1.711 A.

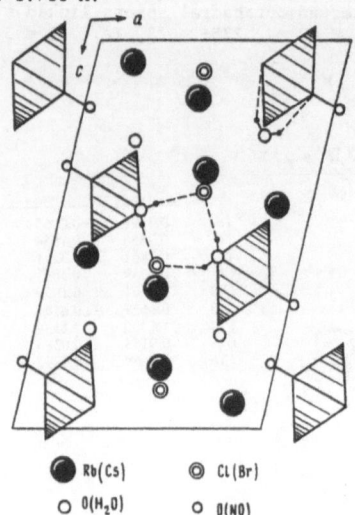

● Rb(Cs)	◎ Cl(Br)
○ O(H₂O)	○ O(NO)

COBALT(II) CHLORIDE TETRAHYDRATE and HEXAHYDRATE
$CoCl_2 \cdot nH_2O$ (n = 4, 6)

Bull. Chem. Soc. Japan, <u>63</u>, 3426-3433.

Tetrahydrate, $P2_1/a$, 11.598 [in text, 11.548 in Abstract], 9.342, 6.056, 110.79,
Z = 4, R = 0.025. Hexahydrate, C2/m, 10.380, 7.048, 6.626, 122.01, Z = 2, R =
0.058. The tetrahydrate is isostructural with the Mn salt (<u>29</u>, 273; <u>37</u>A, 195),
and contains cis-octahedral $[CoCl_2(H_2O)_4]$ molecular units, linked by a system of
O-H...Cl and O-H...O hydrogen bonds; Co-Cl = 2.406, 2.422(1), Co-O =
2.094-2.123(2) A. The structure of the hexahydrate is as previously described
(<u>22</u>, 240; <u>24</u>, 280; <u>27</u>, 435; <u>35</u>A, 402), with trans-octahedral $[CoCl_2(H_2O)_4]$ units
and two additional water molecules, linked by O-H...Cl and O-H...O hydrogen
bonds; Co-Cl = 2.445(2), Co-O = 2.081(5) A.

trans-BIS(DICHLOROSULPHAN)TETRACHLOROPLATINUM(IV)
$PtCl_4(SCl_2)_2$ (I)

trans-BIS(DICHLOROSULPHAN)DICHLOROPALLADIUM(II)
$PdCl_2(SCl_2)_2$ (II)

Z. anorg. Chem., <u>588</u>, 69-76.

I, $P2_1/c$, 6.271, 7.177, 12.368, 100.15, Z = 4, R = 0.087. trans-Octahedral
molecules. Pt-S = 2.36, Pt-Cl = 2.31, S-Cl = 1.99(1) A.

II, P1̄, 4.3387, 6.8213, 8.1737, 104.16, 98.15, 94.42, Z = 1, R = 0.050.
trans-Square-planar molecules. Pd-S = 2.292(2), Pd-Cl = 2.297(3), S-Cl =
2.010(4) A.

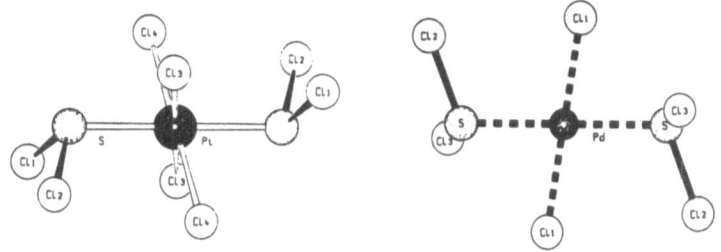

CAESIUM TETRAHALOMETALLATES(II)
Cs_2ZnCl_4, Cs_2CoCl_4, Cs_2CuCl_4, Cs_2ZnBr_4, $Cs_2Cu\ Br_4$

Ferroelectrics, 107, 229-234.

Pnma, a ∿ 10, b ∿ 7.5, c ∿ 13 A, Z = 4, R = 0.023-0.039, β-K_2SO_4-type (2, 86; 38-A, 216; 39A, 184; 40A, 153; 48A, 167; 54A, 112).

SILVER HEXACHLOROYTTRATE
Ag_3YCl_6

Z. anorg. Chem., 582, 143-150.

$R\bar{3}$, 6.8669, 18.305, Z = 3, R = 0.099, 3Ag(1) in 6(c): 0,0,0.2970; 3Ag(1') in 6(c): 0,0,0.2082; 3Ag(2) in 3(b): 0,0,1/2; 3Y in 3(a): 0,0,0; 18 Cl in 18(f): 0.3592,0.3525,0.2510. Isostructural with Na_3GdCl_6 (51A, 147), with similar disorder of some of the Ag+ ions.

SODIUM CERIUM(III) CHLORIDE
$Na_{0.38}(Na_{0.19}Ce_{0.81})Cl_3$

Z. anorg. Chem., 589, 96-100.

$P6_3/m$, 7.5707, 4.3156, Z = 2, R = 0.011. 0.76 Na in 2(b): 0,0,0; (0.38 Na + 1.62 Ce) in 2(c): 1/3,2/3,1/4; Cl in 6(h): 0.38759,0.30279,1/4. $CeCl_3$ variant (UCl_3-type (11, 278; 54A, 107)), with additional and substitutional Na.

SODIUM GADOLINIUM CHLORIDE
$NaGdCl_4$

Z. anorg. Chem., 590, 103-110.

$P\bar{1}$, 7.0281, 6.7625, 6.6672, 100.852, 91.702, 89.760, Z = 2, R = 0.040.
Fluorite-derivative structure with 7-coordinations (monocapped trigonal prisms) for Na (larger black circles in diagram) and Gd (smaller black circles). Na-Cl = 2.77-3.27, Gd-Cl = 2.61-2.82 A.

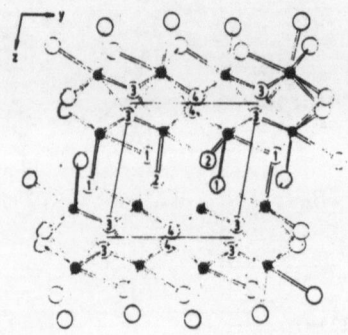

DYSPROSIUM ALUMINUM CHLORIDE
$DyAl_3Cl_{12}$

Z. anorg. Chem., <u>586</u>, 99-105.

$P3_112$, 10.492, 15.636, Z = 3, R = 0.043. $DyCl_8$ square antiprisms linked to $AlCl_4$ tetrahedra.   Dy-Cl = 2.718-2.883(3), Al-Cl = 2.055-2.204(4) A.

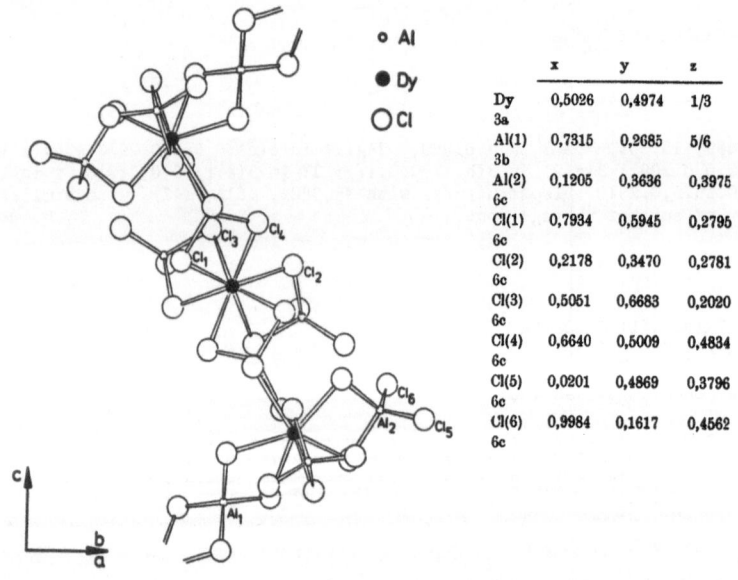

	x	y	z
Dy 3a	0,5026	0,4974	1/3
Al(1) 3b	0,7315	0,2685	5/6
Al(2) 6c	0,1205	0,3636	0,3975
Cl(1) 6c	0,7934	0,5945	0,2795
Cl(2) 6c	0,2178	0,3470	0,2781
Cl(3) 6c	0,5051	0,6683	0,2020
Cl(4) 6c	0,6640	0,5009	0,4834
Cl(5) 6c	0,0201	0,4869	0,3796
Cl(6) 6c	0,9984	0,1617	0,4562

○ Al
● Dy
◯ Cl

INDIUM(I,III) BROMIDE
In_5Br_7

Z. anorg. Chem., <u>582</u>, 128-130.

Reinterpretation of the structure (<u>55</u>A, 111) in C2/c, rather than Cc.

TITANIUM(IV) BROMIDE (CUBIC)   TITANIUM(III) BROMIDE
$TiBr_4$ α-$TiBr_3$

Ž. Neorg. Khim., <u>35</u>, 882-887 [Russ. J. Inorg. Chem., <u>35</u>, 494-498].

TiBr$_4$, Pa3, 11.258, Z = 8, R = 0.070. Ti in 8(c): x,x,x, x = 0.1311; Br(1) in 8(c): x = 0.2480; Br(2) in 24(d): 0.0165,0.0094,0.2482. Structure as previously described (<u>2</u>, 306; cf. monoclinic form in <u>31A</u>, 105).

α-TiBr$_3$, R$\bar{3}$, a = 6.478, c = 18.632 A, Z = 6, R = 0.065. Ti in 6(c): 0,0,0.3336; Br in 18(f): 0.6524 ,0.0006,0.0797. Isostructural with α-TiCl$_3$ (<u>27</u>, 441).

NIOBIUM(V) BROMIDE
α-NbBr$_5$

Z. Naturforsch., <u>45</u>B, 952-956.

Pnma, 12.888, 18.690, 6.149, Z = 8, R = 0.055. Dimeric molecules of two edge-sharing octahedra. Monoclinic (<u>22</u>, 237) and stacking-disordered orthorhombic forms (<u>50A</u>, 136) have been described.

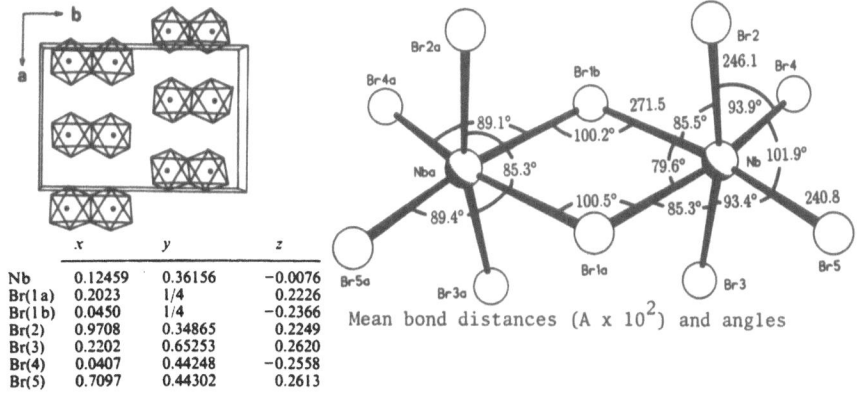

	x	y	z
Nb	0.12459	0.36156	-0.0076
Br(1a)	0.2023	1/4	0.2226
Br(1b)	0.0450	1/4	-0.2366
Br(2)	0.9708	0.34865	0.2249
Br(3)	0.2202	0.65253	0.2620
Br(4)	0.0407	0.44248	-0.2558
Br(5)	0.7097	0.44302	0.2613

Mean bond distances (A x 10^2) and angles

TITANIUM(II) TETRABROMOALUMINATE
Ti(AlBr$_4$)$_2$

Ž. Neorg. Khim, <u>35</u>, 882-887 [Russ. J. Inorg. Chem., <u>35</u>, 494-498].

Pnn2, 6.274, 12.801, 8.676, Z = 2, R = 0.065. Tetrahedral AlBr$_4$$^-$ anions, linked by octahedrally-coordinate Ti^{2+} ions. Al-Br = 2.23-2.40(2), Ti-Br = 2.61-2.78(1) A. A monoclinic form has been described previously (<u>53A</u>, 94).

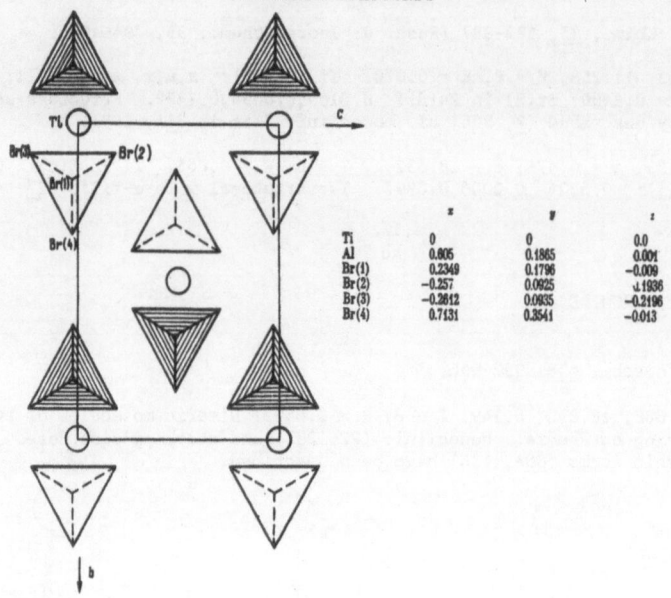

	x	y	z
Tl	0	0	0.0
Al	0.805	0.1865	0.001
Br(1)	0.2349	0.1796	−0.009
Br(2)	−0.257	0.0925	−.1936
Br(3)	−0.2612	0.0935	−0.2196
Br(4)	0.7131	0.3541	−0.013

TETRABROMOGALLATES
$MGaBr_4$ (M = Na, K, In)

Z. anorg. Chem., 585, 38-48.

M = Na, Pnma, 14.044, 7.452, 7.449, Z = 4, R = 0.14. $GaBr_4^-$ tetrahedra, linked by 8-coordinate Na^+ ions.

M = K, R3c, a = 22.161, c = 8.829 A, Z = 18, R = 0.098. β-$GaBr_2$-type (54A, 103).

M = In, P2/a, 13.006, 7.356, 9.061, 106.35, Z = 4, R = 0.155. $GaBr_4^-$ tetrahedra, linked by 10-coordinate In^+ ions.

In the compounds $A(I)M(III)Br_4$ six structure types are found: $NaGaBr_4$, $NaAlCl_4$, $GaCl_2$, β-$GaBr_2$, $KAlBr_4$, $BaSO_4$.

SILVER(I) PENTABROMODIPLUMBATE(II)
$AgPb_2Br_5$

Z. Kristallogr., 191, 135-136.

C2/c, 16.6967, 7.0528, 7.981, 95.95, Z = 4, R = 0.05. [Pb has tetrahedral and Ag linear coordination.]

O = Br
⊕ = Ag
⊕ = Pb

	x	y	z
Pb(1)	0.10358	0.2319	0.0639
Br(1)	0	0.0750	−0.25
Br(2)	0.2206	0.9149	0.3189
Br(3)	0.0876	0.3840	0.4204
Ag(1)	0.25	0.25	0.5

LITHIUM TETRABROMOMANGANATE(II)
Li$_2$MnBr$_4$

Mater. Res. Bull., <u>25</u>, 451-456.

Room temperature, orthorhombic, no details given.

545-758K, Fd3m, 11.2372, at 623K, Z = 8, neutron powder data. 8Li(1) in 16(c):
0,0,0; 8Li(2) + 8Mn in 16(d): 1/2,1/2,1/2; 32Br in 32(e): x,x,x, x = 0.2540.
Modified Zr$_3$S$_4$-type (<u>21</u>, 177; <u>35A</u>, 69), as for LiVO$_2$ (<u>48A</u>, 215).

758-810K, Fm3m, 5.6731, at 773K, Z = 1, neutron powder data. 2Li + 1Mn in 4(b):
1/2,1/2,1/2; 4Br in 4(a): 0,0,0. Disordered, cation-deficient NaCl-type.

CALCIUM MANGANESE(II) BROMIDE OCTAHYDRATE
CaMnBr$_4$.8H$_2$O

Acta Cryst., C<u>46</u>, 538-541.

C2/m, 9.062, 9.582, 9.405, 112.96, Z = 2, R = 0.062. Molecular units with a
MnBr$_4$(H$_2$O)$_2$ octahedron sharing one Br and one H$_2$O with a Ca(H$_2$O)$_4$Br polyhedron;
the unit is disordered about an inversion centre at the Mn atom. The molecular
units are linked by hydrogen bonds. Mn-Br = 2.694, 2.713(2), Mn-O = 2.18(1),
Ca-Br = 3.073(5), Ca-O = 2.31-3.01(4), O-H...O = 2.90, O-H...Br = 3.29-3.62 A.

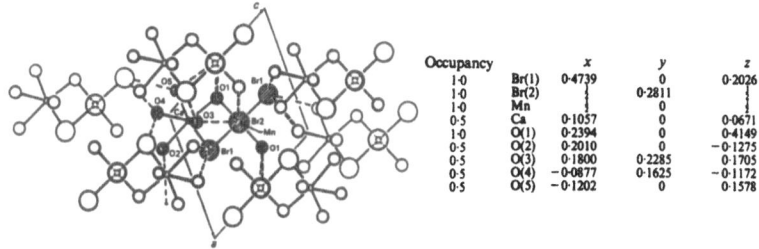

Occupancy		x	y	z
1·0	Br(1)	0·4739	0	0·2026
1·0	Br(2)	½	0·2811	½
1·0	Mn	0	0	0
0·5	Ca	0·1057	0	0·0671
1·0	O(1)	0·2394	0	0·4149
0·5	O(2)	0·2010	0	−0·1275
0·5	O(3)	0·1800	0·2285	0·1705
0·5	O(4)	−0·0877	0·1625	−0·1172
0·5	O(5)	−0·1202	0	0·1578

SCANDIUM IODIDE
Sc$_{0.92}$I$_2$

Inorg. Chem., <u>29</u>, 2030-2032.

P3̄m1, 4.0851, 6.9824, Z = 1, R = 0.020. 0.92 Sc in 1(a): 0,0,0: I in 2(d):
1/3,2/3,0.25025. CdI$_2$-type (<u>1</u>, 161).

ZIRCONIUM(III) IODIDE
ZrI$_3$

Inorg. Chem., <u>29</u>, 2242-2246.

Pmmn, 12.594, 6.679, 7.292, Z = 4, R = 0.022. There is a β-TiCl$_3$-type subcell
(P6$_3$/mcm, Z = 2) (<u>49A</u>, 107), but additional weak reflections indicate
orthorhombic symmetry. The structure contains chains of face-sharing ZrI$_6$
distorted octahedra.

	sym[a]	x	y	z	
I1	4f. m	0.08585	−¼	0.58754	[a]Space group *Pmmn*, origin at 1̄.
I2	4f. m	0.08872	¼	0.91313	
I3	2b. mm	¼	−¼	0.0807	
I4	2a. mm	¼	¼	0.4299	
Zr	4c. m	¼	−0.0125	0.7521	

COPPER(I) IODIDE
CuI

Z. Kristallogr., <u>191</u>, 79-91.

γ-Phase (room temperature), F4̄3m, 6.0337, Z = 4, neutron powder data. Cu in
4(c): 1/4,1/4,1/4; I in 4(a): 0,0,0. Zincblende-type structure.

β-Phase (598K), P6̄m2, 4.289, 7.189, Z = 2, neutron powder data. 0.3Cu(1) in
2(i): 2/3,1/3,0.26; 0.3Cu(2) in 2(i): 2/3,1/3,0.41; 1.4Cu(3) in 2(g): 0,0,0.39;
I(1) in 1(d): 1/3,2/3,1/2; I(2) in 1(a): 0,0,0. H.c.p. anion lattice, with Cu
in tetrahedral and octahedral holes (partial occupancies).

α-Phase (763K), Fm3m, 6.148, Z = 4, neutron powder data. 4Cu in 32(f): x,x,x, x
= 0.302; 4I in 4(a): 0,0,0. F.c.c. anion lattice, with Cu randomly distributed
over all tetrahedral sites.

    There are two phase transitions, the structural changes involving
rearrangement of the anions and of disordered distribution of the cations.

YTTRIUM OSMIUM and IRIDIUM IODIDES
Y_6OsI_{10}, Y_6IrI_{10}

Inorg. Chem., <u>29</u>, 2246-2251.

P1̄, a = 7.6268, 7.6154, b = 9.518, 9.535, c = 9.617, 9.577 A, α = 107.72, 107.67,
β = 97.17, 97.11, γ = 105.13, 105.15°, Z = 1, R = 0.031, 0.037. Isostructural
with the Ru compound (<u>56</u>A, 104), with less tetragonal compression; Y-Os =
2.64-2.75, Y-Ir = 2.69-2.73 A.

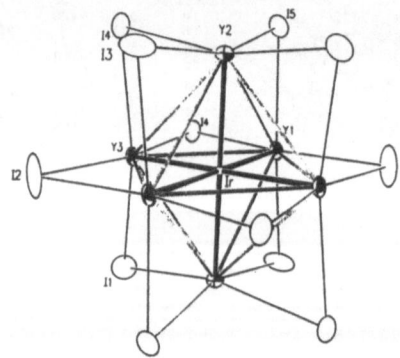

THALLIUM(I) HEXAIODODICUPRATE(I)
$Tl_4Cu_2I_6$

Z. anorg. Chem., <u>587</u>, 23-28.

Pnnm, 9.196, 9.552, 9.336, Z = 2, R = 0.052. Binuclear $Cu_2I_6{}^{4-}$ anions (two
edge-sharing tetrahedra), linked by Tl^+ ions coordinated to 8 iodine atoms
(3.46-3.77 A) and to 1 Cu (3.44 A) and 1 Tl (3.88 A). Cu-I = 2.65-2.68 A,
Cu-I-Cu = 61°.

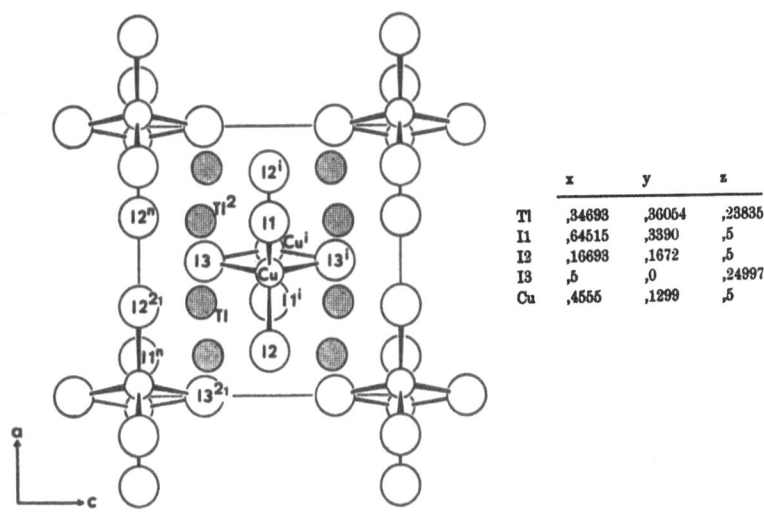

	x	y	z
Tl	,34693	,36054	,23835
I1	,64515	,3390	,5
I2	,16693	,1672	,5
I3	,5	,0	,24997
Cu	,4555	,1299	,5

LITHIUM TETRAIODOZINCATE
Li_2ZnI_4

J. Solid State Chem., 87, 463-466.

Pnma, 14.803, 8.560, 7.012, Z = 4, neutron powder data. Olivine-type structure
(1, 352), with Zn in tetrahedral and Li in octahedral sites. Zn-I = 2.56-2.66,
Li-I = 2.99-3.06(1) A.

		Occupation	x	y	z
Li(1)	4a	4	0	0	0
Li(2)	4c	4	0.269	0.25	0.000
Zn	4c	4	0.0858	0.25	0.4037
I(1)	4c	4	0.0918	0.25	0.769
I(2)	4c	4	0.4228	0.25	0.252
I(3)	8d	8	0.1700	0.0063	0.248

MERCURY PHOSPHIDE IODIDE
$Hg_9P_5I_6$

Z. Kristallogr., 192, 223-231.

$P2_1/c$, 13.112, 12.486, 17.031, 119.90, Z = 4, R = 0.047. (100) Layers of Hg_6
octahedra centred by P_2 groups, with layers linked by Hg-Hg bonds and via
tetrahedral Hg_3P-Hg-P-Hg_3 groups. Hg-Hg = 2.541(7), Hg-P = 2.36-2.49(2), Hg-I =
2.67-3.69(1), P-P = 2.10, 2.16(2) A. Previously described incorrectly as
$Hg_3P_2I_2$ (16, 204).

RHENIUM TELLURIUM OXYFLUORIDES
$O=Re(OTeF_5)_4 \cdot F_2Te(OTeF_5)_2$ (I), $O=Re(OTeF_5)_5$ (II)

Z. anorg. Chem., 590, 37-47.

I, P$\bar{1}$, 10.659, 11.345, 12.371, 81.63, 80.99, 85.31, Z = 2, R = 0.031. ReO$_5$
square pyramid linked to four TeOF$_5$ octahedra, with distorted octahedral
coordination at Re completed by a Re...F contact to a F$_2$Te(OTeF$_5$)$_2$ molecule; the
latter contains a TeF$_2$O$_2$E trigonal bipyramid (E = lone-pair) linked to two TeOF$_5$
octahedra. Re-O = 1.63, 1.89-1.95, Re...F = 2.21 A.

II, P2$_1$/c, 9.017, 18.326, 14.003, 107.50, Z = 4, R = 0.034. Central ReO$_6$
octahedron linked to five TeOF$_5$ octahedra; Re-O = 1.68, 1.87-1.96 A.

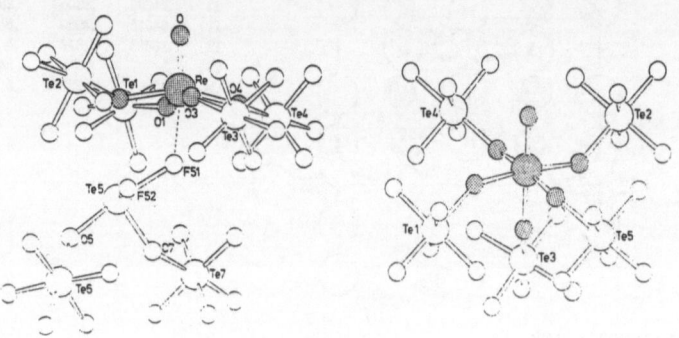

MOLYBDENUM and TUNGSTEN TELLURIUM OXYFLUORIDES
Mo(OTeF$_5$)$_6$ (I), O=Mo(OTeF$_5$)$_4$ (II), O=W(OTeF$_5$)$_4$ (III)

Z. anorg. Chem., <u>590</u>, 23-36.

I, P$\bar{1}$, 8.934, 8.958, 9.053, 68.11, 64.55, 83.39, at 105K, Z = 1, R = 0.03.
Central MoO$_6$ octahedron linked to 6 TeOF$_5$ octahedra; Mo-O = 1.86-1.87, Te-O =
1.89-1.90 A, Mo-O-Te = 146°.

II, P$\bar{1}$, 9.938, 10.220, 9.097, 88.21, 81.37, 71.71, at 107K, Z = 2, R = 0.072.
Central MoO$_5$ square pyramid linked to four TeOF$_5$ octahedra; Mo-O = 1.66 (apical),
1.89-2.01, Te-O = 1.84-1.85 A, Mo-O-Te = 131-154°. The molecules are linked into
pairs by Mo...F interactions which complete distorted octahedral coordination at
Mo. Some Mo and O/F disorder is discussed.

III, P$\bar{1}$, 10.200, 10.507 [only two lattice parameters are listed; from the cell
volume the third parameter is 9.7 A], 75.6, 86.6, 83.5, at 293K, Z = 2, R =
0.063. Structure quite similar to that of II. W-O = 1.63, 1.81-1.89, Te-O =
1.79-1.89 A, W-O-Te = 152-167°, W...F = 2.58 A.

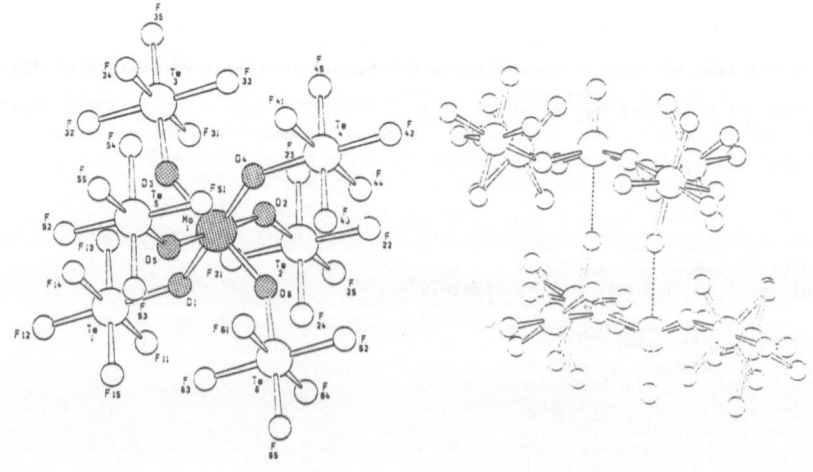

XENON TELLURIUM OXYFLUORIDE HEXAFLUOROARSENATE
[XeO-TeF$_4$-OH.FH]$^+$AsF$_6$$^-$

P$\bar{1}$, 5.0474, 7.5893, 13.8725, 90.14, 95.74, 99.67, at 117K, Z = 2, R = 0.042.
Structure as shown below.

Bond Distances (pm) and Angles (deg) (Standard Deviations in Parentheses)			
Distances			
Xe–F16	245.8 (8)	As–F16	177.1 (7)
Xe–O1	196.2 (9)	Te–F1	183.8 (7)
As–F11	172.2 (8)	Te–F2	182.6 (8)
As–F12	170.2 (8)	Te–F3	182.8 (7)
As–F13	174.2 (7)	Te–F4	182.7 (7)
As–F14	170.6 (7)	Te–O1	194.9 (9)
As–F15	171.2 (7)	Te–O2	182.1 (9)

POTASSIUM TRIFLUOROBIS(PEROXO)TITANATE(IV)
K$_3$Ti(O$_2$)$_2$F$_3$

Ž. Neorg. Khim., 35, 1611-1613 [Russ. J. Inorg. Chem., 35, 915-916].

Cmc2$_1$, 9.307, 8.297, 8.879 [in Abstract; completely different cell parameters are
quoted in the text], Z = 4, R = 0.036. Discrete pentagonal-bipyramidal anions
with 2 F axial and 1 F and 2 peroxo groups equatorial, linked by 7- and
10-coordinate K ions. Ti-F = 1.89-1.98, Ti-O = 1.90, 1.95, O-O = 1.49(1) A.

POTASSIUM µ-FLUORO-µ-PEROXO-BIS(FLUOROXOPEROXOVANADATE(V)) HYDROGEN FLUORIDE
DIHYDRATE
K$_3$[F(O$_2$){VO(O$_2$)F}$_2$].HF.2H$_2$O

Acta Cryst., C46, 1753-1755.

P$\bar{1}$, 8.518, 12.460, 5.981, 92.30, 90.90, 102.70, Z = 2, R = 0.042. Discrete
binuclear anion, with two VO$_5$F$_2$ pentagonal bipyramids sharing a F(O$_2$) face. V-O
= 1.60-2.06, V-F = 1.88-2.12, O-O = 1.44-1.47(1) A. K ions have irregular
coordinations, and HF (78% occupancy) and water molecules form hydrogen bonds to
peroxo oxygen atoms.

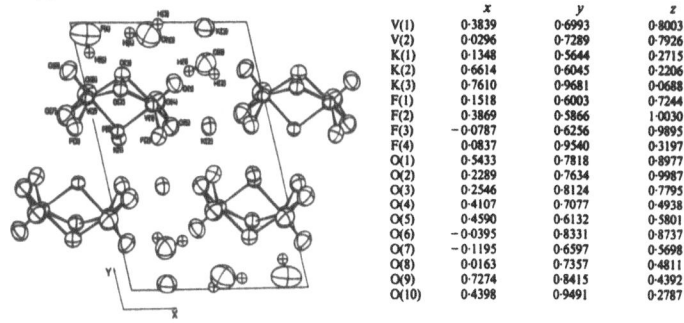

	x	y	z
V(1)	0·3839	0·6993	0·8003
V(2)	0·0296	0·7289	0·7926
K(1)	0·1348	0·5644	0·2715
K(2)	0·6614	0·6045	0·2206
K(3)	0·7610	0·9681	0·0688
F(1)	0·1518	0·6003	0·7244
F(2)	0·3869	0·5866	1·0030
F(3)	−0·0787	0·6256	0·9895
F(4)	0·0837	0·9540	0·3197
O(1)	0·5433	0·7818	0·8977
O(2)	0·2289	0·7634	0·9987
O(3)	0·2546	0·8124	0·7795
O(4)	0·4107	0·7077	0·4938
O(5)	0·4590	0·6132	0·5801
O(6)	−0·0395	0·8331	0·8737
O(7)	−0·1195	0·6597	0·5698
O(8)	0·0163	0·7357	0·4811
O(9)	0·7274	0·8415	0·4392
O(10)	0·4398	0·9491	0·2787

MANGANESE NIOBIUM OXYFLUORIDE
$Mn_{0.17}Nb_6F_{1.66}O_{14.34}$

Z. anorg. Chem., 585, 105-112.

Pmma, 16.677, 3.962, 8.902, Z = 2, R = 0.072. Isostructural with $LiNb_6O_{15}F$ (30A, 334), with one Mn site.

CADMIUM NIOBIUM OXYFLUORIDE
$Cd_{0.25}Nb_6(O,F)_{16}$

Z. anorg. Chem., 580, 95-102.

Pmma, 16.686, 3.951, 8.907, Z = 2, R = 0.033. The structure contains units of one $Nb(O,F)_7$ pentagonal bipyramid and five $Nb(O,F)_6$ octahedra, with channels along c containing Cd (occupancies = 0.39 and 0.11 for the 4(j) and 2(d) sites, respectively). Nb-O = 1.86-2.11, Cd-O = 2.33-3.16 A. The Mn compound is isostructural (preceding report).

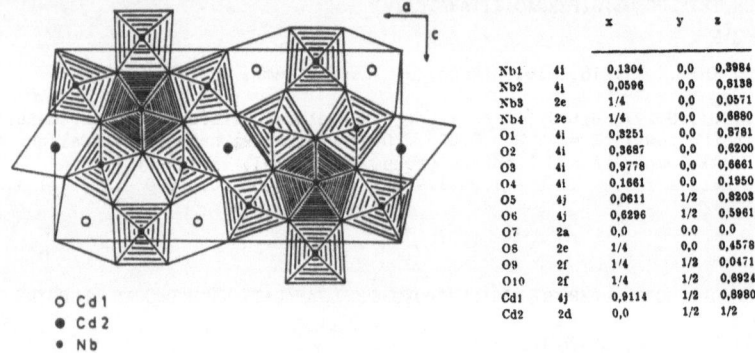

○ Cd 1
● Cd 2
• Nb

		x	y	z
Nb1	4i	0,1304	0,0	0,3984
Nb2	4i	0,0596	0,0	0,8138
Nb3	2e	1/4	0,0	0,0571
Nb4	2e	1/4	0,0	0,6880
O1	4i	0,3251	0,0	0,8781
O2	4i	0,3687	0,0	0,6200
O3	4i	0,9778	0,0	0,6661
O4	4i	0,1661	0,0	0,1950
O5	4j	0,0611	1/2	0,8203
O6	4j	0,6296	1/2	0,5961
O7	2a	0,0	0,0	0,0
O8	2e	1/4	0,0	0,4578
O9	2f	1/4	1/2	0,0471
O10	2f	1/4	1/2	0,8924
Cd1	4j	0,9114	1/2	0,8980
Cd2	2d	0,0	1/2	1/2

POTASSIUM DIFLUOROOCTAMOLYBDATE HEXAHYDRATE
$K_6[Mo_8O_{26}F_2].6H_2O$

Acta Cryst., C46, 2249-2251.

P$\bar{1}$, 10.497, 10.403, 8.001, 107.16, 95.77, 105.53, Z = 1, R = 0.025. Anions consisting of eight edge-sharing MoO_6 octahedra are linked by KO_8F and KO_8 polyhedra, and by hydrogen bonding via the water molecules. Mo-O = 1.70-2.41, Mo-F = 1.97 A.

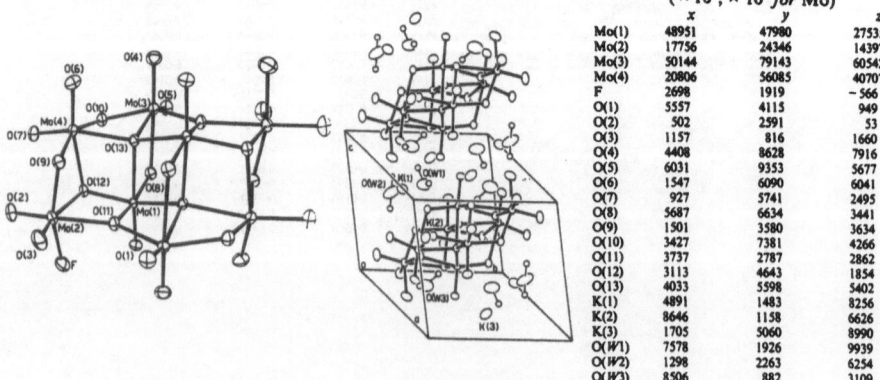

$(\times 10^4, \times 10^5 \text{ for Mo})$

	x	y	z
Mo(1)	48951	47980	27535
Mo(2)	17756	24346	14397
Mo(3)	50144	79143	60542
Mo(4)	20806	56085	40707
F	2698	1919	-566
O(1)	5557	4115	949
O(2)	502	2591	53
O(3)	1157	816	1660
O(4)	4408	8628	7916
O(5)	6031	9353	5677
O(6)	1547	6090	6041
O(7)	927	5741	2495
O(8)	5687	6634	3441
O(9)	1501	3580	3634
O(10)	3427	7381	4266
O(11)	3737	2787	2862
O(12)	3113	4643	1854
O(13)	4033	5598	5402
K(1)	4891	1483	8256
K(2)	8646	1158	6626
K(3)	1705	5060	8990
O(W1)	7578	1926	9939
O(W2)	1298	2263	6254
O(W3)	8506	882	3109

CAESIUM ZINC MOLYBDENUM OXYFLUORIDE
$CsZnMoO_3F_3$ (I)

RUBIDIUM NIOBIUM TUNGSTEN OXIDE
$Rb_{0.3}Nb_{0.3}W_{0.7}O_3$ (II)

RUBIDIUM GALLIUM TUNGSTEN OXIDE
$Rb_{0.3}Ga_{0.1}W_{0.9}O_3$ (III)

Z. anorg. Chem., **582**, 131-142.

I, Fd3m, 10.390, Z = 8, R = 0.059. Cs in 8(b): 3/8,3/8,3/8; Zn/Mo in 16(c):
0,0,0; O/F in 48(f): 0.3165,1/8,1/8. Defect pyrochlore structure, as for
$RbNiCrF_6$ (**38A**, 205).

II, Cmm2, 21.96, 12.66, 3.889, Z = 18, R = 0.037. Hexagonal tungsten bronze
superstructure.

III, P6mm, 7.260, 3.760, Z = 3, R = 0.040. W/Ga in 3(c): 1/2,0,0.5; 0.9 Rb in
1(a): 0,0,0.070; O(1) in 6(e): 0.211,0.789,0.495; O(2) in 3(c): 1/2,0,0.061.
Hexagonal tungsten bronze structure.

SODIUM NEPTUNIUM OXYFLUORIDE
$Na_4Np_{12}F_{46}O_3$

Acta Cryst., C**46**, 2272.

$P6_3/mmc$, 8.022, 16.513, Z = 1, R = 0.054. The material was described previously
as $NaNp_3F_{13}$, (**50A**, 115), but the F(7) site is now reassigned as Na, and F(5) as
oxygen.

	occupation (B_{eq})	x (×10⁴)	y (×10⁴)	z (×10⁴)				
Np		1621	3241	1128	Np—F(3)	2,282 (6) ×2	Np—F(2)	2,307 (10) ×2 (Å)
Na(1)	0·66	3333	6667	9145	Np—F(1)	2,288 (6)	Np—F(5)	2,345 (10) ×2
Na(2)	0·72	6667	3333	2500	Np—O	2,301 (8)	Np—F(4)	2,575 (10)
F(1)		1394	2787	2500	Na(1)—F(4)	2,291 (38)	Na(1)—F(5)	2,540 (15) ×3
F(2)		5376	0751	8361	Na(2)—F(5)	2,886 (11) ×6		
F(3)		3286	0	0				
F(4)		3333	6667	0532				
F(5)		1657	3313	8532				
O	0·75	0	0	0840				

SODIUM OXYCHLORIDE
Na_3OCl

Acta Cryst., C**46**, 736-738.

Pm3m, 4.496, Z = 1, R = 0.027. O in 1(a): 0,0,0; Cl in 1(b): 1/2,1/2,1/2; Na
in 3(d): 1/2,0,0. Antiperovskite structure, as previously described (**55A**,
303).

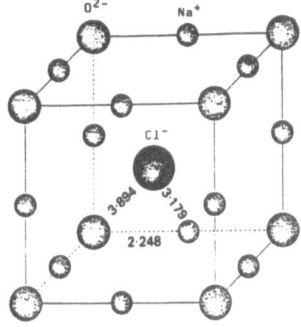

BARIUM INDIUM OXYCHLORIDE
$Ba_3In_2O_5Cl_2$

Z. anorg. Chem., <u>584</u>, 125-128.

I4/mmm, 4.236, 24.929, Z = 2, R = 0.028. Sr_3TiO_7-type (<u>22</u>, 308; <u>24</u>, 440), with
(001) double layers of corner-sharing InO_5Cl octahedra, linked by BaO_{12} and
BaO_4Cl_5 polyhedra.   In-O = 2.05, 2.14, In-Cl = 3.07(1) A.

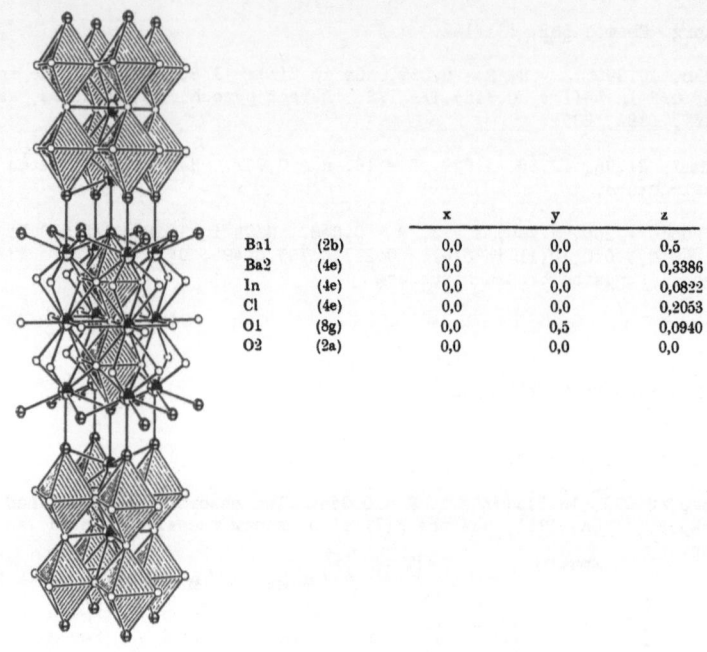

		x	y	z
Ba1	(2b)	0,0	0,0	0,5
Ba2	(4e)	0,0	0,0	0,3386
In	(4e)	0,0	0,0	0,0822
Cl	(4e)	0,0	0,0	0,2053
O1	(8g)	0,0	0,5	0,0940
O2	(2a)	0,0	0,0	0,0

BARIUM BISMUTH(III) OXYCHLORIDE
$Ba_3BiO_3Cl_3$

Z. anorg. Chem., <u>582</u>, 25-29.

Pnma, 7.049, 11.994, 12.089, Z = 4, R = 0.049. Framework of BaO_3Cl_5 and BaO_3Cl_4
polyhedra, with tunnels containing Bi^{3+} ions which have one-sided
(3+2)-coordination (lone-pair). Bi-3 O = 2.02, 2.07 (x 2), Bi-2 Cl = 3.61 A.

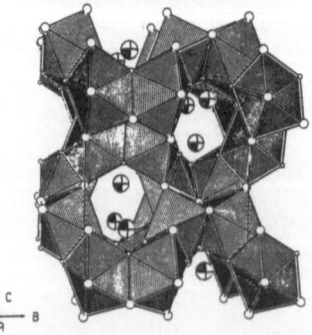

		x	y	z
Bi	(4c)	0,0948	0,25	0,1268
Ba1	(8d)	0,3314	0,5683	0,3872
Ba2	(4c)	0,1451	0,25	0,5245
Cl1	(4c)	0,5867	0,75	0,2669
Cl2	(8d)	0,3902	0,4955	0,1352
O1	(8d)	0,428	0,372	0,445
O2	(4c)	0,781	0,25	0,500

COPPER(I) AND SILVER BISMUTH OXYCHLORIDE
$[M_5Cl][Bi_{48}O_{59}Cl_{30}]$ (M = Cu, Ag)

Acta Chem. Scand., 44, 111-122.

$P\bar{6}2m$ subcells, a = 20.0248, 20.0893, c = 3.8680, 3.8589 A, Z = 1/2 (true cells
have doubled c axes), R = 0.052, 0.058. The Bi-containing part of the structures
resembles the $Bi_4O_5Cl_2$ (42A, 206) and $Bi_{24}O_{31}Cl_{10}$ structures (9, 219); the M_5Cl
parts show disorder, with 3-coordinate Cu but 5-coordinate Ag.

TITANIUM(III) OXYCHLORIDE
TiOCl

Ž. Neorg. Khim., 35, 1945-1946 [Russ. J. Inorg. Chem., 35, 1108-1109].

Pmmn, 3.786, 3.361, 8.045, Z = 2, R = 0.045. Ti in 2(b): 1/4,3/4,0.88068; O in
2(a): 3/4,3/4,0.9446; Cl in 2(a): 1/4,1/4,0.66804. FeOCl-type (3, 67).

LANTHANUM TITANIUM OXYCHLORIDE
$La_3TiO_4Cl_5$

Z. anorg. Chem., 591, 107-117.

Pnma, 16.760, 4.0991, 14.634, Z = 4, R = 0.048. TiO_5 trigonal bipyramids (but Ti
and one O are disordered), linked by three LaO_2Cl_7 tricapped trigonal prisms.

TETRATELLURIUM TETRACHLOROOXOMOLYBDATE(V)
$Te_4(MoOCl_4)_2$

Z. Naturforsch., 45B, 1610-1614.

$P2_1/n$, 11.929, 6.837, 10.498, 90.08, Z = 2, R = 0.013. Square Te_4^{2+} cations and
dimeric $(MoOCl_4)_2^{2-}$ anions (two octahedra sharing a Cl...Cl edge), linked by
Te...Cl and O interactions. Te-Te = 2.695 A.

BARIUM RUTHENIUM(V) OXYCHLORIDE
$Ba_5Ru_2O_9Cl_2$

Z. anorg. Chem., 587, 39-46.

Pnma, 15.310, 5.945, 14.197, Z = 4, R = 0.046. The structure contains Ru_2O_9
groups of two face-sharing octahedra, Cl^- ions, and Ba ions with 8- to
11-coordinations to O and Cl. Ru-O = 1.85-2.10(2) A.

		x	y	z
Ba1	4c	0,0698	0,2500	0,9400
Ba2	4c	0,6262	0,2500	0,9801
Ba3	4c	0,2406	0,2500	0,1459
Ba4	4c	0,5006	0,2500	0,2687
Ba5	4c	0,2936	0,2500	0,8289
Ru1	4c	0,8034	0,2500	0,1252
Ru2	4c	0,8452	0,2500	0,9386
Cl1	4c	0,9937	0,2500	0,6877
Cl2	4c	0,4282	0,2500	0,0528
O1	4c	0,423	0,2500	0,444
O2	4c	0,255	0,2500	0,648
O3	4c	0,183	0,2500	0,329
O4	8d	0,2732	0,478	0,479
O5	8d	0,3412	0,475	0,294
O6	8d	0,4085	0,013	0,615

CALCIUM COPPER OXYCHLORIDE and OXYBROMIDE
$Ca_3Cu_2O_4Cl_2$, $Ca_3Cu_2O_4Br_2$

Mater. Res. Bull., 25, 1085-1090.

I4/mmm, a = 3.861, 3.865, c = 21.349, 23.620 A, Z = 2, powder data. A structure
is derived by adding an additional CuO_2 layer to the $Ca_2CuO_2X_2$ structure (43A,
150). A further neutron diffraction study is planned.

$Ca_3Cu_2O_4Cl_2$

J. Solid State Chem., 84, 178-181.

I4/mmm, 3.863, 21.364, Z = 2, powder data. Structure similar to that of $Sr_3Ti_2O_7$
(22, 308; 24, 440), but with oxygen vacancies, and Cl atoms occupying apical
sites of doubled Cu coordination octahedra.

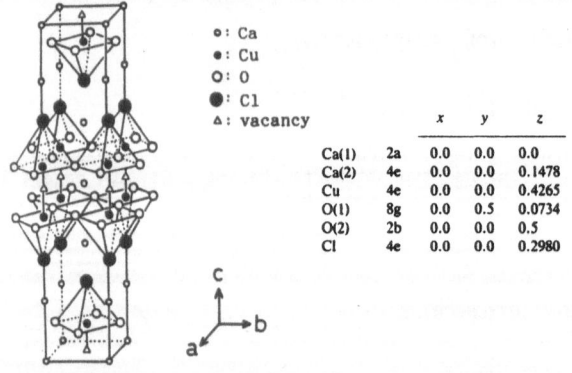

o : Ca
● : Cu
O : O
● : Cl
△ : vacancy

		x	y	z
Ca(1)	2a	0.0	0.0	0.0
Ca(2)	4e	0.0	0.0	0.1478
Cu	4e	0.0	0.0	0.4265
O(1)	8g	0.0	0.5	0.0734
O(2)	2b	0.0	0.0	0.5
Cl	4e	0.0	0.0	0.2980

BARIUM COPPER OXYCHLORIDE
$Ba_{41}Cu_{44}O_{84}Cl_2$

Kristallografija, 35, 324-327 [Soviet Physics - Crystallography, 35, 181-183].

Im3m, 18.27, Z = 2, R = 0.039. Framework with cluster of edge-sharing CuO_4
square planes and CuO_5 square pyramids; Cl and 8- and 9-coordinated Ba are in
cavities.

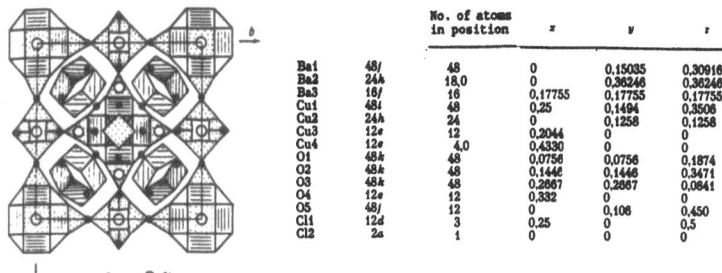

	No. of atoms in position		x	y	z
Ba1	48f	48	0	0,15035	0,30916
Ba2	24h	18,0	0	0,36246	0,36246
Ba3	16f	16	0,17755	0,17755	0,17755
Cu1	48i	48	0,25	0,1494	0,3508
Cu2	24h	24	0	0,1258	0,1258
Cu3	12e	12	0,2044	0	0
Cu4	12e	4,0	0,4330	0	0
O1	48k	48	0,0756	0,0756	0,1874
O2	48k	48	0,1446	0,1446	0,3471
O3	48k	48	0,2667	0,2667	0,0841
O4	12e	12	0,332	0	0
O5	48f	12	0	0,106	0,450
Cl1	12d	3	0,25	0	0,5
Cl2	2a	1	0	0	0

COPPER INDIUM OXYCHLORIDE
$Cu_6O_8 \cdot InCl$

Jpn. J. Appl. Phys., Part 2, 29, L1796-1798.

Fm3m, 9.1661, Z = 4, powder data. Cu in 24(d): 0,1/4,1/4; In in 4(b): 1/2,1/2,1/2; Cl in 4(a): 0,0,0; O in 32(f): x,x,x, x = 0.3590. Structure similar to that of $Ag_6O_8 \cdot AgNO_3$, (11, 348; 17, 510; 28, 178; 30A, 308; 38A, 305), with a framework of face-sharing 26-sided Cu_6O_8 polyhedra, with Cl at the centre, and In in cubic spaces between the polyhedra; Cu has distorted square-planar coordination. Cu-4 O = 1.92, Cu...Cl = 3.24, In-O = 2.24 A.

LANTHANUM CERIUM TANTALUM OXYCHLORIDE
$(La, Ce)_{3.18}TaO_6Cl_{3.27}$

Z. anorg. Chem., 589, 139-157.

Cm, 35.283, 5.429, 9.517, 98.92, Z = 8, R = 0.061. The material was obtained by heating $Ln_2CeTaO_6Cl_3$ (this volume, p. 323), and the main structural features are preserved: chains of face-sharing Cl_6 octahedra, and TaO_6 trigonal prisms.

CADMIUM HYDROXIDE CHLORIDE
$\beta\text{-}Cd_2(OH)_3Cl$

Kristallografija, 35, 995-997 [Soviet Physics-Crystallography, 35, 585-586].

Pnam, 6.800, 9.908, 7.423, Z = 4, R = 0.043. Atacamite-type (12, 220; 53A, 114).

	x	y	z
Cd1	0	0	0
Cd2	0,1915	1/4	0,2610
Cl	0,189	1/4	0,5388
O1(OH)	0,444	0,052(1)	0,283
O2(OH)	0,179	0,25	0,026
H1	0,097	0,074	0,691
H2	0,366	0,25	-0,002

SODIUM OXIDE BROMIDE
Na_4OBr_2

Z. Naturforsch ., 45B, 105-106.

I4/mmm, 4.521, 14.908, Z = 2, R = 0.028. Na(1) in 4(c): 0,1/2,0; Na(2) in 4(e): 0,0,0.1519; O in 2(a): 0,0,0; Br in 4(e): 0,0,0.3481. Anti-K_2NiF_4-type (17, 332; 19, 323), as for the iodide (56A, 111, 276; this volume, p.141). Na-O = 2.26, Na-Br = 2.93-3.20 A.

POTASSIUM OXYBROMIDE
K_4OBr_2

Acta Cryst., C46, 1359-1360.

I4/mmm, 5.145, 16.527, Z = 2, R = 0.043. Anti-K_2NiF_4-type, as for Na_4OI_2 (56A, 111; this volume, p. 141).

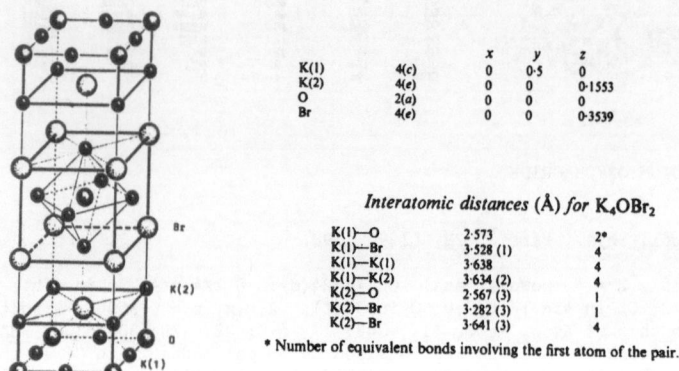

		x	y	z
K(1)	4(c)	0	0·5	0
K(2)	4(e)	0	0	0·1553
O	2(a)	0	0	0
Br	4(e)	0	0	0·3539

Interatomic distances (Å) for K_4OBr_2

K(1)—O	2·573	2*
K(1)—Br	3·528 (1)	4
K(1)—K(1)	3·638	4
K(1)—K(2)	3·634 (2)	4
K(2)—O	2·567 (3)	1
K(2)—Br	3·282 (3)	1
K(2)—Br	3·641 (3)	4

* Number of equivalent bonds involving the first atom of the pair.

BARIUM COPPER OXYBROMIDE and OXYIODIDE
Ba_2CuO_2Br, Ba_2CuO_2I

J. Less-Common Metals, 158, 311-317.

Bromide, R3̄m, 4.281, 29.417, Z = 3, R = 0.029. Ba in 6(c): 0,0,0.7475; Cu in 3(b): 0,0,1/2; O in 6(c): 0,0,0.439; Br in 3(a): 0,0,0. Isostructural with the chloride (43A, 150). Iodide, P2$_1$/m, 11.040, 4.370, 13.980, 91.290, Z = 4, R = 0.055. Both structures contain nearly-linear O-Cu(I)-O^{3-} ions, linked by BaO_3X_3 octahedra. Cu-O = 1.80 A.

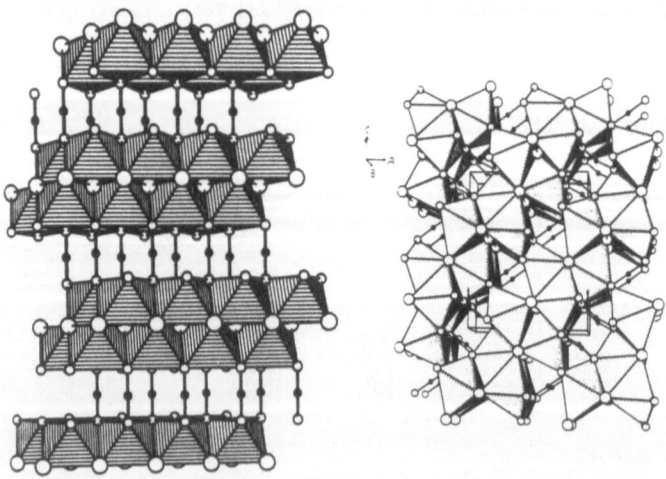

BARIUM COPPER OXIDE BROMIDE IODIDE
$Ba_2CuO_2Br_{0.5}I_{0.5}$

J. Less-Common Metals, 158, L33-L35.

$R\bar{3}m$, a = 4.294, c = 29.973 A, Z = 3, R = 0.069. Ba in 6(c): 0,0,0.7474; Cu in 3(b): 0,0,1/2; O in 6(c): 0,0,0.441; Br/I in 3(a): 0,0,0. Isostructural with Ba_2CuO_2Cl (43A, 150) and related compounds (56A, 110; this volume, preceding report). Cu-2 O = 1.78(6), Ba-3 O = 2.60, Ba-3 Br/I = 3.47 A.

SODIUM OXIDE IODIDE
Na_4OI_2

I4/mmm, 4.655, 15.940, Z = 2, R = 0.045. Anti-K_2NiF_4 type structure (49A, 93), as previouslydescribed (56A, 111, 276; see also the bromide (this volume, page 139)), in contrast to different stoichiometries of Na_3OCl and Na_3OBr (anti-perovskite types (55A, 303)).

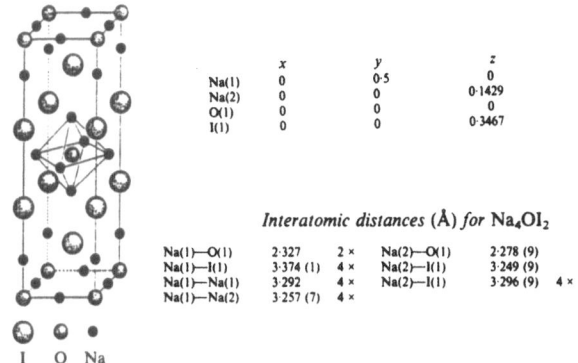

	x	y	z
Na(1)	0	0·5	0
Na(2)	0	0	0·1429
O(1)	0	0	0
I(1)	0	0	0·3467

Interatomic distances (Å) *for* Na_4OI_2

Na(1)—O(1)	2·327	2 ×	Na(2)—O(1)	2·278 (9)	
Na(1)—I(1)	3·374 (1)	4 ×	Na(2)—I(1)	3·249 (9)	
Na(1)—Na(1)	3·292	4 ×	Na(2)—I(1)	3·296 (9)	4 ×
Na(1)—Na(2)	3·257 (7)	4 ×			

I   O   Na

CHROMIUM THIOBROMIDE
CrSBr

Z. anorg. Chem., 585, 157-167.

Pmmn, 4.767, 3.506, 7.965, at 293K, Z = 2, R = 0.026, 0.027, 0.020 at 293, 205, 118K. Cr in 2(b): 3/4,1/4,0.1270; S in 2(a): 3/4,3/4,-0.0782, Br in 2(a): 3/4,3/4,0.3497 (at 293K, similar values at 205, 118K). FeOCl-type (3, 67; 53A, 114), with layers of edge-sharing CrS_4Br_2 distorted octahedra.

POTASSIUM RHENIUM SELENIDE CHLORIDE
$KRe_6Se_5Cl_9$

Mater. Res. Bull., 25, 1227-1234.

Pn3, 13.034, Z = 4, R = 0.040. K in 4(c): 1/2,0,0; Re in 24(h): 0.13121,0.45780,0.53193; (0.54Se + 0.46Cl)   in 8(e): x,x,x, x = 0.1364; (0.62Se + 0.38Cl) in 24(h): 0.4457,0.3044,0.8829; Cl in 24(h): 0.2988,0.4016,0.5755. Isostructural with MMo_6Cl_{14} (37A, 201; 53A, 96), with discrete $Re_6L_8Cl_6{}^-$ clusters (L = Se/Cl) and K^+ ions. Re-Re = 2.61 A (cf. 2.75 A in Re metal).

COPPER TELLURIDE CHLORIDE
CuTeCl

Z. anorg. Chem., **586**, 175-184.

$I4_1/amd$, 15.609, 4.795, Z = 16, R = 0.047. 4Cu(1) in 4(b): 0,1/4,3/8; 12Cu(2)
in 16(g): 0.1273,1/4-x,7/8; 16Te in 16(f): 0.13750,0,1/2; 16Cl in 16(h):
0,0.1197,0.1236. Isostructural with the bromide and iodide (**42**A, 216, 217; **45**A,
207; **49**A, 123), but with different Cu distribution. The structure contains Te_4
helices (Te-Te = 2.758 Å; compare 2.835 Å in elemental Te) and chains of
edge-sharing empty and Cu(1)-filled tetrahedra along $\underline{c}$. Cu(1)-Cl = 2.365 (x 4),
Cu(2)-Cl = 2.318(3) (x 2), Cu(2)-Te = 2.631(1) (x 2) Å.

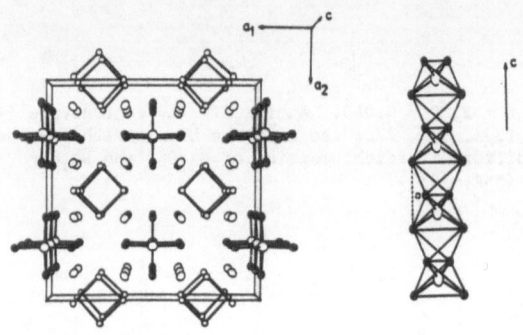

TELLUROIODIDES
Al_3Te_3I, Ga_3Te_3I, In_3Te_3I

Z. Naturforsch., **45**B, 667-678.

Pnma, a = 11.254, 11.168, 11.641, b = 4.100, 4.102, 4.302, c = 19.826, 19.529,
19.974 Å, Z = 4, R = 0.020, 0.045, 0.060. The structures contain ETe_3I tetrahedra
and Te_3-E-E-Te_3 double tetrahedral units (E = Al, Ga, In), linked by sharing
Te...Te edges. Related chlorides and bromides are isostructural.

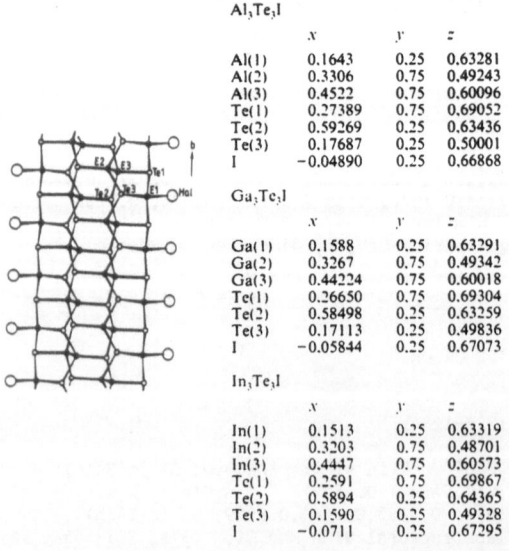

Al_3Te_3I

	x	y	z
Al(1)	0.1643	0.25	0.63281
Al(2)	0.3306	0.75	0.49243
Al(3)	0.4522	0.75	0.60096
Te(1)	0.27389	0.75	0.69052
Te(2)	0.59269	0.25	0.63436
Te(3)	0.17687	0.25	0.50001
I	−0.04890	0.25	0.66868

Ga_3Te_3I

	x	y	z
Ga(1)	0.1588	0.25	0.63291
Ga(2)	0.3267	0.75	0.49342
Ga(3)	0.44224	0.75	0.60018
Te(1)	0.26650	0.75	0.69304
Te(2)	0.58498	0.25	0.63259
Te(3)	0.17113	0.25	0.49836
I	−0.05844	0.25	0.67073

In_3Te_3I

	x	y	z
In(1)	0.1513	0.25	0.63319
In(2)	0.3203	0.75	0.48701
In(3)	0.4447	0.75	0.60573
Te(1)	0.2591	0.75	0.69867
Te(2)	0.5894	0.25	0.64365
Te(3)	0.1590	0.25	0.49328
I	−0.0711	0.25	0.67295

SODIUM OXIDE CYANIDE
Na$_3$O(CN)

Z. Naturforsch., 45B, 407-408.

Pm3m, 4.56, Z = 1, powder data. Na in 3(d): 1/2,0,0; O in 1(a): 0,0,0; CN in
1(b): 1/2,1/2,1/2. Anti-perovskite type, as for the chloride and bromide (55A,
303), with rotating cyanide ion.

Z. anorg. Chem., 591, 41-46.

Pm3m, 4.543, Z = 1, R = 0.046. Na in 3(d): 1/2,0,0; O in 1(a): 0,0,0: 1 C in
6(f): x,1/2,1/2, x = 0.370; 1 N in 6(f): x = 0.387. Anti-perovskite structure
with disordered cyanide ion.

LANTHANUM HEXACYANOCHROMATE(III) PENTAHYDRATE
LaCr(CN)$_6$.5H$_2$O

Acta Cryst., C46, 1994-1996.

P6$_3$/m, 7.7053, 14.8155, Z = 2, R = 0.024. Isostructural with the Fe compound
(39A, 204; 46A, 223), with Cr(CN)$_6$$^{3-}$ octahedra linked by LaN$_6$(H$_2$O)$_3$ tricapped
trigonal prisms, and two uncoordinated water molecules above the triangular faces
of the prisms. Cr-Cr = 2.065(4) A.

		x	y	z
La	2(c)	⅓	⅔	¼
Cr	2(b)	0	0	0
C	12(i)	0·1092	0·2516	0·0807
N	12(i)	0·2199	0·8302	0·1254
O1	4(f)	⅓	⅔	0·9097
O2	6(h)	0·4890	0·4365	¼

NITROPRUSSIC ACID HYDRATE
(H$_3$O)$_2$[Fe(CN)$_5$NO].H$_2$O

J. Phys. Chem. Solids, 51, 381-386.

P2$_1$, 6.327, 11.214, 8.471, 83.12, Z = 2, R = 0.031. Octahedral anions and two
oxonium cations, one of which is hydrogen bonded to the water molecule. Fe-N =
1.619, Fe-C = 1.914-1.976(4) A.

POTASSIUM NITROPRUSSIDE MONOHYDRATE
K$_2$Fe(CN)$_5$NO.H$_2$O

Acta Cryst., C46, 1577-1578.

The structure is now described in Pnam [as previously suggested in 56A, 114] as a
monohydrate (rather than 0.8 hydrate). [But see following report.]

POTASSIUM SODIUM NITROPRUSSIDE HYDRATE
$K_9Na[Fe(CN)_5NO]_5 \cdot 4 \cdot 8H_2O$

Acta Cryst., C46, 1761-1763.

Reinvestigation of the structure of 'potassium nitroprusside 0.8 hydrate' (56A, 114). New data: Pnma, 30.000, 16.053, 11.272, Z = 4, R = 0.039. One K site has been replaced by Na, water molecules added, and occupancies of water sites adjusted.

CAESIUM POTASSIUM HEXACYANOFERRATE(III)
$Cs_2K[Fe(CN)_6]$

J. Chem. Soc., Dalton, 3597-3604.

$P2_1/n$, a = 11.041, 11.145, b = 8.146, 8.131, c = 7.596, 7.664 A, β = 90.49, 90.16°, at 85, 295K, Z = 2, R = 0.023, 0.032. Structure as previously described (43A, 162; 45A, 209). Charge-density study.

BARIUM NITROPRUSSIDE TRIHYDRATE
$Ba[Fe(CN)_5NO].3H_2O$

J. Solid State Chem., 89, 23-30.

Pbcm, 7.620, 19.394, 8.631, Z = 4, R = 0.060. Octahedral anions linked by 9-coordinate Ba ions; two water sites are disordered. The material was previously described as a dihydrate in $Pbc2_1$ ($Pca2_1$), (39A, 204).

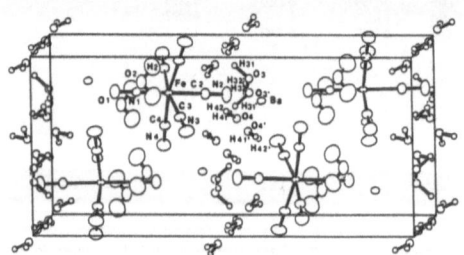

SAMARIUM HEXACYANOFERRATE(III) TETRAHYDRATE
ERBIUM HEXACYANOFERRATE(III) TETRAHYDRATE
$LnFe(CN)_6.4H_2O$ (Ln = Sm, Er)

Acta Cryst., C46, 724-726.

The structures are orthorhombic (56A, 115) and not monoclinic (55A, 137). In addition the material described as the Er compound (56A, 115) is probably the Sm compound (55A, 137).

ERBIUM HEXACYANOFERRATE(III) TETRAHYDRATE
$ErFe(CN)_6 \cdot 4H_2O$

Acta Cryst., C46, 1992-1994.

Cmcm, 7.3212, 12.7576, 13.5636, Z = 4, R = 0.020.   Structure as previously
described (56A, 115; this volume, preceding report).

		x	y	z
Er	4(c)	0	0·3234	¼
Fe	4(a)	0	0	0
C1	16(h)	0·3112	0·4517	0·0900
C2	8(f)	0	0·1366	0·0602
N1	16(h)	0·1987	0·4208	0·1417
N2	8(f)	0	0·2179	0·1004
O1	8(g)	0·2593	0·2190	¼
O2	8(f)	0	0·6561	0·0997

TETRAMMINEPLATINUM(II) TETRAMMINEBIS(HEXACYANOFERRATE(II))PLATINATE(IV)
NONAHYDRATE
$[Pt(NH_3)_4]_2[(NC)_5FeCNPt(NH_3)_4NCFe(CN)_5] \cdot 9H_2O$

Inorg. Chem., 29, 2456-2460.

$P2_1/c$, 15.313, 8.5353, 16.206, 100.52, Z = 2, R = 0.050.   Square-planar cations
and trinuclear cyanide-bridged anions, linked by N-H...NC hydrogen bonding and by
an elaborate system of hydrogen bonds involving the water molecules.

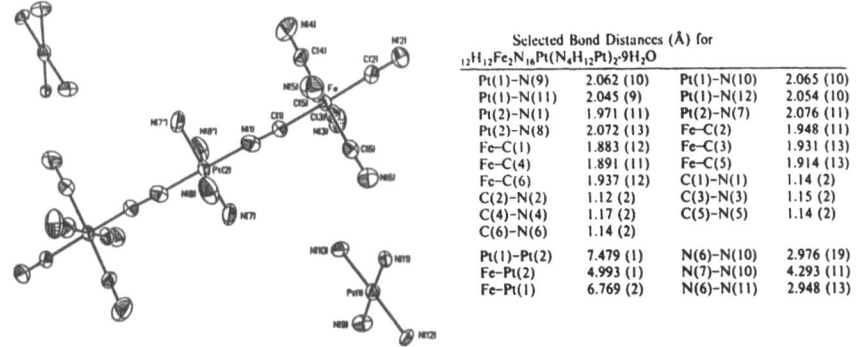

Selected Bond Distances (Å) for
$_{12}H_{12}Fe_2N_{16}Pt(N_4H_{12}Pt)_2 \cdot 9H_2O$

Pt(1)-N(9)	2.062 (10)	Pt(1)-N(10)	2.065 (10)
Pt(1)-N(11)	2.045 (9)	Pt(1)-N(12)	2.054 (10)
Pt(2)-N(1)	1.971 (11)	Pt(2)-N(7)	2.076 (11)
Pt(2)-N(8)	2.072 (13)	Fe-C(2)	1.948 (11)
Fe-C(1)	1.883 (12)	Fe-C(3)	1.931 (13)
Fe-C(4)	1.891 (11)	Fe-C(5)	1.914 (13)
Fe-C(6)	1.937 (12)	C(1)-N(1)	1.14 (2)
C(2)-N(2)	1.12 (2)	C(3)-N(3)	1.15 (2)
C(4)-N(4)	1.17 (2)	C(5)-N(5)	1.14 (2)
C(6)-N(6)	1.14 (2)		
Pt(1)-Pt(2)	7.479 (1)	N(6)-N(10)	2.976 (19)
Fe-Pt(2)	4.993 (1)	N(7)-N(10)	4.293 (11)
Fe-Pt(1)	6.769 (2)	N(6)-N(11)	2.948 (13)

AMMINECOPPER(II) TETRACYANONICKELATE
$[Cu(NH_3)_4]_2[Cu(NH_3)_2]_2[Ni(CN)_4]_4$

Chem. Pap., 44, 13-19.

Pnnm, 15.772, 16.838, 7.333, Z = 2, R = 0.062.   The structure contains
octanuclear centrosymmetric molecules.   There is a central ring of two $Cu(NH_3)_2$
and two $Ni(CN)_4$ groups, with four of the CN groups bridging; the ring Cu atoms
are each bonded in addition to a terminal $-NC-Ni(CN)_2-CN-Cu(NH_3)_4$ group.   Ni
atoms have square-planar coordinations, the ring Cu has distorted
trigonal-bypyramidal and the terminal Cu distorted square-pyramidal
coordination.

SILVER DICYANAMIDE (ORTHORHOMBIC)
$AgN(CN)_2$

Acta Cryst., C46, 2297-2299.

Pnma, 16.133, 3.612, 5.983, Z = 4, R = 0.086. The structure contains infinite
-Ag-N≡C-N-C≡N-Ag- chains along a, similar to those in the trigonal form (43A,
170). Ag-N = 2.08, 2.14(2) A.

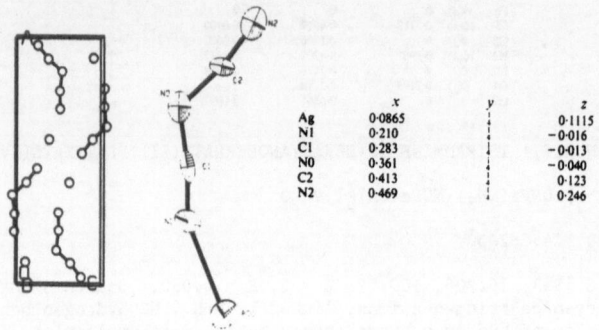

	x	y	z
Ag	0·0865	¼	0·1115
N1	0·210	¼	−0·016
C1	0·283	¼	−0·013
N0	0·361	¼	−0·040
C2	0·413	¼	0·123
N2	0·469	¼	0·246

RUBIDIUM DICYANOAURATE(I)
$RbAu(CN)_2$

Z. Naturforsch., 45B, 629-634.

C2/c, 18.450, 8.998, 13.418, 108.0, Z = 16, R = 0.035. Layers of linear $Au(CN)_2^-$
anions arranged in 4- and 8-membered rings, with holes containing 7- and
8-coordinate Rb ions. Au-C = 1.98-2.06(2) A.

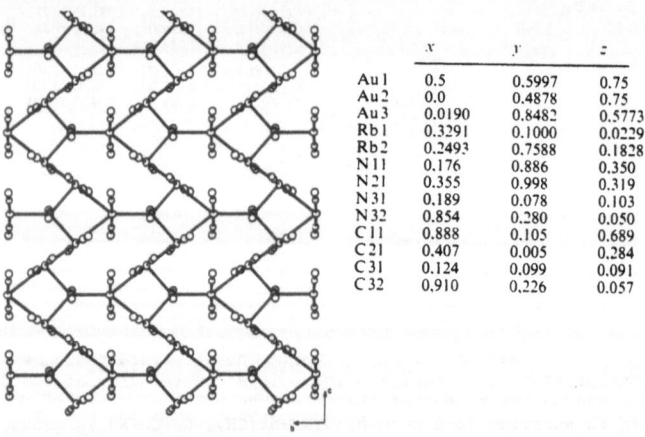

	x	y	z
Au1	0.5	0.5997	0.75
Au2	0.0	0.4878	0.75
Au3	0.0190	0.8482	0.5773
Rb1	0.3291	0.1000	0.0229
Rb2	0.2493	0.7588	0.1828
N11	0.176	0.886	0.350
N21	0.355	0.998	0.319
N31	0.189	0.078	0.103
N32	0.854	0.280	0.050
C11	0.888	0.105	0.689
C21	0.407	0.005	0.284
C31	0.124	0.099	0.091
C32	0.910	0.226	0.057

ZINC CYANIDE   CADMIUM CYANIDE
$Zn(CN)_2$ $Cd(CN)_2$

J. Amer. Chem. Soc., 112, 1546-1554.

$P\bar{4}3m$, a = 5.900, 6.301 A, Z = 2, R = 0.028, 0.036. M(1) in l(a): 0,0,0; M(2) in l(b): 1/2,1/2,1/2; C in 4(e): x,x,x, x = 0.1882,0.1923; N in 4(e): x = 0.3007, 0.2988. Structures as previously described (10, 92), with M(1) coordinated to 4C, M(2) to 4N, and linear M(1)CNM(2) groupings. Zn-C = 1.923(6), Zn-N = 2.037(5); Cd-C = 2.099(5), Cd-N = 2.196(4) A.

ICE I_h
H_2O

J. Chem. Phys., 93, 1412-1417.

$P6_3/mmc$, a = 4.511, 4.506, c = 7.349, 7.346 A, for two crystals, Z = 4, R = 0.007, 0.016. 4 O in 4(f): 1/3,2/3,z, z = 0.0618, 0.0632; 2 H(1) in 4(f): z = 0.173, 0.198; 6 H(2) in 12(k): 0.437,0.873,0.024 and 0.437,0.874,0.028. Structure as previously described (1, 174; 52, 170). O-H = 0.85, 0.82(3) A. Electron-density study.

CLATHRATE HYDRATE
$3.5Xe.8CC\ell_4.136D_2O$

Acta Cryst., B46, 390-399.

$Fd3m$, a = 17.192, 17.240 A, at 13, 100K, Z = 1, neutron radiation, R = 0.041, 0.047. Clathrate hydrate II structure (30B, 268), with disordered hydrogen bonds. Xe atoms occupy statistically 22% of the dodecahedra, and the $CC\ell_4$ molecules exhibit large librations about the centre of the hexakaidecahedron.

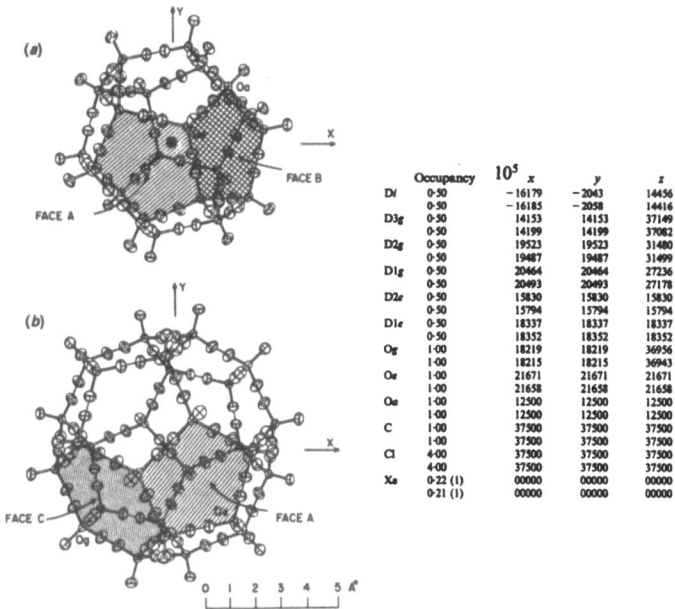

	Occupancy	10^5 x	y	z
Df	0·50	-16179	-2043	14456
	0·50	-16185	-2058	14416
D3g	0·50	14153	14153	37149
	0·50	14199	14199	37082
D2g	0·50	19523	19523	31480
	0·50	19487	19487	31499
D1g	0·50	20464	20464	27236
	0·50	20493	20493	27178
D2f	0·50	15830	15830	15830
	0·50	15794	15794	15794
D1e	0·50	18337	18337	18337
	0·50	18352	18352	18352
Og	1·00	18219	18219	36956
	1·00	18215	18215	36943
Oe	1·00	21671	21671	21671
	1·00	21658	21658	21658
Oa	1·00	12500	12500	12500
	1·00	12500	12500	12500
C	1·00	37500	37500	37500
	1·00	37500	37500	37500
Cl	4·00	37500	37500	37500
	4·00	37500	37500	37500
Xe	0·22 (1)	00000	00000	00000
	0·21 (1)	00000	00000	00000

The dodecahedron of $(D_2O)_{20}$ (a) and hexakaidecahedron of $(D_2O)_{28}$ (b), with D atoms shown in disordered positions.

ALUMINA
Al_2O_3

Mater. Res. Bull., <u>25</u>, 611-621.

γ-Form, slight tetragonal distortion of the cubic spinel type (<u>2</u>, 43, 313).

δ-Form, $P\bar{4}m2$, 5.599, 23.657, Z = 16, powder data. A structural model is proposed based on the spinel type with cation vacancies in the octahedral sites.

σ-Al_2O_3

Neues Jb. Miner. Mh., 217-226.

Fd3m, 7.948, Z = 32/3, R = 0.034. 8 Al(1) in 8(a): 0,0,0; 13.3 Al(2) in 16(d): 5/8,5/8,5/8; O in 32(e): x,x,x, x = 0.381. Spinel structure (<u>1</u>, 350), with vacancies in the octahedral site. Most crystals show satellite reflexions suggesting a complex modulation structure.

VANADIUM DIOXIDE (FORM A)
VO_2

J. Solid State Chem., <u>86</u>, 116-124.

$P4_2/nmc$, 8.4336, 7.6782, Z = 16, powder data. Three-dimensional framework of edge- and corner-sharing VO_6 octahedra.

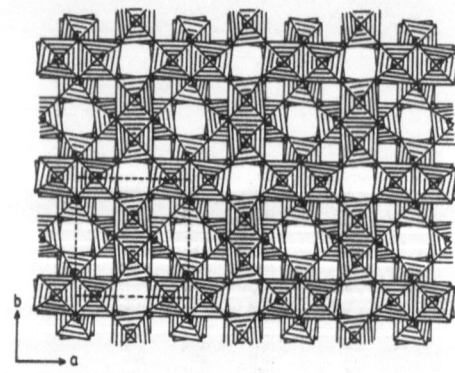

		x	y	z
V	16j	0.18936	0.01764	0.11787
O(1)	16j	0.1674	0.0012	0.3743
O(2)	8i	0.1634		0.3416
O(3)	8i	0.1352		0.8930

INTERATOMIC DISTANCES (Å) AND ANGLES (°) OF VO_6 OCTAHEDRON

V–O(1)^i	1.982(3)	V–O(1)^{ii}	1.885(3)
V–O(1)^{iii}	2.029(2)	V–O(1)^{iv}	2.227(2)
V–O(2)	1.993(2)	V–O(3)	1.784(2)
O(1)^i–V–O(1)^{iii}	86.6(7)	O(1)^i–V–O(1)^{iv}	81.8(8)
O(1)^i–V–O(2)	81.3(8)	O(1)^i–V–O(3)	99.4(9)
O(1)^{ii}–V–O(1)^{iii}	89.9(5)	O(1)^{ii}–V–O(1)^{iv}	84.6(5)
O(1)^{ii}–V–O(2)	99.2(7)	O(1)^{ii}–V–O(3)	94.4(9)
O(1)^{iii}–V–O(1)^{iv}	83.0(5)	O(1)^{iv}–V–O(2)	84.4(4)
O(2)–V–O(3)	85.2(4)	O(3)–V–O(1)^{iii}	107.7(8)

MANGANESE(II, III) OXIDE (HIGH-PRESSURE)
Mn_3O_4

Amer. Min., <u>75</u>, 1249-1252.

Pmab, 9.5564, 9.7996, 3.0240, Z = 4, powder data. Mn(II) in 4(d): 3/4,0.1461,0.6845; Mn(III) in 8(e): 0.0696,0.1147,0.2034; O(1) in 4(c): 0,1/4,0.6204; O(2) in 4(d): 1/4,0.1999,0.1899; O(3) in 8(e): 0.1112,0.9694,0.7956. Isostructural with marokite, $CaMn_2O_4$ (<u>29</u>, 321; <u>31A</u>, 136), with some differences as a result of the substitution of the smaller Mn(II) for Ca. Mn(II)-8 O = 2.12-2.60, Mn(III)-6 O = 1.91-2.32 A.

COPPER(II) OXIDE (TENORITE)   SILVER(I,III) OXIDE
CuO   AgO

SILVER SULPHATE
Ag$_2$SO$_4$

J. Solid State Chem., 89, 184-190.

CuO, C2/c, 4.6833, 3.4208, 5.1294, 99.567, at 11K, Z = 4, neutron powder data.
Cu in 4(c): 1/4,1/4,0; O in 4(e): 0,0.4179,1/4. Structure as previously
described (3, 11, 239; 35A, 207). Cu-O = 1.951, 1.961(1) A (each x 2).

AgO, P2$_1$/c, 5.8517, 3.4674, 5.4838, 107.663, at 9K, Z = 4, neutron powder data
for mixture with Ag$_2$SO$_4$. Ag(I) in 2(a): 0,0,0; Ag(III) in 2(d): 1/2,0,1/2; O
in 4(e): 0.2949,0.3470,0.2187. Structure as previously described (22, 302; 23,
328; 24, 327; 55A, 143). Ag(I)-O = 2.147(3) (x 2), 2.693(3) (x 2), Ag(III)-O =
2.008(3), 2.036(3) (each x 2) A.

Ag$_2$SO$_4$, Fddd, 5.7960, 12.6670, 10.2238, at 9K, Z = 8, neutron powder data for
mixture with AgO. Ag in 16(g): 1/8,1/8,0.4437; S in 8(a): 1/8,1/8,1/8; O in
32(h): -0.0192,0.0571,0.2112. Structure as previously described (2, 89, 425;
44A, 269).

URANIUM DIOXIDE
UO$_{2.11}$, UO$_{2.13}$

J. Solid State Chem., 84, 52-57.

Fm3m, a not given [~5.5 A], Z = 4, neutron radiation, R = 0.031, 0.027.
Fluorite-type structure, as previously described (28, 129), with anion vacancies,
and interstitial oxygen in 1/2,x,x and x,x,x, x = 0.36.

POTASSIUM RUBIDIUM OXIDE
KRbO$_6$

Z. anorg. Chem., 590, 48-54.

P2$_1$/c, 6.370, 5.930, 8.677, 122.83, Z = 2, R = 0.06. Isostructural with RbO$_3$
(40A, 175), with disordered K/Rb.

POTASSIUM TRIBERYLLATE
K$_4$Be$_3$O$_5$

Z. anorg. Chem., 591, 199-208.

C2/c, 10.381, 7.228, 10.788, 118.36, Z = 4, R = 0.051. Infinite -OBeO$_2$BeO$_2$Be-
chains with one tetrahedral and two trigonal planar Be atoms, linked by
6-coordinate K ions. Be-O = 1.68 (tetrahedral), 1.51-1.55 (trigonal) A.

STRONTIUM DIALUMINATE
$Sr_3Al_2O_6$

Inorg. Chem., 29, 4768-4771.

Pa3, 15.8425, Z = 24, neutron powder data. Intricate perovskite superstructure, isostructural with the Ca compound (41A, 221).

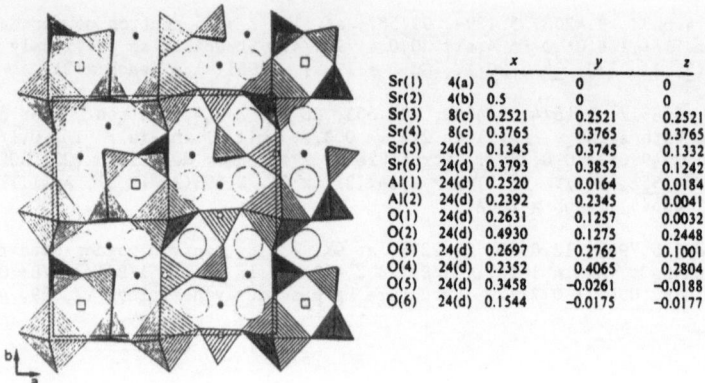

		x	y	z
Sr(1)	4(a)	0	0	0
Sr(2)	4(b)	0.5	0	0
Sr(3)	8(c)	0.2521	0.2521	0.2521
Sr(4)	8(c)	0.3765	0.3765	0.3765
Sr(5)	24(d)	0.1345	0.3745	0.1332
Sr(6)	24(d)	0.3793	0.3852	0.1242
Al(1)	24(d)	0.2520	0.0164	0.0184
Al(2)	24(d)	0.2392	0.2345	0.0041
O(1)	24(d)	0.2631	0.1257	0.0032
O(2)	24(d)	0.4930	0.1275	0.2448
O(3)	24(d)	0.2697	0.2762	0.1001
O(4)	24(d)	0.2352	0.4065	0.2804
O(5)	24(d)	0.3458	-0.0261	-0.0188
O(6)	24(d)	0.1544	-0.0175	-0.0177

STRONTIUM ALUMINATE, GALLATE, and FERRATE
$SrM_{12}O_{19}$ (M = Al, Ga, Fe)

J. Solid State Chem., 87, 186-194.

$P6_3/mmc$, a = 5.5666, 5.7929, 5.8836, c = 22.0018, 22.8123, 23.0376 A, Z = 2, R = 0.033, 0.032, 0.028. Magnetoplumbite-type structures (6, 74; 55A, 145), with the five-coordinate M(2) site probably split into two positions.

BARIUM INDIUM ALUMINATE
Ba_2InAlO_5

Z. anorg. Chem., 591, 174-180.

$P6_3/mmc$, 5.781, 19.625, Z = 4, R = 0.076. New structure type with a framework of tetrahedral $O_3Al-O-AlO_3$ groups and InO_6 octahedra; Ba ions have 12-, 9-, and (6+3)-coordinations.

LEAD HEXAALUMINATE
$PbAl_{12}O_{19}$

J. Solid State Chem., 85, 318-320.

$P6_3/mmc$, 5.5711, 22.045, Z = 2, R = 0.043. Magnetoplumbite structure, as for the Sr compound (41A, 222).

		Number per unit cell	x	$y = 2x$	z
Pb	2d	2	⅓		¼
Al(1)	12k	12	0.8317	0.1080	
Al(2)	4f	4	⅓	0.02809	
Al(3)	4f	4	⅓	0.19013	
Al(4)	2a	2	0	0	
Al(5)	4e	2	0	0.2406	
O(1)	12k	12	0.1545	0.0519	
O(2)	12k	12	0.5016	0.1471	
O(3)	4f	4	⅓	0.0541	
O(4)	4e	4	0	0.1475	
O(5)	6h	6	0.1823	¼	

	Number of bonds	Distance (Å)
Octahedral coordination		
Al(1)—O(1)	2	1.992(7)
—O(2)	2	1.811(7)
—O(3)	1	1.987(6)
—O(4)	1	1.847(6)
Al(3)—O(2)	3	1.880(7)
—O(5)	3	1.966(8)
Al(4)—O(1)	6	1.879(6)
Tetrahedral coordination		
Al(2)—O(1)	3	1.804(7)
—O(3)	1	1.812(10)
Polyhedron 5-coordinated		
Al(5)—O(4)	1	2.044(14)
—O(4)'	1	2.458(14)
—O(5)	3	1.771(11)
Polyhedron 12-coordinated		
Pb —O(2)	6	2.772(5)
—O(5)	6	2.790(9)

BARIUM SCANDIUM ALUMINATE
$\alpha-Ba_2ScAlO_5$

Kristallografija, 35, 213-214 [Soviet Physics - Crystallography, 35, 129-130].

$P6_3/mmc$, 5.7733, 14.530, Z = 3, R = 0.051. Hexagonal packing of six BaO_3 layers (hcchcc), with Sc having octahedral and Sc/Al distorted octahedral coordination; one O site has half-occupancy. Sc-O = 2.12, Sc/Al-O = 1.79, 2.24, Ba-12 O = 2.88-3.01 A.

	Population	x	y	z
Ba (1)	2b 1	0	0	1/4
Ba (2)	4f 1	¹/₃	²/₃	0,58269
Sc	2a 1	0	0	0
O(1)	12k 1	0,169	0,339	0,0879
ScAl(¹/₄Sc+ +³/₄Al)	4f 1	¹/₃	²/₃	0,1385
O(2)	6h 0,5	0,489	0,977	¹/₄

YTTRIUM ALUMINATE (Nd-DOPED)
$Y_3Al_5O_{12}$

Kristallografija, 35, 1488-1491 [Soviet Physics - Crystallography, 35, 876-878].

Ia3d, 12.003, Z = 8, R = 0.019, 0.021 for two laser-irradiated specimens. Garnet structure (52A, 180); oxygen parameters: (0.4695,-0.0507,0.3508).

BARIUM STRONTIUM COPPER β-ALUMINA
$(Ba,Sr)CuAl_{10}O_{17}$

Z. anorg. Chem., 582, 21-24.

$P6_3/mmc$, 5.604, 22.678, Z = 2, R = 0.052. β-Alumina-type structure (5, 72), with blocks of AlO_6 octahedra, and Cu/Al and Al in tetrahedral sites. Cu/Al-O = 1.85, 1.88, Al-O = 1.71, 1.77 (tetrahedral), 1.83-1.98 (octahedral), Ba/Sr-9 O = 2.79, 3.24 A.

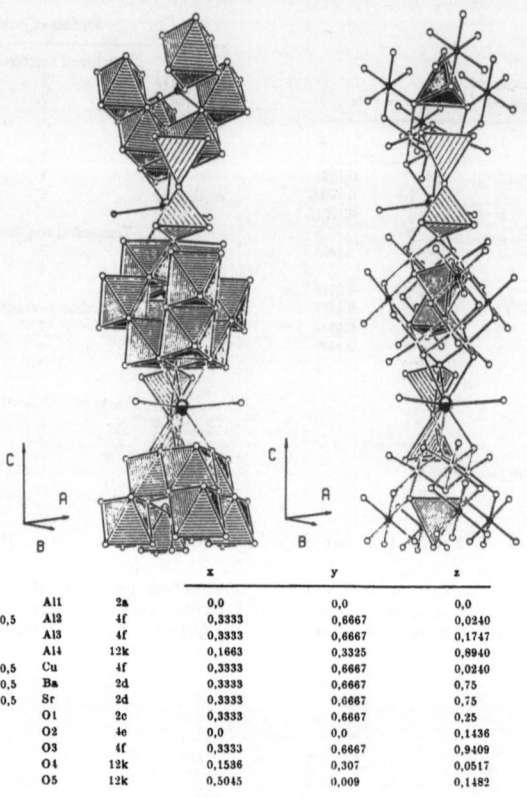

			x	y	z
	Al1	2a	0,0	0,0	0,0
0,5	Al2	4f	0,3333	0,6667	0,0240
	Al3	4f	0,3333	0,6667	0,1747
	Al4	12k	0,1663	0,3325	0,8940
0,5	Cu	4f	0,3333	0,6667	0,0240
0,5	Ba	2d	0,3333	0,6667	0,75
0,5	Sr	2d	0,3333	0,6667	0,75
	O1	2c	0,3333	0,6667	0,25
	O2	4e	0,0	0,0	0,1436
	O3	4f	0,3333	0,6667	0,9409
	O4	12k	0,1536	0,307	0,0517
	O5	12k	0,5045	0,009	0,1482

SILVER β-ALUMINA
$AgA\ell_{11}O_{17}$

J. Phys.: Condens. Matter, 2, 2335-2344.

$P6_3/mmc$, a = 5.5913, 5.6026, 5.6169, c = 22.5293, 22.5489, 22.5973 A at 298, 573, 773K, Z = 2, neutron powder data. Structure as previously described (38A, 247; 46A, 233; 48A, 389), but with increasingly diffuse silver ion distribution as the temperature is increased.

LANTHANUM GADOLINIUM MAGNESIUM ALUMINATE
$La_{0.4}Gd_{0.6}MgA\ell_{11}O_{19}$

J. Chem. Phys., 93, 7076-7084.

$P6_3/mmc$, 5.565, 21.89, Z = 2, R = 0.079. Magnetoplumbite-type structure (6, 74), with Mg/Aℓ disorder, and ordered but partially-occupied La/Gd sites.

		T	$x \times 10^4$	y	z			T	$x \times 10^4$	y	z
La(1)	2d	0.43	6667	3333	2500	O(1)	12j	0.47	1994	3632	2500
Gd(1)	12j	0.077	6727	3755	2500	O(2)	12k	1	1533	3062	533
Al,Mg(1)	2a	1	0000	0000	0000	O(3)	12k	0.94	5053	106	1515
Al,Mg(2)	4f	1	3333	6667	273	O(4)	4e	1	0000	0000	1511
Al,Mg(3)	4f	1	3333	6667	1901	O(5)	4f	1	6667	3333	578
Al,Mg(4)	12k	1	8322	6644	1085						
Al,Mg(5)	4e	0.5	0000	0000	2419						

T: occupancy factors)

CAESIUM LITHOGALLATE
$Cs_2[Li_3GaO_4]$

Z. anorg. Chem., <u>581</u>, 159-172.

Ibam, 5.004, 10.051, 11.885, Z = 4, R = 0.044. Layers of corner-sharing LiO_4 and GaO_4 tetrahedra, linked by (8+2)-coordinate Cs ions. Li-O = 1.99-2.05, Ga-O = 1.87, Cs-O = 3.23-3.31, 3.86 A. The Rb compound is isostructural.

		$x\,10^4$	y	z
Cs	8 j	2390	3720	0
Li(1)	8 g	0	2601	2500
Li(2)	4 b	0	5000	2500
Ga	4 a	5000	5000	2500
O	16 k	2316	1169	1754

POTASSIUM GALLATE
$KGaO_2$

J. Solid State Chem., <u>87</u>, 114-123.

Pcab, 11.07, 5.51, 15.81, Z = 16, re-evaluation of the data of <u>41</u>A, 419, R = 0.105; comparison with the K_2ZnGeO_4 structure (this volume, p. $\overline{1}59$) suggests that <u>x</u> for O(4) should be 0.98 (rather than 0.48).

BARIUM LANTHANUM GALLATE
BARIUM NEODYMIUM GALLATE
$BaLaGaO_4$, $BaNdGaO_4$

STRONTIUM LANTHANUM GALLATE
$SrLaGaO_4$

Z. anorg. Chem., <u>584</u>, 119-124.

Ba compounds, $P2_12_12_1$, a = 10.0438, 9.9514, b = 7.2713, 7.1371, c = 5.9120, 5.8543 A, Z = 4, R = 0.041, 0.038. $\beta-K_2SO_4$-related structures (<u>2</u>, 86), with ordered Ba/Ln distribution.

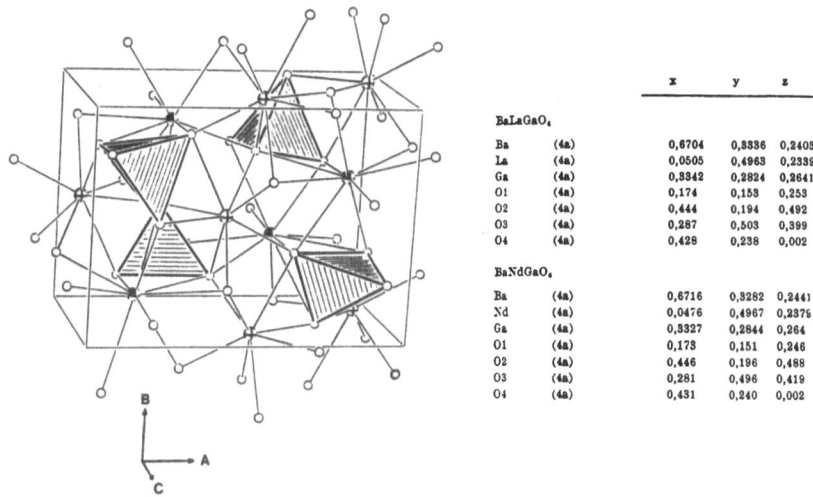

$BaLaGaO_4$		x	y	z
Ba	(4a)	0,6704	0,3336	0,2403
La	(4a)	0,0505	0,4963	0,2339
Ga	(4a)	0,3342	0,2824	0,2641
O1	(4a)	0,174	0,153	0,253
O2	(4a)	0,444	0,194	0,492
O3	(4a)	0,287	0,503	0,399
O4	(4a)	0,428	0,238	0,002

$BaNdGaO_4$				
Ba	(4a)	0,6716	0,3282	0,2441
Nd	(4a)	0,0476	0,4967	0,2379
Ga	(4a)	0,3327	0,2844	0,264
O1	(4a)	0,173	0,151	0,246
O2	(4a)	0,446	0,196	0,488
O3	(4a)	0,281	0,496	0,419
O4	(4a)	0,431	0,240	0,002

Sr compound, I4/mmm, 3.8520, 12.680, Z = 2, R = 0.051. Sr/La in 4(e):
0,0,0.3588; Ga in 2(a): 0,0,0; O(1) in 4(e): 0,0,0.168; O(2) in 4(c): 0,1/2,0.
K_2NiF_4-type (<u>17</u>, 332; <u>19</u>, 323; 49A, 93).

SODIUM INDATE(III)
β-Na_5InO_4

Z. anorg. Chem., <u>583</u>, 24-30.

Pmmn, 7.610, 7.387, 5.478, Z = 2, R = 0.049. Isostructural with β-Li_5AlO_4 (<u>37A</u>,
227; <u>43A</u>, 178). In-4 O = 2.06, 2.08, Na-4 O = 2.31-2.51 A.

		x	y	z
In	2a	,2500	,2500	,2215
Na1	8g	,5563	,5111	,2554
Na2	2b	,2500	,7500	,1970
O1	4e	,2500	,0143	,4212
O2	4f	,0196	,2500	,0188

POTASSIUM SODIUM INDATE(III)
$K_3Na_2InO_4$

Mh. Chem., <u>121</u>, 853-864.

$P2_1/n$, 10.126, 9.699, 7.254, 91.02, Z = 4, R = 0.073. InO_4^{3-} tetrahedra, linked
into a three-dimensional framework by NaO_4 distorted tetrahedra, and K^+ ions with
(4+1)-, (4+2)-, and (5+1)-coordinations. In-O = 2.05-2.09 A.

BARIUM INDATE
$Ba_8In_6O_{17}$

J. Less-Common Metals, <u>157</u>, 71-76.

I4/mmm, 4.1725, 30.487, Z = 1, R = 0.083. The structure can be derived from that
of $Sr_4Ti_3O_{10}$-type (<u>22</u>, 308) by removing some oxygen. In-6 O = 2.05-2.42, Ba-9 or
12 O = 2.54-3.12 A.

		x	y	z
Ba1	(4e)	0,0	0,0	0,4241
Ba2	(4e)	0,0	0,0	0,2990
In1	(2a)	0,0	0,0	0,0
In2	(4e)	0,0	0,0	0,1465
O1	(4e)	0,0	0,0	0,216
O2	(4e)	0,0	0,0	0,079
O3	(4c)[a]	0,0	0,5	0,0
O4	(8g)[a]	0,0	0,5	0,40

[a] Occupancy = 0.75

BARISM STRONTIUM DIINDATES(III)
$Ba_2SrIn_2O_6$ (I), $Ba_{0.07}Sr_{0.93}In_2O_4$ (II)

Z. anorg. Chem., 588, 117-122.

I, I4/mmm, 4.168, 21.290, Z = 2, R = 0.026. M(1) in 2(b): 0,0,1/2; M(2) in
4(e): 0,0,z, z = 0.31789; In in 4(e): z = 0.09875; O(1) in 8(g): 0,1/2,z, z =
0.0791; O(2) in 4(e): z = 0.1962; 2M(1) = 0.5Ba + 1.5Sr, 4M(2) in 3.5Ba + 0.5Sr.
Isostructural with $La_2SrCu_2O_6$ (46A, 291), with Ba/Sr disorder.

II, Pnma, 9.858, 3.273, 11.520, Z = 4, R = 0.040. Atoms in 4(c): x,1/4,z, x =
0.7556, 0.4193, 0.4301, 0.207, 0.119, 0.523, 0.421, z = 0.6524, 0.1064, 0.6110,
0.164, 0.479, 0.781, 0.427, for M(1), In(1), In(2), O(1-4); 4M(1) = 0.28Ba +
3.72Sr. $CaFe_2O_4$-type (21, 290), as for the Sr compound (39A, 213).

LITHIUM PENTAGERMANATE
$Li_4Ge_5O_{12}$

Acta Cryst., C46, 2021-2026.

P$\bar{1}$, 5.120, 9.143, 9.586, 72.95, 77.74, 78.81, Z = 2, R = 0.025. Quasi-close-
packed planes of O atoms with octahedrally-coordinated Li and tetrahedral and
octahedral coordinations for Ge. Li-O = 1.99-2.32, Ge-O = 1.72-1.78
(tetrahedral), 1.82-2.00 (octahedral) A.

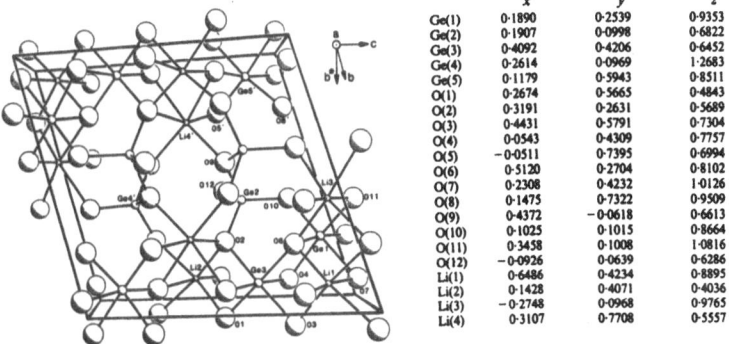

	x	y	z
Ge(1)	0·1890	0·2539	0·9353
Ge(2)	0·1907	0·0998	0·6822
Ge(3)	0·4092	0·4206	0·6452
Ge(4)	0·2614	0·0969	1·2683
Ge(5)	0·1179	0·5943	0·8511
O(1)	0·2674	0·5665	0·4843
O(2)	0·3191	0·2631	0·5689
O(3)	0·4431	0·5791	0·7304
O(4)	0·0543	0·4309	0·7757
O(5)	−0·0511	0·7395	0·6994
O(6)	0·5120	0·2704	0·8102
O(7)	0·2308	0·4232	1·0126
O(8)	0·1475	0·7322	0·9509
O(9)	0·4372	−0·0618	0·6613
O(10)	0·1025	0·1015	0·8664
O(11)	0·3458	0·1008	1·0816
O(12)	−0·0926	0·0639	0·6286
Li(1)	0·6486	0·4234	0·8895
Li(2)	0·1428	0·4071	0·4036
Li(3)	−0·2748	0·0968	0·9765
Li(4)	0·3107	0·7708	0·5557

SODIUM ENNEAGERMANATE
$Na_4Ge_9O_{20}$

Acta Cryst., C46, 1202-1204.

$I4_1/a$, 15.0263, 7.3971, Z = 4, R = 0.026. Structure as previously described (28,
147), with tetrahedrally- and octahedrally-coordinated Ge, and (4+3)-coordinated
Na.

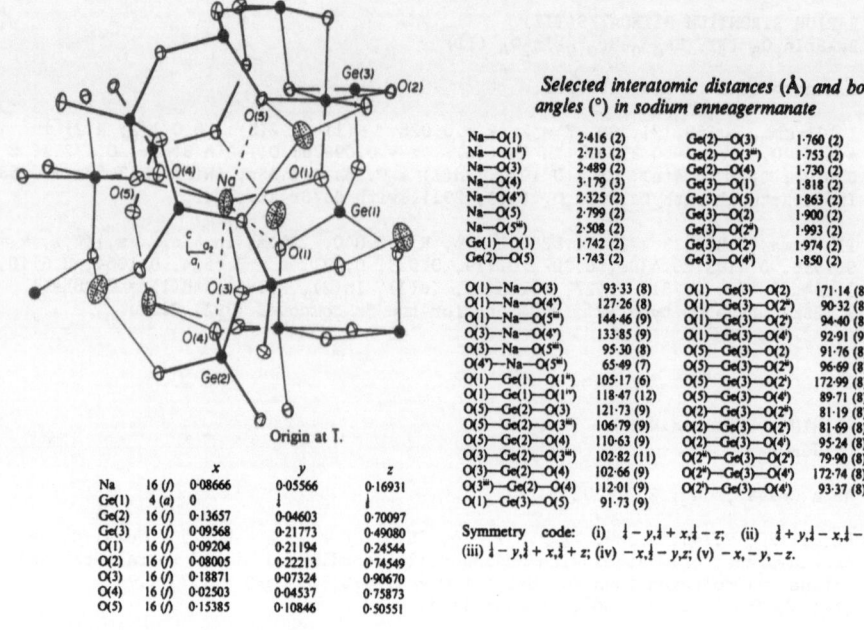

Selected interatomic distances (Å) and bond angles (°) in sodium enneagermanate

Na—O(1)	2·416 (2)	Ge(2)—O(3)	1·760 (2)
Na—O(1^{ii})	2·713 (2)	Ge(2)—O(3^{iii})	1·753 (2)
Na—O(3)	2·489 (2)	Ge(2)—O(4)	1·730 (2)
Na—O(4)	3·179 (3)	Ge(3)—O(1)	1·818 (2)
Na—O(4^{iv})	2·325 (2)	Ge(3)—O(5)	1·863 (2)
Na—O(5)	2·799 (2)	Ge(3)—O(2)	1·900 (2)
Na—O(5^{iii})	2·508 (2)	Ge(3)—O(2^{ii})	1·993 (2)
Ge(1)—O(1)	1·742 (2)	Ge(3)—O(2^i)	1·974 (2)
Ge(2)—O(5)	1·743 (2)	Ge(3)—O(4^i)	1·850 (2)

O(1)—Na—O(3)	93·33 (8)	O(1)—Ge(3)—O(2)	171·14 (8)
O(1)—Na—O(4^{iv})	127·26 (8)	O(1)—Ge(3)—O(2^{ii})	90·32 (8)
O(1)—Na—O(5^{iii})	144·46 (9)	O(1)—Ge(3)—O(2^i)	94·40 (8)
O(3)—Na—O(4^{iv})	133·85 (9)	O(1)—Ge(3)—O(4^i)	92·91 (9)
O(3)—Na—O(5^{iii})	95·30 (8)	O(5)—Ge(3)—O(2)	91·76 (8)
O(4^{iv})—Na—O(5^{iii})	65·49 (7)	O(5)—Ge(3)—O(2^{ii})	96·69 (8)
O(1)—Ge(1)—O(1^{ii})	105·17 (6)	O(5)—Ge(3)—O(2^i)	172·99 (8)
O(1)—Ge(1)—O(1^{iii})	118·47 (12)	O(5)—Ge(3)—O(4^i)	89·71 (8)
O(5)—Ge(2)—O(3)	121·73 (9)	O(2)—Ge(3)—O(2^{ii})	81·19 (8)
O(5)—Ge(2)—O(3^{iii})	106·79 (9)	O(2)—Ge(3)—O(2^i)	81·69 (8)
O(5)—Ge(2)—O(4)	110·63 (9)	O(2)—Ge(3)—O(4^i)	95·24 (8)
O(3)—Ge(2)—O(4)	102·82 (11)	O(2^{ii})—Ge(3)—O(2^i)	79·90 (8)
O(3)—Ge(2)—O(4)	102·66 (9)	O(2^{ii})—Ge(3)—O(4^i)	172·74 (8)
O(3^{iii})—Ge(2)—O(4)	112·01 (9)	O(2^i)—Ge(3)—O(4^i)	93·37 (8)
O(1)—Ge(3)—O(5)	91·73 (9)		

Symmetry code: (i) $\tfrac{1}{4}-y,\tfrac{1}{4}+x,\tfrac{1}{4}-z$; (ii) $\tfrac{1}{4}+y,\tfrac{1}{4}-x,\tfrac{1}{4}-z$; (iii) $\tfrac{1}{4}-y,\tfrac{1}{4}+x,\tfrac{1}{4}+z$; (iv) $-x,\tfrac{1}{4}-y,z$; (v) $-x,-y,-z$.

Origin at Ī.

		x	y	z
Na	16 (*f*)	0·08666	0·05566	0·16931
Ge(1)	4 (*a*)	0	$\tfrac{1}{4}$	$\tfrac{1}{8}$
Ge(2)	16 (*f*)	0·13657	0·04603	0·70097
Ge(3)	16 (*f*)	0·09568	0·21773	0·49080
O(1)	16 (*f*)	0·09204	0·21194	0·24544
O(2)	16 (*f*)	0·08005	0·22213	0·74549
O(3)	16 (*f*)	0·18871	0·07324	0·90670
O(4)	16 (*f*)	0·02503	0·04537	0·75873
O(5)	16 (*f*)	0·15385	0·10846	0·50551

SODIUM POTASSIUM ZINC GERMANATE
(Na,K)$_2$ZnGeO$_4$

Mater. Res. Bull., 25, 371–379.

Pca2$_1$, 10.7399, 5.3616, 14.9796, Z = 8, powder data. Filled cristobalite type, with a framework of corner-sharing GeO$_4$ and ZnO$_4$ tetrahedra, and alkali-metal ions in cavities (Na in two sites, Na/K in two sites).

SODIUM CALCIUM GERMANATE
Na$_2$CaGe$_2$O$_6$

Acta Cryst., C46, 544–546.

R$\bar{3}$m, a = 10.788, c = 13.460 Å, Z = 9, R = 0.051. Ring of six corner-sharing GeO$_4$ tetrahedra, as in the silicate analogue (52A, 322), with disordered Ca/Na sites. There is a cubic 3.8 Å subcell. Ge–O = 1.70–1.79(1) Å.

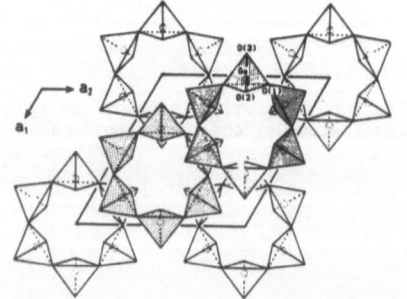

	Occupancy		$\times 10^4$	x	y	z
Ge	1·0			1504·0	−1504·0	5669·8
M(1)	0·35	Ca, 0·65	Na	0	0	2473
M(2)	0·10	Ca, 0·90	Na	5000	0	0
M(3)	0·39	Ca, 0·61	Na	5000	0	5000
M(4)	0·84	Ca, 0·16	Na	0	0	0
O(1)	0·5			2783	235	5267
O(2)	0·5			826	−1451	6799
O(3)	0·5			2697	−2129	5525

BISMUTH GERMANATE
$Bi_4Ge_3O_{12}$

Kristallografija, <u>35</u>, 361-364 [Soviet Physics - Crystallography, <u>35</u>, 204-206].

$I\overline{4}3d$, a = 10.524, 10.540 A, at 293, 573K, Z = 4, neutron radiation, R = 0.021, 0.016. Bi in 16(c): **x,x,x**, x = 0.08751, 0.08763; Ge in 12(a): 3/8,0,1/4; O in 48(e): 0.06951,0.12668,0.28770 and 0.06889,0.12692,0.28766. Eulytite-type structure (<u>2</u>, 122; <u>31</u>A, 227; <u>49</u>A, 177). Ge-4 O = 1.75, Bi-6 O = 2.16, 2.62 A.

CALCIUM ZINC DIGERMANATE
$Ca_2ZnGe_2O_7$ (2 forms), $Ca_2ZnGe_{1.25}Si_{0.75}O_7$

Amer. Min., <u>75</u>, 847-858.

Germanate, high-temperature polymorph, incommensurate structure, average structure is $P\overline{4}2_1m$, 7.950, 5.186 A, Z = 2, R = 0.054. Melilite-type structure (<u>2</u>, 541; <u>17</u>, 574; <u>35</u>A, 455), with layers of ZnO_4 and GeO_4 tetrahedra.

Germanate, low-temperature polymorph, incommensurate structure, average structure is $P2_1$, 8.020, 7.995, 15.506, 89.47°, Z = 4, R = 0.030. Stacking variant of the melilite-type structure.

Germanate/silicate, $P2_1/n$, 9.112, 7.900, 9.380, 114.03, Z = 4, R = 0.017. Tetrahedral sheets with rings of 4, 5, and 6 tetrahedra.

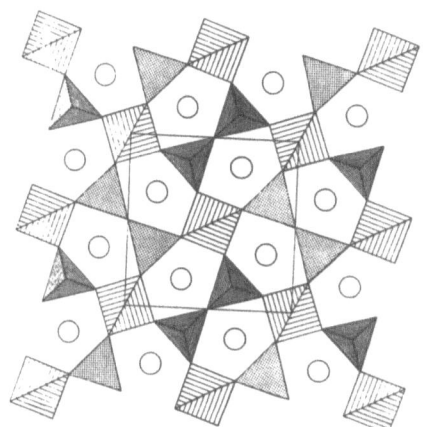

Coordination polyhedron diagram of the melilite structure viewed along the c axis. In the average high-Ca_2ZnGe_2O, structure and the average Ca_2ZnSi_2O, structure, ZnO_4 (T1) tetrahedra are at 0,0,0 and ½,½,0 and connect Ge_2O, and Si_2O, groups (T2). Open circles represent Ca atoms at z = 0.5.

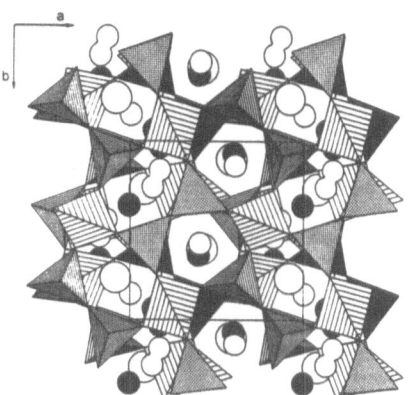

Coordination polyhedron diagram of the average low-Ca_2ZnGe_2O, structure viewed along the c axis. Melilite-type layers are stacked in an ABA'ABA' sequence. Large open circles represent Ca atoms at z ≈ 0, 1, filled small circles Ca atoms at z ≈ ⅓, and open small circles Ca atoms at z ≈ ⅔. The peanut-shaped overlapping circles are due to split Ca atoms with the same z coordinate.

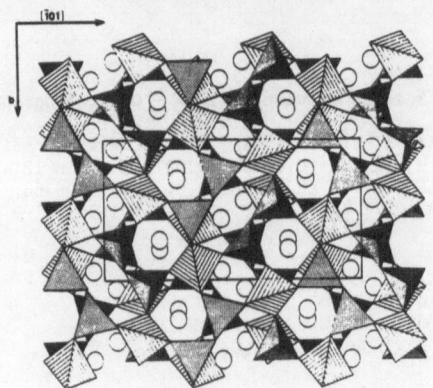

Coordination polyhedron diagram of the structure of $Ca_2ZnGe_{1.25}Si_{0.75}O_7$ (space group $P2_1/n$). The structure is viewed parallel to [101], perpendicular to the tetrahedral layers. The layers (101) are stacked in an ABAB sequence. Pairs of tetrahedra with vertices pointing up or down are occupied by Ge and Si, distorted tetrahedra seen edgewise are occupied by Zn. Open circles represent Ca. Three types of tetrahedral rings are present: five-membered rings, four-membered rings, and extremely elongated six-membered units.

GERMANATES

$NaNbGeO_5$ (I), $NaTaGeO_5$ (II), $LiTaGeO_5$ (III)

Kristallografija, 35, 316-323 [Soviet Physics - Crystallography, 35, 176-180].

I, $P2_1/c$, 6.798, 8.913, 7.523, 115.42, Z = 4, R = 0.026. Isostructural with low-temperature $CaTiSiO_5$ (42A, 416).

II, $C2/c$, 6.843, 8.916, 7.417, 114.77, Z = 4, R = 0.063. Sphene-type structure (2, 117; 6, 111; 42A, 416).

III, $C2/c$, 7.589, 8.130, 7.509, 119.55, Z = 4, R = 0.049. Sphene-type structure.

All three compounds contain chains of MO_6 octahedra (M = Nb or Ta) linked by GeO_4 tetrahedra. Na has 7-coordination and Li 5-coordination.

ZINC GERMANATE (HIGH-PRESSURE)

$ZnGeO_3$

Z. Kristallogr., 191, 93-104.

$R\bar{3}$, a = 4.9568, c = 13.860 A, Z = 6, R = 0.020. Zn in 6(c): 0,0,0.36696; Ge in 6(c): 0,0,0.15597; O in 18(f): 0.3176,0.0297,0.2429. Ilmenite-type structure (3, 69; 44A, 197). Ge-6 O = 1.866, 1.929, Zn-O = 2.001, 2.286(3) A.

POTASSIUM ZINC GERMANATE
K_2ZnGeO_4

J. Solid State Chem., 87, 114-123.

Pca2$_1$, 11.0769, 5.5216, 15.8465, Z = 8, powder data (with a small amount of K_2GeO_3 also present), constrained refinement. The structure derived is a stuffed cristobalite type, with alternating corner-sharing GeO_4 and ZnO_4 tetrahedra, and K ions in cavities.

ERBIUM GERMANATE HYDROXIDE
$Er_{13}Ge_6O_{31}(OH)$

Kristallografija, 35, 642-646 [Soviet Physics - Crystallography, 35, 374-376].

R3 (close to $R\bar{3}$), 15.617, 9.398, Z = 3, R = 0.037. Isostructural with the Gd/F compound (54A, 149).

THULIUM PYROGERMANATE
$Tm_2Ge_2O_7$

J. Phys.: Condens. Matter, 2, 4795-4805.

P4$_1$2$_1$2, 6.764, 12.293, Z = 4, R = 0.028. $O_3Ge-O-GeO_3{}^{6-}$ ions linked by 7-coordinate Tm^{3+} ions. Ge-O = 1.75 (bridging), 1.73-1.77 (terminal), Tm-O = 2.20-2.54 A, Ge-O-Ge = 136°.

CAESIUM STANNATE(IV)
Cs_4SnO_4

Z. anorg. Chem., 587, 145-156.

P2$_1$/c, 11.808, 7.282, 11.667, 111.79, Z = 4, R = 0.070. $SnO_4{}^{4-}$ tetrahedra, linked by 4- to 6-coordinate Cs^+ ions. Sn-O = 1.94-1.97, Cs-O = 2.86-3.62 A.

	x 10^4 y	z
Sn	7584	8754	0083
Cs(1)	4257	1715	7925
Cs(2)	8449	1165	4173
Cs(3)	0633	0892	2042
Cs(4)	6310	1618	6072
O(1)	2158	9882	8396
O(2)	3062	5302	6384
O(3)	6376	8201	4892
O(4)	9139	7329	5251

LEAD(II) TITANIUM(IV) OXIDE
$Pb_{1-x}Ti_xO_{1+x}$ (x = 0-0.08)

Mater. Res. Bull., 25, 979-986.

Mixture of phases: α-PbO-type (1, 89, 93; 52A, 171), P4/nmm, a = 3.978-3.966, c = 5.026-4.977 A, for x = 0-0.08, Z = 2, powder data. O(1) in 2(a): 0,0,0: (1-x)Pb, xTi, and xO(2) in 2(c): 1/2,0,z, z = 0.2345-0.2393, 0.077-0.138, 0.432-0.485, respectively. The additional oxygen is located at the site of the Pb^{2+} lone-pair, and the Ti coordination is similar to that in perovskite-type $PbTiO_3$. β-PbO-type (52A, 171), Pbcm, a = 5.895, 5.895, b = 5.493, 5.491, c = 4.754, 4.755 A, for x = 0.02, 0.04, Z = 4, powder data. Pb, O in 4(d): x,y,1/4, x = 0.2288, 0.2282 for Pb, -0.1346, -0.1383 for O, y = -0.0116, -0.0109 for Pb, 0.0935, 0.0962 for O; TiO_2 does not appear to dissolve in the β-PbO phase.

SILVER PLUMBATE
$Ag_5Pb_2O_6$

J. Less-Common Metals, 161, 17-24.

$P\bar{3}1m$, 5.9324, 6.4105, Z = 1, R = 0.017. Layers of edge-sharing PbO_6 octahedra, with 2- and 3-coordinate Ag ions. Pb-O = 2.219(2), Ag-O = 2.122, 2.286(3) A.

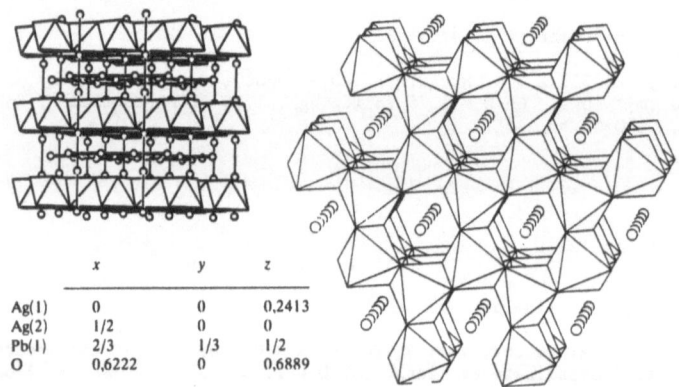

	x	y	z
Ag(1)	0	0	0.2413
Ag(2)	1/2	0	0
Pb(1)	2/3	1/3	1/2
O	0.6222	0	0.6889

LEAD ANTIMONATE
$Pb_2Sb_2O_7$

Kristallografija, 35, 842-846 [Soviet Physics - Crystallography, 35, 494-497].

I2cm, 7.484, 7.857, 10.426, Z = 4, R = 0.033. Deformed weberite type (12, 196; 44A, 142), with SbO_6 octahedra and 8-coordinate Pb.

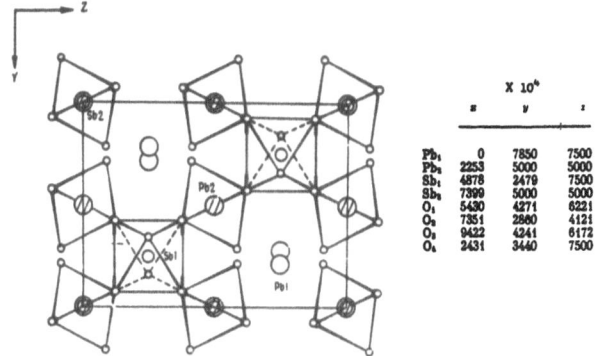

	x	y	z
	\multicolumn{3}{c}{$\times 10^{4}$}	
Pb₁	0	7850	7500
Pb₂	2253	5000	5000
Sb₁	4878	2479	7500
Sb₂	7399	5000	5000
O₁	5430	4271	6221
O₂	7351	2880	4121
O₃	9422	4241	6172
O₄	2431	3440	7500

MANGANESE(II) DIANTIMONATE(V)
$Mn_2Sb_2O_7$

Z. Kristallogr., _190_, 41-46.

$P3_121$, 7.191, 17.402, Z = 6, neutron powder data, structure as previously
proposed (_54_A, 152). Pyrochlore-related structure with 8-coordinate Mn and
6-coordinate Sb. Mn-O = 1.98-2.80, Sb-O = 1.86-2.04(5) A.

		x	y	z
Mn1	3(*a*)	0.856	0	1/3
Mn2	3(*b*)	0.804	0	5/6
Mn3	6(*c*)	0.650	0.138	−0.0058
Sb1	3(*a*)	0.325	0	1/3
Sb2	3(*b*)	0.336	0	5/6
Sb3	6(*c*)	0.516	0.342	0.1642
O1	6(*c*)	0.203	0.228	0.1446
O2	6(*c*)	0.563	0.608	0.1975
O3	6(*c*)	0.195	0.640	0.1464
O4	6(*c*)	−0.049	0.316	0.0527
O5	6(*c*)	−0.036	0.823	0.0547
O6	6(*c*)	0.560	0.400	0.0565
O7	6(*c*)	0.555	0.823	0.0591

BARIUM NICKEL RUTHENIUM ANTIMONY OXIDE
$Ba_3NiRuSbO_9$

Mater. Res. Bull., _25_, 89-93.

$P6_3/mmc$, 5.7856, 14.2321, Z = 2, neutron powder data. Ba(1) in 2(b): 0,0,1/4;
Ba(2) in 4(f): 1/3,2/3,z, z = 0.9097; Ni in 2(a): 0,0,0; Ru/Sb in 4(f): z =
0.1526; O(1) in 6(h): 0.4843,0.9686,1/4; O(2) in 12(k): 0.1705,0.3409,0.4169.
6H-BaTiO₃-type structure(_43_A, 185; _44_A, 195), with Ni in the corner-sharing
octahedra and Ru/Sb in the face-sharing sites. Ni-O = 2.08, Ru/Sb-O = 1.91,
2.05, Ba-12 O = 2.90-2.96(1) A.

LITHIUM ZINC ANTIMONATES
$Li_3Zn_2SbO_6$, $Li_4Zn_{1.5}SbO_6$

Mater. Res. Bull., _25_, 1175-1182.

C2/m, a = 5.259, 5.281, b = 9.036, 9.022, c = 5.209, 5.250 A, β = 110.49,
111.15°, Z = 2, neutron powder data. Cubic close-packed O with the cations
ordered in all the octahedral sites (compare Li_2SbO_6, which has hexagonal
close-packed O, and six of the Li in tetrahedral sites (34A, 273)). Sb-O =
2.00, Zn-O = 2.09-2.14, Li-O = 2.12-2.29 A. The Bi compounds are isostructural.

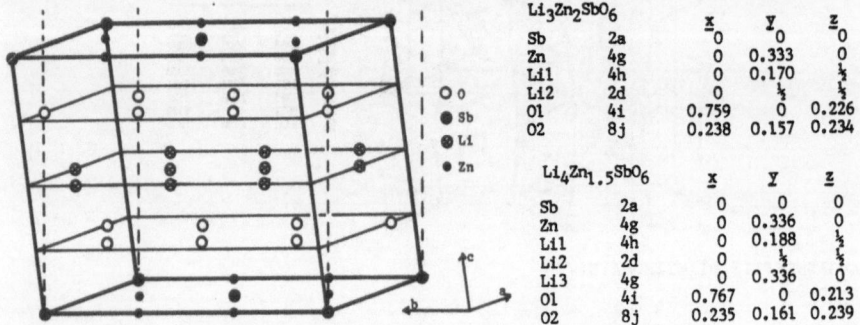

$Li_3Zn_2SbO_6$

		x	y	z
Sb	2a	0	0	0
Zn	4g	0	0.333	0
Li1	4h	0	0.170	½
Li2	2d	0	½	½
O1	4i	0.759	0	0.226
O2	8j	0.238	0.157	0.234

$Li_4Zn_{1.5}SbO_6$

		x	y	z
Sb	2a	0	0	0
Zn	4g	0	0.336	0
Li1	4h	0	0.188	½
Li2	2d	0	½	½
Li3	4g	0	0.336	0
O1	4i	0.767	0	0.213
O2	8j	0.235	0.161	0.239

LANTHANUM LITHIUM ANTIMONATE(V)
La_2LiSbO_6

Mater. Res. Bull., 25, 1271-1277.

$P2_1/n$, 5.6281 [in Abstract, 5.6181 in text], 5.7179, 7.9610, 89.74, Z = 2,
powder data. Monoclinic perovskite (53A, 140; 54A, 190), with ordered cation
distribution. Li-6 O = 1.96-2.07, Sb-6 O = 2.01-2.15, La-8 O = 2.34-2.83(1) A.

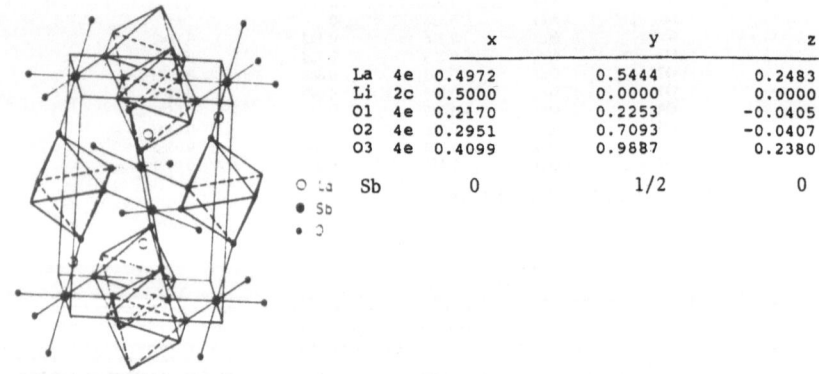

		x	y	z
La	4e	0.4972	0.5444	0.2483
Li	2c	0.5000	0.0000	0.0000
O1	4e	0.2170	0.2253	-0.0405
O2	4e	0.2951	0.7093	-0.0407
O3	4e	0.4099	0.9887	0.2380
Sb		0	1/2	0

CALCIUM BISMUTH OXIDE
$Ca_4Bi_6O_{13}$

Chem. Mater., 2, 454-458.

C2mm, 5.937, 17.356, 7.206, Z = 2, R = 0.029. Chains along c of edge-sharing
BiO_5 square pyramids, linked into (010) sheets by 3-coordinate Bi, with sheets
linked by 7-coordinate Ca ions; there are short Bi...Bi contacts, 3.34 A.

BARIUM BISMUTH OXIDES
$BaBiO_3$, $BaBiO_{3-x}$

Acta Cryst., B$\underline{46}$, 693-698.

$BaBiO_3$, Fm3m, a = 8.7676 A, at 900K, Z = 8, neutron powder data. Ba in 8(c):
1/4,1/4,1/4; Bi(1) in 4(a): 0,0,0; Bi(2) in 4(b): 1/2,1/2,1/2; O in 24(e):
0.2605,0,0. Monoclinic, rhombohedral, and tetragonal forms have been described
($\underline{42}$, 446; $\underline{43A}$, 186; $\underline{44A}$, 195; $\underline{45A}$, 232; $\underline{52A}$, 192).

$BaBiO_{3-x}$, Fm3m, a = 8.7871 A, at 1025K, Z = 8, neutron powder data, structure as
above with no oxygen deficiency, x(O) = 0.2593; and I4/mmm, 6.287, 9.029, Z = 4,
neutron powder data, Ba in 4(d): 0,1/2,1/4; Bi(1) in 2(a): 0,0,0; Bi(2) in
2(b): 0,0,1/2; 6.4 O(1) in x,x,0, x = 0.17; 3.8 O(2) in 4(e): 0,0,0.28.

$BaBiO_3$ (I), $BaBiO_{2.82}$(II)

Mater. Res. Bull., $\underline{25}$, 1467-1476.

I, Fm3m, a = 8.7684 A, at 873K, Z = 8, neutron powder data. Ba in 8(c):
1/4,1/4,1/4; Bi(1) in 4(a): 0,0,0; Bi(2) in 4(b): 1/2,1/2,1/2; O in 24(e):
0.2610,0,0. Double perovskite structure.

II, Pm3m, a = 4.4270 A, at 1083K, Z = 1, neutron powder data. Ba in 1(a):
0,0,0; Bi in 1(b): 1/2,1/2,1/2; 2.82 O in 3(c): 1/2,1/2,0. Perovskite
structure. Further reduction in oxygen content gives a tetragonal phase.

Ba_2BiO_4 3/4 $[Ba_2(Ba_{0.67}Bi_{0.33})BiO_{6-x}]$

Solid State Comm., $\underline{75}$, 759-763.

Fm3m, 8.7670, Z = 16/3, powder data. Ba in 8(c): 1/4,1/4,1/4; Ba/Bi(III) in
4(b): 1/2,1/2,1/2; Bi(V) in 4(a): 0,0,0; O in 24(e): x,0,0, x = 0.20. Double
perovskite.

BARIUM LITHIUM BISMUTHATE
$Ba_4LiBi_3O_{11}$ 4 x $Ba(Bi,Li)O_{2.75}$

J. Solid State Chem., $\underline{84}$, 82-87.

P4/mmm, 6.101, 8.628, Z = 1, powder data. Perovskite-type structure with partial
Bi/Li ordering in the octahedral sites.

Cell occupancy		x	y	z
0.27(2)*	Bi(1)	0.0	0.0	0.0
0.91(1)*	Bi(2)	0.5	0.5	0.0
0.91(1)*	Bi(3)	0.0	0.0	0.5
0.91(1)*	Bi(4)	0.5	0.5	0.5
4.0	Ba	0.5	0.0	0.229(2)
1.83	O(1)	0.0	0.0	0.25
3.67	O(2)	0.25	0.25	0.0
3.67	O(3)	0.25	0.25	0.5
1.83	O(4)	0.5	0.5	0.25

* Occupancy of all Bi sites is completed by Li.

BISMUTH VANADIUM OXIDE
$Bi_{12}(Bi_{0.75}V_{0.05})O_{20}$

J. Solid State Chem., **86**, 59-63.

I23, a = 10.265 A, Z = 2, neutron powder data. Sillenite-type structure (**5**, 70; **44A**, 342), with Bi/V in a tetrahedral site.

		x	y	z	Occupation
(Bi, V)	(2a)	0	0	0	1.44
Bi	(24f)	0.1766	0.3205	0.0096	24
O_I	(24f)	0.1347	0.2502	0.4797	24
O_{II}	(8c)	0.1888	0.1888	0.1888	8
O_{III}	(8c)	0.8876	0.8876	0.8876	8

BISMUTH IRON OXIDE
$Bi_{12}(Bi_{0.5}Fe_{0.5})O_{19.50}$

BISMUTH ZINC OXIDE
$Bi_{12}(Bi_{0.67}Zn_{0.33})O_{19.33}$

Kristallografija, **35**, 1126-1132 [Soviet Physics - Crystallography, **35**, 663-666].

I23, a = 10.184, 10.207 A, Z = 2, neutron radiation, R = 0.019, 0.017. Bi in 24(f): x,y,z, x = 0.17631, 0.17655, y = 0.31786, 0.31821, z = 0.01395, 0.01379; M in 2(a): 0,0,0 (or, more likely, with Bi slightly disordered about this site); O(1) in 24(f): x = 0.13495, 0.13539, y = 0.25148, 0.25132, z = 0.48589, 0.48631; O(2) in 8(c): x,x,x, x = 0.18995, 0.18757; 7 or 6.66 O(3) in 8(c): x = 0.89288, 0.88989. Sillenite-type (**5**, 70; **41A**, 229).

BARIUM PLATINUM SCANDATE and TERBATE
$Ba_8Pt_4Sc_3O_{17.5}$, $Ba_8Pt_4Tb_3O_{17.5}$

J. Less-Common Metals, **159**, 223-229.

Pm3m, a = 8.1531, 8.571 A, Z = 1, R = 0.031 for Sc compounds. Isostructural with the Y compound (following report). Pt(II) has square-planar, Pt(IV) and Sc octahedral, and Ba 9-coordination.

		x	y	z
Ba	8g	0.2432	0.2432	0.2432
Pt^{2+}	3c	0.5	0.5	0.0
Pt^{4+}	1a	0.0	0.0	0.0
Sc	3d	0.5	0.0	0.0
O1	12h	0.256	0.5	0.0
O2	6e	0.0	0.0	0.246

BARIUM PLATINUM YTTRIUM OXIDE
$Ba_8Pt_4Y_3O_{17.5}$

Eur. J. Solid State Inorg. Chem., **25**, 231-235 (1988).

Pm3m, 8.3542, Z = 1. Pt(II) has square-planar, Pt(IV) and Y octahedral, and Ba one-sided 9-coordination.

SODIUM TITANATES
$Na_{0.54}TiO_2$

J. Solid State Chem., 85, 31-37.

$R\bar{3}m$, a = 2.9791, c = 16.928 A, Z = 3, R = 0.057.   1.62 Na in 3(a): 0,0,0; Ti in
3(b): 0,0,1/2; O in 6(c): 0,0,z, z = 0.2294. α-NaFeO$_2$-type (3, 75), with large
thermal parameters for the Na ion.

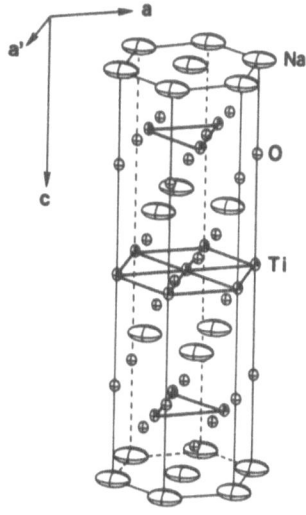

Na_4TiO_4

Z. anorg. Chem., 582, 103-110.

$P\bar{1}$, 8.783, 5.785, 6.476, 124.24, 102.31, 95.87, Z = 2, R = 0.092.   Isostructural
with the Co and Sn compounds (41A, 267; 52A, 183; 56A, 125).   Ti-4 O = 1.83-1.84,
Na-4 or 5 O = 2.30-2.55 A.

SODIUM TETRATITANATE
$Na_2Ti_4O_9$

J. Solid State Chem., 86, 135.

C2/m, 23.108, 2.9392, 10.653, 102.54, Z = 4, rather than the previous $P\bar{1}$
description (56A, 131); cell transformation (021,001,100).

COORDINATES, SPACE GROUP $C2/m$

	x	y	z
Na(1)	0.1366	0.0	0.5974
Na(2)	0.0742	0.0	0.3068
Na(3)	0.1407	0.5	0.1041
Ti(1)	0.0	0.0	0.0
Ti(2)	0.2232	0.5	0.3745
Ti(3)	0.2151	0.5	0.8791
Ti(4)	0.0608	0.5	0.8090
Ti(5)	0.0	0.0	0.5
O(1)	0.0746	0.0	0.9394
O(2)	0.0230	0.5	0.1233
O(3)	0.0308	0.0	0.6947
O(4)	0.1721	0.0	0.3186
O(5)	0.2033	0.0	0.9918
O(6)	0.2418	0.0	0.7812
O(7)	0.0560	0.5	0.4878
O(8)	0.1369	0.5	0.7829
O(9)	0.2152	0.5	0.5528

Occupancy factors (l): Na(1), 0.97 ; Na(2), 0.48. ; Na(3), 0.63

SODIUM HEXATITANATE
$Na_{1.7}Ti_6O_{11}$

J. Solid State Chem., <u>85</u>, 8-14.

$P4_2/mnm$, 11.7456, 2.9866, Z = 2, R = 0.038. Basic units consisting of six edge- and corner-sharing TiO_6 octahedra (random Ti(III)/Ti(IV) occupancy) share edges to form rutile-type chains along $\underline{c}$; sodium ions (occupancy = 0.84) are in tunnels.

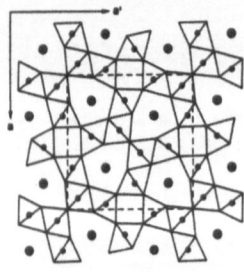

		x	y	z	Occupancy
Na	$4g$	0.6935	-0.6935	0.0	0.84
Ti(1)	$8i$	0.46108	0.19115	0.0	1.0
Ti(2)	$4g$	0.91664	-0.9166	0.0	1.0
O(1)	$8i$	0.2102	0.0206	0.0	1.0
O(2)	$8i$	0.6066	0.1117	0.0	1.0
O(3)	$4f$	0.3291	0.3291	0.0	1.0
O(4)	$2b$	0.5	0.5	0.0	1.0

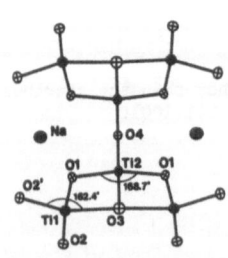

SELECTED INTERATOMIC DISTANCES (Å) AND BOND ANGLES (°)

Na–O(1)	2.528(2) ×4	O(1)–Ti(1)–O(1)	95.62(8)
–O(2)	2.506(2) ×2	O(1)–Ti(1)–O(2)	83.10(6)
–O(3)	2.702(2) ×2	O(1)–Ti(1)–O(2')	88.35(6)
		O(1)–Ti(1)–O(3)	79.90(6)
Mean	2.566	O(2)–Ti(1)–O(2)	97.46(9)
		O(2)–Ti(1)–O(2')	97.91(9)
		O(2)–Ti(1)–O(3)	93.66(6)
Ti(1)–O(1)	2.015(1) ×2	O(2')–Ti(1)–O(3)	162.40(7)
–O(2')	1.947(2)		
–O(2)	1.987(1) ×2		
–O(3)	2.242(1)		
		O(1)–Ti(2)–O(1)	168.70(11)
Mean	2.032	O(1)–Ti(2)–O(3)	86.06(5)
		O(1)–Ti(2)–O(4)	93.84(4)
Ti(2)–O(1)	1.926(2) ×2	O(3)–Ti(2)–O(3)	91.50(6)
–O(3)	2.085(1) ×2	O(3)–Ti(2)–O(4)	87.09(3)
–O(4)	2.0365(4) ×2	O(4)–Ti(2)–O(4)	94.32(2)
Mean	2.016		

POTASSIUM DITITANATE(IV)
$K_6Ti_2O_7$

J. Less-Common Metals, <u>158</u>, 327-337.

P2$_1$/c, 6.582, 9.318, 11.269, 123.46, Z = 2, R = 0.076. Isostructural with the cobaltate (<u>40</u>A, 212) and silicate (<u>49</u>A, 322), with a centrosymmetric $O_3Ti-O-TiO_3{}^{6-}$ ion (refinement in Pc, R = 0.070, gives Ti-O-Ti = 174°). Ti-O = 1.78-1.79 (terminal), 1.88 (bridging) A.

		10^4 x	y	z
Ti	4e	3102	6422	3661
K(1)	4e	4068	2911	2812
K(2)	4e	480	3457	4300
K(3)	4e	7935	229	5470
O(1)	4e	7597	4068	8054
O(2)	4e	5038	2003	5736
O(3)	4e	9624	3315	6382
O(4)	2b	5000	5000	5000

CAESIUM TITANIUM BRONZE
$Cs_{0.13}TiO_2$

Mater. Res. Bull., <u>25</u>, 139-148.

I4$_1$/a, 14.525, 5.943, Z = 32, R = 0.033. Hollandite-type superstructure. Ti-6 O = 1.94-2.02, Cs-12 O = 2.96-3.47 A.

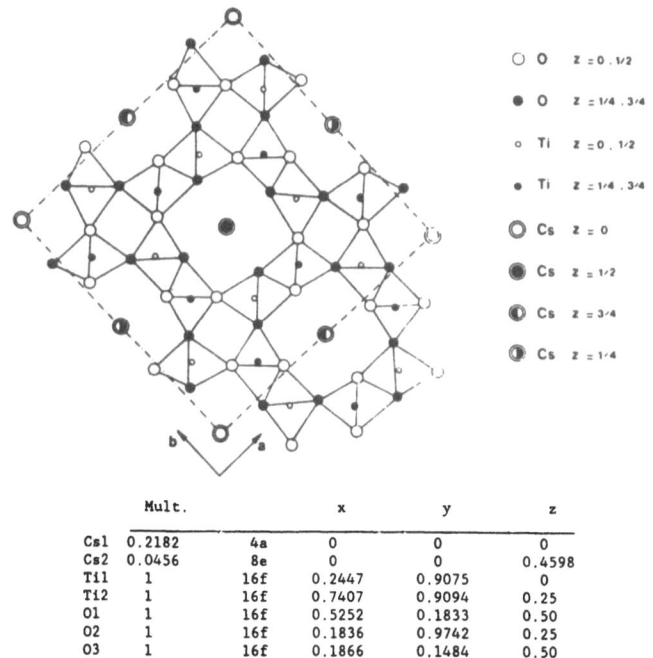

			O	O	$z = 0, 1/2$
			●	O	$z = 1/4, 3/4$
			○	Ti	$z = 0, 1/2$
			●	Ti	$z = 1/4, 3/4$
			◯	Cs	$z = 0$
			◉	Cs	$z = 1/2$
			◑	Cs	$z = 3/4$
			◔	Cs	$z = 1/4$

	Mult.		x	y	z
Cs1	0.2182	4a	0	0	0
Cs2	0.0456	8e	0	0	0.4598
Ti1	1	16f	0.2447	0.9075	0
Ti2	1	16f	0.7407	0.9094	0.25
O1	1	16f	0.5252	0.1833	0.50
O2	1	16f	0.1836	0.9742	0.25
O3	1	16f	0.1866	0.1484	0.50
O4	1	16f	0.3522	0.1853	0.75

HOLLANDITE
$Ba_xTi_8O_{16}$ (x = 1.07, 1.31)

Acta Cryst., B46, 599-609.

I2/m to I4/m transition at 375K (x = 1.07) and 475K (x = 1.31); below each
transition temperature the metastable tetragonal phase coexists with the
monoclinic phase. I2/m, a ∿ 10.3, b ∿ 2.96, c ∿ 10.0 A, β ∿ 91°, at 5-400K;
I4/m, a ∿ 10.1, c ∿ 2.96 A, at 5-500K; neutron powder data. Hollandite-type
structure (13, 190).

SODIUM TITANIUM ARSENIC OXIDE
SODIUM TITANIUM ANTIMONY OXIDE
$Na_2Ti_2As_2O$, $Na_2Ti_2Sb_2O$

Z. anorg. Chem., 584, 150-158.

I4/mmm, a = 4.070, 4.144, c = 15.288, 16.561 A, Z = 2, R = 0.026 for the Sb
compound. Na in 4(e): 0,0,0.3179; Ti in 4(c): 0,1/2,0; Sb in 4(e):
0,0,0.1212; O in 2(b): 0,0,1/2. Anti-K_2NiF_4-type as for Eu_4As_2O (43A, 183).

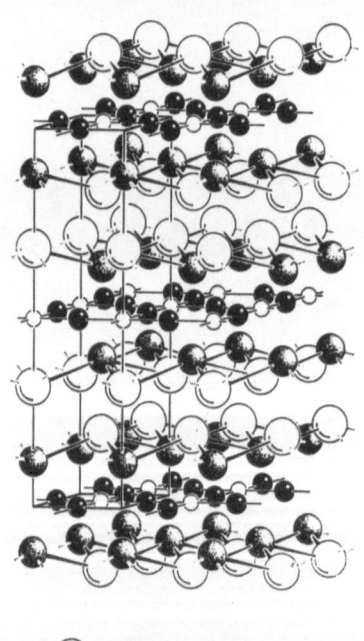

○ Na ◑ Sb ● Ti ○ O

BISMUTH TITANATE
$Bi_4Ti_3O_{12}$

Acta Cryst., B46, 474-487.

B1a1, 5.450, 5.4059, 32.832, 90.00, Z = 4, R = 0.018 (twinned crystal).
Commensurate modulation of an Fmmm parent structure derived from an idealized
I4/mmm structure. Ti^{4+} ions have octahedral coordinations and Bi^{3+} ions have
asymmetric coordinations (lone-pair electrons).

	$10^4\ x$	y	z
Bi(1)	30	5023	5673
Bi(1)′	13	4977	4336
Bi(2)	-21	4793	7113
Bi(2)′	21	5185	2887
Ti(1)	446	-13	5007
Ti(2)	520	-4	6289
Ti(2)′	499	2	3717
O(1)	2990	2760	5102
O(1)′	3548	-2179	4942
O(2)	2704	2442	2495
O(2)′	2736	7571	7489
O(3)	913	-705	5605
O(3)′	918	587	4424
O(4)	552	584	6825
O(4)′	568	-441	3195
O(5)	2904	2800	6121
O(5)′	2962	-2659	3892
O(6)	3677	-1959	6244
O(6)′	3496	2164	3773

ZIRCONIUM TITANATE
ZIRCONIUM TIN TITANATE
$ZrTiO_4$, $ZrSn_{0.5}Ti_{0.5}O_4$

Z. anorg. Chem., 582, 93-102.

Pbcn, a = 4.7855, 4.8145, b = 5.4755, 5.5961, c = 5.0277, 5.1050 A, Z = 2, R =
0.027, 0.054. Zr/Ti/Sn in 4(c): 0,y,1/4, y = 0.2025, 0.1885; O in 8(d): 0.2689,
0.4024, 0.4337; 0.2711, 0.3988, 0.4262. α-PbO_2-type, as previously described for
$ZrTiO_4$ (32A, 267; 53A, 139).

IRON(II) TITANIUM(IV) OXIDES
$FeTi_2O_5$, $(Fe,Mg)Ti_2O_5$

J. Solid State Chem., 88, 334-350; 351-367.

Cmcm, a $\sim$ 3.7, b $\sim$ 9.8, c $\sim$ 10.0 A, for various compositions, Z = 4, neutron
powder data. Pseudobrookite structures (2, 53; 50A, 189).

BISMUTH COPPER TITANIUM OXIDE
$Bi_{0.67}Cu_3Ti_4O_{12}$

Mater. Res. Bull., 25, 477-483.

Im3, 7.4175, Z = 2, powder data. Bi in 2(a): 0,0,0; Cu in 6(b): 0,1/2,1/2; Ti
in 8(c): 1/4,1/4,1/4; O in 24(g): 0.296,0.178,0. Isostructural with
$CaCu_3Mn_4O_{12}$ (41A, 260), with Bi in the icosahedral sites and Ti in the octahedral
sites of a perovskite-type structure; Cu has square-planar coordination, with
four more-distant O neighbours. Bi-12 O = 2.56, Ti-6 O = 1.96, Cu-O = 2.01, 2.83
(each x4) A.

CALCIUM TETRAZIRCONATE
$CaZr_4O_9$

Mater. Res. Bull., 25, 435-442.

C2/c, 17.79, 14.52, 12.00, 119.47, Z = 16, X-ray and neutron powder data.
Isostructural with the Hf compound (41A, 234).

AMMONIUM HEXAVANADATE
$(NH_4)_2V_6O_{16}$

Z. Naturforsch., 45B, 31-38.

$P2_1/m$, 7.858, 8.412, 4.995, 96.43, Z = 1, R = 0.060. Isostructural with the K
salt (18, 443), with (100) layers built from pairs of edge-sharing VO_5 square
pyramids sharing corners with each other and with VO_6 octahedra, with layers
linked by ammonium ions. V-O = 1.60-1.99 (square pyramid, sixth oxygen at
2.92), 1.59-2.29 (octahedron), N-O = 2.82-3.64 A.

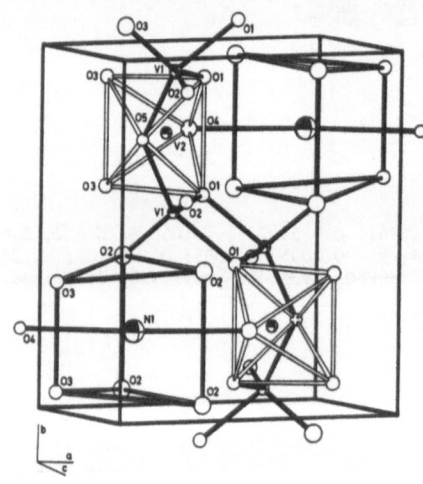

	x	y	z
V(1)	0.6852	0.0546	0.4372
V(2)	0.5742	1/4	0.9159
O(1)	0.5061	0.0874	0.6831
O(2)	0.1706	0.0660	0.4377
O(3)	0.7329	0.0987	0.1161
O(4)	0.4314	1/4	0.1173
O(5)	0.7559	1/4	0.5889
N(1)	0.0614	1/4	0.9452

SODIUM VANADATE
$Na_{0.56}V_2O_5$

Acta Cryst., C46, 536-538.

C2/m, 11.663, 3.6532, 8.92, 90.91, Z = 4, R = 0.056. Isostructural with the Ag
compound (30A, 344), with (001) layers of edge- and corner-sharing VO_6 distorted
octahedra; 7-coordinate Na ions are between the layers. V-O = 1.61-2.54, Na-O =
2.48-2.75 A.

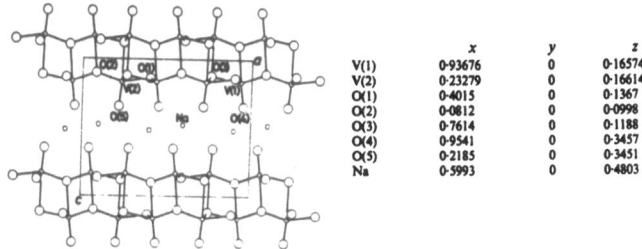

	x	y	z
V(1)	0·93676	0	0·16574
V(2)	0·23279	0	0·16614
O(1)	0·4015	0	0·1367
O(2)	0·0812	0	0·0998
O(3)	0·7614	0	0·1188
O(4)	0·9541	0	0·3457
O(5)	0·2185	0	0·3451
Na	0·5993	0	0·4803

POTASSIUM DIVANADATE
$K_{0.5}V_2O_5$

Acta Cryst., C46, 1590-1592.

Cmcm, 3.6784, 11.6120, 18.6332, Z = 8, R = 0.041. V_2O_5-layers of edge- and corner-sharing VO_6 distorted octahedra, linked by 8-coordinate K ions.

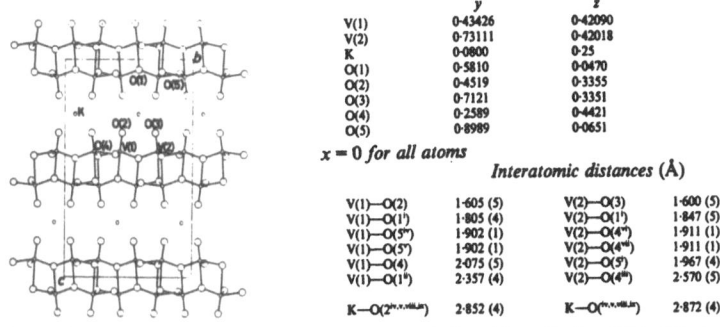

	y	z
V(1)	0·43426	0·42090
V(2)	0·73111	0·42018
K	0·0800	0·25
O(1)	0·5810	0·0470
O(2)	0·4519	0·3355
O(3)	0·7121	0·3351
O(4)	0·2589	0·4421
O(5)	0·8989	0·0651

$x = 0$ for all atoms

Interatomic distances (Å)

V(1)—O(2)	1·605 (5)	V(2)—O(3)	1·600 (5)
V(1)—O(1')	1·805 (4)	V(2)—O(1')	1·847 (5)
V(1)—O(5'')	1·902 (1)	V(2)—O(4'')	1·911 (1)
V(1)—O(5''')	1·902 (1)	V(2)—O(4'''')	1·911 (1)
V(1)—O(4)	2·075 (5)	V(2)—O(5')	1·967 (4)
V(1)—O(1'')	2·357 (4)	V(2)—O(4''')	2·570 (5)

K—O(2⁾	2·852 (4)	K—O⁾	2·872 (4)

Projection of the structure of $K_{0.30}V_2O_5$ onto (100).

Symmetry code: (i) x,y,1/2 − z; (ii) x,1 − y,1/2 + z; (iii) x,1 − y,1 − z; (iv) 1/2 + x,−1/2 + y,1/2 − z; (v) −1/2 + x,−1/2 + y,1/2 − z; (vi) 1/2 + x,1/2 + y,z; (vii) −1/2 + x,1/2 + y,z; (viii) 1/2 + x,−1/2 + y,z; (ix) −1/2 + x, −1/2 + y,z.

POTASSIUM AQUODIOXOTETRAPEROXODIVANADATE(V) TRIHYDRATE
$K_2[V_2O_2(O_2)_4(H_2O)].3H_2O$

Acta Cryst., C46, 738-741.

P1, 6.501, 7.882, 7.501, 107.18, 95.50, 116.20, Z = 1, R = 0.026. Discrete anions with pentagonal pyramidal coordination at each V, linked by 7- and 8-coordinate K ions, and by hydrogen bonding via the water molecules.

	x	y	z
V(1)	0·4072	0·8758	0·7591
V(2)	0·2482	0·5749	0·9630
K(1)	0	0	0
K(2)	0·6284	0·4436	0·5619
O(1)	0·2709	0·9675	0·6635
O(2)	0·4033	0·9578	1·0188
O(3)	0·6294	1·0867	0·9866
O(4)	0·4427	0·6950	0·5456
O(5)	0·6565	0·8968	0·6402
O(6)	0·1288	0·6060	0·7334
O(7)	0·0016	0·6244	0·8835
O(8)	0·5361	0·6587	0·8924
O(9)	0·5688	0·6994	1·0985
O(10)	0·1346	0·3353	0·9029
O(11)	0·1927	0·6579	1·2245
O(12)	0·7097	0·1273	0·4171
O(13)	0·7386	0·4767	0·2144
O(14)	0·1547	0·1496	0·3995

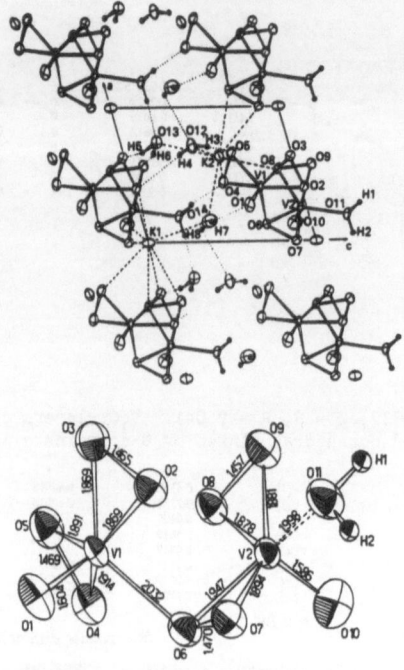

CAESIUM SODIUM VANADATE(V)
Cs_2NaVO_4

Z. anorg. Chem., __587__, 29-38.

$P2_1/m$, 8.399, 6.247, 6.148, 92.66, Z = 2, R = 0.062. $VO_4{}^{3-}$ tetrahedra, linked by 5-coordinate Na and (7+2)- and 10-coordinate Cs ions. V-O = 1.72 A.

		x 10^4 y		z
		x	y	z
V	2e	2422	2500	2907
Na	2e	236	2500	8673
Cs(1)	2e	5038	2500	7994
Cs(2)	2e	8149	2500	3435
O(1)	2e	1756	2500	5503
O(2)	2e	4471	2500	3004
O(3)	4f	8300	5290	8486

POTASSIUM, AMMONIUM, and RUBIDIUM MAGNESIUM DECAVANADATE HEXADECAHYDRATE
$M_2Mg_2V_{10}O_{28} \cdot 16H_2O$ (M = K, NH_4, Rb)

Izv. Akad. Nauk SSSR, Neorg. Mater., __26__, 350-356.

$P\bar{1}$, a = 8.829, 8.874, 8.879, b = 10.738, 10.881, 10.904, c = 11.091, 11.081, 11.052 A, α = 114.23, 114.46, 114.19, β = 74.07, 74.09, 73.70, γ = 108.21, 107.79, 108.81°, Z = 1, R = 0.027, 0.033, 0.042. Isostructural with the Zn compound (__31__A, 142), with decavanadate ions, $Mg(H_2O)_6$ octahedra, and 10-coordinate alkali-metal ions.

STRONTIUM VANADATES
$SrVO_3$, Sr_2VO_4

J. Solid State Chem., <u>86</u>, 101-108.

$SrVO_3$, Pm3m, 3.841, Z = 1, powder data. Perovskite structure.

Sr_2VO_4, I4/mmm, 3.837, 12.576, Z = 2, X-ray and neutron powder data. Sr in 4(e):
0,0,0.35438; V in 2(a): 0,0,0; O(1) in 4(c): 0,1/2,0; O(2) in 4(e):
0,0,0.15778. K_2NiF_4-type structure (<u>17</u>, 332; <u>19</u>, 323; <u>49A</u>, 93).

Sr_2VO_4

J. Solid State Chem., <u>85</u>, 321-325.

I4/mmm, 3.8340, 12.5874, at 200K, Z = 2, neutron powder data. Sr in 4(e):
0,0,0.35438; V in 2(a): 0,0,0; O(1) in 4(c): 0,1/2,0; O(2) in 4(e):
0,0,0.15778. K_2NiF_4-type (<u>17</u>, 332; <u>19</u>, 323; <u>49A</u>, 93). V-6 O = 1.92, 1.99, Sr-9
O = 2.18, 2.48, 2.65 A.

ALVANITE
$(Zn,Ni)Al_4(VO_3)_2(OH)_{12} \cdot 2H_2O$

Neues Jb. Miner. Mh., 385-392.

$P2_1/n$, 17.808, 5.132, 8.881, 92.11, Z = 2, R = 0.056. Brucite-like (100) layers
of edge-sharing $(Zn,Ni)(OH)_6$ and $Al(OH)_6$ octahedra, with one-sixth of the central
positions vacant, linked by chains along <u>b</u> of corner-sharing VO_4 tetrahedra; the
water molecules are in channels along <u>b</u>. (Zn,Ni)-OH = 2.07-2.10, Al-OH =
1.82-1.98, V-O = 1.62, 1.65 (terminal), 1.80, 1.83(1) (bridging) A.

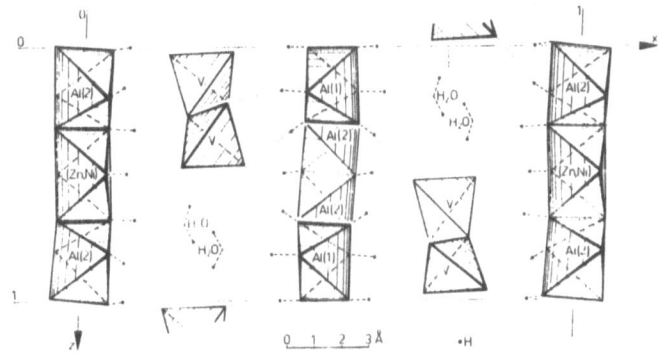

THALLIUM(I) TRIVANADATE
$T\ell V_3 O_8$

Acta Cryst., C$\underline{46}$, 177-179.

P2$_1$/m, 7.780, 8.423, 4.993, 96.48, Z = 2, R = 0.049. Isostructural with the Cs compound ($\underline{31A}$, 141). Chains along $\underline{b}$ of edge- and corner-sharing V(2)O$_5$ square pyramids, linked into (100) sheets by octahedrally-coordinated V(1); sheets are connected by 12-coordinate Tℓ^+ ions. V-O = 1.60-2.29, Tℓ-O = 2.75-3.60 A.

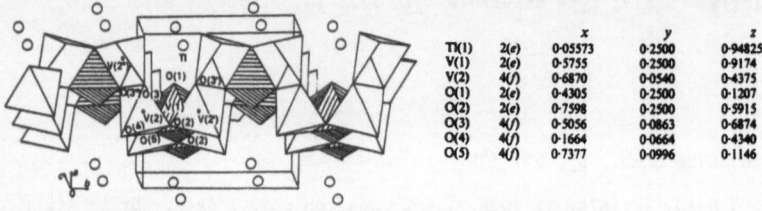

		x	y	z
Tl(1)	2(e)	0·05573	0·2500	0·94825
V(1)	2(e)	0·5755	0·2500	0·9174
V(2)	4(f)	0·6870	0·0540	0·4375
O(1)	2(e)	0·4305	0·2500	0·1207
O(2)	2(e)	0·7598	0·2500	0·5915
O(3)	4(f)	0·5056	0·0863	0·6874
O(4)	4(f)	0·1664	0·0664	0·4340
O(5)	4(f)	0·7377	0·0996	0·1146

LEAD VANADATE BRONZE
$\beta\text{-Pb}_{0.333}V_2O_5$

Acta Cryst., C$\underline{46}$, 1587-1590.

C2/m, 15.463, 3.6477, 10.116, 109.20, Z = 6, R = 0.054. Isostructural with $\beta\text{-Na}_x V_2 O_5$ ($\underline{19}$, 449; $\underline{21}$, 303; $\underline{45A}$, 396), with disorder of the Pb ions resulting in superlattice reflections and diffuse scattering. V-O = 1.60-1.61, 1.81-2.16, 2.27-2.30, Pb-O = 2.35-2.78 A.

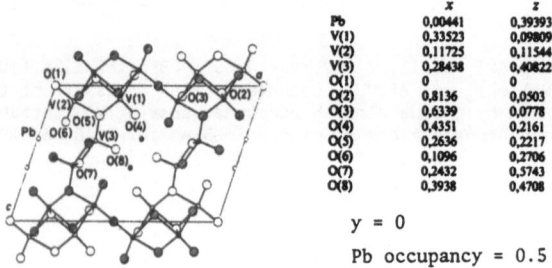

	x	z
Pb	0,00441	0,39393
V(1)	0,33523	0,09809
V(2)	0,11725	0,11544
V(3)	0,28438	0,40822
O(1)	0	0
O(2)	0,8136	0,0503
O(3)	0,6339	0,0778
O(4)	0,4351	0,2161
O(5)	0,2636	0,2217
O(6)	0,1096	0,2706
O(7)	0,2432	0,5743
O(8)	0,3938	0,4708

y = 0

Pb occupancy = 0.5

LEAD GERMANATE VANADATE
$Pb_5(GeO_4)(VO_4)_2$

Ž. Strukt. Khim., $\underline{31}$, No. 4, 80-84.

P6$_3$/m, 10.099, 7.400, Z = 2, powder data. Apatite-type structure.

	x	y	z
Pb(1)	1/3	2/3	0.0064
Pb(2)	0.2545	0.0073	1/4
V/Ge	0.392	0.384	1/4
O(1)	0.299	0.478	1/4
O(2)	0.608	0.480	1/4
O(3)	0.345	0.272	0.067

SODIUM VANADOPHOSPHATE
$Na_{2.44}V_4P_4O_{17}(OH)$

J. Solid State Chem., <u>87</u>, 178-185.

Pnma, 13.723, 6.314, 16.139, Z = 4, R = 0.032. Three complex infinite chains
along <u>b</u> of edge- and corner-sharing VO_6 octahedra and PO_4 tetrahedra, with
tunnels containing Na ions; one Na and one V are disordered. V-O = 1.89-2.23 A.

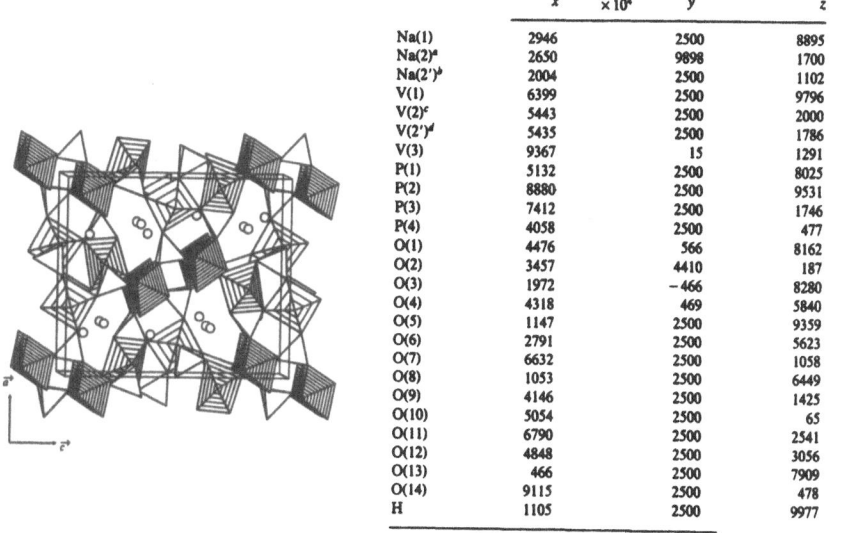

	x $\times 10^4$	y	z
Na(1)	2946	2500	8895
Na(2)[a]	2650	9898	1700
Na(2')[b]	2004	2500	1102
V(1)	6399	2500	9796
V(2)[c]	5443	2500	2000
V(2')[d]	5435	2500	1786
V(3)	9367	15	1291
P(1)	5132	2500	8025
P(2)	8880	2500	9531
P(3)	7412	2500	1746
P(4)	4058	2500	477
O(1)	4476	566	8162
O(2)	3457	4410	187
O(3)	1972	-466	8280
O(4)	4318	469	5840
O(5)	1147	2500	9359
O(6)	2791	2500	5623
O(7)	6632	2500	1058
O(8)	1053	2500	6449
O(9)	4146	2500	1425
O(10)	5054	2500	65
O(11)	6790	2500	2541
O(12)	4848	2500	3056
O(13)	466	2500	7909
O(14)	9115	2500	478
H	1105	2500	9977

Site occupancy (%): [a]65(1), [b]27(1), [c]88(2), [d]12(2)

COBALT(II) METAVANADATE TETRAHYDRATE
$Co(VO_3)_2 \cdot 4H_2O$

Izv. Akad. Nauk SSSR, Neorg. Mater., <u>26</u>, 346-349.

C2/c, 13.040, 9.929, 6.928, 112.35, Z = 4, R = 0.056. Zigzag chains along <u>c</u> of
cis-edge-sharing VO_5 square pyramids, linked by corner sharing with $CoO_4(H_2O)_2$
octahedra, and by hydrogen bonding which includes an additional non-coordinated
water molecule. V-O = 1.65, 1.66, 1.88-2.03, Co-O = 2.05-2.10, O-H...O =
2.71-3.20 A.

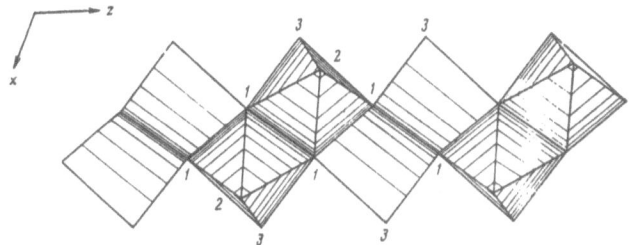

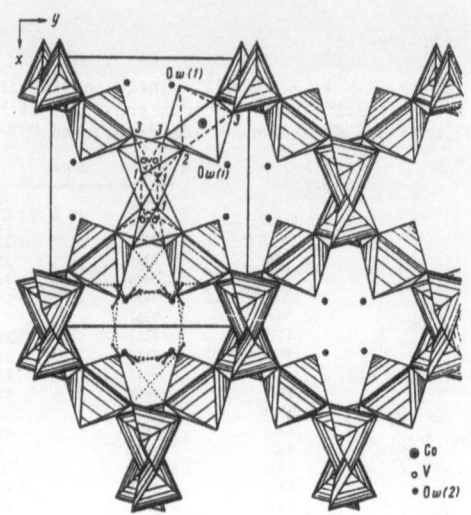

COPPER(II) VANADATE(V) HYDRATES
$Cu_3V_2O_8.H_2O$ (I), $CuV_2O_6.2H_2O$ (II)

Acta Cryst., C46, 15-18.

I, $P2_1/m$, 7.444, 6.658, 7.759, 93.57, Z = 2, R = 0.047. II, P2/c, 5.617, 5.595, 11.333, 91,04, Z = 2, R = 0.045. I contains dimeric units of edge-sharing CuO_5 and $CuO_4(H_2O)$ square pyramids, linked by CuO_4 square planes and by tetrahedral $O_3V-O-VO_3{}^{4-}$ ions. II contains chains along $\underline{a}$ of corner-sharing VO_4 tetrahedra, linked by $CuO_4(H_2O)_2$ distorted octahedra.

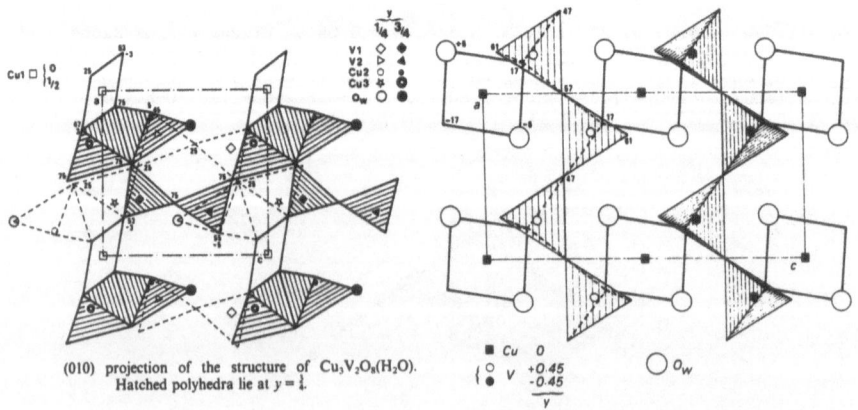

(010) projection of the structure of $Cu_3V_2O_8(H_2O)$. Hatched polyhedra lie at $y = \frac{1}{4}$.

(010) projection of the structure of $CuV_2O_6(H_2O)_2$ (heights of oxygen atoms along $\mathbf{b}$ are in hundreds).

	x	y	z
Cu₃V₂O₈(H₂O)			
Cu(1)	0	0	0
Cu(2)	0·1625	0·25	0·7041
Cu(3)	0·3237	0·25	0·0703
V(1)	0·6613	0·25	0·7841
V(2)	0·7332	0·25	0·3335
O(1)	0·0962	0·25	−0·0595
O(2)	0·2382	−0·0370 (6)	0·1297
O(3)	0·4465	0·25	0·8263
O(4)	0·8613	0·4542 (6)	0·2960
O(5)	0·5396	0·25	0·2142
O(6)	0·6803	0·25	−0·4425
O(w)	0·212	0·25	0·4576
CuV₂O₆(H₂O)₂			
Cu	0	0	0
V	0·2350	0·4519	0·1620
O(1)	0	0·5664	0·2500
O(2)	0·1830	0·1667	0·1272
O(3)	0·5	0·4746	0·25
O(4)	0·2528	0·6146	0·0456
O(w)	0·2612	0·0615	−0·1118

Selected bond lengths (Å) and angles (°) in Cu₃V₂O₈(H₂O) and CuV₂O₆(H₂O)₂, with e.s.d's in parentheses

Cu₃V₂O₈(H₂O)

2 × Cu(1)—O(1)	1·881 (2)	Cu(2)—O(1)	1·929 (5)
2 × Cu(1)—O(2)	1·999 (4)	Cu(2)—O(3)	2·263 (6)
	(d) = 1·940	2 × Cu(2)—O(4)	1·977 (6)
		Cu(2)—O(w)	1·970 (6)
2 × Cu(1)—O(4)	2·594 (6)		(d) = 2·023

Cu(3)—O(1)		1·916 (5)
2 × Cu(3)—O(2)		2·075 (4)
Cu(3)—O(3)		2·153 (5)
Cu(3)—O(5)		1·899 (6)
		(d) = 2·024

2 × V(1)—O(2)	1·719 (6)	2 × V(2)—O(4)	1·696 (4)
V(1)—O(3)	1·652 (5)	V(2)—O(5)	1·664 (6)
V(1)—O(6)	1·773 (6)	V(2)—O(6)	1·806 (5)
	(d) = 1·716		(d) = 1·715

CuV₂O₆(H₂O)₂

2 × Cu—O(2)	1·988 (3)	V—O(1)	1·788 (2)
2 × Cu—O(w)	1·986 (4)	V—O(2)	1·668 (3)
	(d) = 1·987	V—O(3)	1·781 (1)
		V—O(4)	1·607 (3)
2 × Cu—O(4)	2·628 (5)		(d) = 1·711

V—O(1)—V	138·0 (8)	V—O(3)—V	171·8 (8)
Cu—O(4)—V	125·8 (8)	Cu—O(2)—V	134·9 (8)

VOLBORTHITE (SYNTHETIC)
Cu₃V₂O₇(OH)₂·2H₂O

J. Solid State Chem., <u>85</u>, 220–227.

C2/m, 10.606, 5.874, 7.213, 94.90, Z = 2, X-ray and neutron powder data. Structure as previously described (<u>55A</u>, 160), with (001) brucite-type layers, connected by pyrovanadate groups.

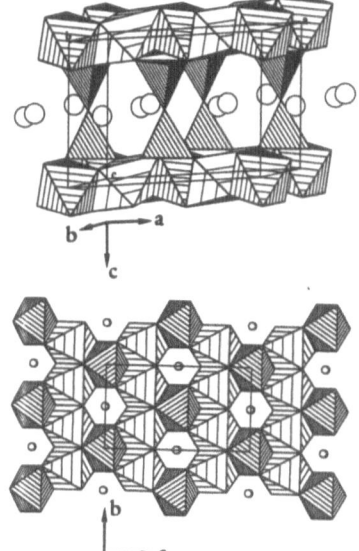

	x	y	z
Cu(1)	0	0	0
	0	**0**	**0**
Cu(2)	1/4	1/4	0
	1/4	**1/4**	**0**
V	0.9959	1/2	0.2516
	fixed	**fixed**	**fixed**
O(1)	0	1/2	1/2
	0	**1/2**	**1/2**
O(2)	0.3424	1/2	0.1143
	0.3428	**1/2**	**0.1095**
O(3)	0.0682	0.2721	0.1846
	0.0756	**0.2635**	**0.1864**
O(4)	0.1548	1/2	0.8464
	0.1622	**1/2**	**0.8381**
O(w)	0.3261	1/2	0.4788
	0.3223	**1/2**	**0.4894**
H(1)	**0.3501**	**1/2**	**0.2376**
H(2)	**0.3536**	**0.3714**	**0.5639**

The values from neutron data appear in bold.

BARIUM COPPER(II) VANADATE
$BaCu_2(VO_4)_2$

Z. anorg. Chem., 591, 167-173.

I$\overline{4}$2d, 12.744, 8.148, Z = 8, R = 0.055. Closely related to $SrNi_2(VO_4)_2$-type (53A, 144), but with distorted octahedral coordination for Cu; Cu-O = 1.94-1.98 (4 distances), 2.31, 2.65(2), V-4 O = 1.68-1.78(2) A.

CERIUM(III) VANADATE(V)
$CeVO_4$

Z. Naturforsch., 45B, 598-602.

Form I, normal pressure, I4_1/amd, 7.383, 6.485, Z = 4, R = 0.020. Zircon-type (1, 345).

Form II, high-pressure, P2_1/n, 7.003, 7.227, 6.685, 105.13, Z = 4, R = 0.024. Monazite-type (9, 236).

Form III, high-pressure, I4_1/a, 5.1645, 11.8482, Z = 4, no structure analysis. Scheelite-type (1, 347).

$CeVO_4$-I

	x	y	z
Ce	0	3/4	1/8
V	0	1/4	3/8
O	0	0,4283	0,2066

$CeVO_4$-II

	x	y	z
Ce	0.22439	0.15657	0.39678
V	0.19948	0.1647	0,8844
O(1)	0.2541	0,9959	0,731
O(2)	0.1124	0.3456	0.0013
O(3)	0.3894	0.2220	0.7710
O(4)	0.0168	0.1052	0.6743

POTASSIUM NEODYMIUM VANADATE
$K_3Nd(VO_4)_2$ (I)

ERBIUM VANADATE
$ErVO_4$ (II)

Vest. Mosk. Univ., Ser. 2: Khim., 31, 266-270.

I, P2_1/m, 7.555, 5.926, 9.161, 90.83, Z = 2, R = 0.023. Isostructural with the analogous phosphate (42A, 352).

II, I4_1/amd, 7.103, 6.305, Z = 4, R = 0.028. Er in 4(a): 0,3/4,1/8; V in 4(b): 0,1/4,3/8; O in 16(h): 0,0.0619,0.1999. Zircon-type (1, 345), as previously described (33A, 309).

ERBIUM VANADATE (HIGH-PRESSURE, PHASE II)
$ErVO_4$

Acta Cryst., C46, 1093-1094.

I4$_1$/a, 5.003, 11.143, Z = 4, R = 0.035. Er in 4(b): 0,1/4,5/8; V in 4(a): 0,1/4,1/8; O in 16(f): 0.1460,0.5050,0.2054 (origin at 1). Scheelite-type (1, 347). V-4 O = 1.722, Er-8 O = 2.333, 2.368(3) A. The normal pressure form has a zircon-type structure (33A, 309; this volume, preceding report).

VIGEZZITE
(Ca,Ce)(Nb,Ta,Ti)$_2$O$_6$

Neues Jb. Miner. Mh., 301-308.

Pnma, 11.065, 7.527, 5.343, Z = 4, R = 0.019. Isostructural with aeschynite (27, 536). (Nb,Ta,Ti)-6 O = 1.89-2.14, (Ca,Ce)-8 O = 2.40-2.53 A.

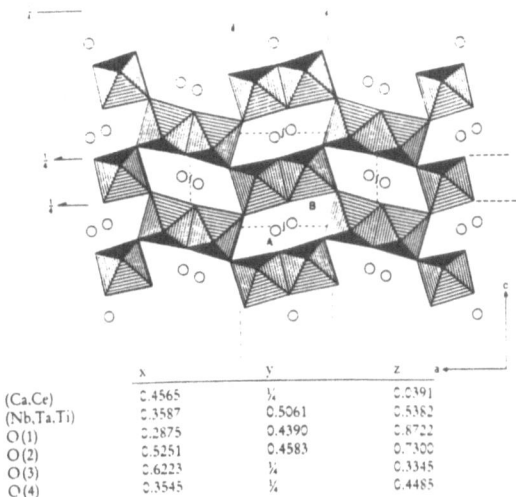

	x	y	z
(Ca,Ce)	0.4565	¼	0.0391
(Nb,Ta,Ti)	0.3587	0.5061	0.5382
O(1)	0.2875	0.4390	0.8722
O(2)	0.5251	0.4583	0.7300
O(3)	0.6223	¼	0.3345
O(4)	0.3545	¼	0.4485

STRONTIUM SODIUM NIOBATE
Sr(Na$_{0.25}$Sb$_{0.75}$)O$_3$

J. Solid State Chem., 84, 16-22.

P2$_1$/n, 8.0913, 8.0871, 8.0918, 89.953, Z = 8, X-ray and neutron powder data. Distorted pervoskite, with an ordered arrangement of SbO$_6$ and NaO$_6$ octahedra. Sb-O = 1.92-2.07, Na-O = 2.16-2.18, Sr-12 O = 2.51-3.21(1) A.

			x	y	z
0.988	Sr1	4(e)	0.2514	¼	0.2401
0.988	Sr2	4(e)	0.2405	¼	0.7678
0.972	Na	2(a)	0	0	0
0.976	Sb1	2(b)	0	0	½
0.976	Sb2	2(c)	½	0	½
0.976	Sb3	2(d)	½	0	0
1	O1	4(e)	0.2646	-0.0316	0.0406
1	O2	4(e)	-0.0188	-0.0259	0.2645
1	O3	4(e)	0.0228	0.2656	0.0287
1	O4	4(e)	0.5304	0.0269	½
1	O5	4(e)	0.4529	½	-0.0253
1	O6	4(e)	½	0.0332	0.4742

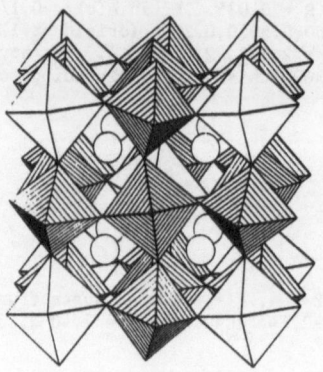

BARIUM NIOBATE(IV)
$BaNbO_3$

Mater. Res. Bull., 25, 9-14.

Pm3m, a = 4.0853, 4.0935 A, for two samples, Z = 1, powder data. Ba in 1(a):
0,0,0; Nb in 1(b): 1/2,1/2,1/2; O in 3(c): 0,1/2,1/2. Perovskite-type
structure, with fully-occupied sites.

BARIUM DINIOBATE (MONOCLINIC)
$BaNb_2O_6$

Ž. Neorg. Khim., 35, 1609-1611 [Russ. J. Inorg. Chem., 35, 914-915].

$P2_1/c$, 3.952, 6.049, 10.434, 90.35, Z = 2, R = 0.09. The structure is related to
those of $CaTa_2O_6$ and α-$SrNb_2O_6$ (28, 201; 48A, 219; 53A, 148), with pairs of
edge-shared NbO_6 octahedra linked by further corner sharing; cavities contain
10-coordinate Ba.

BARIUM DINIOBATE (HIGH-TEMP)
$BaNb_2O_6$

Ž. Neorg. Khim., 35, 2189-2191 [Russ. J. Inorg. Chem., 35, 1246-1248].

$C222_1$, 7.880, 12.215, 10.292, Z = 8, R = 0.063. NbO_6 octahedra and 10- and
12-coordinate Ba; Nb-O = 1.88-2.15(4) A.

SODIUM NIOBIUM PHOSPHATE BRONZE
$Na_4Nb_8P_6O_{35}$

J. Solid State Chem., 89, 75-82.

Pbam, 8.4992, 15.3390, 10.5913, Z = 2, R = 0.036. Framework of edge- and
corner-sharing NbO$_6$ octahedra and PO$_4$ tetrahedra, with Na ions in tunnels.

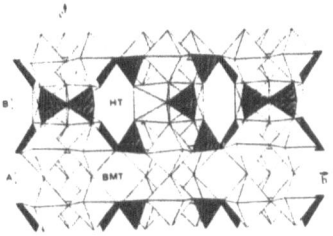

POTASSIUM NIOBIUM PHOSPHATE BRONZES
K$_3$Nb$_6$P$_4$O$_{26}$

J. Solid State Chem., <u>84</u>, 365-374.

Pnma, 14.7484, 31.582, 9.3859, Z = 8, R = 0.044. (100) layers of corner-sharing
NbO$_6$ octahedra and PO$_4$ tetrahedra, with layers linked by further corner sharing,
and tunnels containing K ions.

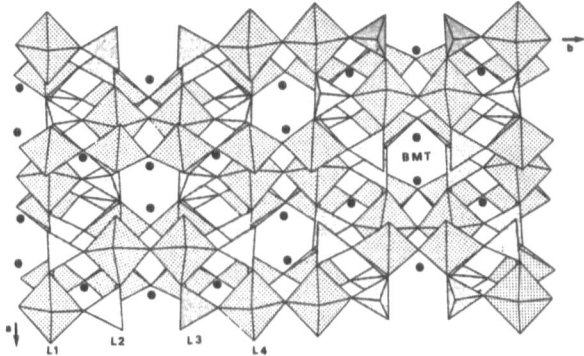

K$_{5-x}$Nb$_8$P$_5$O$_{34}$

J. Solid State Chem., <u>87</u>, 360-365.

P2/c, 13.904, 6.453, 20.64, 125.05, Z = 2, R = 0.029. Nb$_3$P$_2$O$_{13}$ (100) layers of
NbO$_6$ octahedra and PO$_4$ tetrahedra, linked by PO$_4$ tetrahedra and Nb$_2$O$_{11}$ units; K
ions (one disordered over three sites) are in 4- and 6-sided tunnels. Nb-O =
1.81-2.15, P-O = 1.53-1.56, K-O = 2.60-3.31 A.

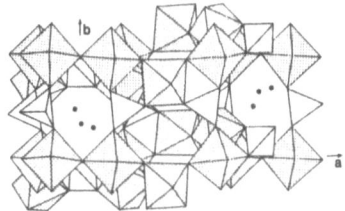

	x	y	z
Nb(1)	0.16613	−0.01720	0.05406
Nb(2)	0.24102	−0.00532	0.26477
Nb(3)	0.53830	0.77407	0.35353
Nb(4)	0.74862	0.77185	0.56941
P(1)	0.000	0.1529	0.250
P(2)	0.22103	0.4980	0.05104
P(3)	0.30488	0.4971	0.31915
K(1)	0.500	0.2743	0.250
K(2)	0.4396	0.7474	0.5162
K(3a)	0.0092	0.6430	0.1360
K(3b)	0.0444	0.504	0.1279
K(3c)	0.0571	0.6856	0.3357
O(1)	0.000	0.000	0.000
O(2)	0.1895	−0.0175	0.1619
O(3)	0.3447	−0.0349	0.1235
O(4)	0.1734	−0.0525	−0.0330
O(5)	0.1732	0.2900	0.0573
O(6)	0.1544	−0.3279	0.0618
O(7)	0.4083	−0.0397	0.3165
O(8)	0.2768	0.0353	0.3778
O(9)	0.0728	0.0114	0.2331
O(10)	0.2509	0.3056	0.2636
O(11)	0.2255	−0.3129	0.2733
O(12)	0.6055	0.7849	0.4697
O(13)	0.4290	0.5329	0.3404
O(14)	0.6475	0.5147	0.3830
O(15)	0.500	0.7214	0.250
O(16)	0.6910	0.5382	0.6059
O(17)	0.8020	0.5173	0.5308
O(18)	0.9181	0.7127	0.6770

CAESIUM PHOSPHONIOBATE BRONZE
$CsNb_3P_3O_{15}$

Mater. Res. Bull., 25, 1155-1160.

Pnnm, 13.4454, 14.8114, 6.4422 A, Z = 4, R = 0.031. Isostructural with the K
compound (56A, 141), with a framework of corner-sharing NbO_6 octahedra and PO_4
tetrahedra, with Cs ions in cavities.

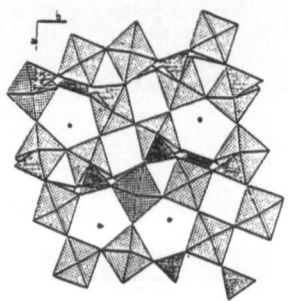

LITHIUM COPPER NIOBATE
$Li_{1-x}Cu_xNbO_3$ (x = 0.08, 0.14, 0.24)

Mater. Res. Bull., 25, 881-889.

R3c, a = 5.155, 5.153, 5.168, c = 13.86, 13.87, 13.89 A, Z = 6, powder data.
Li/Cu in 6(a): 0,0,z, z = 0.283, 0.287, 0.280; Nb in 6(a): z = 0; O in 18(b):
x,y,z, x = 0.065, 0.062, 0.066, y = 0.373, 0.371, 0.369, z = 0.055, 0.053, 0.055.
$LiNbO_3$-type structure (31A, 147; 53A, 147).

LANTHANUM PENTANIOBATE
$LaNb_5O_{14}$

Z. anorg. Chem., 583, 223; 590, 81-92.

Pbcm, 3.8749, 12.4407, 20.2051, Z = 4, R = 0.063. Chains of edge-sharing NbO$_7$ pentagonal bipyramids, linked by corner sharing with NbO$_6$ octahedra; tunnels along a contain 12-coordinate La. Nb-O = 1.77-2.44, La-O = 2.44-2.90 A.

LITHIUM TANTALATE(V)
LiTaO$_3$

Mineral. J., 14, 373-382 (1989).

R3c, a = 5.15329, c = 13.7806 A, Z = 6, R = 0.017. Structure as previously described (49A, 192).

POTASSIUM LITHIUM TANTALATE
K$_{0.73}$Li$_{0.27}$TaO$_3$

Kristallografija, 35, 732-738 [Soviet Physics - Crystallography, 35, 427-431].

P4/mbm, 12.595, 3.936, Z = 10, R = 0.054. Framework of TaO$_6$ octahedra with channels containing K and M(K/Li) ions. Some O atoms may be disordered.

	x	y	z
Ta(1)	0,5	0	0,5
Ta(2)	0,20767	0,07371	0,5
K(1)	0	0	0
K(2)	0,1733	0,3267	0
M	0,3777	0,1223	0
O(1)	0,5	0	0
O(2)	0,210	0,081	0
O(3)	0,344	0,001	0,5
O(4)	0,067	0,140	0,5
O(5)	0,295	0,205	0,5

STRONTIUM LANTHANUM TANTALATE
Sr$_3$LaTa$_3$O$_{12}$

Kristallografija, 35, 630-633 [Soviet Physics - Crystallography, 35, 368-370].

R3̄m, a = 5.6546, c = 27.245 A, Z = 3, R = 0.058. 12-Layer perovskite (46A, 252).

	x	y	z
3Ta1	0	0	0,4229
3Ta2	0	0	0
6A1 *	0	0	0,1397
6A2 *	0	0	0,2856
18O1	0,1667	0,3333	0,0383
18O2	0,1667	0,3333	0,4536

* A = ¹/₂Sr + ¹/₂La.

NEODYMIUM HEPTATANTALATE
NdTa$_7$O$_{19}$

Z. anorg. Chem., <u>588</u>, 43-54.

P$\bar{6}$c2, 6.2229, 19.939, Z = 2, R = 0.034 (0.048 for P6$_3$/mcm, with 2Nd + 2Ta in 4(d)). Isostructural with the La compound (<u>53A</u>, 151), with an ordered cation distribution. Nd-8 O = 2.44, 2.62, Ta(1)-6 O = 1.99, Ta(2)-7 O = 1.90-2.43 A.

		x	y	z
Nd	2c	1/3	2/3	0,0000
Ta1	2e	2/3	1/3	0,0000
Ta2	12l	0,3611	0,3594	0,15597
O1	12l	0,244	0,989	0,1536
O2	12l	0,431	0,051	0,9427
O3	6k	0,371	0,415	1/4
O4	4i	2/3	1/3	0,1660
O5	4h	1/3	2/3	0,1315

THORIUM(IV) DITANTALATE(V)
ThTa$_2$O$_7$

J. Less-Common Metals, <u>158</u>, 275-285.

C2/m, 15.3799, 3.8859, 8.3585, 95.88, Z = 4, R = 0.066. (010) sheets of edge-sharing TaO$_7$ pentagonal bipyramids and TaO$_6$ octahedra, with tunnels along <u>b</u> containing 10-coordinate Th. Ta-O = 1.90-2.17, Th-O = 2.38-2.78 A.

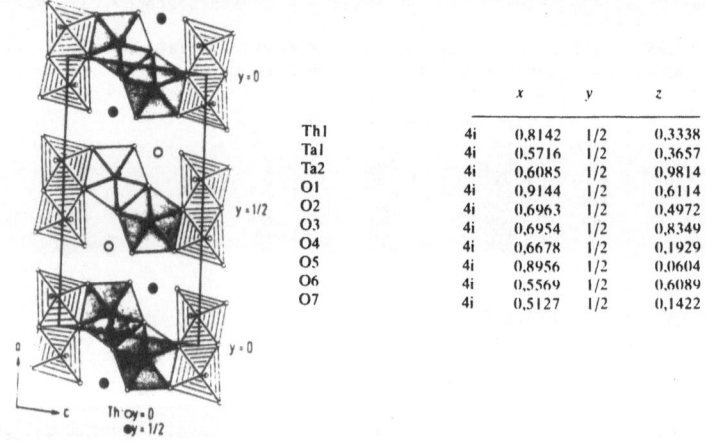

		x	y	z
Th1	4i	0,8142	1/2	0,3338
Ta1	4i	0,5716	1/2	0,3657
Ta2	4i	0,6085	1/2	0,9814
O1	4i	0,9144	1/2	0,6114
O2	4i	0,6963	1/2	0,4972
O3	4i	0,6954	1/2	0,8349
O4	4i	0,6678	1/2	0,1929
O5	4i	0,8956	1/2	0,0604
O6	4i	0,5569	1/2	0,6089
O7	4i	0,5127	1/2	0,1422

POTASSIUM SODIUM CHROMATE
K$_3$Na(CrO$_4$)$_2$

Acta Cryst., C<u>46</u>, 2019-2021.

P$\bar{3}$m1, 5.8580, 7.523, Z = 1, R = 0.030. CrO$_4$ tetrahedra linked by NaO$_6$ octahedra and KO$_{10}$ and KO$_{12}$ polyhedra. There is a phase transition at 239K.

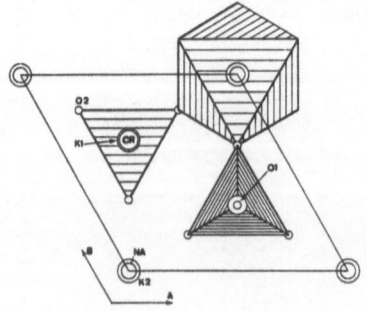

	x	y	z
O(2)	−0·8188	0·8188	0·3025
O(1)			0·0114
Cr			0·22657
K(1)			0·66767
Na	0	0	
K(2)	0	0	0

Interatomic distances (Å) and angles (°) of
$K_3Na(CrO_4)_2$

The standard deviations of the last significant digits are given in parentheses. For each polyhedron, symmetry equivalent distances can be identified by their multiplicity. Distances, corrected for thermal-riding motion, are given in square brackets.

CrO_4 tetrahedron

Cr—O(1)	1·619 (3)	O(1)—Cr—O(2)	110·31 (6)
	[1·646]		
Cr—O(2) × 3	1·6459 (8)	O(2)—Cr—O(2)	108·63 (5)
	[1·661]		
Mean distance	1·639 (1)		

NaO_6 octahedron

Na—O(2) × 6   2·364 (1)

$K(1)O_{10}$ polyhedron		$K(2)O_{12}$ polyhedron	
K(1)—O(1)	2·586 (3)	K(2)—O(2) × 6	2·926 (2)
K(1)—O(2) × 6	2·941 (1)	K(2)—O(1) × 6	3·3834 (4)
K(1)—O(2) × 3	3·151 (2)	Mean distance	3·155 (1)
Mean distance	2·969 (2)		

AMMONIUM BISMUTH CHROMATE DICHROMATE MONOHYDRATE
$NH_4Bi(CrO_4)(Cr_2O_7).H_2O$

Acta Cryst., C46, 1915-1916.

P$\bar{1}$, 6.916, 7.249, 10.798, 80.28, 90.52, 86.49, Z = 2, R = 0.052. Isostructural with the K salt (49A, 197).

CAESIUM NEODYMIUM CHROMATE
$Cs_3Nd(CrO_4)_3$

Kristallografija, 35, 1275-1277 [Soviet Physics - Crystallography, 35, 750-752].

C2/c, 27.798, 10.363, 10.453, 107.27, Z = 8, R = 0.038. (100) Layers of edge- and corner-sharing CrO_4 tetrahedra and NdO_8 distorted square antiprisms, linked by 8- and 10-coordinate Cs ions. Cr-O = 1.62-1.67, Nd-O = 2.36-2.55(1) A.

POTASSIUM TERBIUM CHROMATE
$KTb(CrO_4)_2$

J. Solid State Chem., 85, 83-87.

P$2_12_12_1$, 13.804, 5.735, 9.029, Z = 4, R = 0.073. The structure differs from that of the La and Eu analogues (55A, 172). It contains corner-sharing CrO_4 tetrahedra and TbO_8 bicapped trigonal prisms, with K in large channels along b. Cr-O = 1.62-1.67, Tb-O = 2.25-2.74, K-O = 2.71-3.10(3) A.

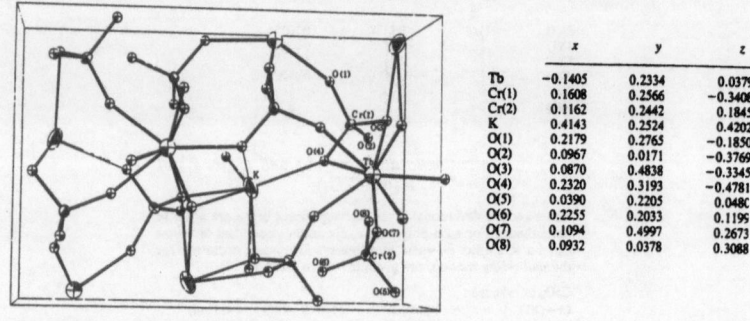

	x	y	z
Tb	−0.1405	0.2334	0.0379
Cr(1)	0.1608	0.2566	−0.3408
Cr(2)	0.1162	0.2442	0.1845
K	0.4143	0.2524	0.4202
O(1)	0.2179	0.2765	−0.1850
O(2)	0.0967	0.0171	−0.3769
O(3)	0.0870	0.4838	−0.3345
O(4)	0.2320	0.3193	−0.4781
O(5)	0.0390	0.2205	0.0480
O(6)	0.2255	0.2033	0.1195
O(7)	0.1094	0.4997	0.2673
O(8)	0.0932	0.0378	0.3088

POTASSIUM URANYL CHROMATE HYDRATE
$K[UO_2CrO_4(OH)].1.5H_2O$

Koord. Khim., $\underline{16}$, 1288-1291.

$P2_1/c$, 13.292, 9.477, 13.137, 104.12, Z = 4, R = 0.041. $[UO_2(OH)]_2{}^{2+}$ dimers are connected by tridentate bridging $CrO_4{}^{2-}$ ions into layers, which are connected by K^+ and water molecules.

AMMONIUM MOLYBDENUM BRONZE
$(NH_4)Mo_3O_9$

Acta Cryst., C$\underline{46}$, 2007-2009.

C2/m, 14.819, 7.708, 6.386, 93.56, Z = 4, R = 0.074. Units of six edge-sharing MoO_6 octahedra share vertices to form (001) layers, which are linked by 8-coordinate ammonium ions. Mo-O = 1.67-2.38, N...O = 2.74-3.14 A.

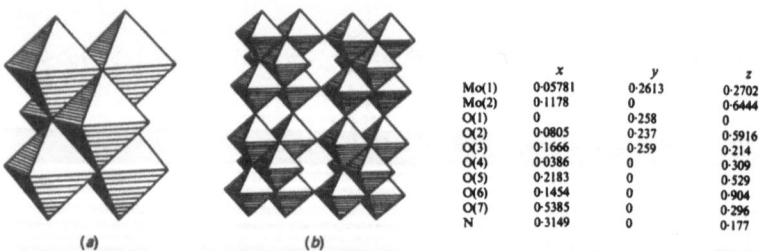

	x	y	z
Mo(1)	0·05781	0·2613	0·2702
Mo(2)	0·1178	0	0·6444
O(1)	0	0·258	0
O(2)	0·0805	0·237	0·5916
O(3)	0·1666	0·259	0·214
O(4)	0·0386	0	0·309
O(5)	0·2183	0	0·529
O(6)	0·1454	0	0·904
O(7)	0·5385	0	0·296
N	0·3149	0	0·177

(a) The unit of six MoO_6 octahedra in $(NH_4)Mo_3O_9$.
(b) Infinite layers formed by vertex sharing of the units shown in (a).

AMMONIUM TRIMOLYBDATE(VI)
$(NH_4)_2Mo_3O_{10}$

Acta Cryst., C$\underline{46}$, 488-489.

Pnma, 13.182, 7.589, 9.286, Z = 4, R = 0.043. Chains along $\underline{b}$ of face- and edge-sharing MoO_6 distorted octahedra, with chains linked by 8- and 10-coordinate ammonium ions. Mo-O = 1.71-2.32, N-O = 2.77-3.39(1) A.

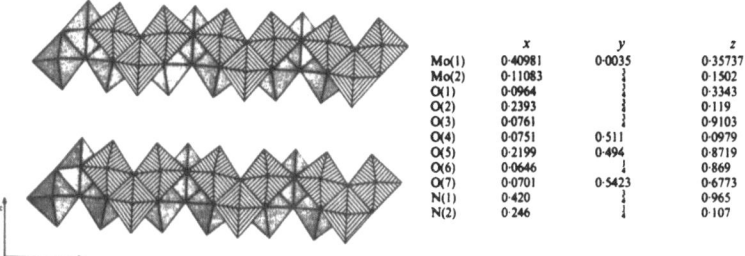

	x	y	z
Mo(1)	0·40981	0·0035	0·35737
Mo(2)	0·11083	¼	0·1502
O(1)	0·0964	¼	0·3343
O(2)	0·2393	¾	0·119
O(3)	0·0761	¾	0·9103
O(4)	0·0751	0·511	0·0979
O(5)	0·2199	0·494	0·8719
O(6)	0·0646	¼	0·869
O(7)	0·0701	0·5423	0·6773
N(1)	0·420	¾	0·965
N(2)	0·246	¼	0·107

AMMONIUM DECAMOLYBDATE
THALLIUM(I) DECAMOLYBDATE
$(NH_4)_8Mo_{10}O_{34}$, $Tl_8Mo_{10}O_{34}$

Acta Cryst., C46, 728.

A (100/001/Ī10) transformation of the ammonium salt cell (55A, 173) indicates isotypism with the thallium salt (51A, 232).

LITHIUM AMMONIUM MOLYBDATE MONOHYDRATE
$LiNH_4MoO_4.H_2O$

Koord. Khim., 16, 793-795.

Pcab, 7.865, 10.880, 12.956, Z = 8, R = 0.068. Isostructural with the LiRb compound (51A, 230), with tetrahedral MoO_4^{2-} ions linked by $LiO_3(H_2O)$ tetrahedra, ammonium ions, and hydrogen bonds.

LITHIUM SODIUM MOLYBDATE DIHYDRATE
$LiNaMoO_4.2H_2O$

Koord. Khim., 16, 616-618.

Pbca, 8.219, 10.372, 13.277, Z = 8, R = 0.024. Isostructural with the sodium compound (34A, 276, 281; 41A, 248), with MoO_4^{2-} ions linked by $NaO_4(H_2O)_2$ octahedra and $LiO_3(H_2O)_2$ trigonal bipyramids; the water molecules are hydrogen bonded to molybdate oxygen atoms. Mo-O = 1.754-1.775, Na-O = 2.354-2.476, Li-O = 2.058-2.176, O-H...O = 2.74-2.86 A.

	x	$\times 10^4$ y	z
Mo	4888	1919	4748
O(1)	6367	1517	3828
O(2)	4347	3554	4568
O(3)	5665	1728	5988
O(4)	3178	923	4614
$O_{w(1)}$	5664	4213	6890
$O_{w(2)}$	7815	3605	2808
Na	8473	15	4084
Li	7668	4385	1380
H(1)	9290	5960	2480
H(2)	9290	6800	1510
H(3)	7390	2880	2980
H(4)	8330	4050	3220

○ O_w ● Na • Li

POTASSIUM TETRAMOLYBDATE
KMo_4O_6

J. Less-Common Metals, _161_, 279-293.

Form I, Pbam, 9.7939, 9.4764, 2.8732, Z = 2, R = 0.087. Form II (high-pressure),
P4/mbm, 9.612, 2.950, Z = 2, R = 0.038. Both forms contain chains of
edge-sharing Mo_6 octahedra, connected by oxygen atoms to form channels along _c_,
which contain the K^+ ions.

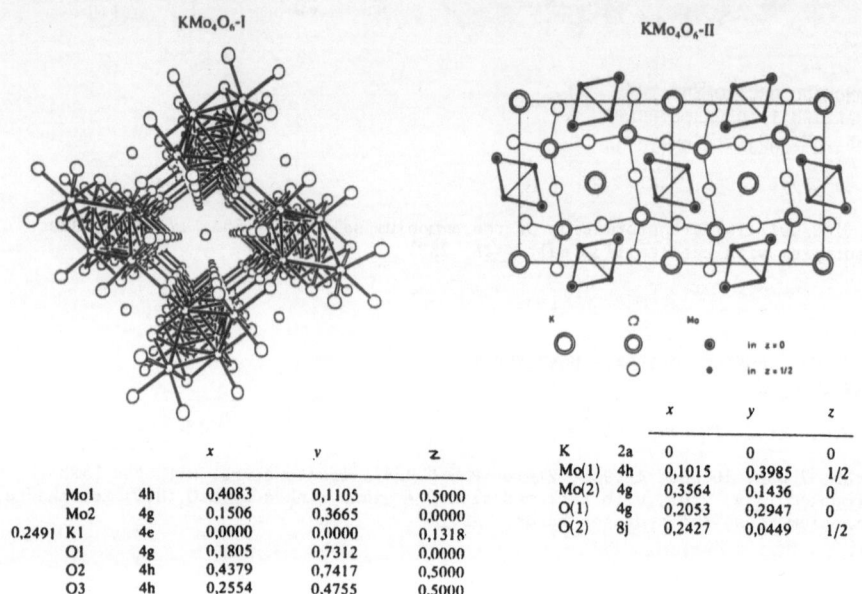

KMo_4O_6-I

KMo_4O_6-II

		x	y	z
Mo1	4h	0,4083	0,1105	0,5000
Mo2	4g	0,1506	0,3665	0,0000
0,2491 K1	4e	0,0000	0,0000	0,1318
O1	4g	0,1805	0,7312	0,0000
O2	4h	0,4379	0,7417	0,5000
O3	4h	0,2554	0,4755	0,5000

		x	y	z
K	2a	0	0	0
Mo(1)	4h	0,1015	0,3985	1/2
Mo(2)	4g	0,3564	0,1436	0
O(1)	4g	0,2053	0,2947	0
O(2)	8j	0,2427	0,0449	1/2

SODIUM RUBIDIUM MOLYBDATE ENNEAHYDRATE
SODIUM POTASSIUM TUNGSTATE ENNEAHYDRATE
$Na_3Rb(MoO_4)_2 \cdot 9H_2O$, $Na_3K(WO_4)_2 \cdot 9H_2O$

Kristallografija, _35_, 1094-1098 [Soviet Physics - Crystallography, _35_, 643-646].

$P6_3/m$, a = 9.648, 9.515, c = 12.157, 12.231 A, Z = 2, R = 0.043, 0.061. MoO_4
octahedra linked by $NaO_2(H_2O)_4$ octahedra (three sharing faces), $Rb(H_2O)_6$ or
$K(H_2O)_6$ trigonal prisms, and hydrogen bonds.

TIN PENTAMOLYBDATE
$SnMo_5O_8$

Acta Cryst., C_46_, 1188-1190.

$P2_1/c$, 7.533, 9.268, 9.970, 109.73, Z = 4, R = 0.033. Isostructural with the La
compound (this volume, p. 194), with chains of $Mo_{10}O_{18}$ clusters, cross-linked to
form channels which contain Sn.

	x	*y*	*z*
Sn	0·26701	0·51884	0·52159
Mo(1)	0·61071	−0·11470	0·48243
Mo(2)	0·38293	−0·11723	0·68149
Mo(3)	0·17715	−0·12160	0·38412
Mo(4)	0·80961	−0·13067	0·29737
Mo(5)	−0·00029	−0·12236	0·58542
O(1)	0·4018	0·0051	0·8498
O(2)	0·0047	−0·2317	0·9306
O(3)	−0·0152	0·0036	0·2358
O(4)	0·3977	−0·2563	0·0186
O(5)	0·7999	−0·2503	0·1197
O(6)	0·2032	−0·2520	0·2201
O(7)	0·5994	−0·2435	0·8247
O(8)	0·3874	−0·0003	0·3340

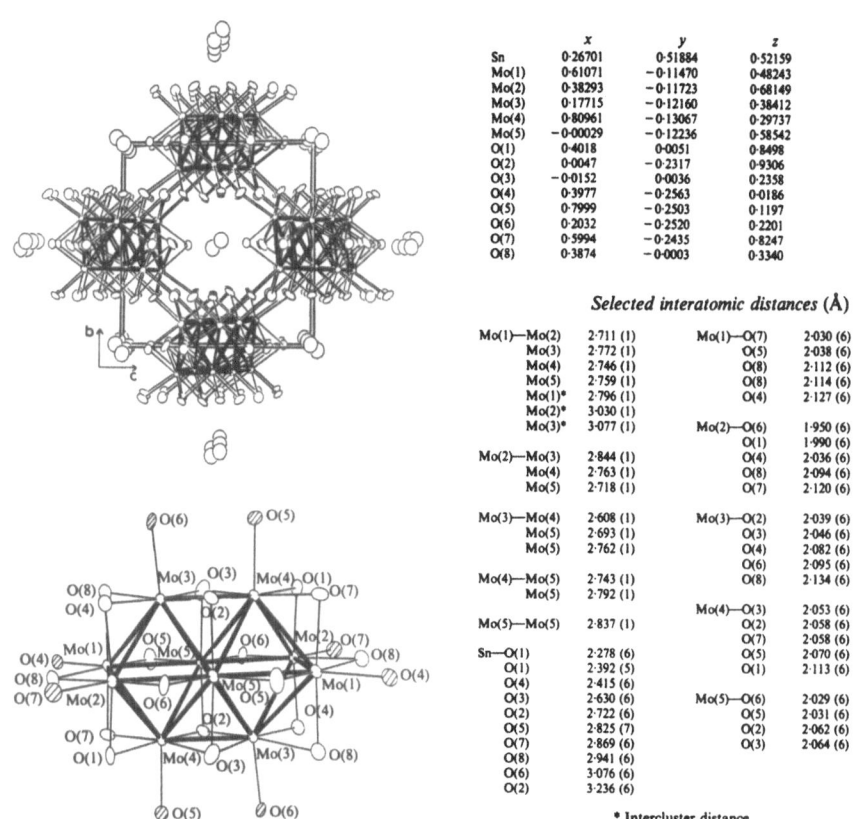

Selected interatomic distances (Å)

Mo(1)—Mo(2)	2·711 (1)	Mo(1)—O(7)	2·030 (6)
Mo(3)	2·772 (1)	O(5)	2·038 (6)
Mo(4)	2·746 (1)	O(8)	2·112 (6)
Mo(5)	2·759 (1)	O(8)	2·114 (6)
Mo(1)*	2·796 (1)	O(4)	2·127 (6)
Mo(2)*	3·030 (1)		
Mo(3)*	3·077 (1)	Mo(2)—O(6)	1·950 (6)
		O(1)	1·990 (6)
Mo(2)—Mo(3)	2·844 (1)	O(4)	2·036 (6)
Mo(4)	2·763 (1)	O(8)	2·094 (6)
Mo(5)	2·718 (1)	O(7)	2·120 (6)
Mo(3)—Mo(4)	2·608 (1)	Mo(3)—O(2)	2·039 (6)
Mo(5)	2·693 (1)	O(3)	2·046 (6)
Mo(5)	2·762 (1)	O(4)	2·082 (6)
		O(6)	2·095 (6)
Mo(4)—Mo(5)	2·743 (1)	O(8)	2·134 (6)
Mo(5)	2·792 (1)		
		Mo(4)—O(3)	2·053 (6)
Mo(5)—Mo(5)	2·837 (1)	O(2)	2·058 (6)
		O(7)	2·058 (6)
Sn—O(1)	2·278 (6)	O(5)	2·070 (6)
O(1)	2·392 (5)	O(1)	2·113 (6)
O(4)	2·415 (6)		
O(3)	2·630 (6)	Mo(5)—O(6)	2·029 (6)
O(2)	2·722 (6)	O(5)	2·031 (6)
O(5)	2·825 (7)	O(2)	2·062 (6)
O(7)	2·869 (6)	O(3)	2·064 (6)
O(8)	2·941 (6)		
O(6)	3·076 (6)		
O(2)	3·236 (6)		

* Intercluster distance.

POTASSIUM ANTIMONY(III) MOLYBDATE
$KSbMo_2O_8$

J. Solid State Chem., **86**, 188–194.

P$\bar{1}$, 5.015, 7.422, 10.304, 90.45, 100.29, 107.79, Z = 2, R = 0.028.   (010) Layers
of Sb_2O_{10} pairs of two edge-sharing SbO_6 octahedra, linked by MoO_5 polyhedra and
by chains of edge-sharing MoO_6 octahedra; 9-coordinate K ions are between the
layers.   Sb-O = 1.99–2.96, Mo-O = 1.70–2.64 A.

	x	*y*	*z*
K	0.40273	0.03602	0.31144
Sb	0.05281	0.47011	0.19223
Mo(1)	0.21191	0.36463	0.54004
Mo(2)	0.35783	0.24446	0.92693
O(1)	0.3589	0.7079	0.1738
O(2)	0.3740	0.4695	0.3862
O(3)	0.2655	0.1493	0.5387
O(4)	0.1351	0.3793	0.7127
O(5)	0.1980	0.4275	0.9662
O(6)	0.5079	0.8100	0.9192
O(7)	0.8950	0.9582	0.1449
O(8)	0.1336	0.6741	0.5524

SODIUM PHOSPHOMOLYBDATE HYDRATE
$Na_3Mo_2P_2O_{11}(OH).2H_2O$

Inorg. Chem., 29, 2879-2881.

$P2_1/n$, 8.128, 13.230, 11.441, 108.84, Z = 4, R = 0.046, atomic positional
parameters not listed. Pairs of edge-shared MoO_6 octahedra are linked into (101)
layers by corner- and edge-sharing with PO_4 tetrahedra. Layers are linked by Na
ions which have distorted octahedral coordination (one Na has a seventh oxygen
neighbour) and by hydrogen bonding via the water molecules.

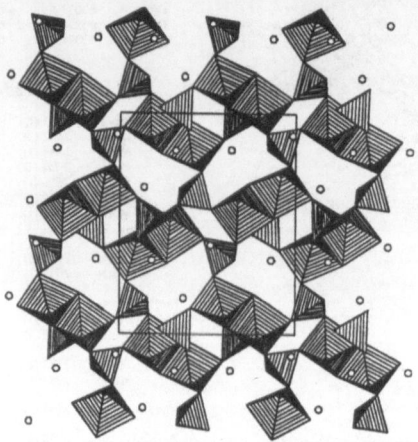

SODIUM MOLYBDENUM(V) PHOSPHATE
$\epsilon-NaMo_2P_3O_{13}$

J. Solid State Chem., 89, 10-15.

$P\bar{1}$, 6.352, 7.448, 10.991, 75.08, 85.33, 79.10, Z = 2, R = 0.033. Framework of
corner-sharing MoO_6 octahedra, PO_4 tetrahedra, and P_2O_7 groups, with Na ions in
cages. Mo-O = 1.66-2.21, P-O = 1.48-1.61 A.

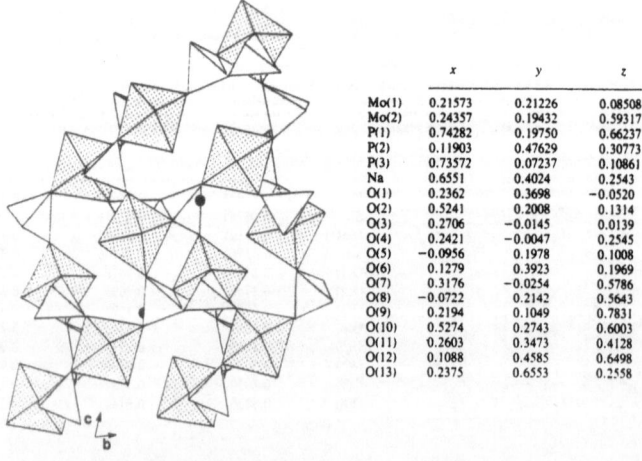

	x	y	z
Mo(1)	0.21573	0.21226	0.08508
Mo(2)	0.24357	0.19432	0.59317
P(1)	0.74282	0.19750	0.66237
P(2)	0.11903	0.47629	0.30773
P(3)	0.73572	0.07237	0.10861
Na	0.6551	0.4024	0.2543
O(1)	0.2362	0.3698	−0.0520
O(2)	0.5241	0.2008	0.1314
O(3)	0.2706	−0.0145	0.0139
O(4)	0.2421	−0.0047	0.2545
O(5)	−0.0956	0.1978	0.1008
O(6)	0.1279	0.3923	0.1969
O(7)	0.3176	−0.0254	0.5786
O(8)	−0.0722	0.2142	0.5643
O(9)	0.2194	0.1049	0.7831
O(10)	0.5274	0.2743	0.6003
O(11)	0.2603	0.3473	0.4128
O(12)	0.1088	0.4585	0.6498
O(13)	0.2375	0.6553	0.2558

ε-NaMo$_2$P$_3$O$_{13}$

ζ-NaMo$_2$P$_3$O$_{13}$

J. Solid State Chem., 89, 31-38.

P2$_1$/c, 6.3682, 22.2546, 8.6172, 126.139, Z = 4, R = 0.062. Framework of
corner-sharing MoO$_6$ octahedra, PO$_4$ tetrahedra, and P$_2$O$_7$ groups, with Na ions in
large cavities. Mo-O = 1.67-2.23, P-O = 1.50-1.62 A.

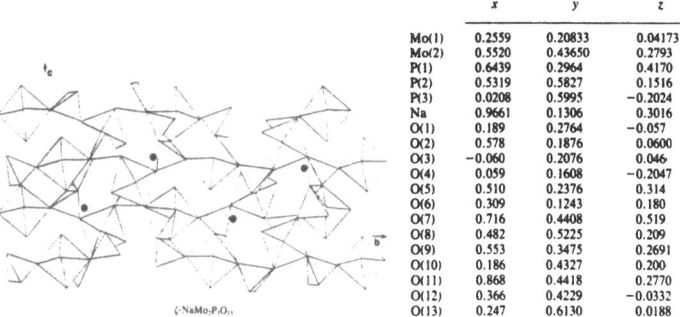

	x	y	z
Mo(1)	0.2559	0.20833	0.04173
Mo(2)	0.5520	0.43650	0.2793
P(1)	0.6439	0.2964	0.4170
P(2)	0.5319	0.5827	0.1516
P(3)	0.0208	0.5995	-0.2024
Na	0.9661	0.1306	0.3016
O(1)	0.189	0.2764	-0.057
O(2)	0.578	0.1876	0.0600
O(3)	-0.060	0.2076	0.046
O(4)	0.059	0.1608	-0.2047
O(5)	0.510	0.2376	0.314
O(6)	0.309	0.1243	0.180
O(7)	0.716	0.4408	0.519
O(8)	0.482	0.5225	0.209
O(9)	0.553	0.3475	0.2691
O(10)	0.186	0.4327	0.200
O(11)	0.868	0.4418	0.2770
O(12)	0.366	0.4229	-0.0332
O(13)	0.247	0.6130	0.0188

POTASSIUM MOLYBDENYL PHOSPHATE DIPHOSPHATE
β-KMo$_2$P$_3$O$_{13}$ K(MoO)$_2$(PO$_4$)(P$_2$O$_7$)

Acta Cryst., C46, 2009-2011.

P2$_1$/c, 9.701, 18.848, 6.389, 106.96, Z = 4, R = 0.026. Framework of
corner-sharing MoO$_6$, PO$_4$, and P$_2$O$_7$ groups, more symmetrical than in the Cs
compound (55A, 177), with K ions (two half-occupancy sites) in tunnels.
[Orthorhombic form described in 55A, 176.]

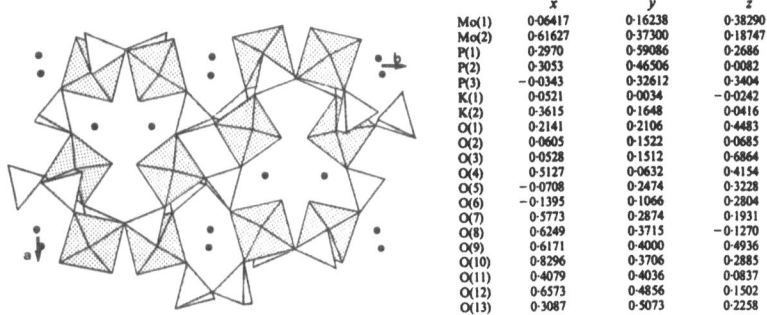

	x	y	z
Mo(1)	0·06417	0·16238	0·38290
Mo(2)	0·61627	0·37300	0·18747
P(1)	0·2970	0·59086	0·2686
P(2)	0·3053	0·46506	0·0082
P(3)	-0·0343	0·32612	0·3404
K(1)	0·0521	0·0034	-0·0242
K(2)	0·3615	0·1648	0·0416
O(1)	0·2141	0·2106	0·4483
O(2)	0·0605	0·1522	0·0685
O(3)	0·0528	0·1512	0·6864
O(4)	0·5127	0·0632	0·4154
O(5)	-0·0708	0·2474	0·3228
O(6)	-0·1395	0·1066	0·2804
O(7)	0·5773	0·2874	0·1931
O(8)	0·6249	0·3715	-0·1270
O(9)	0·6171	0·4000	0·4936
O(10)	0·8296	0·3706	0·2885
O(11)	0·4079	0·4036	0·0837
O(12)	0·6573	0·4856	0·1502
O(13)	0·3087	0·5073	0·2258

CAESIUM PHOSPHOMOLYBDATE
Cs$_4$Mo$_6$P$_{10}$O$_{38}$

J. Solid State Chem., 89, 215-219.

P2$_1$/m, 9.659, 14.404, 6.446, 105.58, Z = 1, R = 0.070. Framework of pairs of
corner-sharing Mo(IV)O$_6$ octahedra, and corner-sharing Mo(III)O$_6$ octahedra, PO$_4$
tetrahedra, and P$_2$O$_7$ groups, with tunnels containing Cs ions. Removal of one Cs
gives the triclinic Cs$_3$ compound (55A, 178).

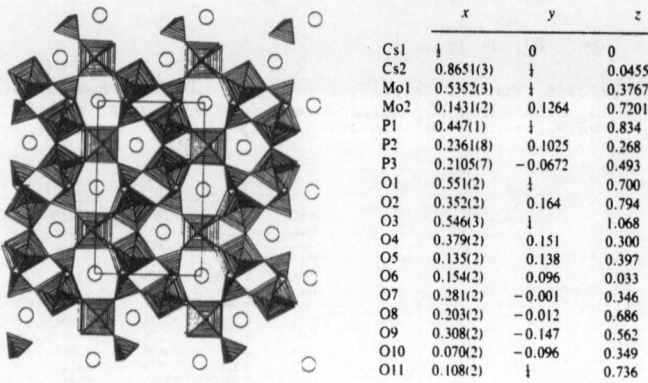

	x	y	z
Cs1	¼	0	0
Cs2	0.8651(3)	¼	0.0455
Mo1	0.5352(3)	¼	0.3767
Mo2	0.1431(2)	0.1264	0.7201
P1	0.447(1)	¼	0.834
P2	0.2361(8)	0.1025	0.268
P3	0.2105(7)	-0.0672	0.493
O1	0.551(2)	¼	0.700
O2	0.352(2)	0.164	0.794
O3	0.546(3)	¼	1.068
O4	0.379(2)	0.151	0.300
O5	0.135(2)	0.138	0.397
O6	0.154(2)	0.096	0.033
O7	0.281(2)	-0.001	0.346
O8	0.203(2)	-0.012	0.686
O9	0.308(2)	-0.147	0.562
O10	0.070(2)	-0.096	0.349
O11	0.108(2)	¼	0.736

BARIUM MOLYBDENUM(V) DIPHOSPHATE
$BaMo_2P_4O_{16}$

J. Solid State Chem., 89, 83-87.

$P2_1/c$, 6.4394, 12.378, 9.1613, 123.92, Z = 2, R = 0.031. Framework of corner-sharing MoO_6 octahedra and $O_3P-O-PO_3$ groups, with 10-coordinate Ba ions in cages. Mo-O = 1.66-2.18, P-O = 1.47-1.59 A.

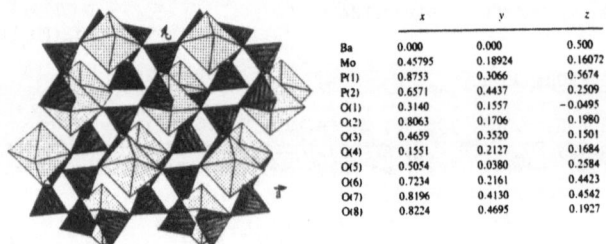

	x	y	z
Ba	0.000	0.000	0.500
Mo	0.45795	0.18924	0.16072
P(1)	0.8753	0.3066	0.5674
P(2)	0.6571	0.4437	0.2509
O(1)	0.3140	0.1557	-0.0495
O(2)	0.8063	0.1706	0.1980
O(3)	0.4659	0.3520	0.1501
O(4)	0.1551	0.2127	0.1684
O(5)	0.5054	0.0380	0.2584
O(6)	0.7234	0.2161	0.4423
O(7)	0.8196	0.4130	0.4542
O(8)	0.8224	0.4695	0.1927

POTASSIUM ALUMINUM PHOSPHOMOLYBDATE
$K_2AlMo_3P_8O_{28}$

Acta Cryst., C46, 1368-1370.

$P\bar{1}$, 4.8171, 7.133, 7.998, 90.53, 92.95, 105.18, Z = 0.5, R = 0.061. Framework of edge- and corner-sharing P_2O_7 groups, MoO_6 and $(Mo,Al)O_6$ octahedra, similar to that in $Na_{0.3}MoP_2O_7$ (55A, 249), with K^+ ions in tunnels. P-O = 1.50-1.53 (terminal), 1.60, 1.61 (bridging), Mo-O = 1.99-2.02, Mo,Al-O = 1.93-2.04, K-8 O = 2.52-3.22(1) A.

	x	y	z
Mo(1)	0·000	0·000	0·000
Mo.Al	0·500	0·500	0·500
K	0·000	0·000	0·500
P(1)	-0·3795	0·2417	0·1953·
P(2)	0·0161	-0·3816	0·2532
O(1)	-0·190	0·102	0·186
O(2)	0·094	-0·201	0·153
O(3)	0·368	0·192	0·071
O(4)	0·155	0·351	0·597
O(5)	0·278	0·550	0·305
O(6)	0·544	0·260	0·375
O(7)	-0·197	0·450	0·138

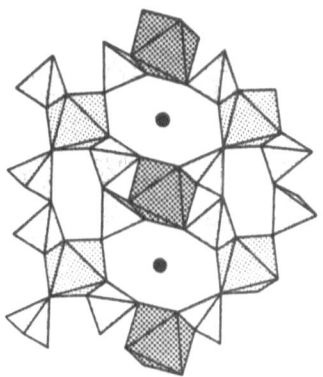

POTASSIUM MOLYBDONIOBOPHOSPHATE
$K_{0.75}NbMoP_3O_{12}$

Acta Cryst., C46, 1381-1383.

Pbcm, 8.8518, 9.1453, 12.5174, Z = 4, R = 0.029. Isostructural with $RbMo_2P_3O_{12}$ (55A, 176), with a framework of $(Mo,Nb)O_6$ octahedra and PO_4 tetrahedra, and tunnels occupied by K ions (partial occupancy (0.75) in the present structure). (Mo,Nb)-O = 1.91-2.11, P-O = 1.52, K-8 O = 2.75-3.28(1) A.

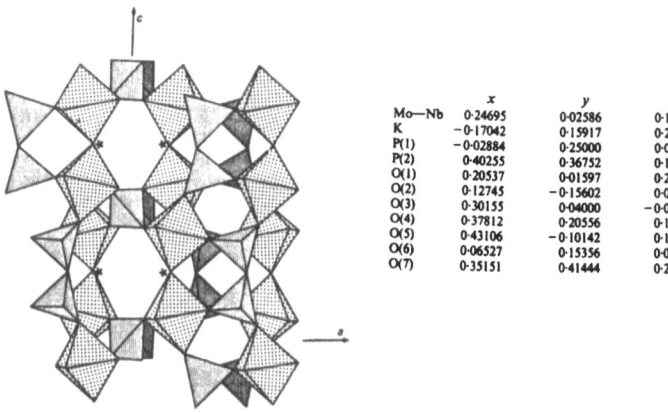

	x	y	z
Mo—Nb	0·24695	0·02586	0·10063
K	−0·17042	0·15917	0·25000
P(1)	−0·02884	0·25000	0·00000
P(2)	0·40255	0·36752	0·13221
O(1)	0·20537	0·01597	0·25000
O(2)	0·12745	−0·15602	0·07173
O(3)	0·30155	0·04000	−0·06273
O(4)	0·37812	0·20556	0·12177
O(5)	0·43106	−0·10142	0·12188
O(6)	0·06527	0·15356	0·07317
O(7)	0·35151	0·41444	0·25000

Projection of $K_{0.75}MoNbP_3O_{12}$ along b.

RUBIDIUM SODIUM SCANDIUM MOLYBDATE
$Rb_3Na_3Sc_2(MoO_4)_6$

Kristallografija, 35, 625-629 [Soviet Physics - Crystallography, 35, 365-367].

Pcca, 10.795, 12.497, 18.652, Z = 4, R = 0.063. Framework of MoO_4 tetrahedra and ScO_6 and MO_6 octahedra (M = Sc/Na), with channels along b which contain 6-coordinate Na and 9- and 10-coordinate Rb ions.

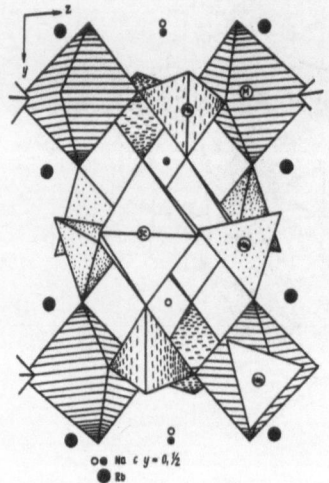

AMMONIUM MOLYBDOSCANDOURANATE HYDRATE
$(NH_4)_2[Sc_2UMo_{12}O_{42}].26H_2O$

Koord. Khim., 16, 207-211.

P$\bar{1}$, 12.000, 12.230, 13.001, 60.06, 117.75, 114.63, Z = 1, R = 0.029.
Centrosymmetric complex anion with Sc atoms having dodecahedral coordination.

LANTHANUM PENTAMOLYBDATE
$LaMo_5O_8$

J. Amer. Chem. Soc., 110, 3295-3296 (1988).

P2_1/a, 9.912, 9.093, 7.575, 109.05, Z = 4, synchrotron and neutron powder data.
Mo_{10} clusters of two edge-sharing Mo_6 octahedra, linked into chains by two
additional Mo-Mo bonds. All Mo (except for those in the shared edge) have
square-pyramidal coordinations to oxygen; Mo-Mo = 2.68-3.07, Mo-O = 1.98-2.16 A.
Chains are connected by bridging O, with tunnels containing 11-coordinate La.

LANTHANUM MOLYBDATE
$LaMo_{7.70}O_{14}$

J. Solid State Chem., 87, 35-43.

C2ca, 11.1708, 9.9848, 9.1960, Z = 4, R = 0.030. Mo_8 clusters (Mo_6 octahedron
with two adjacent faces capped; the Mo(4) site has occupancy 0.85), formed from
MoO_5 polyhedra and MoO_6 octahedra; the clusters are connected to form infinite
Mo_8O_{20} (001) layers, which are conected by strong O-Mo-O bonds, with large
cavities containing La. Mo-Mo = 2.62-2.82, Mo-O = 1.93-2.14, La-11 O = 2.49-3.05
A.

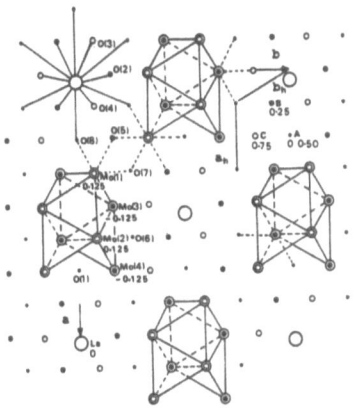

Projection of the idealized LaMo$_{7.70}$O$_{14}$ structure along [001]. The oxygen layers with $z \sim 0$, 1/4, 1/2, 3/4 are indicated by A, B, A, and C, respectively. The La at $z \sim 1/2$ and the Mo at $z \sim 3/8$ and 5/8 have been omitted for purposes of clarity.

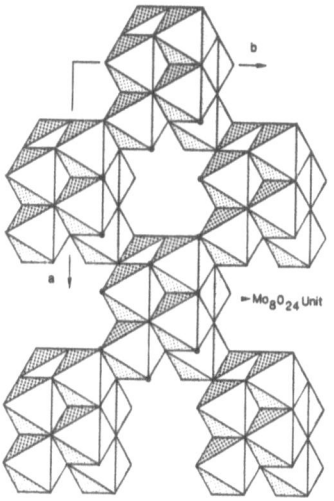

Stacking of the Mo$_8$O$_{24}$ units in the (001) layers. The circles symbolize corners in which the oxygens are shared by the adjoining upper layer.

	x	y	z
Mo(1)	-0.12154	0.07458	0.37944
Mo(2)	0.12254	0.08311	0.37586
Mo(3)	0	-0.16009	0.38115
Mo(4)	0.23430	-0.16269	0.37616
La	0.03038	0	0
O(1)	0.2665	0	1/2
O(2)	-0.0025	-0.1569	-0.2345
O(3)	-0.1258	-0.0815	0.2456
O(4)	0.1188	-0.0779	0.2355
O(5)	-0.2548	0.1703	0.2580
O(6)	0.1200	0.2609	0.4917
O(7)	-0.1230	0.2482	0.4860
O(8)	-0.2465	0	1/2

SODIUM DITUNGSTATE(VI) (PHASE II, HIGH-PRESSURE)
Na$_2$W$_2$O$_7$

Acta Cryst., C46, 317-318.

Cmc2$_1$, 3.7777, 26.6067, 5.4290, Z = 4, R = 0.038. Perovskite-type slabs of WO$_6$ octahedra and 12-coordiante Na(1) ions, with 7-coordinate Na(2) ions between the slabs. W-O = 1.72-2.29, Na-O = 2.39-2.95 A. The normal-pressure form has been described previously (41A, 255).

	x	y	z
W(1)	0	0·44470	0·10218†
W(2)	0	0·34024	0·5745
Na(1)	½	0·5508	0·1047
Na(2)	½	0·2874	0·0523
O(1)	0	0·3936	0·9047
O(2)	0	0·5076	0·3457
O(3)	0	0·4084	0·3942
O(4)	½	0·3517	0·6240
O(5)	0	0·2910	0·7847
O(6)	½	0·4525	0·1483
O(7)	0	0·3059	0·3062

† Fixed.

194 INORGANIC COMPOUNDS

POTASSIUM DITUNGSTATE(VI)
$K_2W_2O_7$

Z. Naturforsch., 45B, 107-110.

$P2_1/c$, 3.883, 13.653, 5.960, 90.40, Z = 2, R = 0.054. (010) Layers of edge- and corner-sharing WO_6 distorted octahedra, linked by 7-coordinate K ions. The Rb compound is isostructural.

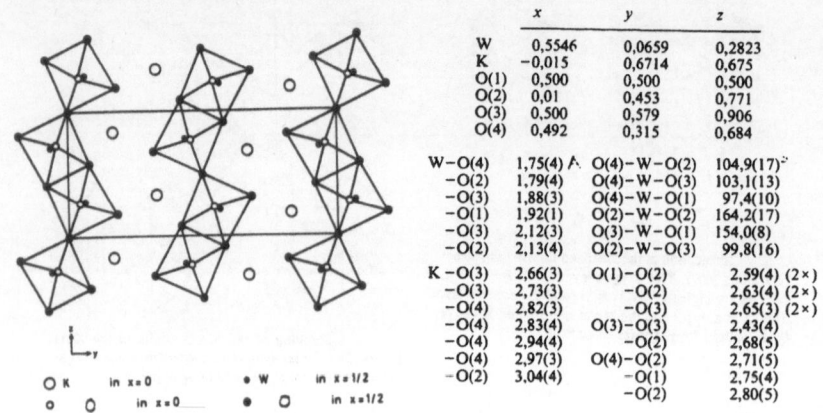

	x	y	z
W	0,5546	0,0659	0,2823
K	−0,015	0,6714	0,675
O(1)	0,500	0,500	0,500
O(2)	0,01	0,453	0,771
O(3)	0,500	0,579	0,906
O(4)	0,492	0,315	0,684

W−O(4)	1,75(4) Å	O(4)−W−O(2)	104,9(17)°
−O(2)	1,79(4)	O(4)−W−O(3)	103,1(13)
−O(3)	1,88(3)	O(4)−W−O(1)	97,4(10)
−O(1)	1,92(1)	O(2)−W−O(2)	164,2(17)
−O(3)	2,12(3)	O(3)−W−O(1)	154,0(8)
−O(2)	2,13(4)	O(2)−W−O(3)	99,8(16)
K −O(3)	2,66(3)	O(1)−O(2)	2,59(4) (2×)
−O(3)	2,73(3)	−O(2)	2,63(4) (2×)
−O(4)	2,82(3)	−O(3)	2,65(3) (2×)
−O(4)	2,83(4)	O(3)−O(3)	2,43(4)
−O(4)	2,94(4)	−O(2)	2,68(5)
−O(4)	2,97(3)	O(4)−O(2)	2,71(5)
−O(2)	3,04(4)	−O(1)	2,75(4)
		−O(2)	2,80(5)

O K in x = 0 • W in x = 1/2 o O in x = 0 • O in x = 1/2

POTASSIUM 11-TUNGSTOBORATE
$K_8H[BW_{11}O_{39}].13H_2O$

Yingyong Huaxue, 6, No. 5, 16-20 (1989).

$P\bar{4}3m$, 10.710, Z = 1, R = 0.045. α-Keggin type anion with a BO_4 tetrahedron.

SODIUM OCTADECATUNGSTOHEXAPHOSPHATE HYDRATE
$Na_{14}H_6P_6W_{18}O_{79}.37.5H_2O$

J. Amer. Chem. Soc., 112, 9386-9387.

C2/c, 16.754, 18.762, 35.898, 99.80, Z = 4, R = 0.043, atomic positional parameters not listed. The heteropolyanion contains two β_3-A-PW_8O_{33} units linked by a central $P_4W_2O_{21}$ unit; the latter contains two corner-sharing WO_6 octahedra and four externally attached PO_4 tetrahedra, which share oxygens with the PW_8 units.

POLYTUNGSTATES

$K_{8.3}Na_{1.7}[Cu_4(H_2O)_2(PW_9O_{34})_2] \cdot 24H_2O$ (I), $Na_{14}Cu[Cu_4(H_2O)_2(P_2W_{15}O_{56})_2] \cdot 53H_2O$ (II)

Inorg. Chem., 29, 1235-1241.

I, P$\bar{1}$, 12.369, 16.957, 11.736, 108.64, 98.47, 82.42, Z = 1, R = 0.043. II, P$\bar{1}$, 13.399, 25.017, 13.339, 104.84, 114.49, 82.61, Z = 1, R = 0.068. Two B-α-PW$_9$O$_{34}$$^{9-}$ units in I, and two α-P$_2$W$_{15}$O$_{56}$$^{12-}$ units in II are linked by a set of four Cu^{2+} ions which have CuO$_4$(H$_2$O)$_2$ distorted octahedral coordinations.

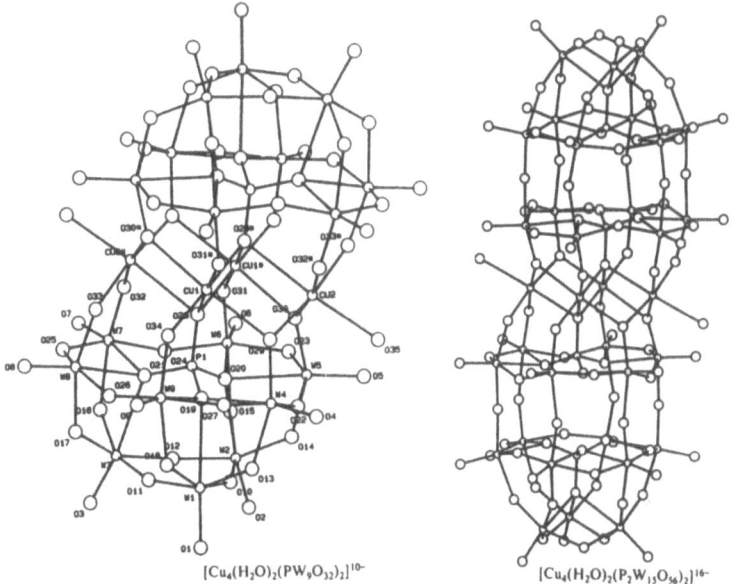

$[Cu_4(H_2O)_2(PW_9O_{32})_2]^{10-}$ $[Cu_4(H_2O)_2(P_2W_{15}O_{56})_2]^{16-}$

COPPER(I) LANTHANON DITUNGSTATES

CuLnW$_2$O$_8$ (Ln = La, Sm)

J. Less-Common Metals, 162, 141-147.

P$\bar{1}$, a = 7.593, 7.407, b = 8.081, 7.944, c = 7.270, 7.174 A, α = 114.97, 114.99, β = 116.30, 116.31, γ = 56.49, 56.89°, Z = 2, R = 0.058, 0.065. α-LiPrW$_2$O$_8$-type (45A, 264), with WO$_6$ octahedra linked to form a W$_4$O$_{16}$$^{8-}$ ion; these anions are connected by Ln^{3+} ions which have 8-coordination and Cu$^+$ ions which have (2+2)-coordination. W-O = 1.72-2.34, La-O = 2.45-2.68, Sm-O = 2.33-2.67, Cu-O = 1.89-1.96, 2.36-2.70 A, O-Cu-O = 146, 135°.

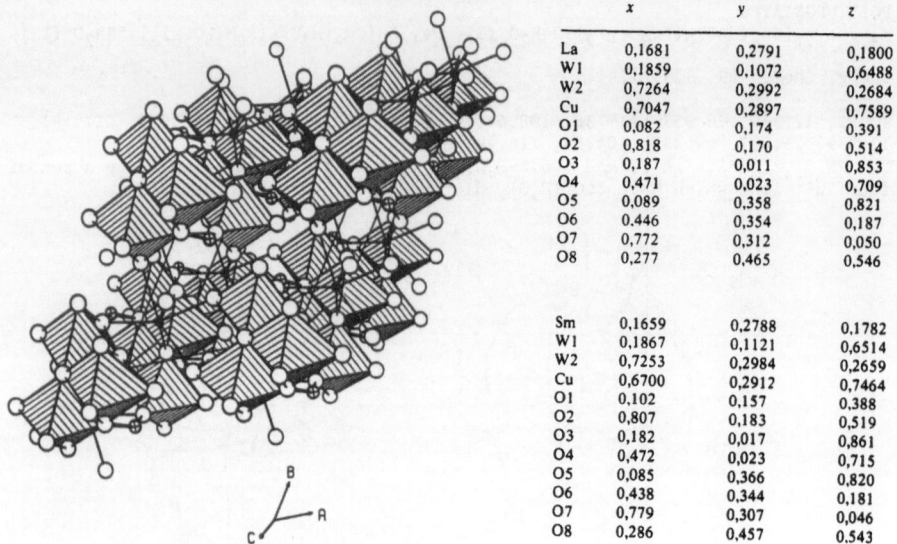

	x	y	z
La	0,1681	0,2791	0,1800
W1	0,1859	0,1072	0,6488
W2	0,7264	0,2992	0,2684
Cu	0,7047	0,2897	0,7589
O1	0,082	0,174	0,391
O2	0,818	0,170	0,514
O3	0,187	0,011	0,853
O4	0,471	0,023	0,709
O5	0,089	0,358	0,821
O6	0,446	0,354	0,187
O7	0,772	0,312	0,050
O8	0,277	0,465	0,546
Sm	0,1659	0,2788	0,1782
W1	0,1867	0,1121	0,6514
W2	0,7253	0,2984	0,2659
Cu	0,6700	0,2912	0,7464
O1	0,102	0,157	0,388
O2	0,807	0,183	0,519
O3	0,182	0,017	0,861
O4	0,472	0,023	0,715
O5	0,085	0,366	0,820
O6	0,438	0,344	0,181
O7	0,779	0,307	0,046
O8	0,286	0,457	0,543

POTASSIUM EUROPIUM ANTIMONY TUNGSTATE
$K_{15}H_3[Eu_3 (H_2O)_3(SbW_9O_{33})(W_5O_{18})_3].25 \cdot 5H_2O$

J. Chem. Soc., Dalton, 1687-1696.

C2/m, 30.250, 18.568, 22.101, 109.19, Z = 4, R = 0.125 (poor crystal quality).
The polyanion contains a central $Eu_3(H_2O)_3$ core coordinated to a B-type α-SbW_9O_{33}
unit and three W_5O_{18} units; each Eu^{3+} has 8-coordination.

LITHIUM PERMANGANATE TRIHYDRATE
$LiMnO_4.3H_2O$

Z. anorg. Chem., _590_, 18-22.

$P6_3mc$, 7.794, 5.427, Z = 2, R = 0.033. Isostructural with the perchlorate (_3_,
117; _49A_, 315), with MnO_4^- tetrahedra and chains of face-sharing $Li(H_2O)_6^+$
octahedra, linked by hydrogen bonds. Mn-O = 1.61, 1.62, Li-O = 2.09, 2.18 A.

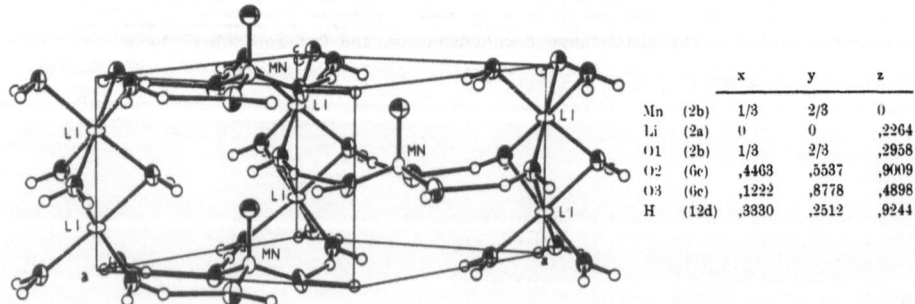

		x	y	z
Mn	(2b)	1/3	2/3	0
Li	(2a)	0	0	,2264
O1	(2b)	1/3	2/3	,2958
O2	(6c)	,4463	,5537	,9009
O3	(6c)	,1222	,8778	,4898
H	(12d)	,3330	,2512	,9244

LITHIUM MANGANESE(IV) OXIDE
$Li_2Mn_4O_9$

Mater. Res. Bull., <u>25</u>, 657-664.

Fd3m, 8.174, Z~3.5, neutron powder data. 7Li in 8(a): 1/8,1/8,1/8; 14Mn in
16(d): 1/2,1/2,1/2; 32 O in 32(e): x,x,x, x = 0.2634. Spinel-type (<u>1</u>, 350).

SODIUM TETRA(PERMANGANATO)DILITHATE
$Na_{10}[Li_2(MnO_4)_4]$

Angew. Chem., <u>102</u>, 835-836 [Angew. Chem. Int. Edn. Engl., <u>29</u>, 800-801]; Z.
anorg. Chem., <u>590</u>, 7-17.

Pnma, 10.481, 15.184, 10.444, Z = 4 [not 16], R = 0.086. Chains along <u>b</u> of
alternating 6- and 8-membered rings built from corner-sharing MnO_4 and LiO_4
tetrahedra, linked by 5- to 7-coordinate Na ions.

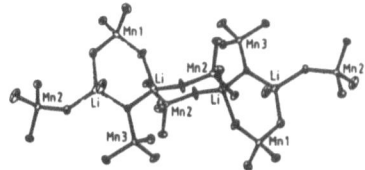

POTASSIUM LITHIUM MANGANATE(V)
$K_{11}Li(MnO_4)_4$ $K_{11}[Li(OMnO_3)_4]$

Z. anorg. Chem., <u>586</u>, 106-114.

I$\bar{4}$2m, 7.8718, 17.509, Z = 2, R = 0.039. The structure contains
tetrakis(manganato(V))lithate groupings, with Li coordinated tetrahedrally to four
O atoms of four tetrahedral manganate groups; K ions have irregular
8-coordinations. Li-O = 1.92, Mn-O = 1.68-1.69 A.

BIRNESSITE (Na, Mg, and K)
$M_xMn_2O_4 \cdot yH_2O$ (M_x = $Na_{0.58}$, $Mg_{0.29}$, $K_{0.46}$, y = 1.5, 1.7, 1.4)

Amer. Min., <u>75</u>, 477-489.

C2/m subcells, a = 5.175, 5.049, 5.149, b = 2.850, 2.845, 2.843, c = 7.337,
7.051, 7.176 A, β = 103.18, 96.65, 100.76°, for Na-, Mg-, and K-rich phases, Z =
1, powder data. Structure similar to that of chalcophanite (<u>17</u>, 429; <u>19</u>, 454).
Some of the interlayer water molecules and cations share the same sites;
superstructures result from water/cation ordering.

		x	y	z	Occ*
Mn	Na	0	0	0	
	Mg	0	0	0	
	K	0	0	0	
	Ch	0	0	0	
O1	Na	0.376	0	0.133	
	Mg	0.356	0	0.141	
	K	0.365	0	0.136	
	Ch	0.28	0	0.15	
O2	Na	0.595	0	0.500	2.0
	Mg	0.703	0	0.506	1.5
	K	0.723	0	0.522	1.3
	Ch	½	0	½	
O3	Na	0	0	½	0.3
	K	0	0	½	0.3
Mg	Mg	0.023	0	0.282	0.5
	Ch**	0.099	0	0.304	

Note: Na, Mg, and K indicate Na-, Mg-, and K-birn, respectively; Ch indicates chalcophanite *C2/m* subcell.

* Occupancy factors are atoms/unit-cell.
** Zn position for chalcophanite monoclinic subcell.

TETRAAMMINEPLATINUM(II) BIS[PERTECHNETATE(VII)]
$[Pt(NH_3)_4][TcO_4]_2$

Acta Cryst., C46, 8-10.

$P\bar{1}$, 5.178, 7.725, 7.935, 69.33, 79.74, 77.41, Z = 1, R = 0.026. Square-planar cations and tetrahedral anions, linked by N-H...O hydrogen bonds. Pt-N = 2.05, Tc-O = 1.71, N-H...O = 2.96-3.24(1) A.

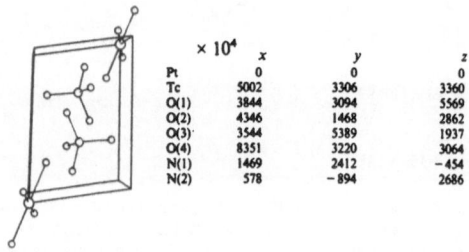

$\times 10^4$	x	y	z
Pt	0	0	0
Tc	5002	3306	3360
O(1)	3844	3094	5569
O(2)	4346	1468	2862
O(3)	3544	5389	1937
O(4)	8351	3220	3064
N(1)	1469	2412	-454
N(2)	578	-894	2686

SODIUM FERRITE
$NaFeO_2$

Z. Kristallogr., 193, 51-69.

Below 1273K, β-form, $Pn2_1a$, a = 5.675-5.681, b = 5.386-5.471, c = 7.144-7.354 A, at 298-1223K, Z = 4 neutron powder data. Above 1273K, γ-form, $P4_12_12$, a = 5.621-5.631, c = 7.365-7.377 A, at 1273-1373K, Z = 4, neutron powder data. Structure as previously described (18, 422; 28, 133); the phase change involves a rotation of FeO_4 tetrahedra.

SODIUM β"-FERRITE
$Na_{1.5}(H_3O)_{0.5}Fe_{10}ZnO_{17}\cdot0\cdot3H_2O$

J. Solid State Chem., 87, 298-307.

$R\bar{3}m$, a = 5.940, c = 35.731 A, Z = 3, R = 0.037. β"-Alumina structure, as for the Na/K ferrite (56A, 160), with water molecules localized in the conduction region.

		Occupancy	x	z
Fe(1)	18h	0.999	0.16949	−0.06972
Fe(2)	6c	0.990	0	0.35066
Fe(3)	6c	0.990	0	0.44981
Fe(4)	3a	0.996	0	0
O(1)	18h	0.984	0.15482	0.03398
O(2)	18h	0.984	0.16413	0.23612
O(3)	6c	0.990	0	0.29680
O(4)	6c	0.990	0	0.09631
O(5)	18h	0.237	0.02269	0.49851
O(6)	9d	0.160	0.5	0.5
O(7)	18h	0.050	0.73320	0.16322
Na(1)	18h	0.118	0.75380	0.16740
Na(2)	6c	0.450	0	0.17254

$$y = -x$$

POTASSIUM FERRATE(V)
K_3FeO_4

Z. anorg. Chem., **586**, 115-124.

Pnma, 7.7016, 9.0920, 7.8370, Z = 4, R = 0.032. FeO_4 distorted tetrahedra, linked by (7+1)- and 7-coordinate K ions. Fe-O = 1.714-1.724, K-O = 2.643-3.147, 3.449 A.

	x	y	z
Fe	0.2746	1/4	0.0087
K(1)	0.0363	0.4629	0.2805
K(2)	0.3279	1/4	0.5623
O(1)	0.1824	0.0835	0.9553
O(2)	0.9789	1/4	0.5779
O(3)	0.2847	1/4	0.2277

POTASSIUM TETRAFERRATE(III)
$K_{14}Fe_4O_{13}$

Z. anorg. Chem., **580**, 57-70.

$P2_1/c$, 6.779, 29.562, 6.721, 120.31, Z = 2, R = 0.045. Isostructural with $Na_{14}Al_4O_{13}$ (**50A**, 173). Fe-4 O = 1.86-1.95 A.

		x 10^5	y	z
K(1)	4e	27020	99911	04042
K(2)	4e	28181	88583	04223
K(3)	4e	72410	78024	94851
K(4)	4e	14796	94967	49876
K(5)	4e	87016	84879	50207
K(6)	4e	27466	77047	05440
K(7)	4e	74005	88812	95706
Fe(1)	4e	65423	94560	49937
Fe(2)	4e	35251	83626	49912
O(1)	4e	47248	89641	50106
O(2)	4e	32559	05665	76663
O(3)	4e	33137	67151	26540
O(4)	4e	07107	82954	23161
O(5)	4e	06192	05683	23535
O(6)	4e	56286	70599	00165
O(7)	2d	50000	0	50000

SODIUM SILICATE FERRITE
$Na_{1-x}(Fe_{1-x}Si_x)O_2$ (x = 0-0.1, 0.125-0.200)

J. Solid State Chem., 85, 202-219.

x = 0-0.1, Pn2₁a, a = 5.6731-5.5727, b = 5.3853-5.3615, c = 7.1440-7.1845 A, Z = 4, powder data. β-NaFeO₂-type (18, 422; 28, 133).

x = 0.125-0.200, Pbca, a = 5.3923-5.3749, b = 10.7658-10.7280, c = 14.6468-14.5826 A, Z = 16, powder data. KGaO₂-type (41A, 419, but with some errors in oxygen parameters corrected [see also this volume, p. 153]).

REVISED ATOMIC COORDINATES FOR KGaO₂

	x	y	z
Ga(1)	0.258(1)	0.0088(3)	0.1890
Ga(2)	0.280(1)	0.2633(3)	0.0644
K(1)	0.758(3)	0.0140(7)	0.0613
K(2)	0.799(3)	0.2645(7)	0.1852
O(1)	0.569(8)	0.302(2)	0.011
O(2)	0.158(6)	0.406(2)	0.103
O(3)	0.311(7)	0.164(2)	0.155
O(4)	0.059(9)	0.479(3)	0.276

CALCIUM FERRITE
$Ca_2Fe_{22}O_{33}$

Kristallografija, 35, 1105-1109 [Soviet Physics - Crystallography, 35, 650-652].

R32, a = 6.028, c = 62.224 A, Z = 3, powder data. Structure proposed.

BISMUTH IRON OXIDE
$BiFeO_3$

Acta Cryst., B46, 698-702.

R3c, a = 5.57874, c = 13.8688 A, at 294K, Z = 6 (rhombohedral cell, a = 5.6343 A, α = 59.348°, Z = 2), R = 0.024. Bi in 6(a): 0,0,z, z = 0; Fe in 6(a): z = 0.22077; O in 18(b): 0.4428,0.0187,0.9520. Rhombohedrally-distorted perovskite.

STRONTIUM BISMUTH LEAD FERRATE
$Sr_2BiPbFeO_6$

J. Less-Common Metals, 159, 337-341.

Fmmm subcell, 5.425, 5.490, 23.22, Z = 4. Perovskite-like slabs of FeO_6 octahedra between (Bi,Pb)O double sheets.

			x	y	z
0.4875	Bi1	8i	0	1/2	0.1893
0.4875	Pb1	8i	0	1/2	0.1893
0.025	Sr1	8i	0	1/2	0.1893
0.975	Sr2	8i	0	0	0.0781
0.0125	Bi2	8i	0	0	0.0781
0.0125	Pb2	8i	0	0	0.0781
1.0000	Fe	4b	1/2	0	0
1.0000	O1	8e	1/4	1/4	0
1.0000	O2	8i	1/2	0	0.0793
1.0000	O3	8i	0	0	0.1900

β"-FERRITE (POTASSIUM, CADMIUM-STABILIZED)
$K_{1.33}Fe_{10.27}Cd_{0.73}O_{16.87}$

Solid State Ionics, <u>40-41</u>, 95-98.

$R\bar{3}m$, a = 6.000, c = 35.908 A, Z = 3, R = 0.040. β"-Alumina structure, but with K
only in the BR site (4K in 18(h): 0.0289,2x,-0.1713).

STRONTIUM LANTHANUM FERRATES
$Sr_2LaFe_3O_8$

J. Solid State Chem., <u>84</u>, 237-244.

Pmma, 5.5095, 11.8845, 5.6028, Z = 2, X-ray and neutron powder data.
Oxygen-deficient perovskite-type structure, with layers of octahedrally- and
tetrahedrally-coordinated Fe along <u>b</u> (OOTOOT). Fe-O = 1.93-2.37 (octahedral),
1.85-1.91 (tetrahedral), Sr/La-12 or 8 O = 2.47-3.05(1) A.

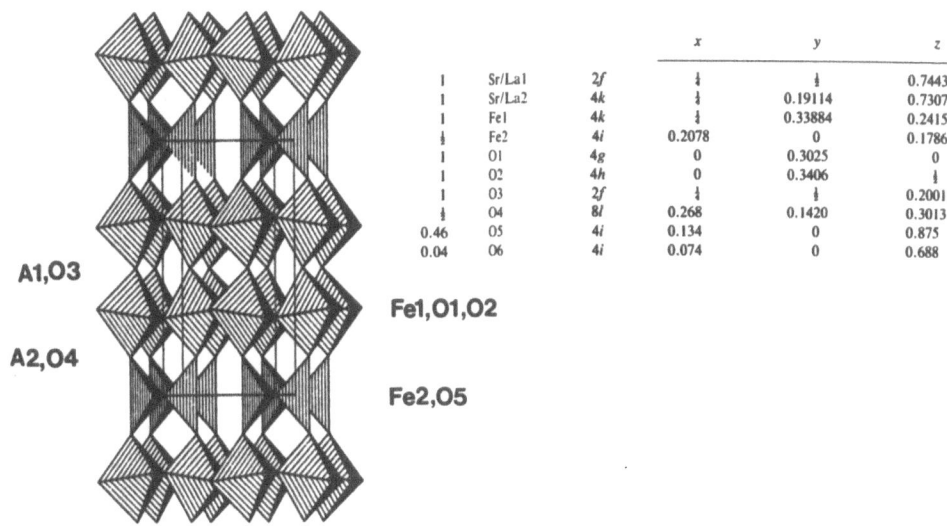

			x	y	z
1	Sr/La1	2f	¼	¼	0.7443
1	Sr/La2	4k	¼	0.19114	0.7307
1	Fe1	4k	¼	0.33884	0.2415
½	Fe2	4i	0.2078	0	0.1786
1	O1	4g	0	0.3025	0
1	O2	4h	0	0.3406	½
1	O3	2f	¼	¼	0.2001
½	O4	8l	0.268	0.1420	0.3013
0.46	O5	4i	0.134	0	0.875
0.04	O6	4i	0.074	0	0.688

A1,O3

A2,O4

Fe1,O1,O2

Fe2,O5

$Sr_2LaFe_3O_{8.94}$

J. Solid State Chem., <u>84</u>, 271-279.

$R\bar{3}c$, a = 5.4784, c = 13.3928 A, Z = 2 (rhombohedral cell has a = 5.4712 A, α =
60.09°, Z = 2/3), neutron powder data. La/Sr in 6(a): 0,0,1/4; Fe in 6(b):
0,0,0; O in 18(e): x,0,1/4, x = 0.5190 (0.5266 at 50K, with a $P\bar{3}m1$ magnetic
structure). Distorted perovskite.

BARIUM LANTHANUM FERRATES
$BaLaFe_2O_{5.91}$

J. Solid State Chem., **85**, 38-43.

Pm3m, 3.9268, at 5K, neutron powder data. Ba/La in 1(b): 1/2,1/2,1/2; Fe in 1(a): 0,0,0; O in 3(d): 1/2,0,0. Perovskite structure (1, 300).

$Ba_6La_2Fe_4O_{15}$

J. Less-Common Metals, **157**, 173-178.

$P6_3mc$, 11.904, 7.111, Z = 2, R = 0.032. The structure contains FeO_4 tetrahedra, FeO_6 octahedra, chains of face-sharing BaO_6 octahedra, and further 6-, 10-, and 12-coordinate Ba ions (one site is occupied by Ba+La). Fe-O = 1.81-1.92 (tetrahedral), 1.97, 2.20 (octahedral) A.

		x	y	z
Ba1	(2a)	0,0	0,0	0,0
Ba2	(2b)	0,3333	0,6667	0,4784
Ba3	(6c)	0,3471	0,1735	0,6648
2/6 Ba and 4/6 La	(6c)	0,9555	0,4778	0,3312
Fe1	(6c)	0,1778	0,8222	0,6622
Fe2	(2b)	0,6667	0,3333	0,5238
O1	(6c)	0,415	0,585	0,166
O2	(6c)	0,752	0,248	0,340
O3	(12d)	0,674	0,066	0,024
O4	(6c)	0,806	0,903	0,762

BARIUM LANTHANUM ALUMINUM IRON OXIDE
$Ba_6La_2Al_{1.5}Fe_{2.5}O_{15}$

Z. anorg. Chem., **588**, 97-101.

$P6_3mc$, 11.814, 7.1003 A, Z = 2, R = 0.046. Isostructural with related materials (56A, 120; this volume, preceding report), with partial Al/Fe ordering (Fe in the tetrahedral site, Fe/Al in the octahedral site).

		x	y	z
Ba1	(2a)	0,0	0,0	0,0
Ba2	(6c)	0,1716	0,8284	0,1749
Ba3	(2b)	0,33333	0,66667	0,4900
2Ba/4La	(6c)	0,47862	0,52138	0,8445
Fe1	(2b)	0,33333	0,66667	0,035
3Fe/3Al	(6c)	0,1789	0,8211	0,674
O1	(12d)	0,678	0,071	0,042
O2	(6c)	0,248	0,752	0,843
O3	(6c)	0,414	0,586	0,176
O4	(6c)	0,903	0,097	0,274

BARIUM CALCIUM LANTHANUM IRON OXIDE
$Ba_5CaLa_2Fe_4O_{15}$

Z. anorg. Chem., $\underline{585}$, 82-86.

$P6_3mc$, 11.770, 7.039, Z = 2, R = 0.041. $Ba_6Nd_2Al_4O_{15}$-type ($\underline{56}$A, 120), with Ca
partially ordered in the 2(a) Ba and 6(c) La sites.

LANTHANON IRON OXIDES
$Ba_6Nd_2Fe_4O_{15}$, $Ba_5SrLa_2Fe_4O_{15}$, $Ba_5SrNd_2Fe_4O_{15}$

J. Less-Common Metals, $\underline{158}$, 147-152.

$P6_3mc$, a = 11.815, 11.819, 11.730, c = 7.078, 7.052, 7.015 A, Z = 2, R = 0.042,
0.052, 0.047. Isostructural with the aluminate ($\underline{56}$A, 120), with face-sharing
BaO_6 octahedra, with Fe in octahedral and tetrahedral sites, and 8-coordinate
M/Ln.

		x	y	z
Ba1	(2a)	0.0	0.0	0.0
Ba2	(2b)	0.3333	0.6667	0.4850
Ba3	(6c)	0.3465	0.1732	0.6671
2/6 Ba + 4/6 Nd	(6c)	0.9572	0.4786	0.3302
Fe1	(6c)	0.1774	0.8226	0.6572
Fe2	(2b)	0.6667	0.3333	0.5198
O1	(6c)	0.415	0.585	0.169
O2	(6c)	0.755	0.245	0.335
O3	(12d)	0.665	0.058	0.039
O4	(6c)	0.813	0.906	0.782
Ba1	(2a)	0.0	0.0	0.0
Ba2	(2b)	0.3333	0.6667	0.4809
Ba3	(6c)	0.3461	0.1730	0.6663
2/6 Sr + 4/6 La	(6c)	0.9552	0.4776	0.3292
Fe1	(6c)	0.1776	0.8224	0.6635
Fe2	(2b)	0.6667	0.3333	0.525
O1	(6c)	0.415	0.585	0.167
O2	(6c)	0.751	0.249	0.344
O3	(12d)	0.672	0.064	0.034
O4	(6c)	0.816	0.907	0.768
Ba1	(2a)	0.0	0.0	0.0
Ba2	(2b)	0.3333	0.6667	0.4843
Ba3	(6c)	0.3455	0.1728	0.6674
2/6 Sr + 4/6 Nd	(6c)	0.9564	0.4782	0.3312
Fe1	(6c)	0.1779	0.8221	0.6617
Fe2	(2b)	0.6667	0.3333	0.5233
O1	(6c)	0.415	0.585	0.166
O2	(6c)	0.750	0.250	0.340
O3	(12d)	0.674	0.064	0.027
O4	(6c)	0.816	0.908	0.769

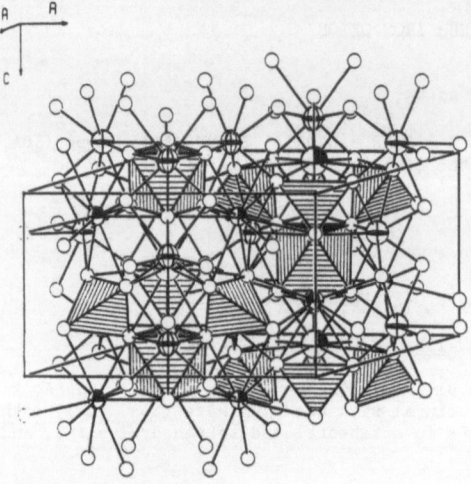

BARIUM CALCIUM LANTHANON FERRATES
$Ba_{4.5}Ca_{1.5}Nd_2Fe_4O_{15}$, $Ba_5CaSm_2Fe_4O_{15}$

J. Less-Common Metals, 162, 175-180.

$P6_3mc$, a = 11.6244, 11.655, c = 6.9205, 6.987 A, Z = 2, R = 0.065, 0.045.
$Ba_6Nd_2Al_4O_{15}$-type (56A, 120), with Ca in one Ba and one Ba/Ln site.

		x	y	z
0.85Ba1/1.15Ca1	(2a)	0,0	0,0	0,0
Ba2	(6c)	0,1713	0,8287	0,173
Ba3	(2b)	0,33333	0,66667	0,490
0,15Ba/1,85Ca/4Nd	(6c)	0,4778	0,5222	0,840
Fe1	(2b)	0,33333	0,66667	0,033
Fe2	(6c)	0,1764	0,8236	0,666
O1	(12d)	0,670	0,063	0,031
O2	(6c)	0,250	0,750	0,847
O3	(6c)	0,417	0,583	0,167
O4	(6c)	0.910	0.090	0.265
(II)				
1,3Ba1/0,7Ca1	(2a)	0,0	0,0	0,0
Ba2	(6c)	0,1718	0,8282	0,1694
Ba3	(2b)	0,33333	0,66667	0,487
0,7Ba/1,3Ca/4Sm	(6c)	0,4783	0,5217	0,8386
Fe1	(2b)	0,33333	0,66667	0,027
Fe2	(6c)	0,1780	0,8220	0,664
O1	(12d)	0,669	0,066	0,026
O2	(6c)	0,248	0,752	0,843
O3	(6c)	0,419	0,581	0,170
O4	(6c)	0,904	0,096	0,252

TERBIUM IRON GARNET
$Tb_3Fe_5O_{12}$

J. Solid State Chem., 84, 39-51.

290K, Ia3d, 12.4339, Z = 8, neutron powder data. Tb in 24(c): 1/8,0,1/4; Fe(1)
in 24(d): (3/8,0,1/4); Fe(2) in 16(a): 0,0,0; O in 96(h):
-0.02752,0.05570,0.15018. Garnet structure.

39, 13, 5K, R$\bar{3}$, a = 10.7478, 10.7444, 10.7442 A, α = 109.43, 109.41, 109.41°, Z = 4, neutron powder data. Slight distortion of the cubic structure.

LUTETIUM IRON COBALT OXIDE
LUTETIUM IRON OXIDE
$LuFeCoO_4$, $LuFe_2O_4$

Acta Cryst., C$\underline{46}$, 1917-1918.

R$\bar{3}$m, a = 3.4180, 3.4406, c = 25.28, 25.28 A, Z = 3, R = 0.016, 0.029. Lu in 3(a): 0,0,0; Fe/Co or Fe in 6(c): 0,0,z, z = 0.21485, 0.21518; O(1) in 6(c): z = 0.1284, 0.1281; O(2) in 6(c): 0.2923, 0.2926. Essentially In_2ZnS_4-type ($\underline{27}$, 246; $\underline{37A}$, 99), as for Yb_2FeO_4 ($\underline{41A}$, 265; $\underline{46A}$, 282).

CAESIUM RUTHENATE(VI)
Cs_2RuO_4

Z. anorg. Chem., $\underline{591}$, 87-94.

Pnma, 8.512, 6.475, 11.458, Z = 4, R = 0.058. β-K_2SO_4-type ($\underline{2}$, 86; $\underline{48A}$, 304); Ru-4 O = 1.75-1.77, Cs-9 or 11 O = 3.02-3.97 A.

CALCIUM RUTHENATE(IV)   STRONTIUM RUTHENATE(IV)
$CaRuO_3$ $SrRuO_3$

Solid State Ionics, $\underline{43}$, 171-177.

Ca, Pnma, 5.524, 7.649, 5.354, Z = 4, R = 0.031. Ca in 4(c): 0.9448,1/4,0.0139; Ru in 4(b): 0,0,1/2; O(1) in 4(c): 0,0258,1/4,0.5920; O(2) in 8(d): 0.2021,0.4518,0.1973. Sr, Pm3m, 3.910, Z = 1, R = 0.032. Sr in 1(a): 0,0,0; Ru in 1(b): 1/2,1/2,1/2; O in 3(c): 0,1/2,1/2. Distorted (Ca) and ideal (Sr) perovskites.

STRONTIUM RUTHENATE(IV)   STRONTIUM DIRUTHENATE(IV)
Sr_2RuO_4 (I) $Sr_3Ru_2O_7$ (II)

Z. anorg. Chem., $\underline{591}$, 161-166.

I, I4/mmm, 3.871, 12.702, Z = 2, R = 0.043. K_2NiF_4-type ($\underline{17}$, 332; $\underline{19}$, 323; $\underline{49A}$, 93); Ru-O = 1.94, 2.07 A.

II, I4/mmm, 3.890, 20.552, Z = 2, R = 0.089. $Sr_3Ti_2O_7$-type ($\underline{22}$, 308; $\underline{24}$, 440); Ru-O = 1.95-2.01 A.

BARIUM RUTHENIUM MANGANESE OXIDE
$Ba_4Ru_{1.1}Mn_{1.9}O_{10}$

Mh. Chem., 121, 635-640.

$Cmc2_1$, 5.735, 13.148, 12.855, Z = 4, R = 0.061. Groups of three face-sharing MO_6 octahedra are connected into sheets; Ba ions have 10- and 11-coordinations. Ru/Mn-O = 1.85-2.17(4), Ba-O = 2.52-3.28(4), Ru/Mn...Ru/Mn = 2.48, 2.68(2) A.

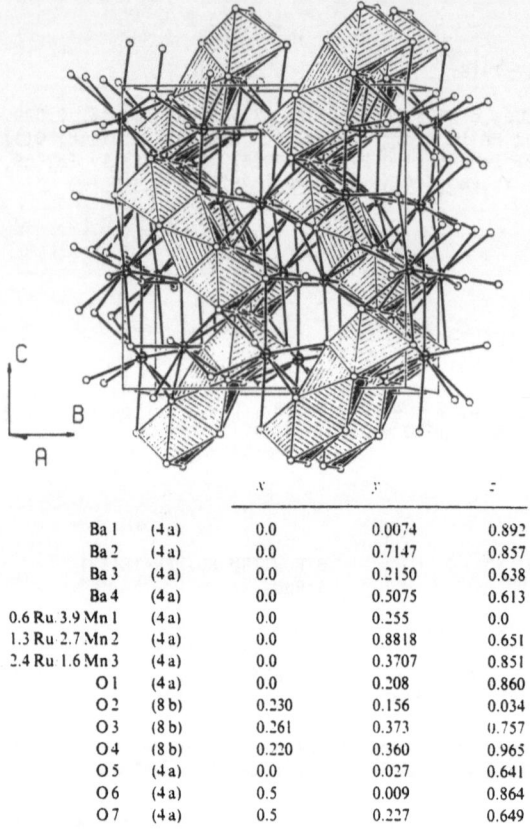

		x	y	z
Ba 1	(4 a)	0.0	0.0074	0.892
Ba 2	(4 a)	0.0	0.7147	0.857
Ba 3	(4 a)	0.0	0.2150	0.638
Ba 4	(4 a)	0.0	0.5075	0.613
0.6 Ru. 3.9 Mn 1	(4 a)	0.0	0.255	0.0
1.3 Ru 2.7 Mn 2	(4 a)	0.0	0.8818	0.651
2.4 Ru 1.6 Mn 3	(4 a)	0.0	0.3707	0.851
O 1	(4 a)	0.0	0.208	0.860
O 2	(8 b)	0.230	0.156	0.034
O 3	(8 b)	0.261	0.373	0.757
O 4	(8 b)	0.220	0.360	0.965
O 5	(4 a)	0.0	0.027	0.641
O 6	(4 a)	0.5	0.009	0.864
O 7	(4 a)	0.5	0.227	0.649

BARIUM NICKEL DIRUTHENATE(V)
BARIUM COBALT(II) DIRUTHENATE(V)
BARIUM ZINC DIRUTHENATE(V)
$Ba_3MRu_2O_9$ (M = Ni, Co, Zn)

J. Solid State Chem., 89, 174-183.

Ni, Zn, $P6_3/mmc$, a = 5.7256, 5.7549, c = 14.0596, 14.1328 A, at 5K, Z = 2, neutron powder data. Co, Cmcm, 5.7456, 9.9177, 14.0862, at 2K (hexagonal at room temperature), Z = 4, neutron powder data. 6H-perovskites, with pairs of face-sharing RuO_6 octahedra, and corner-sharing MO_6 octahedra (M = Ni, Co, Zn), as in $Ba_3NiSb_2O_9$ (44A, 195).

$Ba_3NiRu_2O_9$		x	y	z
Ba1	2b	0	0	1/4
Ba2	4f	1/3	2/3	0.9110
Ni	2a	0	0	0
Ru	4f	1/3	2/3	0.1546
O1	6h	0.4867	-0.0266	1/4
O2	12k	0.1708	0.3415	0.4170

$Ba_3CoRu_2O_9$		x	y	z
Ba1	4c	0	0.000	1/4
Ba2	8f	0	0.3342	0.0897
Co	4a	0	0	0
Ru	8f	0	0.3284	0.8450
O1	4c	0	0.5136	1/4
O2	8g	0.270	0.2477	1/4
O3	8f	0	0.8335	0.0833
O4	16h	0.2547	0.0892	0.0833

$Ba_3ZnRu_2O_9$		x	y	z
Ba1	2b	0	0	1/4
Ba2	4f	1/3	2/3	0.9101
Ni	2a	0	0	0
Ru	4f	1/3	2/3	0.1551
O1	6h	0.4864	-0.0273	1/4
O2	12k	0.1715	0.3430	0.41571

SODIUM RUBIDIUM COBALTATE(II)
$Na_7Rb(CoO_3)_2$

Z. anorg. Chem., 588, 7-18.

C2/m, 10.847, 4.376, 10.720, 91.04, Z = 2, R = 0.099. Planar CoO_3^{4-} groups, linked by 4-coordinate Na and 10-coordinate Rb ions. Co-O = 1.82-1.91 A.

POTASSIUM RUBIDIUM COBALT(II) OXIDE
$K_4Rb_2Co_2O_5$

Z. anorg. Chem., 591, 67-76.

$P4_2/mnm$, 6.742, 11.722, Z = 2, R = 0.10. Isostructural with related materials (55A, 194), with $O_2Co-O-CoO_2^{6-}$ ions; Co-O = 1.93 (bridging), 1.81 (terminal) A.

BARIUM LANTHANUM COBALTATE
$Ba_6La_2Co_4O_{15}$

Z. anorg. Chem., 584, 114-118.

$P6_3mc$, 11.8082, 7.0019, Z = 2, R = 0.056. Co_4O_{15} groups of one CoO_6 octahedron sharing corners with three CoO_4 tetrahedra, linked by Ba ions (6-, 10-, and 12-coordinations) and one Ba/La site (8-coordination). Co-O = 1.76-1.93 (tetrahedral); 2.00, 2.12(4) A (octahedral).

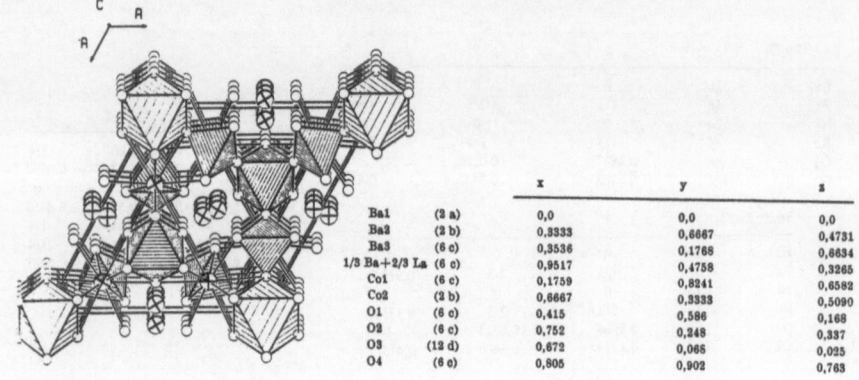

		x	y	z
Ba1	(2 a)	0,0	0,0	0,0
Ba2	(2 b)	0,3333	0,6667	0,4731
Ba3	(6 c)	0,3536	0,1768	0,6634
1/3 Ba+2/3 La	(6 c)	0,9517	0,4758	0,3265
Co1	(6 c)	0,1759	0,8241	0,6582
Co2	(2 b)	0,6667	0,3333	0,5090
O1	(6 c)	0,415	0,586	0,168
O2	(6 c)	0,752	0,248	0,337
O3	(12 d)	0,672	0,065	0,025
O4	(6 c)	0,805	0,902	0,763

STRONTIUM IRIDATE(IV)
$SrIrO_3$

Z. Kristallogr., 191, 239-247.

Im3, 9.340, Z = 12, R = 0.026. Perovskite-type structure, with rows of pairs of edge-sharing IrO_6 octahedra, linked into a three-dimensional framework; Sr is distributed over four sites, each coordinated to six oxygens. Ir-O = 1.96-1.98, Sr-O - 2.41-3.34 A.

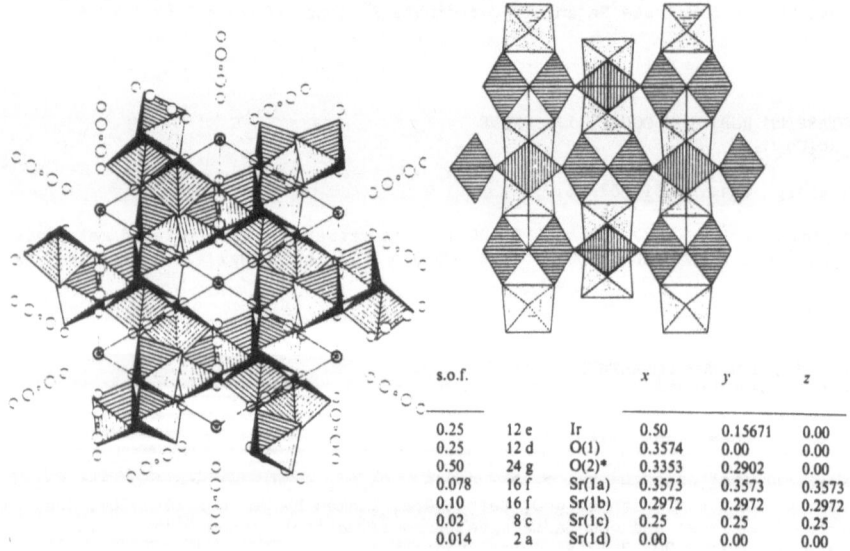

s.o.f.			x	y	z
0.25	12 e	Ir	0.50	0.15671	0.00
0.25	12 d	O(1)	0.3574	0.00	0.00
0.50	24 g	O(2)*	0.3353	0.2902	0.00
0.078	16 f	Sr(1a)	0.3573	0.3573	0.3573
0.10	16 f	Sr(1b)	0.2972	0.2972	0.2972
0.02	8 c	Sr(1c)	0.25	0.25	0.25
0.014	2 a	Sr(1d)	0.00	0.00	0.00

BARIUM IRIDATE TITANATE
$Ba_4Ir_{1.45}Ti_{1.55}O_{10}$

Z. anorg. Chem., 586, 87-92.

$Cmc2_1$, 5.783, 13.362, 13.033, Z = 4, R = 0.063. Isostructural with $Ba_4(Ti,Pt)_2PtO_{10}$ (48A, 214), with groups of three face-sharing $(Ir,Ti)O_6$ octahedra linked into layers, and 10- and 11-coordinated Ba ions. Ir-O = 1.94-2.06, Ir, Ti-O = 1.77-2.26(3) A.

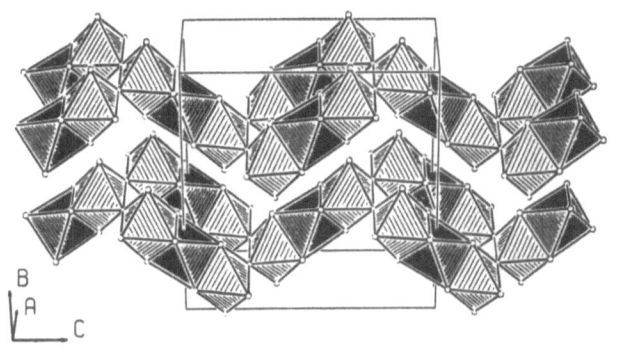

		x	y	z
Ba1	(4a)	0,0	0,0122	0,8986
Ba2	(4a)	0,0	0,7182	0,8619
Ba3	(4a)	0,0	0,2171	0,6423
Ba4	(4a)	0,0	0,5118	0,6107
Ir	(4a)	0,0	0,2500	0,0
0,5Ir/3,5Ti(1)	(4a)	0,0	0,8859	0,6576
1,3Ir/2,7Ti(2)	(4a)	0,0	0,3758	0,8519
O1	(4a)	0,0	0,211	0,847
O2	(4a)	0,0	0,018	0,658
O3	(4a)	0,5	0,018	0,863
O4	(4a)	0,5	0,217	0,645
O5	(8b)	0,220	0,139	0,043
O6	(8b)	0,236	0,373	0,754
O7	(8b)	0,236	0,356	0,969

BARIUM IRIDATE MANGANATE
Ba(Ir,Mn)O$_3$ BaIr$_{0.36}$Mn$_{0.64}$O$_3$

J. Less-Common Metals, 157, 301-306.

R$\bar{3}$m, a = 5.709, c = 21.319, Z = 9, R = 0.036. 9-Layer perovskite, sequence
(hhc)$_3$, with three face-sharing (In,Mn)O$_6$ linked by further corner-sharing.
In,Mn-6 O = 1.94-1.96, Ba-12 O = 2.85-2.96(1) Å.

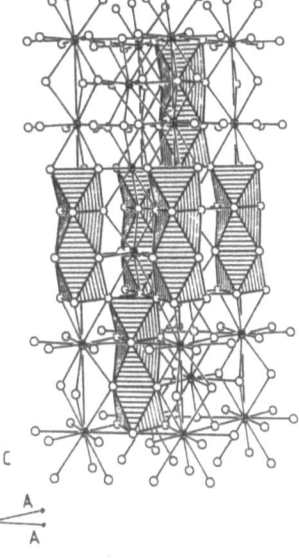

		x	y	z
Ba1	(3a)	0,0	0,0	0,0
Ba2	(6c)	0,0	0,0	0,2189
0,8Ir/2,2Mn	(3b)	0,0	0,0	0,5
2,4Ir/3,6Mn	(6c)	0,0	0,0	0,3830
O1	(18h)	0,151	0,849	0,5584
O2	(9e)	0,5	0,0	0,0

BARIUM IRIDATE RUTHENATE
$Ba_5(Ir,Ru)_3O_{12}$

Z. anorg. Chem., **580**, 71-77.

Pnma, 10.853, 5.897, 19.819, Z = 4, R = 0.056. Groups of three face-sharing MO_6 octahedra, linked by 7- to 12-coordinate Ba ions. M-O = 1.87-2.09(3) A.

		x	y	z
3.3 Ir1/0.7 Ru1	(4c)	0.7847	0.25	0.5565
1.4 Ir2 2.6 Ru2	(4c)	0.8810	0.25	0.6740
1.3 Ir3 2.7 Ru3	(4c)	0.6949	0.25	0.4302
Ba1	(4c)	0.0271	0.25	0.4268
Ba2	(4c)	0.1860	0.25	0.6305
Ba3	(4c)	0.5477	0.25	0.6757
Ba4	(4c)	0.3875	0.25	0.4686
Ba5	(4c)	0.1959	0.25	0.2539
O1	(8d)	0.309	0.478	0.0182
O2	(8d)	0.513	0.982	0.207
O3	(8d)	0.095	0.015	0.105
O4	(8d)	0.262	0.496	0.869
O5	(4c)	0.779	0.25	0.755
O6	(4c)	0.306	0.25	0.142
O7	(4c)	0.467	0.25	0.918
O8	(4c)	0.111	0.25	0.976

BARIUM EUROPIUM(III) IRIDATE(V)
Ba_2EuIrO_6

J. Less-Common Metals, **161**, 1-6.

$P6_3/mmc$, 5.9159, 14.7440, Z = 3, R = 0.053. 6-L Perovskite with sequence $(hcc)_2$. There are pairs of face-sharing MO_6 octahedra (M=Eu/Ir) connected by sharing corners with isolated EuO_6 octahedra to form a three-dimensional framework; Ba ions have 12-coordination.

		x	y	z
Ba1	(2b)	0,0	0,0	0.25
Ba2	(4f)	0.3333	0.6666	0.0955
0.75Ir/0.25Eu1	(4f)	0.3333	0.6666	0.8366
Eu2	(2a)	0,0	0,0	0,0
O1	(6h)	0.511	1.022	0.25
O2	(12k)	0.823	1.647	0.087

(Å)

Ba1-O1	2.960(28)	(6x)	Ir/Eu1-O2	1.961(27) (3x)
Ba1-O2	3.008(24)	(6x)	Ir/Eu1-O1	2.044(27) (3x)
Ba2-O1	2.916(22)	(3x)	Eu2-O2	2.219(27) (6x)
Ba2-O2	2.962(25)	(6x)		
Ba2-O2	3.134(22)	(3x)	Ir/Eu1-Ir/Eu1	2.553(2)

LITHIUM NICKEL OXIDE
$Li_{1-x}Ni_{1+x}O_2$

Mater. Res. Bull., **25**, 623-630.

$R\bar{3}m$, a = 5.787-5.810 A, α = 59.78-59.21°, Z = 1, powder data. In hexagonal setting, Ni mainly in 3(a), Li mainly in 3(b), as in $LiNiO_2$ (**18**, 424), but with disorder of these sites as x increases; O in 6(c): 0,0,z, z = 0.258.

NICKELATES 213

LANTHANUM STRONTIUM NICKELATES
$La_{2-x}Sr_xNiO_4$ (x = 0, 0.6, 1.0, 1.4, 1.6)

Mater. Res. Bull., 25, 293-306.

I4/mmm, a = 3.87, 3.80, 3.82, 3.82, 3.82, c = 12.62, 12.70, 12.42, 12.32, 12.33 A
(i.e. a minimum and maximum at x = 0.6), Z = 2, powder data. La/Sr in 4(e):
0,0,z, z = 0.3622,0.3610,0.3612,0.3606,0.3597; Ni in 2(a): 0,0,0; O(1) in 4(c):
1/2,0,0; O(2) in 4(e): z = 0.179,0.166,0.165,0.159,0.162. K_2NiF_4-type (17,
332; 19, 323; 49A, 93). The distortion of the NiO_6 octahedron increases
gradually with increasing x, while one La-O bond changes length, but only in the
range x = 0-0.6.

$La_{1.6}Sr_{0.4}NiO_{3.47}$

J. Solid State Chem., 84, 165-170.

Immm, 3.8728, 3.7242, 12.767, Z = 2, powder data. Oxygen-deficient K_2NiF_4-type
(17, 332; 19, 323; 49A, 93).

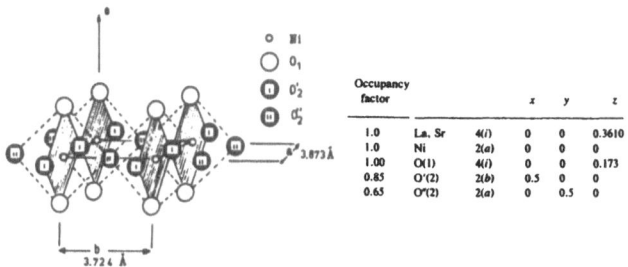

Occupancy factor			x	y	z
1.0	La, Sr	4(i)	0	0	0.3610
1.0	Ni	2(a)	0	0	0
1.00	O(1)	4(i)	0	0	0.173
0.85	O'(2)	2(b)	0.5	0	0
0.65	O''(2)	2(a)	0	0.5	0

PRASEODYMIUM NICKELATE
PrNiO₃

Mater. Res. Bull., 25, 1091-1098.

Below 773K, Pbnm, a = 5.4146, 5.4294, 5.4456, b = 5.3757, 5.3849, 5.3952, c =
7.6199, 7.6362, 7.6520 A, at 293, 473, 673K, Z = 4, powder data. Pr in 4(c):
0.9957,0.0291,1/4; Ni in 4(b): 1/2,0,0; O(1) in 4(c): 0.0680,0.4926,1/4; O(2)
in 8(d): 0.7175,0.2844,0.0351 (parameters at room temperature, similar values at
the other temperatures). Orthorhombic perovskite (20, 273; 37A, 266). Ni-6 O =
1.94 A.

Above 773K, R3̄c, a = 5.4577, 5.4614, c = 13.1058, 13.1319 A, at 773, 873K, Z = 6,
powder data. Pr in 6(a): 0,0,1/4; Ni in 6(b): 0,0,0; O in 18(e): x,0,1/4, x =
-0.5554, -0.5558. Rhombohedral perovskite.Ni-6 O = 1.94 A.

BARIUM PLATINUM LANTHANON OXIDES
$Ba_8Pt_4Ln_3O_{17.5}$ (Ln = Er, Yb, Tm)

Z. anorg. Chem., <u>584</u>, 7-11.

Pm3m, a = 8.3441, 8.3145, 8.3296 A, Z = 1, R = 0.046, 0.042, 0.048.
Isostructural with the Y compound (this volume, p. 164) with square-planar
coordination for Pt(II), octahedral for Pt(IV) and Ln, and 9-coordination for
Ba.

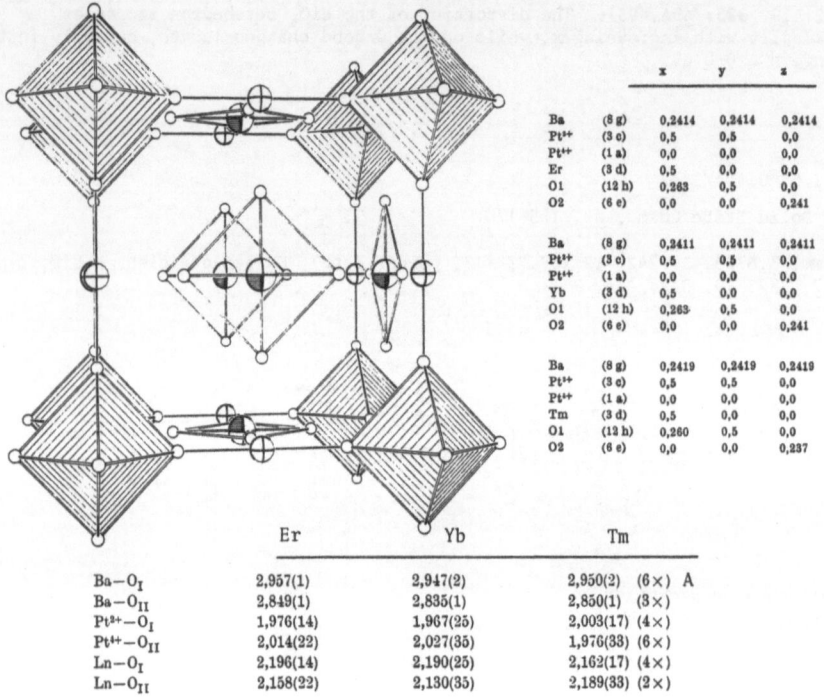

		x	y	z
Ba	(8 g)	0,2414	0,2414	0,2414
Pt^{2+}	(3 c)	0,5	0,5	0,0
Pt^{4+}	(1 a)	0,0	0,0	0,0
Er	(3 d)	0,5	0,0	0,0
O1	(12 h)	0,263	0,5	0,0
O2	(6 e)	0,0	0,0	0,241
Ba	(8 g)	0,2411	0,2411	0,2411
Pt^{2+}	(3 c)	0,5	0,5	0,0
Pt^{4+}	(1 a)	0,0	0,0	0,0
Yb	(3 d)	0,5	0,0	0,0
O1	(12 h)	0,263	0,5	0,0
O2	(6 e)	0,0	0,0	0,241
Ba	(8 g)	0,2419	0,2419	0,2419
Pt^{2+}	(3 c)	0,5	0,5	0,0
Pt^{4+}	(1 a)	0,0	0,0	0,0
Tm	(3 d)	0,5	0,0	0,0
O1	(12 h)	0,260	0,5	0,0
O2	(6 e)	0,0	0,0	0,237

	Er	Yb	Tm	
$Ba-O_I$	2,957(1)	2,947(2)	2,950(2)	(6×) A
$Ba-O_{II}$	2,849(1)	2,835(1)	2,850(1)	(3×)
$Pt^{2+}-O_I$	1,976(14)	1,967(25)	2,003(17)	(4×)
$Pt^{4+}-O_{II}$	2,014(22)	2,027(35)	1,976(33)	(6×)
$Ln-O_I$	2,196(14)	2,190(25)	2,162(17)	(4×)
$Ln-O_{II}$	2,158(22)	2,130(35)	2,189(33)	(2×)

LITHIUM COPPER OXIDES
$LiCu_2O_2$ (I), $LiCu_3O_3$ (II)

J. Solid State Chem., <u>88</u>, 534-542.

I, $P4_2/nmc$, 5.719, 12.401, Z = 8, R = 0.044. II, P4/mmm, 2.810, 8.889, Z = 1, R
= 0.023. Both structures contain $Cu(I)O_2$ linear groups, $Cu(II)O_4$ square planes
(with one or two more-distant O), and LiO_5 square pyramids. In II, Li and Cu(II)
are completely disordered, and some crystals of I exhibit orthohombic symmetry
and partial Li/Cu(II) disorder.

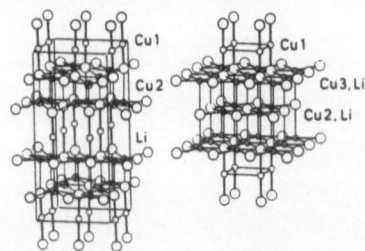

Projections of the crystal structures of Li-
Cu_2O_2 (I) and $LiCu_3O_3$ (II). O represents O atoms and
o Cu^I atoms in (I) and (II); ● represents Cu^{II} and ⊕ Li
in (I); in (II), O and ● represent sites for both kind of
atoms, Li and Cu^{II}. Cu^IO_2-dumbbells are graphically
emphasized.

I   II

$LiCu_2O_2$		x	y	z	
Cu1	8g	0.4947	3/4	0.9949	
Cu2	8g	1/4	0.5004	0.1557	
O2	8g	0.011	1/4	0.1551	
O1	8g	0.487	3/4	0.1436	
Li1	8g	3/4	0.989	0.179	

$LiCu_3O_3$		x	y	z	Occupation
Cu1	1a	0	0	0	0.0625
Cu2	1b	0	0	1/2	0.0505
Cu3	2h	1/2	1/2	0.2283	0.0743
O1	1d	1/2	1/2	1/2	0.0625
O2	2g	0	0	0.2088	0.1250

Lithium atoms presumed to complete the occupancy of the Cu2 and Cu3 sites

SODIUM CUPRATE(III)
$NaCuO_2$

J. Solid State Chem., 89, 308-314.

C2/m, 6.363, 2.753, 6.110, 120.78, Z = 2, R = 0.030 and neutron powder data. Cu in 2(a): 0,0,0; Na in 2(d): 0,1/2,1/2; O in 4(i): 0.3330,0,0.7762. Structure as previously described (56A, 169).

CALCIUM CUPRATE(II)
Ca_2CuO_3

Acta Chem. Scand., 44, 516-518.

Immm, 3.2781, 3.7870, 12.277, for $Ca_{1.82}Sr_{0.18}CuO_3$, Z = 2, R = 0.019. Ca in 4(j): 1/2,0,0.15016; Cu in 2(d): 0,1/2,0; O(1) in 4(i): 0,0,0.3402; O(2) in 2(a): 0,0,0. Structure as previously described (35A, 246), with CuO_4 square planes linked by CaO_7 monocapped trigonal prisms. Cu-O = 1.89, 1.96, Ca-O = 2.33-2.51 A.

CALCIUM STRONTIUM CUPRATE(II)
$(Ca,Sr)_2CuO_3$

Izv. Akad. Nauk Mold. SSR, Ser. Fiz.-Tekh. Mat. Nauk, No. 3, 26-28 (1989).

Imm2, 12.4, 3.8, 3.25, Z = 2, R = 0.063. Structure essentially as previously described (35A, 246; this volume, preceding report).

BARIUM ALUMINUM COPPER OXIDE
$Ba_{46}Al_6Cu_{24}O_{84}$

Inorg. Chem., 29, 2837-2841.

$P6_3mc$, 13.1524, 17.3122, $Z = 1$, $R = 0.018$. Layer structure with all the Cu in bowl-shaped rings of composition Cu_6O_{15} built from edge- and corner-sharing CuO_4 square planes. Aℓ ions have tetrahedral coordinations and Ba ions have 7- to 9-coordinations.

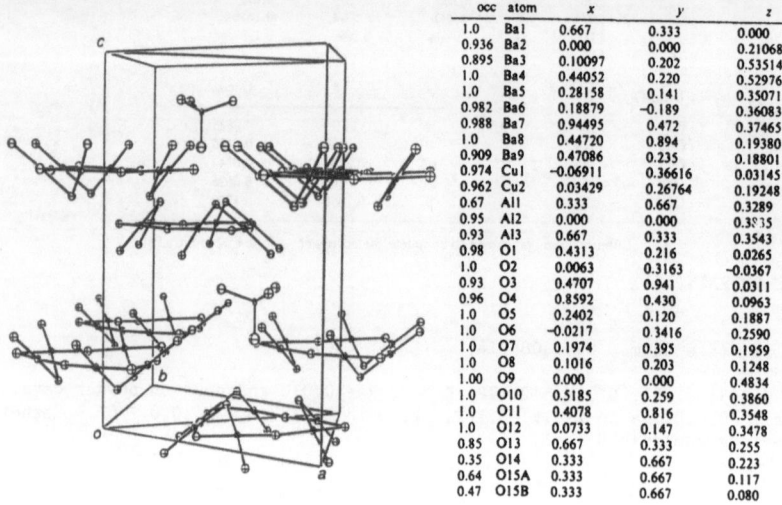

occ	atom	x	y	z
1.0	Ba1	0.667	0.333	0.000
0.936	Ba2	0.000	0.000	0.21068
0.895	Ba3	0.10097	0.202	0.53514
1.0	Ba4	0.44052	0.220	0.52976
1.0	Ba5	0.28158	0.141	0.35071
0.982	Ba6	0.18879	−0.189	0.36083
0.988	Ba7	0.94495	0.472	0.37465
1.0	Ba8	0.44720	0.894	0.19380
0.909	Ba9	0.47086	0.235	0.18801
0.974	Cu1	−0.06911	0.36616	0.03145
0.962	Cu2	0.03429	0.26764	0.19248
0.67	Al1	0.333	0.667	0.3289
0.95	Al2	0.000	0.000	0.3835
0.93	Al3	0.667	0.333	0.3543
0.98	O1	0.4313	0.216	0.0265
1.0	O2	0.0063	0.3163	−0.0367
0.93	O3	0.4707	0.941	0.0311
0.96	O4	0.8592	0.430	0.0963
1.0	O5	0.2402	0.120	0.1887
1.0	O6	−0.0217	0.3416	0.2590
1.0	O7	0.1974	0.395	0.1959
1.0	O8	0.1016	0.203	0.1248
1.00	O9	0.000	0.000	0.4834
1.0	O10	0.5185	0.259	0.3860
1.0	O11	0.4078	0.816	0.3548
1.0	O12	0.0733	0.147	0.3478
0.85	O13	0.667	0.333	0.255
0.35	O14	0.333	0.667	0.223
0.64	O15A	0.333	0.667	0.117
0.47	O15B	0.333	0.667	0.080

STRONTIUM ZINC COPPER(I) OXIDE
$Sr_9Zn_4Cu_2O_{14}$

Z. anorg. Chem., **583**, 17-23.

$C2/m$, 22.217, 3.612, 11.286, 98.63, $Z = 2$, $R = 0.067$. Three-dimensional framework of SrO_6 octahedra and trigonal prisms, with tetrahedrally coordinated Zn^{2+} and linearly-coordinated Cu^+ ions. Cu–O = 1.81, 1.86, Zn–O = 1.92-2.07, Sr–O = 2.43-2.85(1) A.

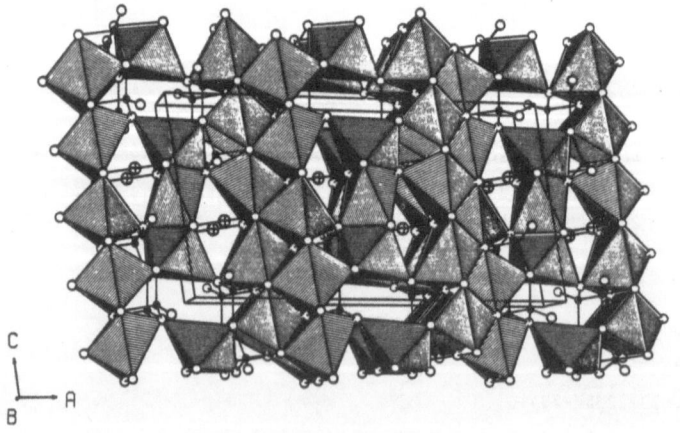

		x	y	z
Sr1	(4i)	0,3132	0	0,1029
Sr2	(4i)	0,4902	0	0,1920
Sr3	(2d)	0	0,5	0,5
Sr4	(4i)	0,6938	0	0,5704
Sr5	(4i)	0,6867	0	0,2255
Zn1	(4i)	0,8905	0	0,2976
Zn2	(4i)	0,1024	0	0,0379
Cu	(4i)	0,1061	0	0,3755
O1	(4i)	0,197	0	0,069
O2	(4i)	0,809	0	0,598
O3	(4i)	0,406	0	0,359
O4	(4i)	0,797	0	0,269
O5	(4i)	0,581	0	0,094
O6	(4i)	0,024	0	0,350
O7	(4i)	0,911	0	0,134

LANTHANON DICUPRATES(II)
Ln$_2$Cu$_2$O$_5$ (Ln = Ho, Tm, Yb)

Solid State Comm., 75, 785-788.

Pn2$_1$a, a = 10.8126, 10.7406, 10.7244, b = 12.4485, 12.3715, 12.3372, c = 3.4969,
3.4588, 3.4324 A, Z = 4, neutron powder data. Structures as previously described
for the Ho compound (43A, 217).

BARIUM YTTRIUM CUPRATE
BaY$_2$CuO$_5$

Acta Cryst., C46, 1986-1988.

Pbnm, 7.1342, 12.1811, 5.6580, Z = 4, neutron powder data. Structure essentially
as previously described (49A, 221; 54A, 194; 55A, 198; 56A, 175).

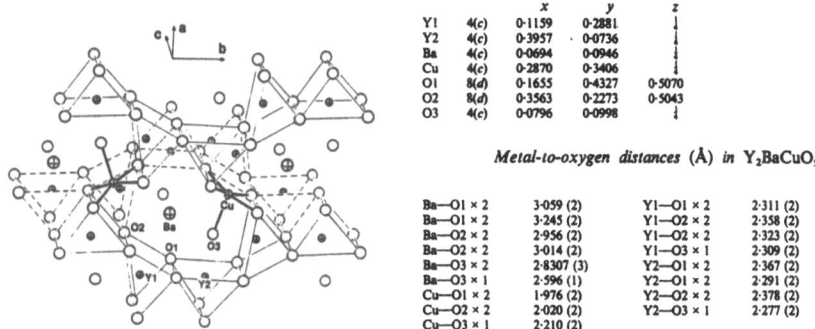

		x	y	z
Y1	4(c)	0·1159	0·2881	¼
Y2	4(c)	0·3957	0·0736	¼
Ba	4(c)	0·0694	0·0946	¼
Cu	4(c)	0·2870	0·3406	¼
O1	8(d)	0·1655	0·4327	0·5070
O2	8(d)	0·3563	0·2273	0·5043
O3	4(c)	0·0796	0·0998	¼

Metal-to-oxygen distances (Å) in Y$_2$BaCuO$_5$

Ba—O1 × 2	3·059 (2)	Y1—O1 × 2	2·311 (2)
Ba—O1 × 2	3·245 (2)	Y1—O2 × 2	2·358 (2)
Ba—O2 × 2	2·956 (2)	Y1—O2 × 2	2·323 (2)
Ba—O2 × 2	3·014 (2)	Y1—O3 × 1	2·309 (2)
Ba—O3 × 2	2·8307 (3)	Y2—O1 × 2	2·367 (2)
Ba—O3 × 1	2·596 (1)	Y2—O1 × 2	2·291 (2)
Cu—O1 × 2	1·976 (2)	Y2—O2 × 2	2·378 (2)
Cu—O2 × 2	2·020 (2)	Y2—O3 × 1	2·277 (2)
Cu—O3 × 1	2·210 (2)		

BARIUM YTTERBIUM COPPER OXIDE
BARIUM LUTETIUM COPPER OXIDE
BaYb$_2$CuO$_5$, BaLu$_2$CuO$_5$

J. Solid State Chem., 89, 385-388.

Pnma, a = 12.0588, 12.0314, b = 5.6123, 5.5989, c = 7.0535, 7.0380 A, Z = 4,
neutron powder data. Isostructural with the Y compound (49A, 221; 54A, 194; 55A,
198), with a framework of LnO$_7$ units, and cavities occupied by 11-coordinate Ba
and 5-coordinate Cu (distorted square pyramid). [See also preceding report.]

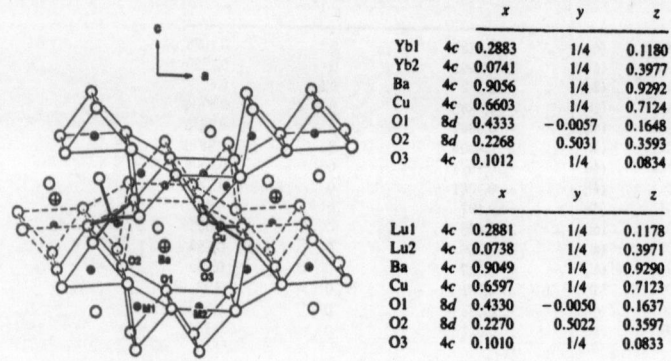

		x	y	z
Yb1	4c	0.2883	1/4	0.1180
Yb2	4c	0.0741	1/4	0.3977
Ba	4c	0.9056	1/4	0.9292
Cu	4c	0.6603	1/4	0.7124
O1	8d	0.4333	−0.0057	0.1648
O2	8d	0.2268	0.5031	0.3593
O3	4c	0.1012	1/4	0.0834

		x	y	z
Lu1	4c	0.2881	1/4	0.1178
Lu2	4c	0.0738	1/4	0.3971
Ba	4c	0.9049	1/4	0.9290
Cu	4c	0.6597	1/4	0.7123
O1	8d	0.4330	−0.0050	0.1637
O2	8d	0.2270	0.5022	0.3597
O3	4c	0.1010	1/4	0.0833

COPPER OXIDE SUPERCONDUCTORS

A great deal of work continues to appear on high temperature (90K)
superconductors (not all the materials studied have superconducting properties):

STRONTIUM CUPRATE
$Sr_2CuO_{3.9}$

J. Solid State Chem., $\underline{88}$, 513-519.

I4/mmm, 3.7907, 12.417, Z = 1.8, neutron powder data. 3.6 Sr in 4(e):
0,0,0.3539; 1.8 Cu in 2(a): 0,0,0; 2.8 O(1) in 4(c): 0,1/2,0; 4 O(2) in
4(e): 0,0,0.1528. K_2NiF_4-type ($\underline{17}$, 332; $\underline{19}$, 323; $\underline{49A}$, 93). Cu-6 O = 1.90,
Sr-9 O = 2.50-2.68 A.

COPPER OXIDES
M_2CuO_4, $La_2(Cu,M')O_4$, $LnSrBaCu_3O_7$, $LnCaBaCu_3O_7$, $BaM''O_3$

Acta Chem., Scand., $\underline{44}$, 769-776, 902-906.

$La_{1.8}Ba_{0.2}CuO_4$, $La_{1.9}Ca_{0.1}CuO_4$, $La_{1.8}Ca_{0.2}CuO_4$, $La_2Cu_{0.5}Ni_{0.5}O_4$, I4/mmm, a ∿ 3.8,
c ∿ 13.2 A, Z = 2, X-ray and neutron powder data confirm the K_2NiF_4-type
structure ($\underline{54A}$, 195).

$La_2Cu_{0.8}Zn_{0.2}O_4$, Abma, 5.404, 13.155, 5.456, Z = 4, X-ray and neutron powder
data. La_2CuO_4-type ($\underline{43A}$, 216).

$HoSrBaCu_3O_7$, $NdSrBaCu_3O_7$, $LaSrBaCu_3O_7$, $NdCaBaCu_3O_7$, $LaCaBaCu_3O_7$, Pmmm, a ∿ 3.9, b
∿ 3.9, c ∿ 11.7 A, Z = 1, X-ray and neutron powder data confirm the structure
($\underline{54A}$, 193; $\underline{55A}$, 199).

$Nd_{1.85}Ce_{0.15}CuO_4$, I4/mmm, 3.968, 12.135, Z = 2, X-ray powder data, Nd_2CuO_4-type
($\underline{41A}$, 269).

$BaBiO_3$, $BaBi_{0.5}La_{0.5}O_3$, $BaBi_{0.5}Cu_{0.5}O_3$, I2/m, X-ray powder data. $BaBiO_3$-type
($\underline{42A}$, 446; $\underline{44A}$, 195; $\underline{52A}$, 192).

LANTHANUM STRONTIUM CUPRATE
$La_{1.83}Sr_{0.17}CuO_4$

Kristallografija, $\underline{35}$, 498-500 [Soviet Physics - Crystallography, $\underline{35}$, 289-290].

I4/mmm, 3.764, 13.175, Z = 2, neutron powder data. La, Sr in 4(e): 0,0,0.3610;
Cu in 2(a): 0,0,0; O(1) in 4(c): 0,1/2,0; O(2) in 4(e): 0,0,0.1827.
K_2NiF_4-type, as previously described ($\underline{39A}$, 252; $\underline{54A}$, 195).

$(La, Sr)_2CuO_4$

Kristallografija, $\underline{35}$, 861-868 [Soviet Physics - Crystallography, $\underline{35}$, 506-510].

12 and 10 atomic % Sr, Pbma, a = 5.363, 5.363, b = 5.338, 5.348, c = 13.167,
13.194 A, at 5, 15K, Z = 4, R = 0.022, 0.023; 3 atomic % Sr, Abma, 5.365, 5.353,
13.223, Z = 4, R = 0.015 (twinned crystals). Distorted K_2NiF_4-types ($\underline{54A}$, 195),
with the lower symmetry resulting from some La/Sr ordering.

LANTHANUM STRONTIUM COPPER TITANIUM OXIDE
LANTHANUM STRONTIUM COPPER IRIDIUM OXIDE
$La_{1.5}Sr_{0.5}Cu_{0.75}M_{0.25}O_4$ (M = Ti, Ir)

J. Less-Common Metals, $\underline{167}$, 179-183.

Abma, a = 5.3886, 5.3991, b = 5.3921, 5.4070, c = 13.1333, 13.1042 A, Z = 4, R =
0.042, 0.040. La_2CuO_4-type ($\underline{39A}$, 252; $\underline{43A}$, 216; $\underline{56A}$, 172), with Cu/M disorder
giving a less-distorted octahedral coordination; Cu/Ti-O = 1.91 (x 4), 2.32 (x
2), Cu/Ir-O = 1.91 (x 4), 2.33 (x 2) A.

		x	y	z
La/Sr	(8f)	0.0051	0.0	0.3598
Cu/Ti	(4a)	0.0	0.0	0.0
O1	(8e)	0.25	0.25	0.0078
O2	(8f)	0.0142	0.0	0.1768
La/Sr	(8f)	0.9871	0.0	0.3599
Cu/Ir	(4a)	0.0	0.0	0.0
O1	(8e)	0.25	0.25	0.0089
O2	(8f)	0.9766	0.0	0.1773

NEODYMIUM CERIUM COPPER OXIDE
$Nd_{2-x}Ce_xCuO_4$ (x = 0.05-0.30)

Solid State Comm., $\underline{73}$, 791-795.

I4/mmm, a = 3.94, c = 12.12-12.04 A, for x = 0.05-0.30, at 1.5 and 295K, Z = 2,
X-ray and neutron powder data. Structure as previously described ($\underline{41A}$, 269).

LANTHANUM DYSPROPSIUM COPPER OXIDES
$La_{1.25}Dy_{0.75}CuO_{3.75}F_{0.5}$ (I), $La_{1.25}Dy_{0.75}CuO_4$ (II)

I, Cmma, 5.4609, 5.5089, 12.4842, Z = 2, neutron powder data. II, P4/nmm,
3.8661, 12.4475, Z = 2, neutron powder data. Rocksalt and fluorite layers, with
Dy in only one Ln site, and excess anions in I incorporated as an interstitial
defect in the rocksalt layers.

LANTHANUM STRONTIUM COPPER ALUMINUM OXIDE
LaSrCuAlO$_5$

J. Solid State Chem., 88, 250-260.

Pbcm, 7.9219, 11.020, 5.4235, Z = 4, R = 0.029. Perovskite-related structure,
but the lower oxygen content results in tetrahedral coordination for Al and
square-pyramidal coordination for Cu. Al-O = 1.71-1.75, Cu-O = 1.93, 1.95 (each
x2), 2.43 (next nearest 2.95) A.

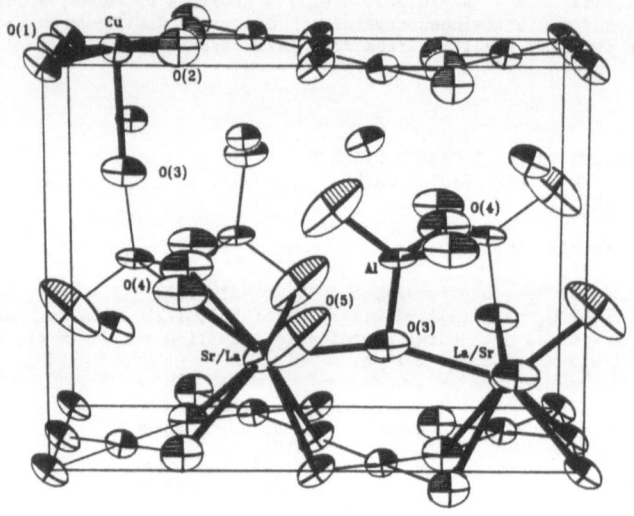

		Population	x	y	z
La(1)	4d	0.57	0.19497	0.13578	0.2500
Sr(2)	4d	0.61	-0.25266	0.09786	0.2500
La(2)	4d	0.39	-0.25266	0.09786	0.2500
Sr(1)	4d	0.43	0.19497	0.13578	0.2500
Cu	4d	1.0	-0.0234	0.12623	-0.2500
Al	4d	1.0	0.5014	0.1573	-0.2500
O(1)	4a	1.0	0.0000	0.0000	0.0000
O(2)	4c	1.0	-0.0429	0.2500	0.5000
O(3)	4d	1.0	0.2823	0.1375	-0.2500
O(4)	4c	1.0	-0.454	0.2500	0.5000
O(5)	4d	1.0	0.372	-0.0317	0.2500

CALCIUM STRONTIUM COPPER PLATINUM OXIDE
Ca$_{1.75}$Sr$_{1.5}$Cu$_{0.75}$PtO$_6$

Kristallografija, 35, 869-873 [Soviet Physics - Crystallography, 35, 511-513].

R$\bar{3}$c, a = 9.442, c = 11.125 A, Z = 6, R = 0.041 (twinned crystal). 1.5Ca in 6(a):
0,0,1/4; (9Ca + 9Sr) in 18(e): 0.3633,0,1/4; 4.5Cu in 18(e): 0.055,0.055,1/4;
6Pt in 6(b): 0,0,0; 36O in 36(f): 0.1808,0.0237,0.1112. Isostructural with
Sr$_4$PtO$_6$ (23, 360).

LANTHANUM STRONTIUM COPPER OXIDE
$La_2SrCu_2O_6$

Mater. Res. Bull., <u>25</u>, 199-204.

I4/mmm, 3.8647, 19.9410, Z = 2, neutron powder data. La/Sr(1) in 2(a): 0,0,0;
La/Sr(2) in 4(e): 0,0,0.1787; Cu in 4(e): 0,0,0.5919; O(1) in 8(g):
0,1/2,0.0845; O(2) in 4(e): 0,0,0.2973; [2(b) site essentially vacant].
Structure as previously described (<u>46A</u>, 291). The material is not a
superconductor.

LANTHANUM BARIUM STRONTIUM COPPER OXIDE
$(La,Ba)_2SrCu_2O_6$

Mater. Res. Bull., <u>25</u>, 1279-1286.

I4/mmm, a = 3.85-3.86, c = 19.96-20.22 A, for four La/Ba compositions with O
content 5.96-6.30, Z = 2, neutron powder data. Structure as previously described
(<u>46A</u>, 291) with statistical La/Ba/Sr distribution over two sites, and some oxygen
in the 1/2,1/2,0 site, even in the oxygen-deficient sample. The material is not
a superconductor.

LANTHANUM NEODYMIUM STRONTIUM CUPRATE
$LaNdSrCu_2O_6$ (I)

NEODYMIUM STRONTIUM CUPRATE
$Nd_{1.4}Sr_{1.6}Cu_2O_{5.79}$ (II)

J. Solid State Chem., <u>85</u>, 88-99.

I, I4/mmm, 3.8540, 19.7688, Z = 2, neutron powder data. II, Immm, 3.7780,
11.3712, 20.1315, Z = 6, neutron powder data. I is isostructural with $La_2SrCu_2O_6$
(<u>46A</u>, 291; this volume, above), but with unusual Nd/La ordering. II has a
related structure with a tripled <u>b</u> axis, and ordered oxygen vacancies giving a
one-dimensional sub-lattice of corner-sharing CuO_5 square pyramids.

I

		x'	y	z	Site occupancy
A(1)	2a	0	0	0	1.0 Nd
A(2)	4e	0	0	0.1790	0.5 La, 0.5 Sr
Cu(1)	4e	0	0	0.5913	
O(1)	8g	0	½	0.0814	
O(2)	4e	0	0	0.7025	

II

		x	y	z	Site occupancy
A(1)	2a	0	0	0	Nd 1.0
A(2)	4g	0	0.3073	0	Nd 0.6; Sr 0.4
A(3)	4i	0	0	0.1948	Nd 1.0
A(4)	8l	0	0.3256	0.1763	Sr 1.0
Cu(1)	4i	0	0	0.5873	
Cu(2)	8l	0	0.3524	0.6013	
O(1)	8l	0	0.1621	0.4163	
O(2)	8l	0	0.1356	0.0955	
O(4)	4j	½	0	0.4242	0.603
O(5)	4i	0	0	0.6940	
O(6)	8l	0	0.3451	0.6989	
O(7)	2c	½	½	0	0.174
O(8)	4h	0	0.3732	½	

O(3) site is unoccupied

NEODYMIUM BISMUTH STRONTIUM CALCIUM COPPER OXIDE
$Nd_{1.7}Bi_{0.1}Sr_{0.9}Ca_{0.3}Cu_2O_6$

Physica C, <u>169</u>, 169-173.

I4/mmm, 3.833, 19.539, Z = 2, R = 0.051. M(1) in 4(e): 0,0,0.1796; M(2) in 2(a): 0,0,0; Cu in 4(e): 0,0,0.4100; O(1) in 4(e): 0,0,0.291; O(2) in 8(g): 1/2,0,0.0813. Isostructural with $La_2SrCu_2O_6$ (<u>46A</u>, 291; this volume, p. 221).

BARIUM YTTRIUM COPPER OXIDE
$Ba_2YCu_3O_{6.7}$

Phys. Rev. B, <u>41</u>, 4220-4223.

Pmmm, a = 3.8336, 3.8222, b = 3.8828, 3.8801, c = 11.7503, 11.7193 A, at 292, 115K, Z = 1, R = 0.032, 0.038. Structure as previously described (<u>54A</u>, 193; <u>55A</u>, 200), with the O(5) site vacant and O(4) disordered.

LANTHANUM CALCIUM BARIUM ALUMINUM CUPRATE
$(La,Ca)Ba_2Cu_2(A\ell,Cu)O_{6.78}$

Acta Cryst., C<u>46</u>, 2001-2003.

P4/mmm, 3.908, 11.647, Z = 1, R = 0.030. Variant of the $YBa_2Cu_3O_{7-x}$ structure (<u>54A</u>, 193; <u>55A</u>, 199,200).

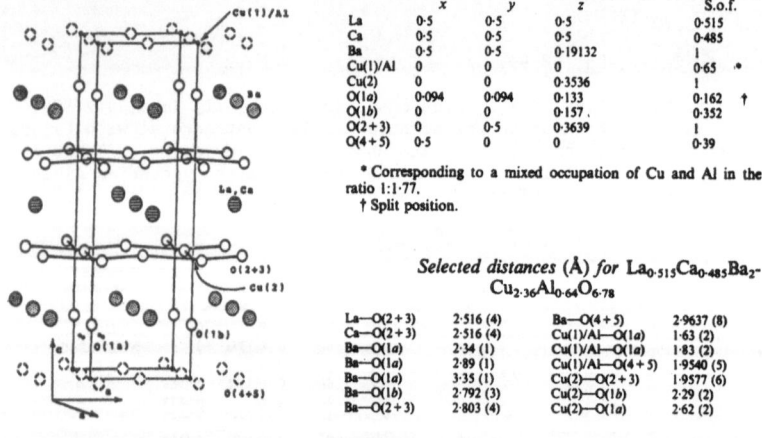

	x	y	z	S.o.f.
La	0·5	0·5	0·5	0·515
Ca	0·5	0·5	0·5	0·485
Ba	0·5	0·5	0·19132	1
Cu(1)/Al	0	0	0	0·65 *
Cu(2)	0	0	0·3536	1
O(1a)	0·094	0·094	0·133	0·162 †
O(1b)	0	0	0·157	0·352
O(2+3)	0	0·5	0·3639	1
O(4+5)	0·5	0	0	0·39

* Corresponding to a mixed occupation of Cu and Al in the ratio 1:1·77.
† Split position.

Selected distances (Å) for $La_{0.515}Ca_{0.485}Ba_2$-$Cu_{2.36}Al_{0.64}O_{6.78}$

La—O(2+3)	2·516 (4)	Ba—O(4+5)	2·9637 (8)
Ca—O(2+3)	2·516 (4)	Cu(1)/Al—O(1a)	1·63 (2)
Ba—O(1a)	2·34 (1)	Cu(1)/Al—O(1a)	1·83 (2)
Ba—O(1a)	2·89 (1)	Cu(1)/Al—O(4+5)	1·9540 (5)
Ba—O(1a)	3·35 (1)	Cu(2)—O(2+3)	1·9577 (6)
Ba—O(1b)	2·792 (3)	Cu(2)—O(1b)	2·29 (2)
Ba—O(2+3)	2·803 (4)	Cu(2)—O(1a)	2·62 (2)

BARIUM LANTHANON ALUMINUM COPPER OXIDES
$Ba_2Gd(Cu,A\ell)_3O_{6.88}$ (I), $Ba_2Er(Cu,A\ell)_3O_{6.6}$ (II)

Z. Phys. B: Condens. Matter, <u>80</u>, 177-180.

P4/mmm, a = 3.884, 3.855, c = 11.685, 11.789 A, Z = 1, R = 0.025, 0.031, and neutron data for I. Structures similar to those of related materials (<u>54A</u>, 194; <u>55A</u>, 199).

BARIUM YTTRIUM CUPRATE
Ba$_2$YCu$_3$O$_7$

Acta Cryst., C$\underline{46}$, 354–358.

Pmmm, 3.836, 3.883, 11.686, Z = 1, R = 0.050. Structure as previously described (54A, 193; 55A, 200).

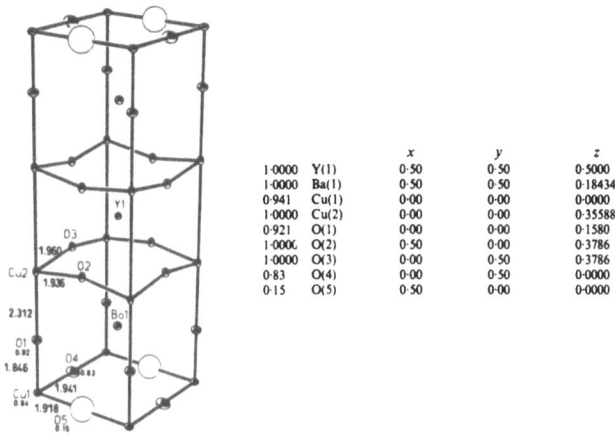

		x	y	z
1·0000	Y(1)	0·50	0·50	0·5000
1·0000	Ba(1)	0·50	0·50	0·18434
0·941	Cu(1)	0·00	0·00	0·0000
1·0000	Cu(2)	0·00	0·00	0·35588
0·921	O(1)	0·00	0·00	0·1580
1·0000	O(2)	0·50	0·00	0·3786
1·0000	O(3)	0·00	0·50	0·3786
0·83	O(4)	0·00	0·50	0·0000
0·15	O(5)	0·50	0·00	0·0000

Ba$_2$YCu$_3$O$_{7-x}$

J. Less-Common Metals, $\underline{162}$, 181–195.

Structure of the orthorhombic phase as previously described (54A, 193; 55A, 200).

LEAD STRONTIUM YTTRIUM CUPRATE
Pb$_{0.5}$Sr$_{2.5}$YCu$_2$O$_{7-x}$

J. Solid State Chem., $\underline{84}$, 375–385.

P4/mmm, 3.8253, 11.891, Z = 1, powder data. Structure similar to those of related materials (54A, 193; 55A, 199), with 80% occupancy for the O(2) site.

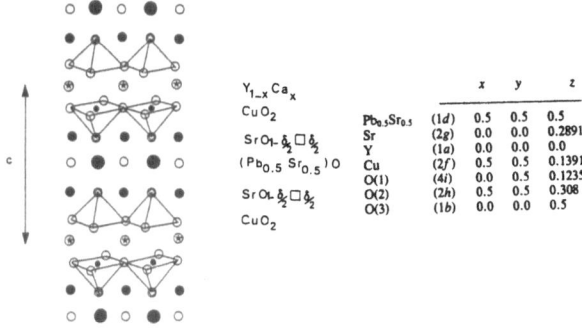

			x	y	z
	Pb$_{0.5}$Sr$_{0.5}$	(1d)	0.5	0.5	0.5
	Sr	(2g)	0.0	0.0	0.2891
	Y	(1a)	0.0	0.0	0.0
	Cu	(2f)	0.5	0.5	0.1391
	O(1)	(4i)	0.0	0.5	0.1235
	O(2)	(2h)	0.5	0.5	0.308
	O(3)	(1b)	0.0	0.0	0.5

LEAD STRONTIUM YTTRIUM COPPER NICKEL OXIDE
$Pb_2Sr_2YCu_{2.2}Ni_{0.8}O_8$

Physica C, 165, 499-504.

Cmmm, 5.4035, 5.4386, 15.6965, at 255K. Z = 2, neutron powder data at 255 and 1.5K. Structure approximately as previously described in P4/mmm (56A, 177; Physica C, 157, 272); Ni substitutes in one Cu site.

LEAD STRONTIUM NEODYMIUM CUPRATE
$Pb_2(Sr,Nd)_3Cu_3O_8$

J. Solid State Chem., 84, 144-152.

Cmmm, 5.437, 5.472, 15.797, Z = 2, R = 0.067. Double layers of CuO_5 square pyramids are separated by (Nd,Sr) oxygen-deficient layers which are stacked between (PbO)-Cu-(PbO) slabs.

		x	y	z
Pb	4l	0.5	0.0	0.3885
Sr, Nd	4k	0.0	0.0	0.2217
Nd, Sr	2a	0.0	0.0	0.0
Cu1	2d	0.0	0.0	0.5
Cu2	4l	0.5	0.0	0.1103
O1	4l	0.5	0.0	0.257
O2	16r	0.075	−0.075	0.389
O3	8m	0.25	0.25	0.0985

BARIUM HOLMIUM COPPER PLATINUM OXIDE
$Ba_2Ho_2Cu_{1.1}Pt_{0.9}O_8$

Acta Cryst., C46, 970-972.

Pcmn, 10.303, 5.668, 13.178, Z = 4, R = 0.064. Isostructural with the Er compound (55A, 204).

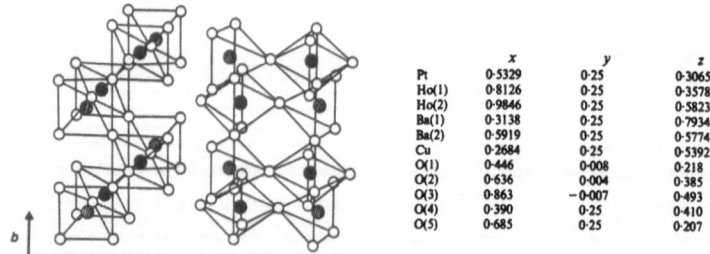

	x	y	z
Pt	0·5329	0·25	0·3065
Ho(1)	0·8126	0·25	0·3578
Ho(2)	0·9846	0·25	0·5823
Ba(1)	0·3138	0·25	0·7934
Ba(2)	0·5919	0·25	0·5774
Cu	0·2684	0·25	0·5392
O(1)	0·446	0·008	0·218
O(2)	0·636	0·004	0·385
O(3)	0·863	−0·007	0·493
O(4)	0·390	0·25	0·410
O(5)	0·685	0·25	0·207

Column-like units containing the double zigzag chains of Cu, O and Pt ions. Atom key: open circles O, filled circles Pt, shaded circles Cu.

BISMUTH LEAD CALCIUM STRONTIUM COPPER OXIDE
$Bi_{1.7}Pb_{0.3}Ca_2Sr_2Cu_3O_{9.9}$

Physica C, <u>167</u>, 291-296.

Amaa, 5.399, 5.413, 37.13, Z = 4, neutron powder data. Layer structure, with 30%
vacancies in the Bi sites, and Pb in the Ca sites.

STRONTIUM LEAD CUPRATE
$Sr_{4.79}Pb_{3.21}Cu_{0.66}O_{11.12}$

J. Solid State Chem., <u>85</u>, 44-50.

$P\bar{6}2m$, 10.072, 3.542, Z = 1, R = 0.040. Separate chains along <u>c</u> of edge-sharing
PbO_6 octahedra and edge-sharing SrO_9 tricapped trigonal prisms, linked into a
three-dimensional framework by corner sharing. The Sr/Pb site has
7-coordination, two Cu ions partially occupy tetrahedral sites, and one oxygen
site is partially occupied.

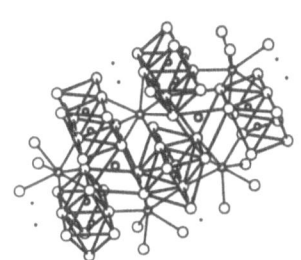

	x	y	z
Pb1	0.3368	0	0
Sr1	⅓	⅔	½
Sr2ᵃ	0.7005	0	½
Pb2	0.7005	0	½
Cu1ᶜ	0	0	0.371
Cu2ᶜ	0	0	0.112
O1ᵈ	0.170	0	½
O2	0.4604	0	½
O3	0.2412	0.4446	0

ᵃ The site occupancy is 0.930(13) and Pb2 is 1–0.930.
ᶜ Site occupancy for Cu1 is 0.197(19) and for Cu2 it is 0.134(15).
ᵈ Site occupancy is 0.71(9).

BARIUM COPPER YTTRIUM TUNGSTEN OXIDE
$Ba_4CuYW_2O_{12}$

Acta Chem. Scand., <u>44</u>, 855-856.

Fm3m, 8.3065, Z = 2, powder data. W in 4(a): 0,0,0; Cu,Y in 4(b): 1/2,1/2,1/2;
Ba in 8(c): 1/4,1/4,1/4; O in 24(e): 0.232,0,0. Perovskite-type structure
(<u>40</u>A, 206).

LANTHANUM STRONTIUM CUPRATE
$La_{6.16}Sr_{1.84}Cu_{7.66}O_{20}$

J. Solid State Chem., <u>84</u>, 335-341.

P4/mbm, 10.7468, 3.8633, Z = 1, R = 0.032. Framework of corner-sharing CuO_6
octahedra, CuO_5 square pyramids, and CuO_4 square planes, with 10-coordinate
La/Sr.

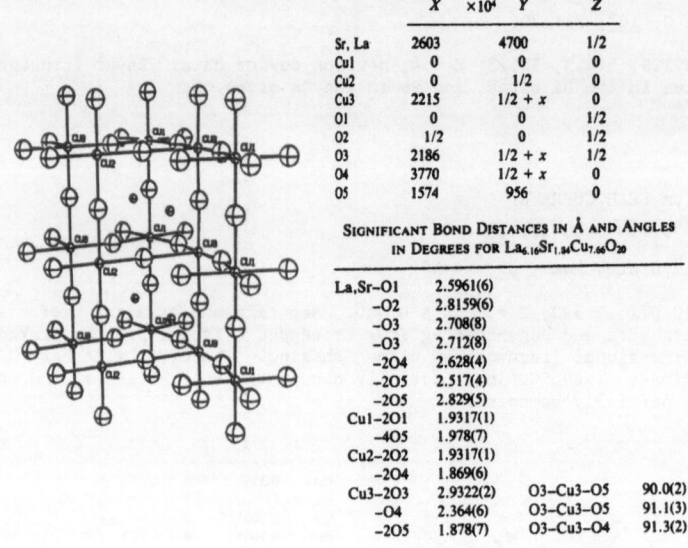

	$X \times 10^4$	Y	Z
Sr, La	2603	4700	1/2
Cu1	0	0	0
Cu2	0	1/2	0
Cu3	2215	$1/2 + x$	0
O1	0	0	1/2
O2	1/2	0	1/2
O3	2186	$1/2 + x$	1/2
O4	3770	$1/2 + x$	0
O5	1574	956	0

SIGNIFICANT BOND DISTANCES IN Å AND ANGLES IN DEGREES FOR $La_{6.16}Sr_{1.84}Cu_{7.66}O_{20}$

La,Sr–O1	2.5961(6)		
–O2	2.8159(6)		
–O3	2.708(8)		
–O3	2.712(8)		
–2O4	2.628(4)		
–2O5	2.517(4)		
–2O5	2.829(5)		
Cu1–2O1	1.9317(1)		
–4O5	1.978(7)		
Cu2–2O2	1.9317(1)		
–2O4	1.869(6)		
Cu3–2O3	2.9322(2)	O3–Cu3–O5	90.0(2)
–O4	2.364(6)	O3–Cu3–O5	91.1(3)
–2O5	1.878(7)	O3–Cu3–O4	91.3(2)

Additional data on superconductors have been given in the following (probably not exhaustive) list of structural papers:

Chin. Sci. Bull., 34, 1345-1347.

Diwen Wuli Xuebao, 12, 272-276.

Jpn. J. Appl. Phys., Part 2, 29, L57-L59, L423-L425, L572-L575, L588-L590, L1092-L1095, L1422-L1424, L1799-L1802, L1803-L1806, L1856-L1858.

J. Less-Common Metals, 157, 233-244; 160, L5-L8, 309-322; 164/165 [Symposium proceedings].

J. Mater. Res., 5, 46-52, 731-736.

J. Phys. (Paris), 51, 579-586.

Kenkyu Hokoku - Asahi Garasu Kogyo Gijutsu Shoreikai, 55, 37-47.

Mater. Res. Bull., 25, 465-476.

Mater. Res. Soc. Symp. Proc., 156, 317-328; 166, 181-186, 187-192.

Physica C, 165, 161-165; 166, 79-86; 168, 1-7, 426-429, 546-548; 169, 179-183, 217-226, 377-380; 170, 87-94, 139-152; 171, 19-24, 339-343, 468-478, 561-566; 172, 138-142, 183-189.

Phys. Lett. B, 4, 791-794.

Phys. Rev. B, 41, 1863-1877, 1889-1893; 42, 138-149, 387-392, 4228-4239.

Prog. High Temp. Supercond., 22, 409-411, 429-431, 432-433.

Res. Dev. Rev.-Mitsubishi Kasei Corp., 4, 59-67.

Solid State Comm., 73 , 683-685.

Supercond. Sci. Technol., <u>3</u>, 194-198.

Sverkhprovodimost: Fiz. Khim., Tekh., <u>2</u>,. No. 3, 57-59; <u>2</u>, No. 5, 25-28 (1989).

ALKALI-METAL SILVER OXIDES
MAgO (M = Na, K, Rb)

Z. anorg. Chem., <u>585</u>, 75-81.

I4/mmm, a = 9.520, 9.925, 10.025, c = 4.617, 5.458, 5.670, Z = 8, R = 0.033,
0.021, 0.15. M in 8(h): x,x,0, x = 0.1558,0.1478,0.1457; Ag in 8(j): x,1/2,0,
x = 0.1595,0.1772,0.1837; O in 8(i): x,0,0, x = 0.3068,0.2944,0.2893; revised
parameters also given for CsAgO and MCuO (M = Li, Na, K, Rb). Structures
essentially as previously described (<u>52A</u>, 240), but in higher-symmetry space
group, and with planar Ag_4O_4 rings.

BARIUM MERCURATE(II)
$BaHgO_2$

J. Less-Common Metals, <u>162</u>, 169-174.

$P6_322$, 6.904, 11.970, Z = 6, R not given. Ba(1) in 4(f): 1/3,2/3,0.5353; Ba(2)
in 2(a): 0,0,0; Hg in 6(h): 0.1716,0.3432,1/4; O in 12(i): 0.658,0.971,0.396.
(001) Layers of BaO_6 trigonal prisms are connected along <u>c</u> by linear O-Hg-O
units. Hg-O = 2.00(4), Ba-O = 2.59-2.74(5) A.

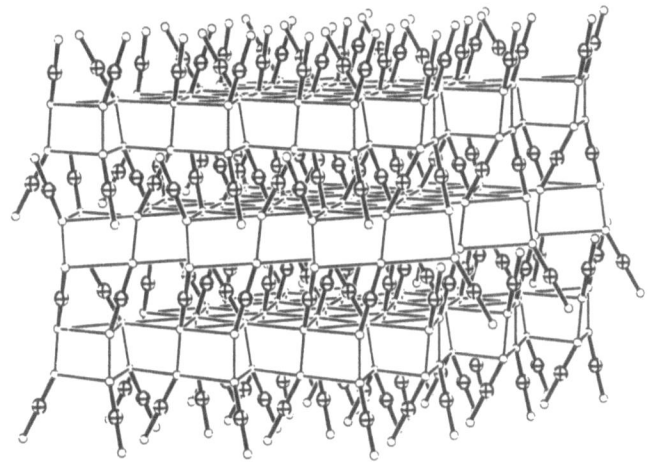

BARIUM STRONTIUM MERCURATE(II)
$Ba_{0.75}Sr_{0.25}HgO_2$

Mh. Chem., 121, 787-791.

$P6_322$, 6.897, 11.986, Z = 6, R = 0.068. Ba in 4(f): 1/3,2/3,0.5350; (0.5Ba +
1.5Sr) in 2(a): 0,0,0; Hg in 6(h): 0.1715,0.3431,1/4; O in 12(i):
0.664,0.967,0.394. Linear $[O-Hg-O]^{2-}$ anions linked by MO_6 trigonal prisms; only
one Ba site is substituted by Sr. Hg-O = 1.96(3), M-) = 2.56-2.76 A.

BARIUM YTTRIUM NICKEL OXIDE
BaY_2NiO_5

J. Solid State Chem., 88, 291-302.

Immm, 3.7703, 5.7760, 11.3581, Z = 2, R = 0.026. Ba in 2(a): 0,0,0; Y in 4(j):
0,1/2,0.2061; Ni in 2(c): 1/2,1/2,0; O(1) in 2(d): 0,1/2,0; O(2) in 8(ℓ):
0,0.2394,0.3510. Isostructural with the Nd compound (53A, 176), with chains of
corner-sharing NiO_6 compressed octahedra, 7-coordinate Y, and 10-coordinate Ba.

BARIUM BISMUTH LANTHANON OXIDES
Ba_2BiLnO_6 (Ln = Y, Dy)

J. Less-Common Metals, 161, L16-L17.

Fm3m, a = 8.5681, 8.5831 A, Z = 4, R = 0.070, 0.051. Ba in 8(c): 1/4,1/4,1/4;
Bi in 4(a): 0,0,0; Ln in 4(b): 1/2,0,0; O in 24(e): x,0,0, x = 0.246,0.242.
Ordered perovskite structure. Bi-6 O = 2.10, 2.09(3), Ln-6 O = 2.18, 2.21(3),
Ba-12 O = 3.030, 3.035(1) A.

BARIUM PLATINUM LANTHANUM OXIDE
BARIUM NICKEL YTTRIUM OXIDE
$BaPtLa_2O_5$ (I), $BaNiY_2O_5$ (II)

J. Less-Common Metals, 166, L7-L9.

I, P4/mbm, 6.9061, 5.9378, Z = 2, R = 0.061. II, Immm, 3.7610, 5.7633, 11.3167,
Z = 2, R = 0.043. In I, Pt has square-planar coordination (Pt-O = 2.00(1) A), La
8-coordination (La-O = 2.36-2.65 A), and Ba 10-coordination (Ba-O = 2.97, 3.05
A); in II Ni has flattened octahedral coordination (Ni-O = 2.18 (x 4), 1.88 (x 2)
A), Y 7-coordination (Y-O = 2.25-2.41 A), and Ba 10-coordination (Ba-O = 2.88,
2.94 A).

		x	y	z
(I)				
Ba	(2a)	0,0	0,0	0,0
Pt	(2d)	0,0	0.5	0,0
La	(4h)	0.1745	0.6745	0.5
O1	(8k)	0.3611	0.8611	0.7528
O2	(2b)	0,0	0,0	0.5
(II)				
Ba	(2a)	0,0	0,0	0,0
Ni	(2c)	0.5	0.5	0,0
Y	(4j)	0.5	0,0	0.7025
O1	(8l)	0,0	0.7598	0.3512
O2	(2d)	0.5	0,0	0.5

STRONTIUM BISMUTH(V) NEODYMIUM OXIDE
Sr_2BiNdO_6

J. Less-Common Metals, <u>161</u>, 141-146.

$P2_1/n$, 5.948, 6.101, 8.490, 90.19, Z = 2, R = 0.048. Perovskite-type structure, with ordered Bi/Nd distribution. Bi-6 O = 2.09-2.12, Nd-6 O = 2.32-2.35, Sr-12 O = 2.48-3.80(1) A.

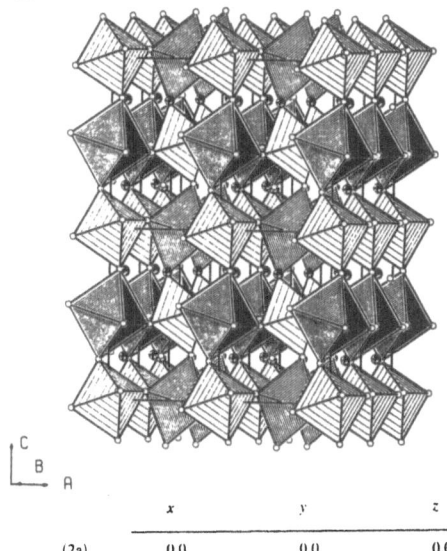

		x	y	z
Bi	(2a)	0.0	0.0	0.0
Nd	(2b)	0.0	0.0	0.5
Sr	(4e)	0.9893	0.4542	0.7495
O1	(4e)	0.397	0.538	0.734
O2	(4e)	0.312	0.784	0.445
O3	(4e)	0.215	0.311	0.451

BARIUM NEODYMIUM ZINC OXIDE
$BaNd_2ZnO_5$

J. Solid State Chem., <u>86</u>, 233-237; J. Less-Common Metals, <u>167</u>, 193-198.

I4/mcm, 6.747, 11.537, Z = 4, R = 0.030, 0.064. (001) Layers of ZnO_4 tetra-hedra and 10-coordinate Ba ions, with an oxide ion (O(1)) and 8-coordinate Nd between the layers.

		x	y	z
Ba	4(a)	0	0	¼
Nd	8(h)	0.17420	¼ + x	0
Zn	4(b)	0	½	¼
O_1	4(c)	0	0	0
O_2	16(l)	0.3557	¼ + x	0.1311

M–O	Distance (Å)
Ba–O_1 × 2	2.884
Ba–O_2 × 8	2.931(3)
Nd–O_1 × 2	2.493(1)
Nd–O_2 × 2	2.299(4)
Nd–O_2 × 4	2.635(4)
Zn–O_2 × 4	1.944

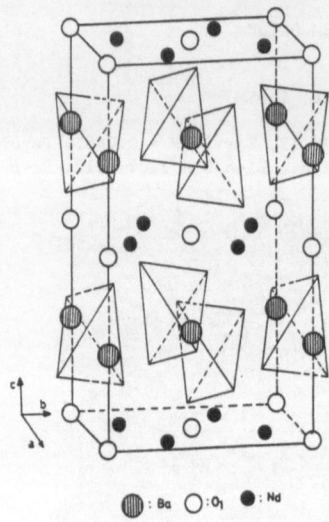

BARIUM CALCIUM DYSPROSIUM ZIRCONIUM OXIDE
$Ba_2Ca_5Dy_{18}ZrO_{36}$

J. Less-Common Metals, 163, 347-351.

$P6_3$, 17.6947, 3.3658, Z = 1, R = 0.092. Isostructural with $M_3Ln_{10}O_{18}$ compounds
(54A, 200; 56A, 182).

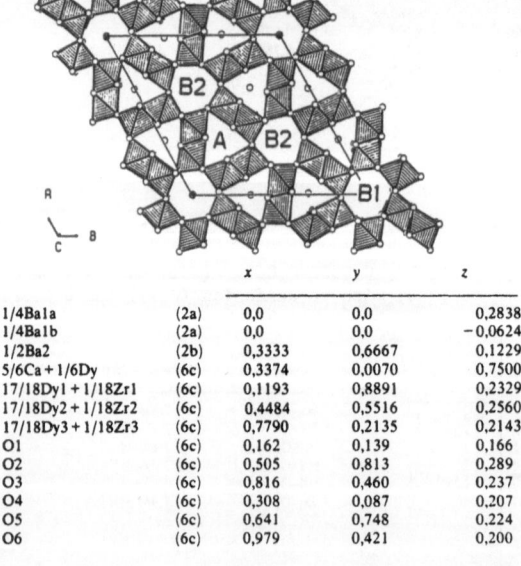

		x	y	z
1/4Ba1a	(2a)	0,0	0,0	0,2838
1/4Ba1b	(2a)	0,0	0,0	−0,0624
1/2Ba2	(2b)	0,3333	0,6667	0,1229
5/6Ca + 1/6Dy	(6c)	0,3374	0,0070	0,7500
17/18Dy1 + 1/18Zr1	(6c)	0,1193	0,8891	0,2329
17/18Dy2 + 1/18Zr2	(6c)	0,4484	0,5516	0,2560
17/18Dy3 + 1/18Zr3	(6c)	0,7790	0,2135	0,2143
O1	(6c)	0,162	0,139	0,166
O2	(6c)	0,505	0,813	0,289
O3	(6c)	0,816	0,460	0,237
O4	(6c)	0,308	0,087	0,207
O5	(6c)	0,641	0,748	0,224
O6	(6c)	0,979	0,421	0,200

BARIUM NICKEL TERBATE
$BaNiTb_2O_5$ (I)

BARIUM COPPER LANTHANUM SAMARIUM OXIDE
$BaCuLaSmO_5$ (II)

J. Less-Common Metals, <u>167</u>, 185-192.

I, Immm, 3.783, 5.80, 11.407, Z = 2, R = 0.056. II, P4/mbm, 6.743, 5.836 Z = 2,
R = 0.039. The structures are related to those of other $BaMLn_2O_5$ compounds, with
octahedral coordination for Ni in I, and in II square-planar coordination for Cu
and La/Sm disorder (see e.g. <u>54</u>A, 199).

BARIUM THULIUM OXIDE   BARIUM HOLMIUM OXIDE
$Ba_3Tm_4O_9$ $Ba_3Ho_4O_9$

Z. anorg. Chem., <u>591</u>, 181-187.

R3, a = 6.056, 6.098, c = 24.957, 25.136 A, Z = 3, R = 0.047, 0.066. Isostructural with the Yb compound (<u>50</u>A, 223).

LANTHANON OXIDES
$BaCa_2Sc_5Yb_5O_{18}$ (I), $BaCa_2Ho_5Y_5O_{18}$ (II), $BaCa_2La_5Y_5O_{18}$ (III)

Z. anorg. Chem., <u>589</u>, 89-95.

$P6_3$, a = 17.138, 17.765, 18.000, c = 3.217, 3.377, 3.446 A, Z = 2, R = 0.061,
0.091, 0.091. $M_3Ln_{10}O_{18}$-type (<u>56</u>A, 182), with an Ln_9O_{18} octahedral framework
(Ln = Sc/Yb, Ho/Y, La/Y), Ba in a large tunnel, and Ca/Ln in a smaller tunnel.

SODIUM HYDROXIDE TETRAHYDRATE
β-NaOH.4H_2O

Z. anorg. Chem., **582**, 162-168.

$P2_12_12_1$, 6.237, 6.288, 13.121, at 118K, Z = 4, R = 0.027. Columns along **a** of face-sharing Na(H_2O)$_6$$^+$ octahedra, linked by hydrogen bonds via the OH$^-$ ion and an uncoordinated water molecule. Na-O = 2.34-2.44, O-H...O = 2.66-2.97 A.

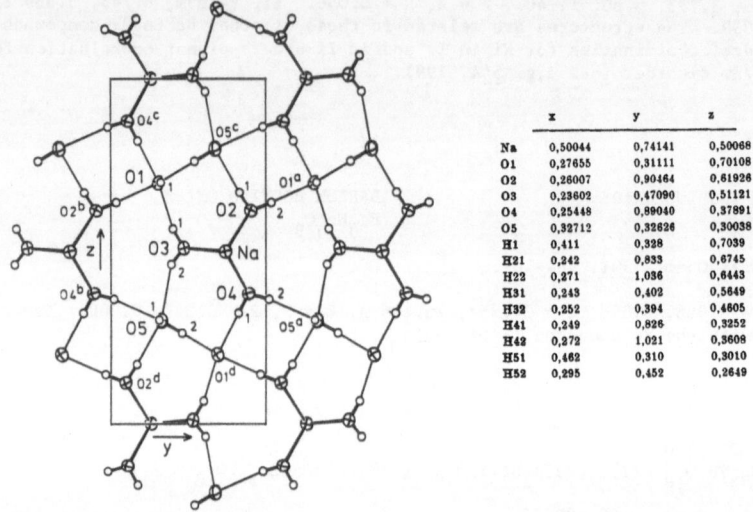

	x	y	z
Na	0,50044	0,74141	0,50068
O1	0,27655	0,31111	0,70108
O2	0,26007	0,90464	0,61926
O3	0,23602	0,47090	0,51121
O4	0,25448	0,89040	0,37891
O5	0,32712	0,32626	0,30038
H1	0,411	0,328	0,7039
H21	0,242	0,833	0,6745
H22	0,271	1,036	0,6443
H31	0,243	0,402	0,5649
H32	0,252	0,394	0,4605
H41	0,249	0,826	0,3252
H42	0,272	1,021	0,3608
H51	0,462	0,310	0,3010
H52	0,295	0,452	0,2649

CAESIUM HYDROXIDE MONOHYDRATE
CsOH.H_2O

J. Chem. Phys., **93**, 5972-5978.

Room-temperature phase, 295K, P3m1, cell parameters not given, for H and D compounds, Z = 1, neutron powder data. 383 and 414K, P6/mmm, cell parameters not given, Z = 1, neutron powder data. Structure as previously described (<u>49A</u>, 227), with O and H disorder at the higher temperatures.

CAESIUM SODIUM HYDROXIDE HEXAHYDRATE
CsNa$_2$(OH)$_3$.6H_2O

Angew. Chem., **102**, 949-950 [Angew. Chem. Int. Edn. Engl., **29**, 904-905].

$Pca2_1$, 13.951, 6.089, 12.508, Z = 4, R = 0.026. Columns along **b** of face-sharing Na(H_2O)$_6$ octahedra linked by hydrogen bonding via OH$^-$ ions, with small cavities (into which the H atoms of the OH$^-$ ions probably project) and large cavities containing Cs$^+$ ions coordinated to 9 H_2O and 3 OH$^-$. Na-O = 2.275-2.688(3), O-H...O = 2.633, 2.798(4), Cs-O = 3.287-3.981(3) A.

BARIUM HYDROXIDE TRIHYDRATE
Ba(OH)$_2$.3H_2O

Acta Cryst., <u>C46</u>, 361-363.

Pnma, 7.640, 11.403, 5.965, Z = 4, neutron radiation, R = 0.044. Structure as previously described (53A, 180), with H positions now determined more accurately.

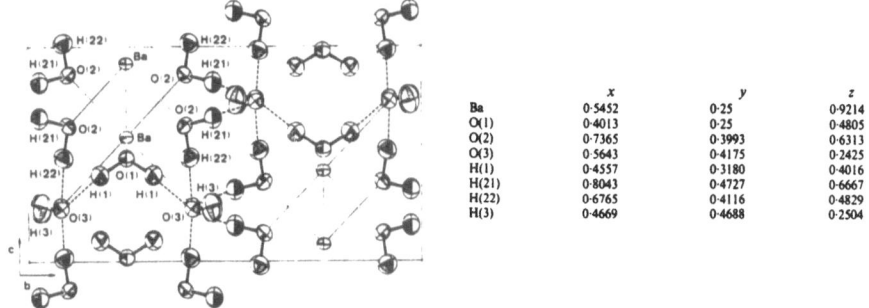

	x	y	z
Ba	0·5452	0·25	0·9214
O(1)	0·4013	0·25	0·4805
O(2)	0·7365	0·3993	0·6313
O(3)	0·5643	0·4175	0·2425
H(1)	0·4557	0·3180	0·4016
H(21)	0·8043	0·4727	0·6667
H(22)	0·6765	0·4116	0·4829
H(3)	0·4669	0·4688	0·2504

BARIUM HYDROXIDE OCTAHYDRATE
$Ba(OH)_2 \cdot 8H_2O$

Z. Kristallogr., 192, 111–118.

$P2_1/n$, 9.274, 9.260, 11.817, 98.95, Z = 4, R = 0.024. Structure as previously described (28, 119; 29, 297), with H atoms now located. $Ba(H_2O)_8^{2+}$ cations and anion pairs, linked by hydrogen bonds. Ba–O = 2.726–2.814(1), O–H...O = 2.628–2.985(2) A.

	x	y	z
Ba	0.00035	0.14806	0.24566
Oh(1)	0.0258	−0.3478	0.3926
Oh(2)	−0.0200	−0.3551	0.1398
Ow(1)	−0.2001	0.0529	0.3832
Ow(2)	−0.0825	0.3873	0.3537
Ow(3)	0.1236	−0.0838	0.3739
Ow(4)	0.2550	0.2477	0.3839
Ow(5)	−0.2576	0.2497	0.1205
Ow(6)	−0.1202	−0.0832	0.1124
Ow(7)	0.0680	0.3749	0.1044
Ow(8)	0.2130	0.0485	0.1267
Hh(1)	0.033	−0.355	0.463
Hh(2)	−0.005	−0.348	0.199
Hw(11)	−0.292	0.097	0.382
Hw(12)	−0.175	0.049	0.446
Hw(21)	−0.166	0.392	0.356
Hw(22)	−0.064	0.448	0.356
Hw(31)	0.200	−0.093	0.366
Hw(32)	0.094	−0.181	0.387
Hw(41)	0.349	0.214	0.377
Hw(42)	0.252	0.245	0.448
Hw(51)	−0.261	0.324	0.128
Hw(52)	−0.354	0.222	0.120
Hw(61)	−0.133	−0.072	0.049
Hw(62)	−0.092	−0.174	0.114
Hw(71)	0.053	0.443	0.117
Hw(72)	0.050	0.366	0.044
Hw(81)	0.296	0.089	0.118
Hw(82)	0.219	−0.028	0.128

HYDROGARNET (BARIUM INDIUM)
$Ba_3In_2(OD)_{12}$

J. Solid State Chem., 87, 173–177.

Ia3d, 14.0658, Z = 8, neutron powder data. Ba in 24(c): 1/8,0,1/4: In in 16(a): 0,0,0; O in 96(h): −0.02960,0.05343,0.14041; H in 96(h): −0.0912,0.0424,0.1567. Hydrogarnet structure (54A, 207).

ALUMINUM HYDROXIDE GERMANATE
$Al_2(OH)_2GeO_4$

Oesterreiche Akad. Wissenschaften, Math.-Naturwiss. Klasse, Anzeiger, __127__, 51-52.

C2/c, 5.418, 8.247, 9.303, 111.00, Z = 4, synchrotron radiation, R = 0.13. New
structure type, with cubic close-packing of O and OH, Al in octahedral and Ga
in tetrahedral holes.

POTASSIUM HEXAHYDROXOPLUMBATE(IV)
$K_2Pb(OH)_6$

Kristallografija, __35__, 491-492 [Soviet Physics - Crystallography, __35__, 284-285];
Ž. Neorg. Khim., __35__, 2285-2289 [Russ. J. Inorg. Chem., __35__, 1301-1304].

R$\bar{3}$, a = 6.621, c = 12.975 A, Z = 3, R = 0.066. Pb in 3(a): 0,0,0; K in 6(c):
0,0,0.7132; O in 18(f): 0.217,0.301, 0.0933. Isostructural with the Pt compound
(__9__, 213; __49A__, 229), with $Pb(OH)_6^{2-}$ octahedra linked by 9-coordinate K^+ ions and
hydrogen bonds. Pb-O = 2.16, K-O = 2.81-3.08(1) A.

VANADIUM HYDROXIDE OXIDE
$H_2V_3O_8$

J. Solid State Chem., __89__, 372-377.

Pnam, 16.9298, 9.3589, 3.6443, Z = 4, powder data.   (100) Layer of edge-sharing
VO_6 octahedra and VO_5 trigonal bipyramids. V-O = 1.58-2.47 A.

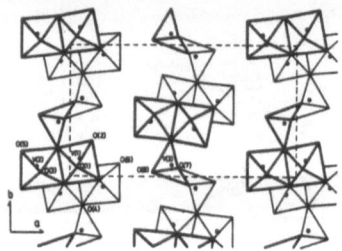

COPPER(II) HYDROXIDE
$Cu(OH)_2$

Acta Cryst., C__46__, 2279-2284.

$Cmc2_1$, 2.9471, 10.593, 5.2564, Z = 4, R = 0.042. Structure generally as previously described in Cmcm (__26__, 358), but the Cu coordination is square-pyramidal,
Cu-O = 1.948 (x 2), 1.972 (x 2), 2.356(5) A, with a sixth oxygen at 2.915(5) A.
These polyhedra share edges to form corrugated (010) layers, which are linked by
hydrogen bonds.

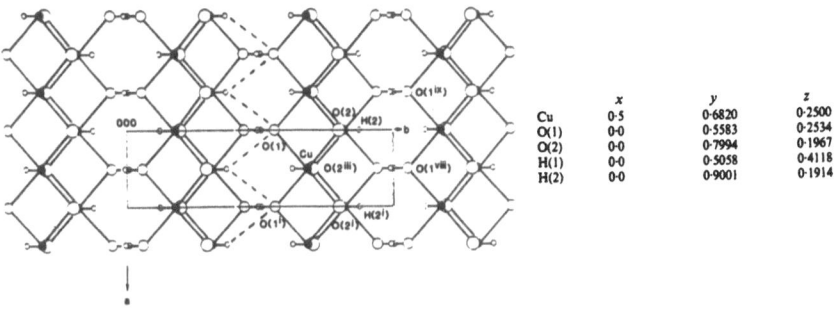

	x	y	z
Cu	0·5	0·6820	0·2500
O(1)	0·0	0·5583	0·2534
O(2)	0·0	0·7994	0·1967
H(1)	0·0	0·5058	0·4118
H(2)	0·0	0·9001	0·1914

STRONTIUM TETRAHYDROXOZINCATE MONOHYDRATE
SrZn(OH)$_4$·H$_2$O

Koord. Khim., 16, 1255-1259.

P$\bar{1}$, 6.245, 6.298, 7.838, 105.39, 65.15, 108.63, Z = 2, R = 0.071. Zn(OH)$_4^{2-}$ tetrahedra linked by 7-coordinate Sr. Zn-O = 1.94-1.98, Sr-O = 2.54-2.67 A.

CADMIUM HYDROXIDE
γ-Cd(OH)$_2$

Mater. Res. Bull., 25, 987-996.

Im, 5.664, 10.223, 3.404, β = 91.52°, at 133K, Z = 4, R = 0.027. The hydrogen positions indicate that the compound is Cd$_2$O(OH)$_2$·H$_2$O. Cd has distorted octa-hedral coordination, Cd-O = 2.25-2.44(1) A.

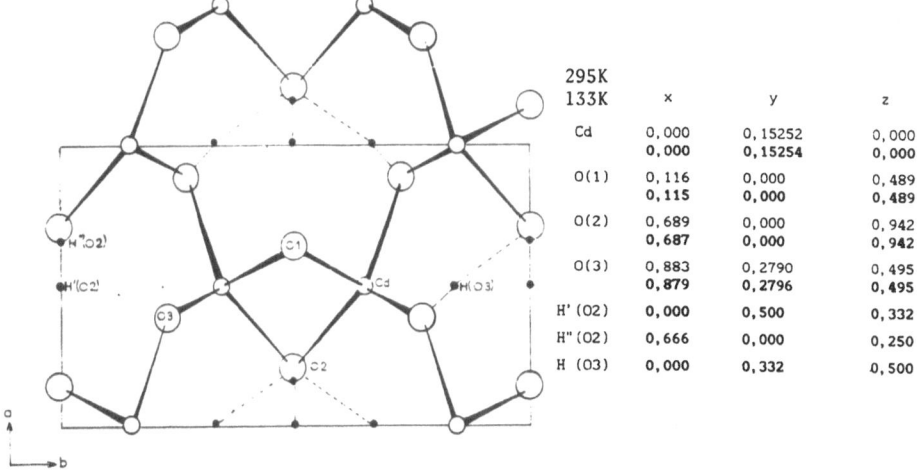

	295K 133K	x	y	z
Cd		0,000	0,15252	0,000
		0,000	0,15254	0,000
O(1)		0,116	0,000	0,489
		0,115	0,000	0,489
O(2)		0,689	0,000	0,942
		0,687	0,000	0,942
O(3)		0,883	0,2790	0,495
		0,879	0,2796	0,495
H'(O2)		0,000	0,500	0,332
H"(O2)		0,666	0,000	0,250
H(O3)		0,000	0,332	0,500

DEUTERIUM SULPHIDE
D_2S

Z. Kristallogr., <u>193</u>, 1-19.

Phase I, Fm3m, 5.8486, at 160K, Z = 4, neutron powder data. S in 4(a): 0,0,0;
D in sperical shell around S.

Phase II, Pa3, 5.7647, at 120K, Z = 4, neutron powder data. S in 4(a): 0,0,0;
D again in a shell around S.

Phase III, Pbcm, 4.0760, 13.3801, 6.7215, at 1.5K, Z = 8, neutron powder data.
S(1,2), D(1,2) in 4(d): x,y,1/4, x = -0.042, 0.460, 0.173, 0.194, y = 0.1449,
0.4017, 0.2168, 0.0736; D(3) in 8(e): 0.6724,0.3797,0.1057. Hexagonal close-
packed S, with distortions to accommodate the H atoms.

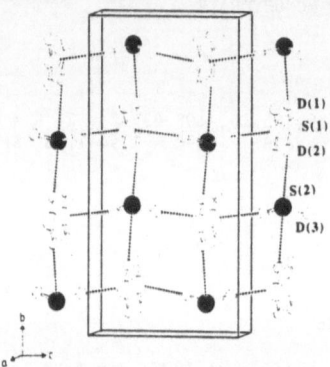

POTASSIUM BIS(DITHIOCARBONATO)NITRIDOTECHNETATE(V) DIHYDRATE
$K_2[TcN(S_2CO)_2].2H_2O$

J. Chem. Soc., Dalton, 2923-2925.

$P2_1/n$, 8.353, 15.630, 9.230, 90.94, Z = 4, R = 0.035. Anions in which Tc has
distorted square-pyramidal 5-coordination, linked by K ions which have distorted
tetrahedral coordinations to N, O, and H_2O. Tc-N = 1.621(6), Tc-S = 2.386-2.392(2)
A.

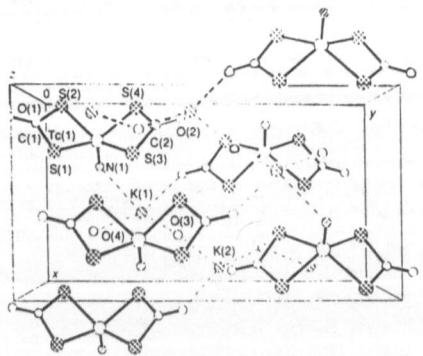

TETRAPHOSPHORUS TRISELENIDE
α'-P_4Se_3, α-P_4Se_3

Inorg. Chem., 29, 2889-2894.

α', Pnma, 10.997, 9.845, 13.803, at 133K, Z = 8, R = 0.057. α, Pnma, 11.788, 9.720, 26.254, at 265K, Z = 16, R = 0.043. The α'-form is isostructural with α-P_4S_3 (21, 253; 50A, 55). The structure of the α-form is as previously described (23, 402).

SODIUM SELENOARSENATE DODECAHYDRATE
$Na_3AsO_3Se.12H_2O$ (I)

SODIUM TETRASELENOARSENATE ENNEAHYDRATE
$Na_3AsSe_4.9H_2O$ (II)

Z. anorg. Chem., 581, 141-152.

I, $P2_12_12_1$, 9.220, 13.018, 14.048, Z = 4, R = 0.041. II, $P2_13$, 12.149, Z = 4, R = 0.045. I is isostructural with the S analogue (this volume, p. 275), and II is isostructural with $Na_3SbS_4.9H_2O$ (13, 282; 45A, 276, 399). Both structures contain tetrahedral anions linked by hydrogen bonds via the water molecules, and by 6-coordinate Na ions. O—H...Se = 3.36-3.69 A.

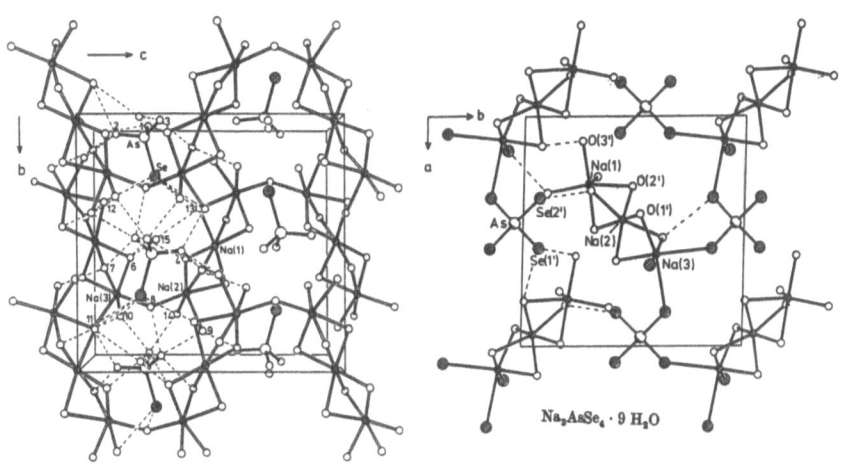

Na₂AsO₃Se · 12 H₂O   Na₃AsSe₄ · 9 H₂O

LITHIUM AMMONIUM TITANIUM SULPHIDES
$Li_{0.23}(ND_3)_{0.63}TiS_2$ (I), $Li_{0.11}(ND_4)_{0.11}(ND_3)_{0.54}TiS_2$ (II)

Chem. Mater., 2, 75-81.

I, R$\bar{3}$m, a = 3.4234, 3.4121, c = 26.824, 26.492 A, at 300, 12K; II, R$\bar{3}$m, a = 3.4206, 3.4093, c = 26.775, 26.447 A, at 300, 12K. 3R TiS_2 polytypes, with Li coordinated to 3 ND_3 molecules and ammonium ions with trigonal prismatic coordination between the TiS_2 sheets.

LANTHANUM OXYSULPHIDE
$La_2O_2S_2$

Acta Cryst., C46, 1376–1378.

Cmca, 13.215, 5.943, 5.938, Z = 4, R = 0.017. La in 8(d): 0.3392,0,0,; S in 8(f): 0,0.3750,0.3747; O in 8(e): 1/4,0.2446,1/4. (100) Layers of edge-sharing LaO_4S_4 square antiprisms, linked along **a** by sharing oxygen atoms or S_2 pairs.

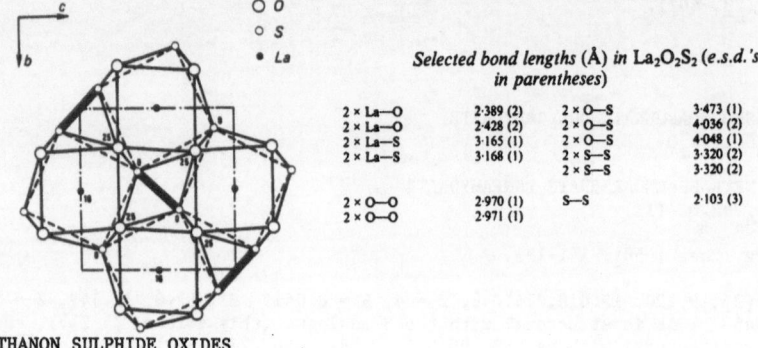

○ O
○ S
● La

Selected bond lengths (Å) in $La_2O_2S_2$ (e.s.d.'s in parentheses)

2 × La—O	2·389 (2)	2 × O—S	3·473 (1)
2 × La—O	2·428 (2)	2 × O—S	4·036 (2)
2 × La—S	3·165 (1)	2 × O—S	4·048 (1)
2 × La—S	3·168 (1)	2 × S—S	3·320 (2)
		2 × S—S	3·320 (2)
2 × O—O	2·970 (1)	S—S	2·103 (3)
2 × O—O	2·971 (1)		

LANTHANON SULPHIDE OXIDES
Ln_2S_2O　　(Ln = Er, Tm, Yb)

J. Less-Common Metals, 158, 137–145.

$P2_1/c$, a = 8.1920, 8.1556, 8.1181, b = 6.8326, 6.7960, 6.7616, c = 6.8050, 6.7739, 6.7403 A, β = 99.631, 99.844, 99.922°, Z = 4, R = 0.032, 0.027, 0.056. Ln ions are 7-coordinate.

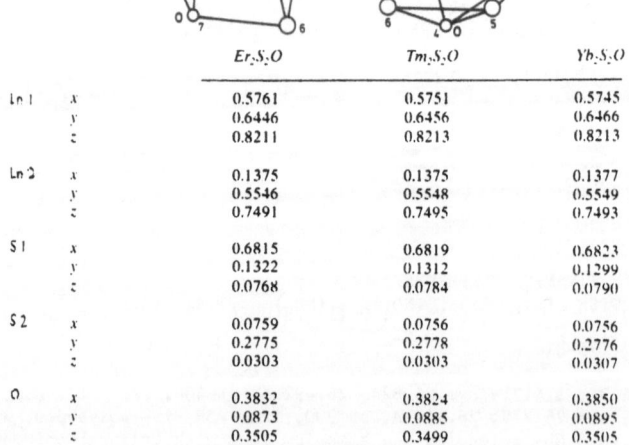

		Er_2S_2O	Tm_2S_2O	Yb_2S_2O
Ln 1	x	0.5761	0.5751	0.5745
	y	0.6446	0.6456	0.6466
	z	0.8211	0.8213	0.8213
Ln 2	x	0.1375	0.1375	0.1377
	y	0.5546	0.5548	0.5549
	z	0.7491	0.7495	0.7493
S 1	x	0.6815	0.6819	0.6823
	y	0.1322	0.1312	0.1299
	z	0.0768	0.0784	0.0790
S 2	x	0.0759	0.0756	0.0756
	y	0.2775	0.2778	0.2776
	z	0.0303	0.0303	0.0307
O	x	0.3832	0.3824	0.3850
	y	0.0873	0.0885	0.0895
	z	0.3505	0.3499	0.3505

LUTETIUM URANIUM OXYSULPHIDE
$(UOS)_4LuS$

Acta Cryst., C46, 1205–1207.

I4/mmm, 3.8014, 34.20, Z = 2, R = 0.079. Two (UOS)$_2$ sheets alternate with one (LuS) sheet. U atoms have 8- and 9-coordinations (formal oxidation states 3.5 and 4, respectively), and Lu has octahedral coordination.

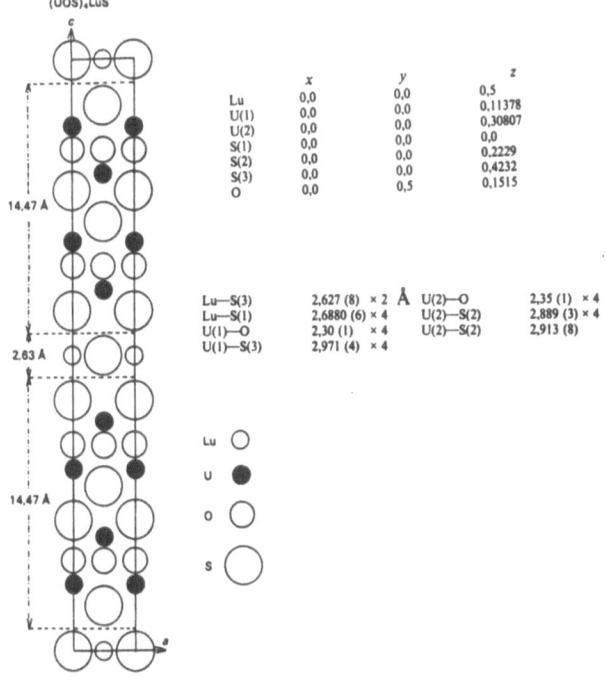

(UOS)₄LuS

	x	y	z
Lu	0,0	0,0	0,5
U(1)	0,0	0,0	0,11378
U(2)	0,0	0,0	0,30807
S(1)	0,0	0,0	0,0
S(2)	0,0	0,0	0,2229
S(3)	0,0	0,0	0,4232
O	0,0	0,5	0,1515

Lu—S(3)	2,627 (8) × 2 Å	U(2)—O	2,35 (1) × 4	
Lu—S(1)	2,6880 (6) × 4	U(2)—S(2)	2,889 (3) × 4	
U(1)—O	2,30 (1) × 4	U(2)—S(2)	2,913 (8)	
U(1)—S(3)	2,971 (4) × 4			

Lu ○
U ●
O ○
S ○

LITHIUM HEPTABORATE
$Li_3B_7O_{12}$

Acta Cryst., C46, 1999-2001.

P$\bar{1}$, 6.487, 7.840, 8.510, 92.11, 104.85, 99.47, Z = 2, R = 0.061. Framework of B_3O_7, B_3O_8, and BO_3 groups connected by sharing oxygen atoms, with 4- and (4+1)-coordinated Li ions in holes. B-O = 1.35-1.50, Li-O = 1.93-2.18, 2.30(1) A.

	x	y	z
O1	0·4104	0·3898	0·3137
O2	0·3280	0·6576	0·2118
O3	0·5607	0·8011	0·0695
O4	0·6845	0·5850	0·2421
O5	0·2618	0·0565	0·1079
O6	0·0601	0·2390	0·4741
O7	0·2252	0·0281	0·3687
O8	0·7294	0·2960	0·3100
O9	-0·0501	-0·1365	0·1449
O10	-0·0359	0·1705	0·1825
O11	0·0661	0·4675	0·2920
O12	0·6394	0·8627	0·3576
B1	0·2676	0·5033	0·2718
B2	0·0938	0·0360	0·2021
B3	0·6156	0·4269	0·2872
B4	-0·0424	0·2906	0·3136
B5	0·5586	0·7245	0·2230
B6	0·2149	0·1378	0·4942
B7	0·2491	0·1319	-0·0365
Li1	-0·0119	0·6948	0·3146
Li2	0·5178	0·0756	0·3062
Li3	0·2145	0·7977	0·0292

SODIUM BORATE HYDRATE
β-Na$[B_5O_6(OH)_4]$.3H$_2$O

Z. Naturforsch., 45B, 1155-1166.

C2/c, 11.097, 16.435, 13.786, 115.02, Z = 8, R = 0.035. The structure contains
the same pentaborate anion as in the α-form (sborgite, 38A, 292), with a slightly
different orientation in the cell; Na ions have octahedral and tetrahedral coordinations.

STRONTIUM BERYLLIUM BORATE
SrBe$_2$(BO$_3$)$_2$

J. Solid State Chem., 85, 270-274.

P2$_1$/n, 9.247, 4.492, 11.561, 112.17, Z = 4, R = 0.034. Layers of corner-sharing
BO$_3$ triangles, BeO$_4$ distorted tetrahedra, and a pair of edge-sharing BeO$_4$ tetrahedra, with layers linked by 9-coordinate Sr ions. B-O = 1.36-1.39, Be-O = 1.58-
1.70, Sr-O = 2.55-2.96 A.

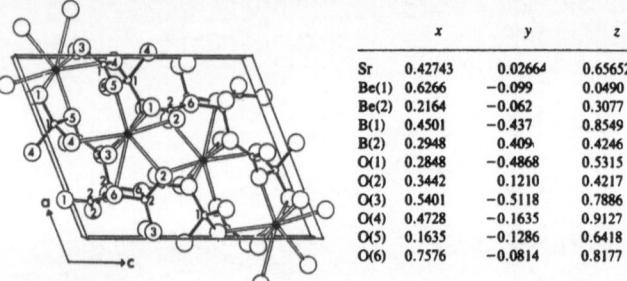

	x	y	z
Sr	0.42743	0.02664	0.65652
Be(1)	0.6266	−0.099	0.0490
Be(2)	0.2164	−0.062	0.3077
B(1)	0.4501	−0.437	0.8549
B(2)	0.2948	0.409	0.4246
O(1)	0.2848	−0.4868	0.5315
O(2)	0.3442	0.1210	0.4217
O(3)	0.5401	−0.5118	0.7886
O(4)	0.4728	−0.1635	0.9127
O(5)	0.1635	−0.1286	0.6418
O(6)	0.7576	−0.0814	0.8177

PINAKIOLITE (ANTIMONY)
(Mg,Mn,Sb)$_3$O$_2$BO$_3$

Z. Kristallogr., 191, 105-116.

C2/m, 21.773, 6.153, 5.327, 94.38, Z = 8, R = 0.047. Structure as previously
described (13, 355; 40A, 222).

	Atom	x ×10^4	y	z
100% Mn	M(1)	0	0	0
100% Mn	M(2)	0	0	5000
38% Sb + 32% Mn + 30% Mg	M(3)	2508	0	5043
52%Mn + 48% Mg	M(4)	−60	5000	450
52%Mn + 48% Mg	M(5)	−66	5000	4391
52%Mn + 48% Mg	M(45)	−180	5000	2597
100% Mg	M(6)	2500	2500	0
100% Mg	M(7)	3880	2416	7168
	B(1)	1308	0	7814
	B(2)	3636	0	2011
	O(1)	1075	0	10131
	O(2)	1003	0	5473
	O(3)	1952	0	8152
	O(4)	4006	0	−33
	O(5)	3933	0	4383
	O(6)	2990	0	1912
	O(7)	5151	2812	2566
	O(8)	2953	2666	6809

BARIUM BORATE
β-BaB$_2$O$_4$

Rep. Res. Lab. Eng. Mater., Tokyo Inst. Technol., 15, 1-12.

R3c, a = 12.5316, c = 12.7285 A, Z = 18, R = 0.013. Structure as previously
described (49A, 241; 51A, 265; 55A, 302), with two independent planar B$_3$O$_6$$^{3-}$
anions, linked by Ba^{2+} cations.

ALUMINOBORATES

$Co_{2.1}Al_{0.9}BO_5$ (I), Ni_2AlBO_5 (II), Cu_2AlBO_5 (III)

J. Solid State Chem., **84**, 289-298.

I, II, Pbam, a = 12.010, 12.013, b = 9.197, 9.111, c = 2.993, 2.942 A, Z = 4, R =
0.044, 0.041. III, $P2_1/a$, 9.365, 11.778, 3.072, 97.71, Z = 4, R = 0.046. I and
II are isostructural with ludwigite (**13**, 353), and III is a distorted variant; all
three structures exhibit partial cation ordering.

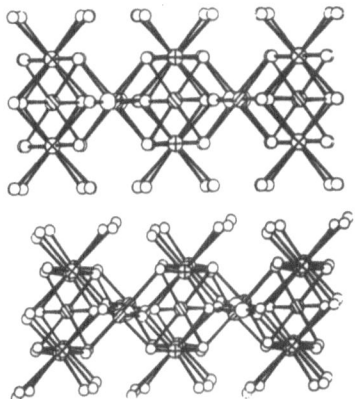

Comparative views of Ni_2AlBO_3 (top) and
Cu_2AlBO_3 (bottom).

NICKEL SCANDIUM BORATE

$NiScBO_4$

Z. anorg. Chem., **585**, 87-92.

Pnma, 9.415, 3.173, 9.461, Z = 4, R = 0.043. Warwickite-type (**13**, 351; **40A**,
222; **44A**, 234), with Ni/Sc disordered in the two octahedral sites.

AMMONIUM COBALT HEXABORATE HYDRATE

$(NH_4)_2[Co(H_2O)_2\{B_6O_7(OH)_6\}_2]\cdot 2H_2O$

Latv. PSR Zinat. Akad. Vestis, Kim. Ser., No. 2, 173-180.

$P\bar{1}$, 11.053, 7.867, 7.535, 90.13, 94.86, 107.95, Z = 1, R = 0.022. Isostructural
with the K/Co and K/Mg compounds (**39A**, 269; **51A**, 265), with centrosymmetric complex anions with Co coordinated octahedrally to 2 H_2O and 4 O.

NICKEL ALUMINUM BORATE
NICKEL IRON(III) BORATE
NICKEL GALLIUM BORATE

Ni_2MBO_5 (M = Al, Fe, Ga)

Z. anorg. Chem., **582**, 15-20.

Pbam, a = 9.012, 9.200, 9.200, b = 12.011, 12.210, 12.154, c = 2.944, 3.006, 2.986
A, Z = 4, R = 0.055, 0.059, 0.042. Ludwigite-type structure, as for related materials (**39A**, 269; **52A**, 255; **56A**, 190), with Ni/Al disorder in the Al compound. The

Cr compound is isostructural.

NICKEL ANTIMONY BORATE
$Ni_{5.33}Sb_{0.67}B_2O_{10}$

J. Less-Common Metals, 158, 339-345.

P2/m subcell, a = 5.376, 5.355, b = 3.031, 3.021, c = 10.603, 10.563 A, β = 94.47,
94.49°, at 293, 105K, Z = 1, R = 0.079, 0.071. Hulsite-type (42A, 310).

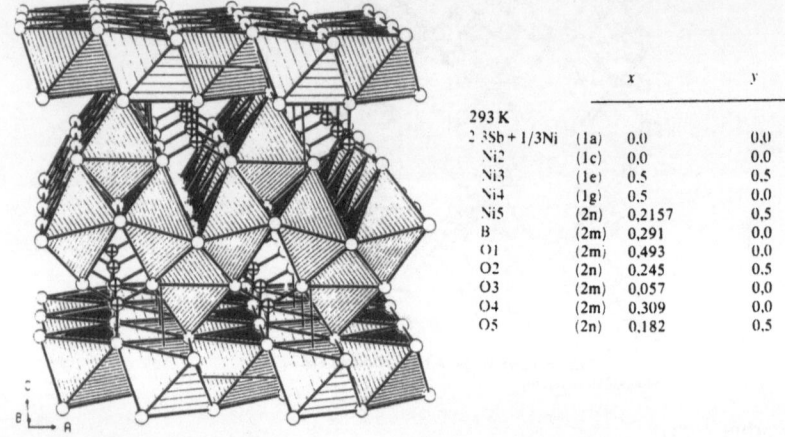

293 K		x	y	z
2/3Sb + 1/3Ni	(1a)	0.0	0.0	0.0
Ni2	(1c)	0.0	0.0	0.5
Ni3	(1e)	0.5	0.5	0.0
Ni4	(1g)	0.5	0.0	0.5
Ni5	(2n)	0.2157	0.5	0.2815
B	(2m)	0.291	0.0	0.764
O1	(2m)	0.493	0.0	0.300
O2	(2n)	0.245	0.5	0.468
O3	(2m)	0.057	0.0	0.699
O4	(2m)	0.309	0.0	0.892
O5	(2n)	0.182	0.5	0.096

NICKEL TANTALUM BORATE   NICKEL NIOBIUM BORATE
$Ni_{5.33}Ta_{0.67}B_2O_{10}$ $Ni_{5.33}Nb_{0.67}B_2O_{10}$

Solid State Ionics, 43, 1-5.

C2/c, a = 10.525, 10.507, b = 6.159, 6.162, c = 21.642, 21.657 A, β = 101.86,
101.84°, Z = 8, R = 0.065, 0.087. (001) Layers of edge-sharing NiO_6 and NbO_6
or TaO_6 octahedra, with layers linked by zigzag ribbons of edge-sharing NiO_6
octahedra; small straight tunnels contain three-coordinated B. Two (of seven)
Ni sites have partial Ta or Nb occupancy. The Sb compound has a slightly different structure (preceding report).

BARIUM COPPER(II) BORATE
$Ba_2Cu(BO_3)_2$

Acta Cryst., C46, 370-372.

Pnma, 8.023, 11.290, 13.889, Z = 8, R = 0.020. $Cu_2(BO_3)_4^{8-}$ units of two CuO_4
distorted square planes, linked by Ba^{2+} ions. Cu-O = 1.91-2.00, B-O = 1.33-
1.41 A.

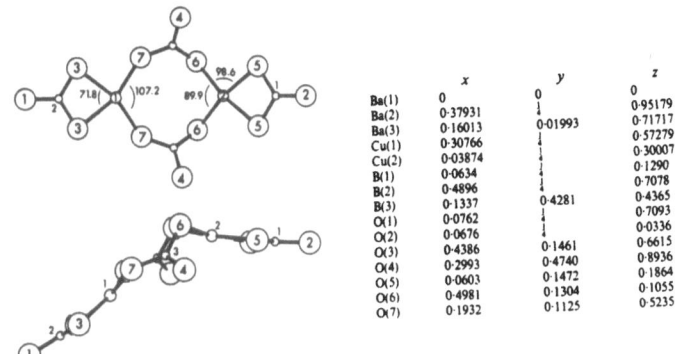

	x	y	z
Ba(1)	0	0	0
Ba(2)	0·37931	¼	0·95179
Ba(3)	0·16013		0·71717
Cu(1)	0·30766	0·01993	0·57279
Cu(2)	0·03874	¼	0·30007
B(1)	0·0634	¼	0·1290
B(2)	0·4896		0·7078
B(3)	0·1337	0·4281	0·4365
O(1)	0·0762	¼	0·7093
O(2)	0·0676		0·0336
O(3)	0·4386	0·1461	0·6615
O(4)	0·2993	0·4740	0·8936
O(5)	0·0603	0·1472	0·1864
O(6)	0·4981	0·1304	0·1055
O(7)	0·1932	0·1125	0·5235

Sketches of the $Cu_2(BO_3)_4$ unit. Top: view along [100]. Angles in °. Bottom: view along direction orthogonal to mirror plane at $y = \frac{1}{4}$.

SILVER DODECABORATE TRIHYDRATE
$Ag_6[B_{12}O_{18}(OH)_6].3H_2O$

$3 \times [Ag_2O.2B_2O_3.2H_2O]$

Z. Kristallogr., 190, 85-96.

$P2_1/c$, 11.784, 10.654, 10.116, 112.10, Z = 2, R = 0.056. Isolated dodecaborate anions, linked by Ag^+ ions (two of which have partial occupancy).

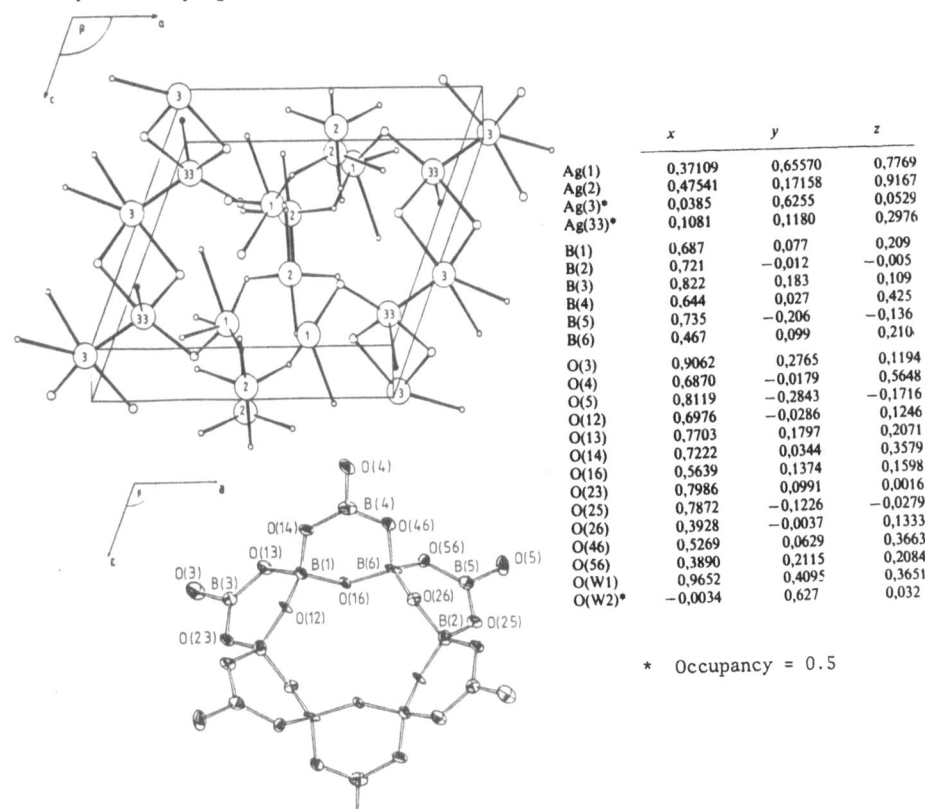

	x	y	z
Ag(1)	0,37109	0,65570	0,7769
Ag(2)	0,47541	0,17158	0,9167
Ag(3)*	0,0385	0,6255	0,0529
Ag(33)*	0,1081	0,1180	0,2976
B(1)	0,687	0,077	0,209
B(2)	0,721	-0,012	-0,005
B(3)	0,822	0,183	0,109
B(4)	0,644	0,027	0,425
B(5)	0,735	-0,206	-0,136
B(6)	0,467	0,099	0,210
O(3)	0,9062	0,2765	0,1194
O(4)	0,6870	-0,0179	0,5648
O(5)	0,8119	-0,2843	-0,1716
O(12)	0,6976	-0,0286	0,1246
O(13)	0,7703	0,1797	0,2071
O(14)	0,7222	0,0344	0,3579
O(16)	0,5639	0,1374	0,1598
O(23)	0,7986	0,0991	0,0016
O(25)	0,7872	-0,1226	-0,0279
O(26)	0,3928	-0,0037	0,1333
O(46)	0,5269	0,0629	0,3663
O(56)	0,3890	0,2115	0,2084
O(W1)	0,9652	0,4095	0,3651
O(W2)*	-0,0034	0,627	0,032

* Occupancy = 0.5

BARIUM LANTHANON BORATES
$Ba_3Ln_2(BO_3)_4$ (Ln = La, Pr)

Kristallografija, 35, 856-860 [Soviet Physics - Crystallography, 35, 503-505]

Pnma, a = 7.734, 7.733, b = 17.043, 16.843, c = 9.056, 9.009 A, Z = 4, R = 0.033, 0.039. Isostructural with related materials (39A, 270; 54A, 214), but with disordered Ba/Ln sites.

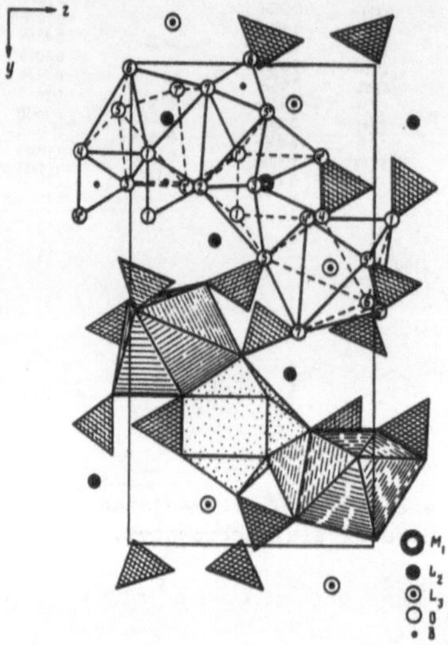

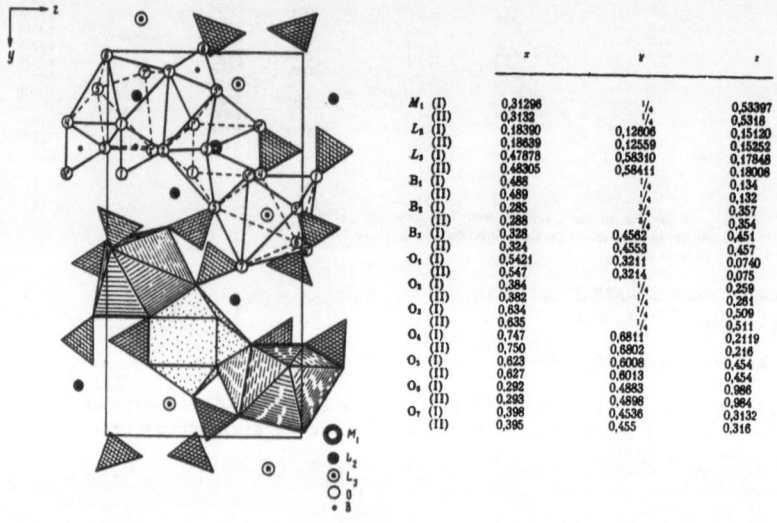

		x	y	z
M_1	(I)	0,31296	¼	0,53397
	(II)	0,3132	¼	0,5318
L_2	(I)	0,18390	0,12606	0,15120
	(II)	0,18639	0,12559	0,15252
L_3	(I)	0,47878	0,58310	0,17848
	(II)	0,48305	0,58411	0,18008
B_1	(I)	0,488	¼	0,134
	(II)	0,489	¼	0,132
B_2	(I)	0,285	¾	0,357
	(II)	0,288	¾	0,354
B_3	(I)	0,328	0,4562	0,451
	(II)	0,324	0,4553	0,457
O_1	(I)	0,5421	0,3211	0,0740
	(II)	0,547	0,3214	0,075
O_2	(I)	0,384	¼	0,259
	(II)	0,382	¼	0,261
O_3	(I)	0,634	¼	0,509
	(II)	0,635	¼	0,511
O_4	(I)	0,747	0,6811	0,2119
	(II)	0,750	0,6802	0,216
O_5	(I)	0,623	0,6008	0,454
	(II)	0,627	0,6013	0,454
O_6	(I)	0,292	0,4883	0,986
	(II)	0,293	0,4898	0,984
O_7	(I)	0,398	0,4536	0,3132
	(II)	0,395	0,455	0,316

LANTHANUM BOROGERMANATE
$LaBGeO_5$

Izv. Akad. Nauk SSSR, Neorg. Mater., 26, 1105-1107.

$P3_1$ (pseudo $P3_121$), 7.020, 6.879, Z = 3, R = 0.03. Isostructural with stillwellite, $CeBSiO_5$ (29, 411; 32A, 433). Ge-4 O = 1.72-1.79, B-4 O = 1.45-1.50, La-9 O = 2.41-2.74(1) A.

LITHIUM BOROURANATE
$LiBUO_5$

Acta Cryst., C46, 372-374.

$P2_1/c$, 5.767, 10.574, 6.835, 105.04, Z = 4, R = 0.062. The structure resembles

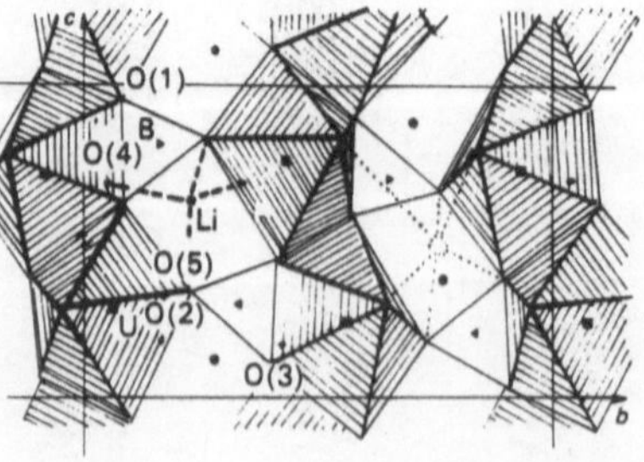

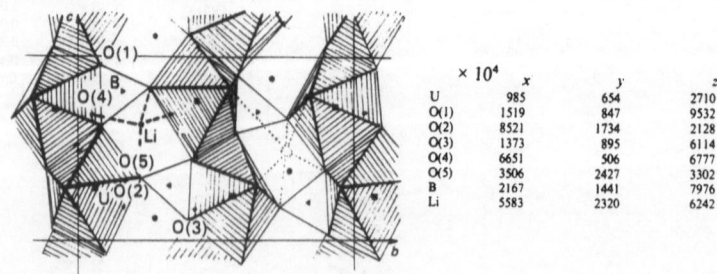

$\times 10^4$	x	y	z
U	985	654	2710
O(1)	1519	847	9532
O(2)	8521	1734	2128
O(3)	1373	895	6114
O(4)	6651	506	6777
O(5)	3506	2427	3302
B	2167	1441	7976
Li	5583	2320	6242

that of (orthorhombic) NaBUO$_5$ (55A, 220), but with tetrahedral coordination for
Li (octahedral for Na); U has pentagonal bipyramidal coordination. U-O = 1.79,
1.80, 2.28-2.41(2), B-O = 1.36-1.41(4), Li-O = 1.93-2.09(6) A.

SODIUM TRIHYDROGENTETRACARBONATE
$Na_5H_3(CO_3)_4$

Acta Cryst., B46, 466-474.

P$\bar{1}$, 3.4762, 10.0393, 15.5969, 107.770, 95.589, 95.028, Z = 2, R = 0.029. The
carbonate groups are linked by hydrogen bonds to form two independent $[H_3(CO_3)_4]^{5-}$
fragments; there are two asymmetric (O...O = 2.572 and 2.597(1) A) and two symmetric hydrogen bonds (O...O = 2.492 and 2.507(1) A). Na ions have 5- to 7-
coordinations. Electron-density study.

	x	y	z
Na1	0·78632	0·67317	0·04873
Na2	0·81130	0·18185	0·09932
Na3	0·18047	0·71800	0·27916
Na4	0·42761	0·21298	0·30787
Na5	0·77185	0·74803	0·47762
O1	0·28749	0·55431	0·09748
O2	0·37640	0·34347	0·10976
O3	0·62239	0·54136	0·22131
O4	0·94197	0·38381	0·30290
O5	0·18035	0·60058	0·39492
O6	0·30491	0·40970	0·43290
O7	0·30385	0·85333	0·05650
O8	0·30195	0·06961	0·15427
O9	0·07345	0·87712	0·18646
O10	0·88991	0·06120	0·32907
O11	0·73764	0·86380	0·36291
O12	0·64812	0·07382	0·45689
C1	0·41972	0·47544	0·13898
C2	0·13807	0·46837	0·37668
C3	0·23680	0·93709	0·12978
C4	0·75976	0·99593	0·38122

POTASSIUM BICARBONATE
$KHCO_3$

J. Solid State Chem., 86, 180-187.

301K, P2$_1$/a, 15.181, 5.629, 3.713, 104.64, Z = 4, R = 0.026.   353K, C2/m, 15.194,
5.640, 3.734, 104.86, Z = 4, R = 0.023. The room-temperature structure is as
previously described (16, 327; 17, 516; 33A, 433; 40A, 226), with disordered H
sites. In the high-temperature phase disorder of the carbonate groups results in
higher symmetry.

room temperature

high temperature

POTASSIUM and RUBIDIUM CARBONATE FLUORIDE
K_3CO_3F, Rb_3CO_3F

Z. Naturforsch., $\underline{45}$B, 943-946.

R$\bar{3}$c, a = 7.4181, 7.761, c = 16.3918, 17.412 A, Z = 6, R = 0.04, 0.07. K or Rb in
18(e): x,x,3/4, x = 0.5669, 0.5680; C in 6(a): 0,0,3/4; O in 18(e): x = 0.1737,
0.1663; F in 6(b): 0,0,0. The structure contains carbonate and fluoride anions,
linked by alkali-metal cations which are coordinated to 5 O and 2 F. C-O =
1.289(1), 1.291(7), K-O = 2.80-2.92, K-F = 2.59, Rb-O = 2.93-3.11, Rb-F = 2.27 A.

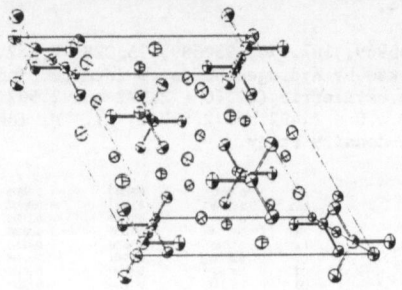

CALCITE (MAGNESIAN)
(Ca,Mg)CO$_3$

Amer. Min., $\underline{75}$, 1151-1158.

R$\bar{3}$c, a = 4.9673, 4.9382, c = 16.9631, 16.382 A, for 6.4 and 12.9 mol % Mg, Z = 6,
R = 0.023, 0.037. Ca/Mg in 6(b): 0,0,0; C in 6(a): 0,0,1/4; O in 18(e), x,0,1/4,
x = 0.2575, 0.2587. Calcite structure ($\underline{1}$, 292; $\underline{52}$A, 258), with disordered Ca/Mg
distribution.

TETRAAMMINECARBONATOCOBALT(III) NITRATE HYDRATE
[Co(NH$_3$)$_4$(CO$_3$)](NO$_3$).0.5H$_2$O

Struct. Chem., $\underline{1}$, 227-234.

P2$_1$/n, 7.4960, 22.673, 10.513, 91.41, Z = 8, R = 0.033. Octahedral cations,
linked to each other and to the nitrate anions and water molecules by an intricate system of hydrogen bonds. Co-O = 1.896-1.916(1), Co-N = 1.942-1.969(1) A.

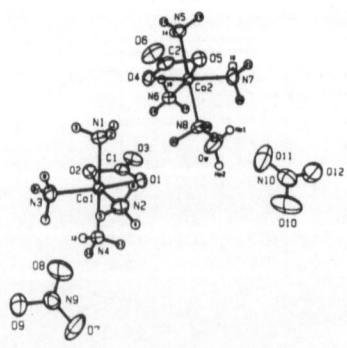

POTASSIUM NICKEL HYDROGENBISCARBONATE TETRAHYDRATE
$KNiH(CO_3)_2 \cdot 4H_2O$

Acta Cryst., B$\underline{46}$, 458–466.

P$\bar{1}$, 5.3824, 6.6737, 6.9480, 115.881, 90.678, 108.017, Z = 1, R = 0.035. Iso-structural with the Mg compound ($\underline{38A}$, 299; $\underline{55A}$, 221), with two carbonate groups linked by a short symmetric hydrogen bond, O...O = 2.456(2) A. Ni has octahedral coordination, and K has 12-coordination. Electron-density study.

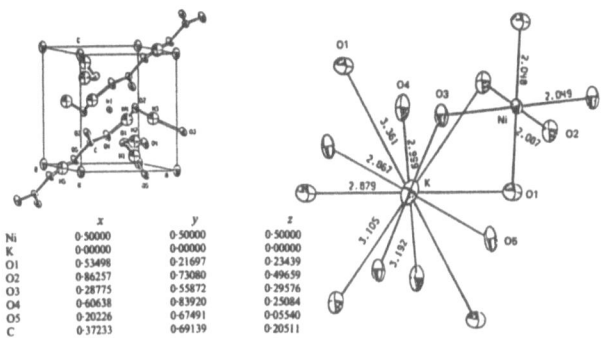

	x	y	z
Ni	0·50000	0·50000	0·50000
K	0·00000	0·00000	0·00000
O1	0·53498	0·21697	0·23439
O2	0·86257	0·73080	0·49659
O3	0·28775	0·55872	0·29576
O4	0·60638	0·83920	0·25084
O5	0·20226	0·67491	0·05540
C	0·37233	0·69139	0·20511

SZYMANSKIITE
$(Hg_2)_8(Ni,Mg)_6(CO_3)_{12}(OH)_2(H_3O)_8 \cdot 3H_2O$

Canad. Miner., $\underline{28}$, 703–707, 709–718.

P6_3, 17.3984, 6.0078, Z = 1, R = 0.035. -O-Hg-Hg-O- chains, face-sharing -Hg-O$_3$-Hg- columns, and (Ni,Mg)O$_6$ distorted octahedra, held together by carbonate groups. Very large tunnels (13 A across) contain disordered carbonate, oxonium, and water, leaving still a 7 A tunnel (the mineral can be regarded as a non-silicate zeolite).

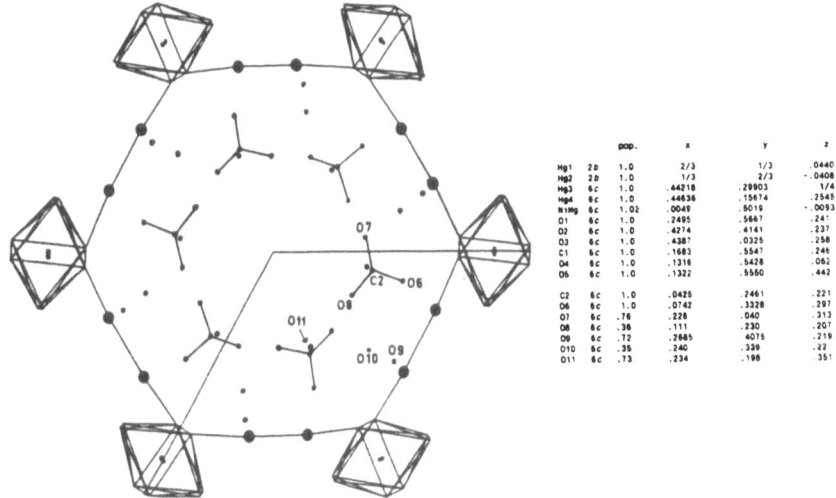

	pop.	x	y	z
Hg1	2b	1.0	2/3	1/3	.0440
Hg2	2b	1.0	1/3	2/3	-.0408
Hg3	6c	1.0	.64218	.29903	1/4
Hg4	6c	1.0	.44636	.15674	.2545
Ni/Mg	6c	1.02	.0049	.5019	-.0093
O1	6c	1.0	.2495	.5667	.241
O2	6c	1.0	.4274	.4141	.237
O3	6c	1.0	.4387	.0325	.258
C1	6c	1.0	.1683	.5547	.246
O4	6c	1.0	.1316	.5428	.062
O5	6c	1.0	.1322	.5550	.442
C2	6c	1.0	.0425	.2461	.221
O6	6c	1.0	.0742	.3328	.297
O7	6c	.76	.226	.040	.313
O8	6c	.36	.111	.230	.207
O9	6c	.72	.2885	4075	.219
O10	6c	.35	.240	.339	.22
O11	6c	.73	.234	.198	351

POTASSIUM URANYL CARBONATE
$K_4[UO_2(CO_3)_3]$

Chin. J. Chem., No. 4, 313-318.

C2/c, 10.240, 9.198, 12.222, 95.12, Z = 4, R = 0.12. U has hexagonal bipyramidal
coordination, with three bidentate carbonate ligands in equatorial sites.

SODIUM NITRITE
$NaNO_2$

Acta Cryst., B46, 343-347.

Im2m, 3.518, 5.535, 5.382, at 120K, Z = 2, R = 0.013. Na in 2(a): 0,0.58814,0;
N in 2(a): 0,0.12244,0; O in 4(d): 0,0,0.19620. Structure as previously described
(2, 65; 41A, 297). Electron-density study.

POTASSIUM NITRITE (CUBIC)
KNO_2

Z. Kristallogr., 190, 63-74.

Fm3m, a = 6.7 A, at 373K, Z = 4, R = 0.075. 4 K in 4(b): 0,0,1/2; 4 N in 32(e):
x,x,x, x = 0.0387; 8 O in 96(k): x,x,z, x = 0.071, z = 0.124 or x = 0.107, z =
0.048. Orientationally-disordered nitrite ions.

SODIUM AMMONIUM HEXANITROIRIDATE(III)
$Na(NH_4)_2[Ir(NO_2)_6]$

Ž. Neorg. Khim., 35, 682-684 [Russ. J. Inorg. Chem., 35, 384-386].

Fm3 (or F23), 10.512, Z = 4, R = 0.023. Ir in 4(a): 0,0,0; Na in 4(b): 1/2,0,0;
N(1) in 8(c): 1/4,1/4,1/4; N(2) in 24(e): 0,0.1960,0; O in 48(h): 0,0.2553,0.1029.
Cubic close-packing of octahedral anions, with Na in octahedral and ammonium in
tetrahedral holes, each coordinated to 12 O. Ir-N = 2.06 A. The Rh compound is
isostructural (10, 137).

POTASSIUM DI-μ-HYDROXO-BIS[DINITROPLATINATE(II)] SESQUIHYDRATE
$K_2[Pt_2(NO_2)_4(OH)_2] \cdot 1 \cdot 5H_2O$ (I)

POTASSIUM cyclo-TRI-μ3-OXO-TRIS[DINITROPLATINATE(II)]TRINITROPLATINATE(IV)
 TRIHYDRATE
$K_5[Pt_4(NO_2)_9O_3] \cdot 3H_2O$ (II)

Inorg. Chem., 29, 73-76, 1394.

I, Pbca, 11.879, 13.094, 32.060, Z = 16, R = 0.065. II, P$\bar{1}$, 9.940, 10.069, 15.206,
72.15, 74.69, 72.24, Z = 2, R = 0.032. I contains dimeric anions linked by hydrogen bonds; II is a mixed-valence tetranuclear complex (as previously described
(55A, 225)).

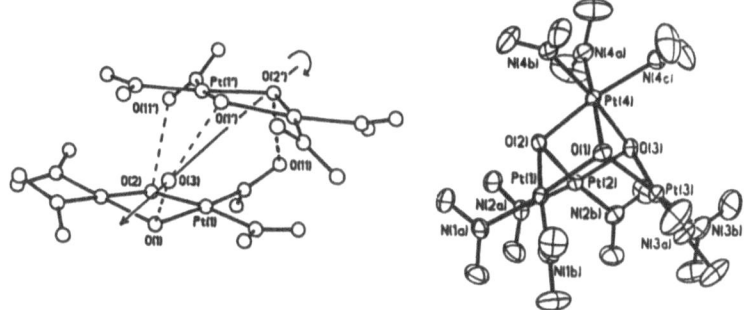

SILVER ACETYLIDE NITRATE
Ag$_2$C$_2$·6AgNO$_3$

Huaxue Xuebao, 48, 232–236.

R$\bar{3}$, a = 7.945 A, α = 106.08°, Z = 1, R = 0.063. Ag$_8$C$_2$$^{6+}$ and 6 NO$_3$$^-$ ions, with
the C$_2$$^{2-}$ ion in a cage of 8 Ag$^+$ ions.

YTTRIUM NITRATE MONOHYDRATE
Y(NO$_3$)$_3$·H$_2$O

Acta Cryst., C46, 525–527.

P$\bar{1}$, 7.388, 7.889, 8.204, 64.43, 70.90, 62.74, Z = 2, R = 0.070. Chains along **b** of
linked YO$_8$(H$_2$O) polyhedra, linked by hydrogen bonds. Y–O = 2.31–2.47(1) A.

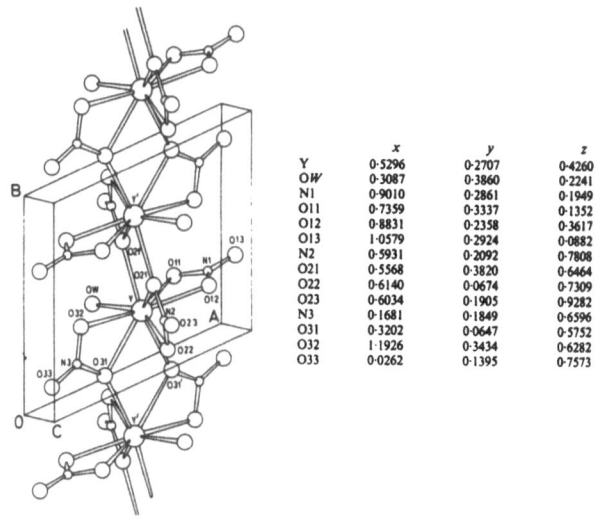

	x	y	z
Y	0·5296	0·2707	0·4260
OW	0·3087	0·3860	0·2241
N1	0·9010	0·2861	0·1949
O11	0·7359	0·3337	0·1352
O12	0·8831	0·2358	0·3617
O13	1·0579	0·2924	0·0882
N2	0·5931	0·2092	0·7808
O21	0·5568	0·3820	0·6464
O22	0·6140	0·0674	0·7309
O23	0·6034	0·1905	0·9282
N3	0·1681	0·1849	0·6596
O31	0·3202	0·0647	0·5752
O32	1·1926	0·3434	0·6282
O33	0·0262	0·1395	0·7573

CAESIUM LANTHANUM NITRATE HYDRATE
Cs$_2$La(NO$_3$)$_5$·2H$_2$O

Kristallografija, 35, 1395–1398 [Soviet Physics – Crystallography, 35, 823–825].

I2/a, 10.868, 9.063, 17.557, 103.20, Z = 4, R = 0.053. Isolated $[La(NO_3)_5(H_2O)_2]^{2-}$ complexes (bidentate nitrate ligands, distorted icosahedral geometry), linked by Cs^+ ions and hydrogen bonds. La–O = 2.55–2.72(1) A.

CAESIUM CERIUM(III) NITRATE HYDRATE
$Cs_2Ce(NO_3)_5 \cdot 2H_2O$

Kristallografija, 35, 1399–1402 [Soviet Physics – Crystallography, 35, 826–828].

C2/c, 11.125, 8.798, 16.503, 103.77, Z = 4, R = 0.039. Isolated $[Ce(NO_3)_5(H_2O)_2]^{2-}$ complexes (three bidentate and two unidentate nitrate ligands, tetracapped trigonal prismatic geometry), linked by Cs^+ ions and hydrogen bonds. Ce–O = 2.52–2.64 A.

AMMONIUM PRASEODYMIUM NITRATE HYDRATE
$(NH_4)_2[Pr(NO_3)_5(H_2O)_2] \cdot 2H_2O$

Z. anorg. Chem., 591, 77–86.

C2/c, 11.047, 8.928, 17.875, 101.78, Z = 4, R = 0.026. Anions with Pr exhibiting 12-coordination to 5 bidentate nitrate and 2 water molecules, linked by ammonium ions and additional water molecules.

SODIUM NEODYMIUM NITRATE HYDRATE
$Na_2[Nd(NO_3)_5] \cdot H_2O$

Izv. Akad. Nauk SSSR, Neorg. Khim., 26, 2357–2362.

P2/a [c unique], 15.147, 21.216, 7.903, γ = 90.74°, Z = 8, R = 0.048. Three independent Nd ions each having icosahedral coordination, with bridging nitrate ligands.

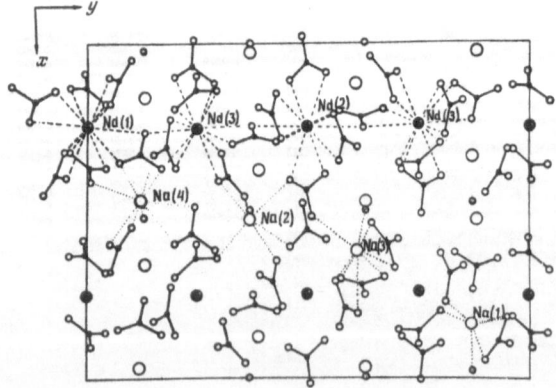

SAMARIUM NITRATE HEXAHYDRATE
$[Sm(H_2O)_4(NO_3)_3] \cdot 2H_2O$

Jiegou Huaxue, 9, 164–167.

P$\bar{1}$, 6.747, 9.156, 11.673, 69.90, 88.88, 69.29, Z = 2, R = 0.032. Sm has 10-coordination, Sm-O(N) = 2.52-2.72, Sm-OH$_2$ = 2.41-2.43 A, with coordination poly-hedra linked by hydrogen bonding via the non-coordinated water molecules.

DYSPROSIUM NITRATE HEXAHYDRATE
[Dy(H$_2$O)$_4$(NO$_3$)$_3$].2H$_2$O

Xibei Daxue Xuebao, Ziran Kexueban, 20, 53-58.

P$\bar{1}$, 6.703, 9.048, 11.535, 70.60, 89.08, 69.23, Z = 2, R = 0.143. Isostructural with the Sm compound (preceding report).

PHOSPHOROUS ACID
H$_3$PO$_3$

Z. anorg. Chem., 591, 17-31.

Pna2$_1$, 7.166, 12.013, 6.743, at 15K, Z = 8, R = 0.032-0.053 for one X-ray and three neutron data sets. Structure as previously described (21, 374), with tetra-hedral H-PO(OH)$_2$ molecules linked by hydrogen bonds. P-H = 1.39, P=O = 1.50, P-O = 1.55, O-H = 1.01, H...O = 1.55-1.60 A.

COBALT(II) PHOSPHITE MONOHYDRATE
Co(HPO$_3$).H$_2$O

J. Appl. Phys., 67, 5998-6000.

Pca2$_1$, 8.984, 7.918, 10.139, Z = 4, R = 0.036. Layers with zigzag chains of edge-sharing CoO$_6$ octahedra connected by corner-sharing and via HPO$_3$ tetrahedra; layers are linked by hydrogen bonds.

COPPER BIS(HYDROGENPHOSPHONATE)
Cu(HPO$_3$H)$_2$

Acta Cryst., C46, 1378-1381.

P2$_1$/a, 7.4748, 9.9406, 7.5175, 99.722, Z = 4, R = 0.053. HPO$_2$(OH)$^-$ tetrahedra hydrogen bonded into tetrameric anionic units, linked by pairs of edge-sharing CuO$_6$ distorted octahedra.

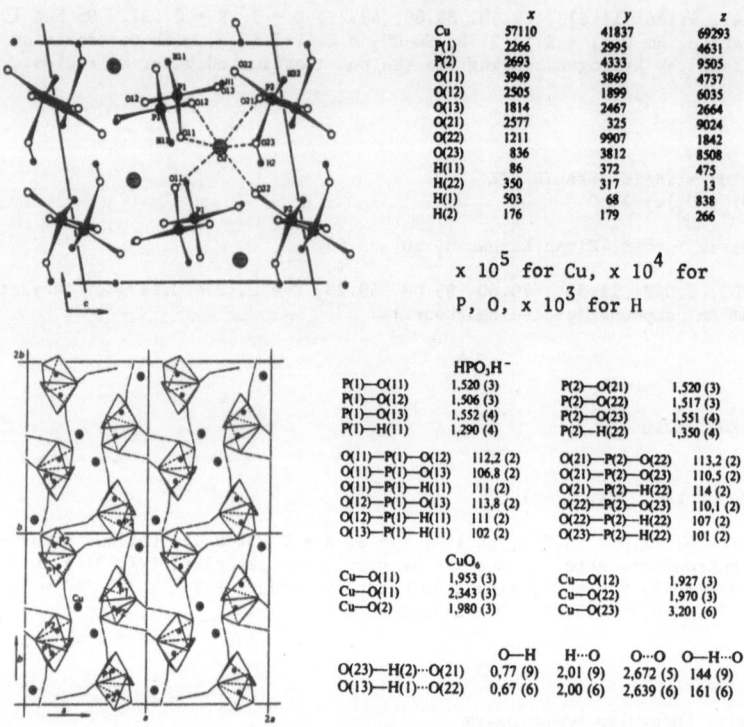

	x	y	z
Cu	57110	41837	69293
P(1)	2266	2995	4631
P(2)	2693	4333	9505
O(11)	3949	3869	4737
O(12)	2505	1899	6035
O(13)	1814	2467	2664
O(21)	2577	325	9024
O(22)	1211	9907	1842
O(23)	836	3812	8508
H(11)	86	372	475
H(22)	350	317	13
H(1)	503	68	838
H(2)	176	179	266

$x \times 10^5$ for Cu, $\times 10^4$ for P, O, $\times 10^3$ for H

HPO_3H^-

P(1)—O(11)	1,520 (3)	P(2)—O(21)	1,520 (3)
P(1)—O(12)	1,506 (3)	P(2)—O(22)	1,517 (3)
P(1)—O(13)	1,552 (4)	P(2)—O(23)	1,551 (4)
P(1)—H(11)	1,290 (4)	P(2)—H(22)	1,350 (4)
O(11)—P(1)—O(12)	112,2 (2)	O(21)—P(2)—O(22)	113,2 (2)
O(11)—P(1)—O(13)	106,8 (2)	O(21)—P(2)—O(23)	110,5 (2)
O(11)—P(1)—H(11)	111 (2)	O(21)—P(2)—H(22)	114 (2)
O(12)—P(1)—O(13)	113,8 (2)	O(22)—P(2)—O(23)	110,1 (2)
O(12)—P(1)—H(11)	111 (2)	O(22)—P(2)—H(22)	107 (2)
O(13)—P(1)—H(11)	102 (2)	O(23)—P(2)—H(22)	101 (2)

CuO_6

Cu—O(11)	1,953 (3)	Cu—O(12)	1,927 (3)
Cu—O(11)	2,343 (3)	Cu—O(22)	1,970 (3)
Cu—O(2)	1,980 (3)	Cu—O(23)	3,201 (6)

	O—H	H···O	O···O	O—H···O
O(23)—H(2)···O(21)	0,77 (9)	2,01 (9)	2,672 (5)	144 (9)
O(13)—H(1)···O(22)	0,67 (6)	2,00 (6)	2,639 (6)	161 (6)

CALCIUM and STRONTIUM ZINC HYDROGEN PHOSPHITE HYDRATE
$MZn(HPO_3)_2 \cdot 2H_2O$ (M = Ca, Sr)

Inorg. Chem., 29, 958-963.

M = Ca, $P2_1/n$, 7.131, 7.766, 14.479, 97.30, Z = 4, R = 0.045. M = Sr, $P\bar{1}$, 7.728, 7.967, 7.448, 99.56, 107.93, 100.51, Z = 2, R = 0.034. Both structures contain layers of HPO_3^{2-} tetrahedra linked by tetrahedrally-coordinated Zn and 8-coordinate M ions, with the interlayer regions lined by water molecules coordinated to the M ions.

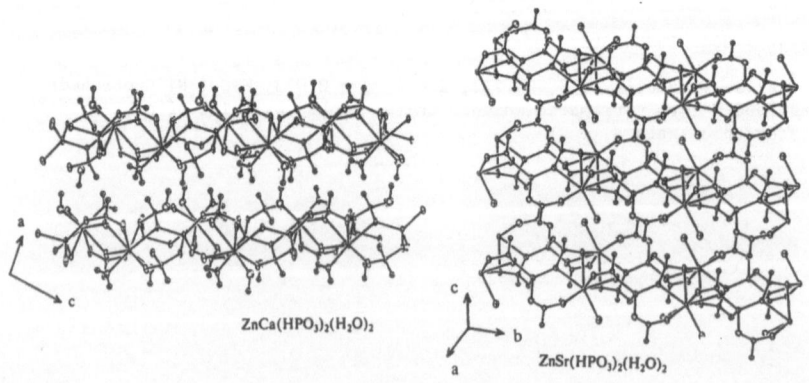

ZnCa(HPO₃)₂(H₂O)₂   ZnSr(HPO₃)₂(H₂O)₂

SILVER NEODYMIUM METAPHOSPHATE
$AgNd(PO_3)_4$

Izv. Akad. Nauk SSSR, Neorg. Mater., 26, 1288-1290.

$P2_1/n$, 9.947, 13.17, 7.271, 90.48, Z = 4, R = 0.048. Isostructural with the Na
compound (42A, 335).

AMMONIUM DIHYDROGEN PHOSPHATE
$NH_4H_2PO_4$

Ferroelectrics, 107, 247-252.

$I\bar{4}2d$, a = 7.467, c = 7.540 A, at 152K, Z = 4, R = 0.030. N in 4(b): 0,0,1/2; P
in 4(a): 0,0,0; O in 16(e): 0.0846,0.1478,0.1153; H(O) in 8(d): 1/4,0.146,1/8;
H(N) in 16(e): 0.010,0.084,0.558. Structure as previously described (1, 363; 54A,
224), with perhaps some disorder of the O site.

SODIUM PHOSPHATE (LOW-TEMPERATURE)
Na_3PO_4

Z. Kristallogr., 192, 233-243.

$P\bar{4}2_1c$, 10.8084, 6.8178, Z = 8, X-ray and neutron powder data. The structure is
similar to that of the high-temperature cubic form (46A, 318), with ordered phosphate tetrahedra.

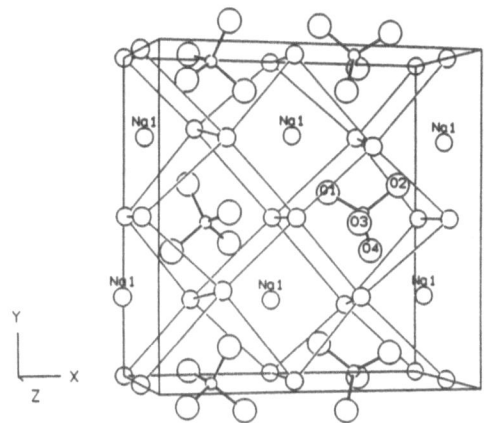

DISODIUM HYDROGEN PHOSPHATE HYDRATES
$Na_2HPO_4 \cdot nH_2O$ (n = 1, 2, 7, 12)

J. Phys. Chem., 94, 7830-7834.

n = 1, $P\bar{1}$, 5.565, 7.949, 5.735, 103.83, 102.15, 107.47, Z = 2, R = 0.020. Tetrahedral O_3POH^{2-} anions linked by $NaO_4(H_2O)_2$ and NaO_5 polyhedra and hydrogen bonds.
P-OH = 1.617, P-O = 1.513-1.532(1) A.

n = 2, Pbca, 10.343, 6.596, 16.833, Z = 8, R = 0.029. Structure as previously
described (43A, 247).

n = 7, P2$_1$/n, 10.432, 11.000, 9.252, 95.62, Z = 4, R = 0.022. Structure as previously described (35A, 335).

n = 12, I2/c, 14.178, 9.021, 12.771, 108.88, Z = 4, R = 0.046. Structure as previously described (44A, 244), with disorder of the phosphate group and of hydrogen bonds.

POTASSIUM DIHYDROGEN PHOSPHATE
KH_2PO_4

Phys. Status Solidi A, 117, K93–K96.

127K, paraelectric, F$\bar{4}$d2 [alternate setting of I$\bar{4}$2d], 10.495, 6.924, Z = 8, R = 0.029. 115K, ferroelectric, Fdd2, 10.467, 10.533, 6.926, Z = 8, R = 0.045. Structures as previously described (17, 478; 49A, 260).

RUBIDIUM DIHYDROGEN PHOSPHATE
RbH_2PO_4

Phys. Status Solidi A, 122, K117–K120.

Ferroelectric phase, Fdd2, 10.696, 10.794, 7.260, at 145K, Z = 8, R = 0.069. Paraelectric phase, F$\bar{4}$d2 [alternate setting of I$\bar{4}$2d], 10.718, 7.246, at 155K, Z = 8, R = 0.057. Structures as previously described for ferroelectric (48A, 280) and paraelectric phases (44A, 245; 45A, 300; 48A, 280; 49A, 261).

APATITES (HEXAGONAL and MONOCLINIC)
$Ca_5(PO_4)_3(F,Cl,OH)$

Amer. Min., 75, 295–304, 1216.

Hexagonal, P6$_3$/m, 9.462, 6.849, Z = 2, R = 0.015. Monoclinic, P2$_1$/b (c unique), 9.488, 18.963, 6.822, γ = 119.97°, Z = 4, R = 0.047. Apatite structures (2, 99). Solid solution is achieved by a 0.4 A shift along c of Cl, relative to its position in chloroapatite, accompanied by splitting of the Ca(2) site. The monoclinic form results from Cl/OH ordering.

		hexagonal	
	x	y	z
Ca(1)	⅔	⅓	0.0018
Ca(2)$_A$	−0.00699	0.23861	¼
Ca(2)$_B$	−0.0018	0.2709	¼
P	0.36968	0.40053	¼
O(1)	0.4862	0.3315	¼
O(2)	0.4851	0.5884	¼
O(3)	0.2597	0.3450	0.0699
F	0	0	¼
O(H)	0	0	0.200
Cl$_B$	0	0	0.368
Cl$_A$	0	0	0.44

BERLINITE (ALPHA, TRIGONAL)
$AlPO_4$

J. Appl. Cryst., 23, 397–400.

P3$_1$21, 4.804, 10.758, at 2.9 GPa, Z = 3, synchrotron radiation, R = 0.063. Al in 3(a): 0.4567,0,1/3; P in 3(b): 0.4544,0,5/6; O(1) in 6(c): 0.4097,0.3089,0.3914;

O(2) in 6(c): 0.4089,0.2774,0.8776. Structure as previously described (3, 424; 45A, 303; 53A, 203).

Z. Kristallogr., 192, 119-136.

P3₁21, a = 4.941-4.605, c = 10.940-10.558 A, at 0.0001-8.51 GPa, Z = 3, R = 0.041-0.26. Al in 3(a): x,0,1/3, x = 0.4663-0.449; P in 3(b): x,0,5/6, x = 0.4667-0.434; O(1) in 6(c): x,y,z, x = 0.4162-0.386, y = 0.2915-0.322, z = 0.3977-0.3866; O(2) in 6(c): x,y,z, x = 0.4153-0.397, y = 0.2570-0.299, z = 0.8838-0.8722. Structure as previously described (3, 424; 53A, 203).

ALUMINUM PHOSPHATE
AlPO₄-25

J. Phys. Chem., 94, 3365-3367.

298K, Ibam, 15.085, 18.761, 8.363, Z = 24; 593K, Acmm, 9.4489, 15.2028, 8.4084, Z = 12, neutron powder data. The 2D net of AlPO₄-21 (52A, 271; see also 55A, 233) is retained, but the double-crankshaft chains disappear. The room-temperature phase has a doubled b-axis, and the detailed structure has not been established.

	sym		x	y	z
T(1)	16o	1	0.3535	0.0982	0.1987
T(2)	8m	m	0.1566	0.2500	0.3142
O(1)	8l	2	0.3017	0.0000	0.2500
O(2)	8k	2	0.5000	0.1275	0.2500
O(3)	16o	1	0.2317	0.1625	0.2497
O(4)	8m	m	0.3308	0.0971	0.0000
O(5)	4e	2/m	0.0000	0.2500	0.2500
O(6)	4g	mm	0.1076	0.2500	0.5000

ALUMINUM PHOSPHATE (FORM CJ₂)
AlPO₄   0.5 x [Al₂(PO₄)₂(H₃O)F.2H₂O]

J. Solid State Chem., 87, 241-244.

P2₁2₁2₁, 9.456, 9.621, 9.965, Z = 8, R = 0.059. Framework of corner-sharing PO₄ tetrahedra, AlO₅ distorted trigonal bipyramids and AlO₅F distorted octahedra, with two types of open channel, along a and c. P-O = 1.51-1.56, Al-O = 1.78-1.92, Al-F = 1.83 A.

	$X \times 10^4$	Y	Z
P(1)	4,237	2086	3574
P(2)	9,200	-149	3611
Al(1)	6,031	2367	6396
Al(2)	6,529	246	1767
O(1)	4,316	3488	2834
O(2)	7,649	234	3361
O(3)	5,502	315	124
O(4)	4,896	2193	4981
O(5)	10,196	952	3013
O(6)	5,440	1556	7923
O(7)	4,951	973	2747
O(8)	2,677	1646	3773
O(9)	6,936	689	5925
O(10)	2,437	856	6852
O(11)	4,775	3381	54
F(1)	7,210	1978	1370

AMBLYGONITE-MONTEBRASITE
LiAlPO$_4$(F,OH)

Amer. Min., <u>75</u>, 992-1008.

C$\overline{1}$, a $\sim$ 6.7, b $\sim$ 7.7, c $\sim$ 7.0 A, α $\sim$ 91, β $\sim$ 117, γ $\sim$ 91°, for 7 samples, Z = 4,
R = 0.021-0.029. Structure as previously described (<u>22</u>, 414; <u>23</u>, 429), but with
a C-centred cell used to emphasize structural relationships with the kieserite
and titanite group minerals. The Al-O,OH,F distances decrease with OH substitution for F; Li is disordered in two sites within a 6-coordinate cavity.

GOYAZITE
SrAl$_3$[PO$_3$(O$_{1/2}$(OH)$_{1/2}$)]$_2$(OH)$_6$

Neues Jb. Miner. Mh., 241-247 (1971); Mineral. J., <u>13</u>, 390-396 (1987).

R$\overline{3}$m, a = 7.015, c = 16.558 A, Z = 3, R = 0.039. Sr in 3(a): 0,0,0; Al in 9(d):
1/6,-1/6,-1/6; P in 6(c): 0,0,0.3083; O(1) in 6(c): 0,0,0.4017; O(2) in 18(h):
0.2158,-0.2158,-0.0566; O(H) in 18(h): 0.1262,-0.1262,0.1366. Alunite-type
structure (<u>5</u>, 92; <u>30A</u>, 376; <u>42A</u>, 369; <u>43A</u>, 279). P-O = 1.536(4), Al-O = 1.889,
1.910(4), Sr-O = 2.$\overline{7}$29, 2.76$\overline{7}$(4) A.

FLORENCITE
CeAl$_3$(PO$_4$)$_2$(OH)$_6$

Neues Jb. Miner. Mh., 227-231.

R$\overline{3}$m, a = 6.972, c = 16.261 A, Z = 3, R = 0.035. Ce in 3(a): 0,0,0; Al in 9(d):
1/6,-1/6,-1/6; P in 6(c): 0,0,0.3119; O(1) in 6(c): 0,0,0.4049; O(2) in 18(h):
0.2126,-0.2126,-0.0548: O(H) in 18(h): 0.1243,-0.1243,0.1338; H in 18(h): 0.193,
-0.193,0.122. Alunite-type structure (<u>5</u>, 92; <u>30A</u>, 376; <u>42A</u>, 369; <u>43A</u>, 279).
P-O = 1.512, 1.557, Al-O = 1.889, 1.901, Ce-O = 2.643, 2.$\overline{7}$18(5) A.

LITHIUM LEAD(II) PHOSPHATE
LiPbPO$_4$

Kristallografija, <u>35</u>, 205-208 [Soviet Physics - Crystallography, <u>35</u>, 124-126]

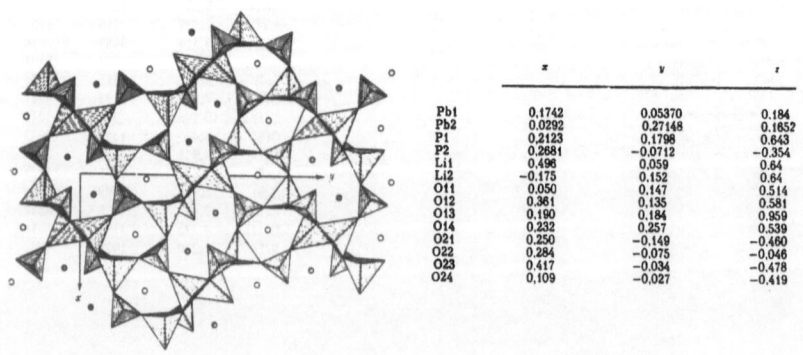

	x	y	z
Pb1	0.1742	0.05370	0.184
Pb2	0.0292	0.27148	0.1652
P1	0.2123	0.1798	0.643
P2	0.2681	−0.0712	−0.354
Li1	0.496	0.059	0.84
Li2	−0.175	0.152	0.84
O11	0.050	0.147	0.514
O12	0.361	0.135	0.581
O13	0.190	0.184	0.959
O14	0.232	0.257	0.539
O21	0.250	−0.149	−0.460
O22	0.284	−0.075	−0.046
O23	0.417	−0.034	−0.478
O24	0.109	−0.027	−0.419

Pna2$_1$, 7.982, 18.626, 4.938, Z = 8, R = 0.051. Framework of corner-sharing PO$_4$ and LiO$_4$ tetrahedra, with 8- and 9-coordinate Pb ions in channels. P-O = 1.48-1.60, Li-O = 1.82-2.06, Pb-O = 2.29-3.37(3) A.

TITANIUM HYDROGEN PHOSPHATES
α-Ti(HPO$_4$)$_2$.H$_2$O (I), γ-Ti(H$_2$PO$_4$)(PO$_4$).2H$_2$O (II)

Acta Chem. Scand., 44, 865-872.

I, P2$_1$/c, 8.630, 5.006, 16.189, 110.20, Z = 4, Powder data. Isostructural with the Zr compound (34A, 328; 43A, 255).

II, P2$_1$, 5.181, 6.347, 11.881, 102.59, Z = 2, powder data. Structural model proposed.

SODIUM TITANYL PHOSPHATES
NaTiOPO$_4$

Kristallografija, 35, 634-637 [Soviet Physics - Crystallography, 35, 370-371].

P2$_1$22, 8.755, 9.124, 10.518, Z = 8, R = 0.076. Framework with chains of corner-sharing TiO$_6$ octahedra linked by PO$_4$ tetrahedra; Na ions are in holes and are disordered over several positions. Ti-O = 1.81-1.86, 1.95-2.13 A.

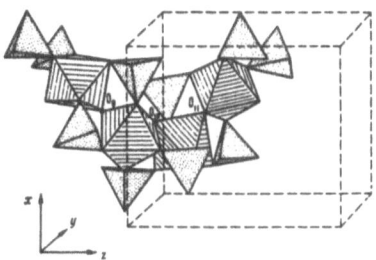

Na$_4$TiO(PO$_4$)$_2$

Kristallografija, 35, 847-851 [Soviet Physics - Crystallography, 35, 497-500].

Ibam, 15.647, 14.989, 7.081, at 573K, Z = 8, R = 0.057. Chains along c of trans-corner-sharing TiO$_6$ octahedra which are also linked by PO$_4$ tetrahedra; Na$^+$ ions are statistically distributed over six sites.

	Population	x	y	z
Ti	1	1/4	1/4	1/4
P$_1$	1	0,1196	0,1236	0
P$_2$	1	0,1228	0,3800	1/2
Na$_1$	0,99	0,2391	0	1/4
Na$_2$	0,98	0	0,2614	1/4
Na$_3$	0,73	0.1107	0,1239	1/2
Na$_4$	0,80	0,1148	0,3871	0
Na$_5$	0,59	0	0	1/4
Na$_6$	0,41	1/2	0	1/4
O$_1$	1	0,2467	0,3001	0
O$_2$	1	0,3331	0,3435	0,3200
O$_3$	1	0,1533	0,3286	0,3230
O$_4$	1	0,1228	0,0235	0
O$_5$	1	0,0276	0,3777	1/2
O$_6$	1	0,1598	0,4716	1/2
O$_7$	1	0,0299	0,1598	0

POTASSIUM TITANYL PHOSPHATE   POTASSIUM STANNYL PHOSPHATE
KTiOPO$_4$ KSnOPO$_4$

Acta Cryst., B46, 333-343, 692.

Pna2$_1$, a = 12.819, 13.145, b = 6.399, 6.526, c = 10.584, 10.738 A, Z = 8, R =
0.024, 0.021. Structures as previously described for the Ti compound (40A, 245;
52A, 275). The SnO$_6$ octahedra (Sn-O = 1.96-2.13 A) are much less distorted than
the TiO$_6$ octahedra.

KSnOPO$_4$

Inorg. Chem., 29, 3245-3247.

Pna2$_1$, 13.170, 6.5387, 10.745, Z = 8, R = 0.031. Isostructural with the Ti compound (preceding report). Sn-6 O = 1.96-2.13(1) A.

MAGNESIUM TITANIUM(III) PHOSPHATE
Mg$_3$Ti$_4$(PO$_4$)$_6$

J. Solid State Chem., 84, 299-307.

P$\bar{1}$, 6.9311, 7.9616, 9.4299, 67.614, 69.348, 79.327, Z = 1, R = 0.021. The
structure contains a framework of corner-sharing (Ti,Mg)O$_6$ octahedra, Ti$_2$O$_{10}$
and (Ti,Mg)$_2$O$_{10}$ octahedral units (two octahedra sharing an edge), and PO$_4$ tetrahedra; additional Mg ions have trigonal bipyramidal coordination.

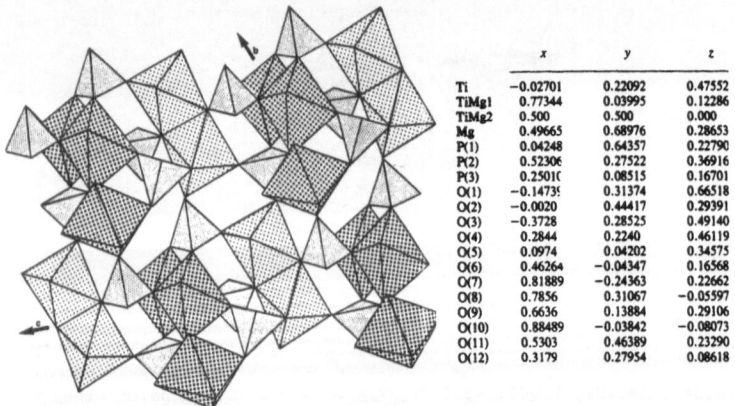

	x	y	z
Ti	−0.02701	0.22092	0.47552
TiMg1	0.77344	0.03995	0.12286
TiMg2	0.500	0.500	0.000
Mg	0.49665	0.68976	0.28653
P(1)	0.04248	0.64357	0.22790
P(2)	0.52306	0.27522	0.36916
P(3)	0.25010	0.08515	0.16701
O(1)	−0.14739	0.31374	0.66518
O(2)	−0.0020	0.44417	0.29391
O(3)	−0.3728	0.28525	0.49140
O(4)	0.2844	0.2240	0.46119
O(5)	0.0974	0.04202	0.34575
O(6)	0.46264	−0.04347	0.16568
O(7)	0.81889	−0.24363	0.22662
O(8)	0.7856	0.31067	−0.05597
O(9)	0.6636	0.13884	0.29106
O(10)	0.88489	−0.03842	−0.08073
O(11)	0.5303	0.46389	0.23290
O(12)	0.3179	0.27954	0.08618

POTASSIUM ZIRCONIUM PHOSPHATE
K$_2$Zr(PO$_4$)$_2$

Z. Kristallogr., 193, 155-159.

P$\bar{3}$, 5.176, 9.011, Z = 1, R = 0.029 (twinned crystal). (001) Layers of ZrO$_6$ octahedra sharing corners with PO$_4$ tetrahedra; layers are linked by K ions. Zr-O =
2.07(1) A.

HAFNIUM(IV) HYDROGEN PHOSPHATE MONOHYDRATE
α-Hf(HPO₄)₂·H₂O

Anal. Sci., 6, 689-693.

$P2_1/n$, 9.0142, 5.2567, 15.4768, 101.636, Z = 4, synchrotron powder data. Isostructural with the Zr compound (34A, 328; 43A, 255).

POTASSIUM VANADYL PHOSPHATE
KVOPO₄

Inorg. Chem., 29, 2158-2163.

$Pna2_1$, 12.816, 6.388, 10.556, Z = 8, R = 0.056. Isostructural with the Ti compound (this volume, p.258). P-4 O = 1.52-1.57, V-6 O = 1.67-2.20, K-8 and 9 O = 2.65-3.12 A.

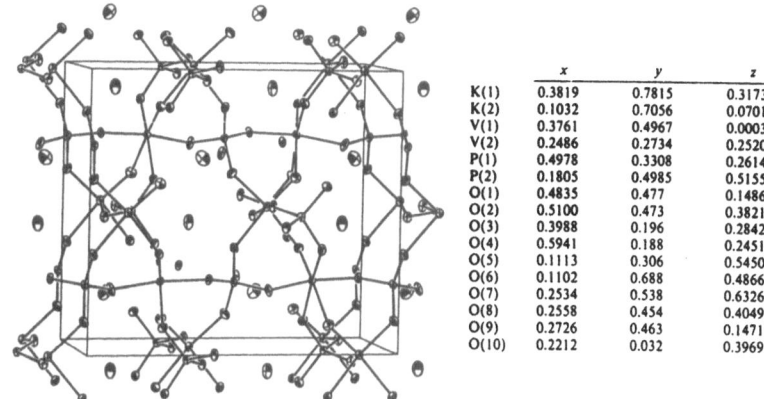

	x	y	z
K(1)	0.3819	0.7815	0.3173
K(2)	0.1032	0.7056	0.0701
V(1)	0.3761	0.4967	0.0003
V(2)	0.2486	0.2734	0.2520
P(1)	0.4978	0.3308	0.2614
P(2)	0.1805	0.4985	0.5155
O(1)	0.4835	0.477	0.1486
O(2)	0.5100	0.473	0.3821
O(3)	0.3988	0.196	0.2842
O(4)	0.5941	0.188	0.2451
O(5)	0.1113	0.306	0.5450
O(6)	0.1102	0.688	0.4866
O(7)	0.2534	0.538	0.6326
O(8)	0.2558	0.454	0.4049
O(9)	0.2726	0.463	0.1471
O(10)	0.2212	0.032	0.3969

SODIUM CHROMIUM(III) FLUORIDE PHOSPHATE
Na₅CrF₂(PO₄)₂

Ž. Neorg. Khim., 35, 839-842 [Russ. J. Inorg. Chem., 35, 470-472].

P3, 10.576, 6.669, Z = 3, R = 0.045. (001) Layers of PO₄ tetrahedra, trans-CrO₄F₂ and Na(O,F)₆ octahedra, linked by additional Na coordination octahedra. Cr-O = 1.95-1.99, Cr-F = 1.91, 1.95(1) A.

	x	y	z
Cr	0.1675	0.3343	0.0
P(1)	0.4823	0.5163	0.2502
P(2)	0.2948	0.1458	-0.2502
Na(1)	0.4932	0.4770	0.7598
Na(2)	0.3520	0.1651	0.2385
Na(3)	0.1692	0.3330	0.4994
Na(4)	0.0	0.0	0.4387
Na(5)	0.3333	0.6667	0.5810
Na(6)	0.6667	0.3333	0.5076
Na(7)	0.6667	0.3333	-0.0068
Na(8)	0.0	0.0	-0.0039
Na(9)	0.3333	0.6667	0.0283
O(1)	0.4124	0.5043	0.4598
O(2)	0.5561	0.4241	0.2551
O(3)	0.5856	0.6844	0.2118
O(4)	0.3694	0.4634	0.0807
O(5)	0.2502	0.2601	-0.2061
O(6)	0.2298	0.0354	-0.0707
O(7)	0.2342	0.0707	-0.4401
O(8)	0.4632	0.2334	-0.2342
F(1)	0.1827	0.4717	-0.1985
F(2)	0.1438	0.1908	0.2014

MOLYBDENUM PHOSPHATE
$Mo_8O_{12}(PO_4)_4(HPO_4)_2.13H_2O$

J. Amer. Chem. Soc., <u>112</u>, 8182-8183.

P$\bar{1}$, 10.466, 12.341, 8.228, 94.75, 111.46, 89.48, Z = 1, R = 0.067, atomic positional parameters not listed. Framework of edge- and corner-sharing MoO_6
octahedra and PO_4 and $PO_3(OH)$ tetrahedra, with two tetrameric Mo_4 units; Mo-Mo =
2.61, 2.66 A. Water molecules are in tunnels.

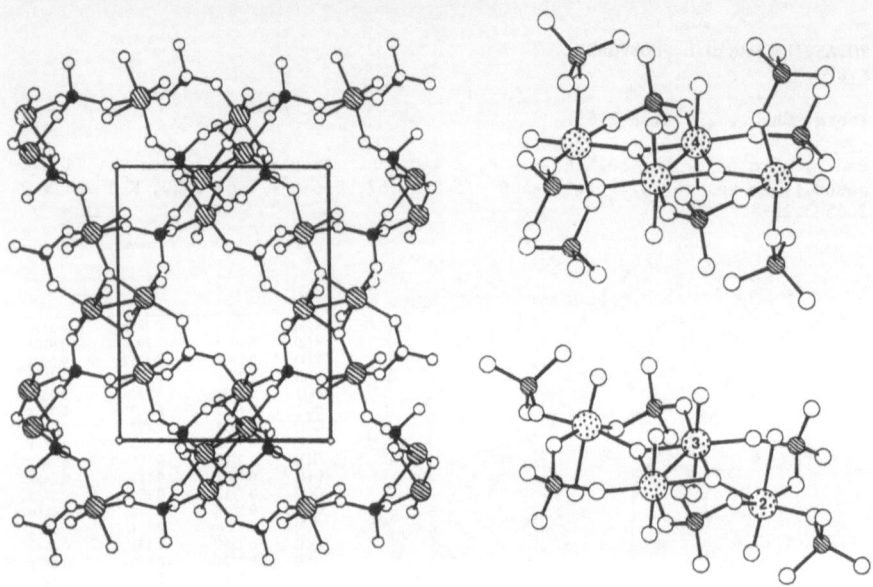

MOLYBDENUM(V) ALUMINOPHOSPHATE
$MoAlP_2O_9$

Z. Kristallogr., <u>190</u>, 135-142.

P4/ncc, 8.8030, 8.6970, Z = 4, R = 0.027. Isostructural with $VSiP_2O_9$ (<u>42A</u>, 343),
with ReO_3-type rows of MoO_6 octahedra and AlP_2O_6 tetrahedral columns. Mo-O =
1.65, 1.98 (x 4), 2.70, Al-O = 1.73 (x 4), P-O = 1.51-1.53 A.

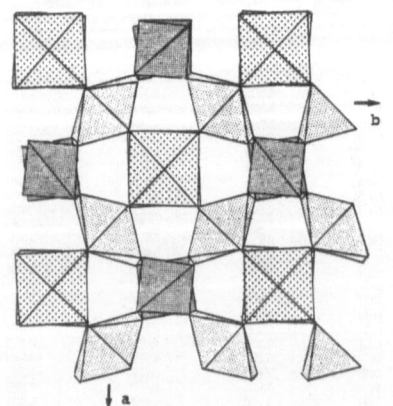

	x	y	z
Al	0.750	0.250	0.000
Mo	0.250	0.250	0.17748
P	0.56514	0.435	0.250
O(1)	0.3929	0.4141	0.2332
O(2)	0.250	0.250	−0.0126
O(3)	0.6472	0.3779	0.1084

BEUSITE (Ca-RICH)
$(Mn,Fe,Ca)_3(PO_4)_2$

Canad. Miner., 28, 141-146.

$P2_1/c$, 8.797, 11.758, 6.170, 99.31, Z = 4, R = 0.033. Isostructural with graftonite (33A, 401), with Ca preferentially in the 7-coordinate M(1) site.

CAESIUM MANGANESE(II) PHOSPHATE
$CsMnPO_4$

Kristallografija, 35, 42-46 [Soviet Physics - Crystallography, 35, 22-25].

$Pbn2_1$, 9.575, 9.128, 5.595, Z = 4, R = 0.043. β-Tridymite-like framework of corner-sharing PO_4 and MnO_4 tetrahedra, with some oxygen disorder, and Cs ions in channels. P-O = 1.50-1.59, Mn-O = 1.89-2.15, Cs-O = 3.07-3.71(1) A.

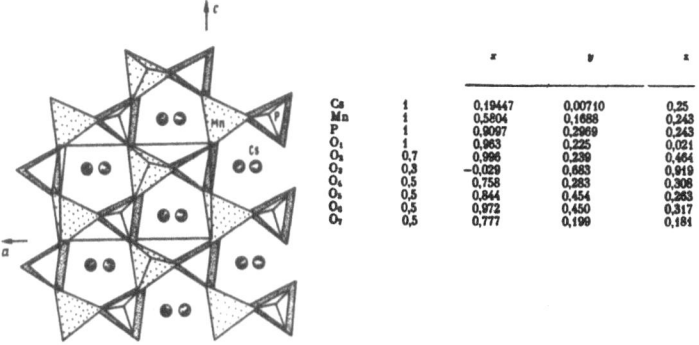

		x	y	z
Ca	1	0,19447	0,00710	0,25
Mn	1	0,5804	0,1688	0,243
P	1	0,9097	0,2969	0,243
O₁	1	0,963	0,225	0,021
O₂	0,7	0,996	0,239	0,464
O₃	0,3	-0,029	0,683	0,919
O₄	0,5	0,758	0,283	0,306
O₅	0,5	0,844	0,454	0,263
O₆	0,5	0,972	0,450	0,317
O₇	0,5	0,777	0,199	0,181

IRON PHOSPHATE OXIDE
$β-Fe_2(PO_4)O$

J. Solid State Chem., 86, 195-205.

$I4_1/amd$, 5.3360, 12.457, Z = 4, R = 0.035. Structure as previously described from powder data (55A, 240); new O(1) parameters are y = 0.4950, z = 0.8075.

POTASSIUM IRON(III) FLUORIDE PHOSPHATE
$KFePO_4$

Izv. Akad. Nauk SSSR, Neorg. Mater., 26, 595-601.

$P2_1nb$, 10.651, 12.88, 6.370, Z = 8, R = 0.016. Isostructural with $KTiOPO_4$ (40A, 245; 52A, 275).

CAESIUM POTASSIUM SODIUM IRON PHOSPHATE FLUORIDE
$(Cs_{0.92}K_{0.08})NaFe_9(PO_4)_6F_2$

Kristallografija, 35, 1122-1125 [Soviet Physics - Crystallography, 35, 660-662].

A2/m, 8.688, 10.645, 12.371, β = 113.51°, Z = 2, R = 0.032. PO_4 tetrahedra and face-, edge-, and corner-sharing FeO_6 and FeO_5F octahedra (one of the latter has one very long Fe-O bond), and 10-coordinate Na and Cs ions (the latter with some off-centre K substitution).

NISSONITE
$[CuMg(PO_4)(OH)(H_2O)_2]_2 \cdot H_2O$

Amer. Min., _75_, 1170-1175

C2/c, 22.523, 5.015, 10.506, 99.62, Z = 4, R = 0.068. Slabs parallel to (100), with a sheet of CuO_6 octahedra sandwiched between sheets of MgO_6 octahedra and PO_4 tetrahedra. One water molecule is not coordinated to a cation, and slabs are connected by hydrogen bonds. Cu-O = 1.94-2.02 (4 distances), 2.36, 2.66, Mg-O = 2.05-2.12, P-O = 1.51-1.60 A.

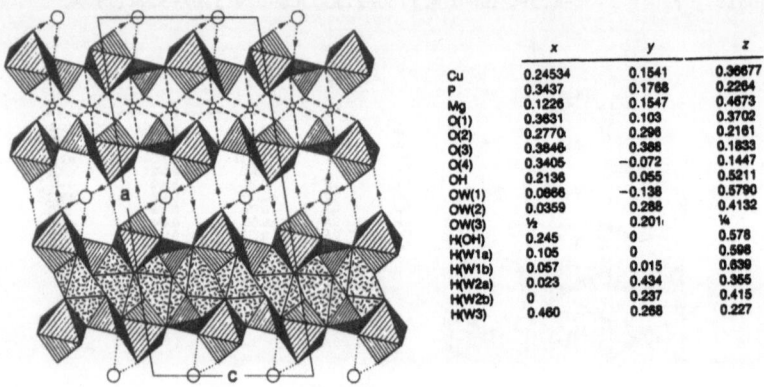

	x	y	z
Cu	0.24534	0.1541	0.36677
P	0.3437	0.1768	0.2264
Mg	0.1226	0.1547	0.4673
O(1)	0.3631	0.103	0.3702
O(2)	0.2770	0.298	0.2161
O(3)	0.3846	0.388	0.1833
O(4)	0.3405	-0.072	0.1447
OH	0.2136	0.055	0.5211
OW(1)	0.0866	-0.138	0.5790
OW(2)	0.0359	0.288	0.4132
OW(3)	½	0.201	¼
H(OH)	0.245	0	0.576
H(W1a)	0.105	0	0.596
H(W1b)	0.057	0.015	0.639
H(W2a)	0.023	0.434	0.355
H(W2b)	0	0.237	0.415
H(W3)	0.460	0.268	0.227

STRONTIUM ZINC PHOSPHATE
$\alpha\text{-}SrZn_2(PO_4)_2$

J. Solid State Chem., _85_, 164-168.

$P2_1/c$, 8.3232, 9.5101, 9.0317, 92.293, Z = 4, R = 0.026. Hurlbutite-type structure (_24_, 406; _40A_, 238; _41A_, 313), with a framework of corner-sharing PO_4 and ZnO_4 tetrahedra, with 7-coordinate Sr in large holes. P-O = 1.525-1.549, Zn-O = 1.914-1.992, Sr-O = 2.535-2.728(1) A.

	x	y	z
Sr	0.2456	0.0819	0.6074
Zn₁	0.5739	0.5711	0.2887
Zn₂	0.9195	0.3197	0.4510
P₁	0.5428	0.2941	0.4314
P₂	0.0576	0.0847	0.2507
O₁	0.4746	0.3866	0.3024
O₂	0.4178	0.2059	0.0528
O₃	0.2276	0.0621	0.3200
O₄	0.0174	0.4490	0.3135
O₅	0.0637	0.3033	0.6287
O₆	0.5658	0.1413	0.3796
O₇	0.6999	0.1407	0.9909
O₈	0.9510	0.3729	0.8793

ZINC IRON PHOSPHATE HYDRATE
$Zn_2(Zn,Fe)Fe(PO_4)_3 \cdot 2H_2O$

Z. Naturforsch., <u>45</u>B, 1255-1261.

P$\bar{1}$, 6.415, 9.144, 9.834, 70.79, 78.32, 73.69, Z = 2, R = 0.044. Framework of
corner- and edge-sharing PO_4 tetrahedra, ZnO_4 tetrahedra and ZnO_5 trigonal bipyramids, $(Zn,Fe)O_5(H_2O)$ octahedra, and $Fe(III)O_5(H_2O)$ octahedra.   P-O = 1.51-
1.56, Zn-O = 1.90-2.01 (tetrahedra), 2.00-2.03 (trigonal bipyramids),
(Zn,Fe)-O = 2.07-2.20, Fe-O = 1.94-2.14 A.

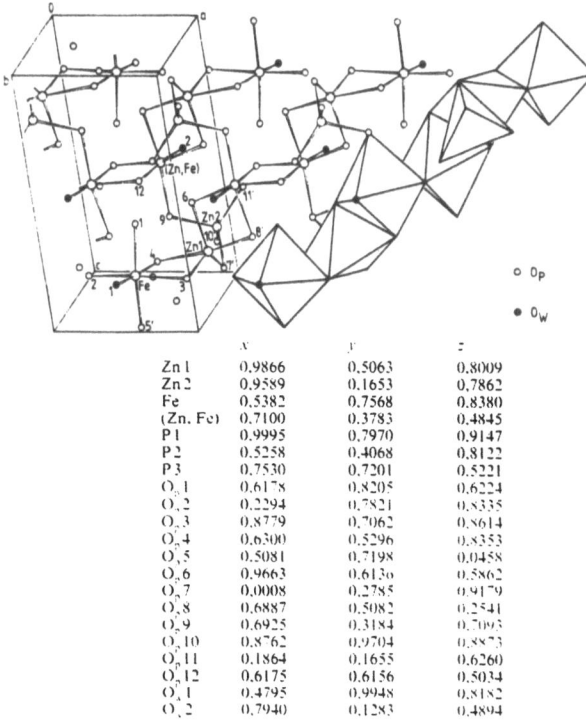

	x	y	z
Zn 1	0.9866	0.5063	0.8009
Zn 2	0.9589	0.1653	0.7862
Fe	0.5382	0.7568	0.8380
(Zn, Fe)	0.7100	0.3783	0.4845
P 1	0.9995	0.7970	0.9147
P 2	0.5258	0.4068	0.8122
P 3	0.7530	0.7201	0.5221
O 1	0.6178	0.8205	0.6224
O 2	0.2294	0.7821	0.8335
O 3	0.8779	0.7062	0.8614
O 4	0.6300	0.5296	0.8353
O 5	0.5081	0.7198	0.0458
O 6	0.9663	0.6136	0.5862
O 7	0.0008	0.2785	0.9179
O 8	0.6887	0.5082	0.2541
O 9	0.6925	0.3184	0.7093
O 10	0.8762	0.9704	0.8873
O 11	0.1864	0.1655	0.6260
O 12	0.6175	0.6156	0.5034
O 1	0.4795	0.9948	0.8182
O 2	0.7940	0.1283	0.4894

LANTHANON PHOSPHATES
$LnPO_4$

Inorg. Chim. Acta, <u>174</u>, 155-159.

I4_1/amd, a = 6.880, 6.914, 6.865, 6.903, c = 6.017, 6.042, 6.004, 6.034 A, for
Ln = Gd/Er (0.5:0.5), Gd/Y (0.5:0.5), Gd/Yb (0.5:0.5), Gd/Yb (0.75:0.25), Z =
4, R = 0.043, 0.041, 0.053, 0.049. Zircon-type (<u>1</u>, 345).

STRONTIUM LANTHANUM PHOSPHATE
$Sr_3La(PO_4)_3$

Eur. J. Solid State Inorg. Chem., <u>27</u>, 855-867.

I$\bar{4}$3d, neutron powder data. Eulytite-type (<u>2</u>, 122; <u>31</u>A, 227), with Sr/La disordered in a single site, and O distributed over three partially-occupied sites.

LEAD(II) PYROPHOSPHATE (HIGH-TEMPERATURE)
$Pb_2P_2O_7$

Z. anorg. Chem., <u>584</u>, 173-177.

$P2_1/n$, a = 12.852, 7.076, 7.096, 95.4, at 1023K, Z = 4, powder data. Monoclinic variation of the triclinic room-temperature form (<u>53A</u>, 217); only sym-
metrized Pb and P positions are given.

SODIUM ZIRCONIUM DIPHOSPHATE
$(Na_{0.67}Zr_{0.33})_2P_2O_7$

Acta Cryst., C<u>46</u>, 2011-2013.

Fddd, 6.867, 12.345, 27.527, Z = 16, R = 0.07. Pyrophosphate anions linked by
(Na,Zr)O_6 octahedra. P-O = 1.508-1.547(5) (terminal), 1.599(3) (bridging) A,
P-O-P = 131.3°, Na,Zr-O = 2.375-2.776(6) A.

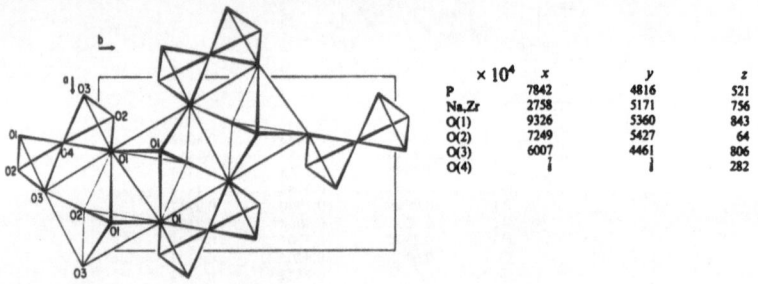

$\times 10^4$	x	y	z
P	7842	4816	521
Na,Zr	2758	5171	756
O(1)	9326	5360	843
O(2)	7249	5427	64
O(3)	6007	4461	806
O(4)	⅛	⅛	282

LITHIUM VANADIUM(III) PYROPHOSPHATE
$LiVP_2O_7$

J. Solid State Chem., <u>86</u>, 143-148.

$P2_1$, 4.8048, 8.113, 6.9393, 109.01, Z = 2, R = 0.022. Isostructural with the Fe
compound (<u>52A</u>, 354), with a framework of corner-sharing VO_6 octahedra and
$O_3P-O-PO_3^{4-}$ ions; tunnels contain 4-coordinate Li^+ ions. V-O = 1.95-2.03, P-O =
1.51-1.52 (terminal), 1.60 (bridging), Li-O = 1.96-2.12 A, P-O-P = 128.6°.

	x	y	z
V	0.21539	0.43900	0.72831
P(1)	0.21012	0.72275	0.08485
P(2)	0.40041	0.11997	0.52131
O(1)	-0.1857	0.4773	0.5189
O(2)	0.3959	0.6172	0.6066
O(3)	0.0524	0.2632	0.8512
O(4)	0.2489	0.2846	0.5075
O(5)	0.1302	0.6337	0.8835
O(6)	0.6031	0.3727	0.9179
O(7)	0.4024	0.5913	0.2458
Li	0.1959	0.8050	0.6816

POTASSIUM VANADIUM(IV) PYROPHOSPHATE
$\beta-K_2V_3P_4O_{17}$

J. Solid State Chem., <u>87</u>, 396-401.

P2$_1$/c, 9.298, 4.879, 17.998, 114.98, Z = 2, R = 0.055. Layers of VO$_5$ square
pyramids (one V is disordered in two sites) and P$_2$O$_7$ groups, with 8-coordinate
K ions. V-O = 1.62-2.12 A. The material is a high-temperature form; a normal
temperature form is also known (56A, 221).

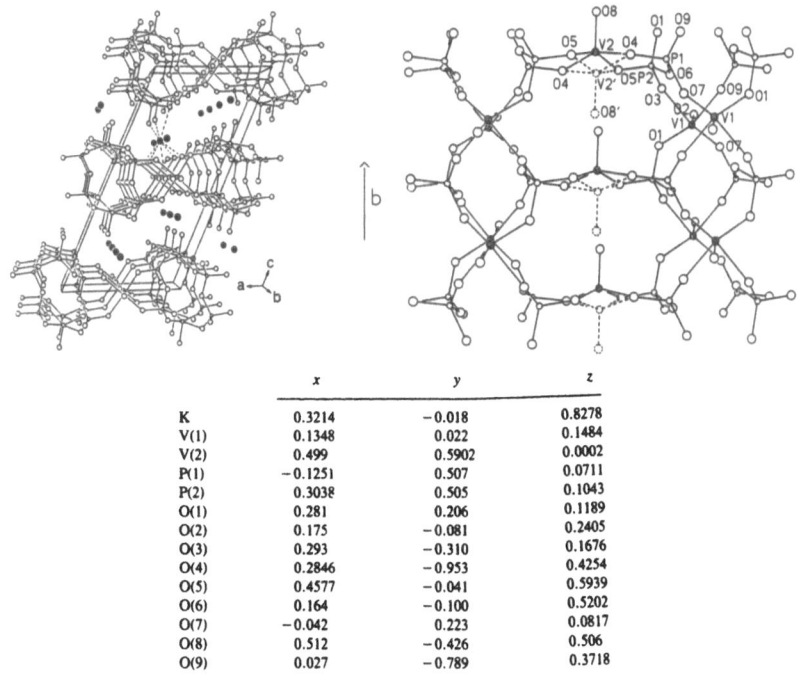

	x	y	z
K	0.3214	−0.018	0.8278
V(1)	0.1348	0.022	0.1484
V(2)	0.499	0.5902	0.0002
P(1)	−0.1251	0.507	0.0711
P(2)	0.3038	0.505	0.1043
O(1)	0.281	0.206	0.1189
O(2)	0.175	−0.081	0.2405
O(3)	0.293	−0.310	0.1676
O(4)	0.2846	−0.953	0.4254
O(5)	0.4577	−0.041	0.5939
O(6)	0.164	−0.100	0.5202
O(7)	−0.042	0.223	0.0817
O(8)	0.512	−0.426	0.506
O(9)	0.027	−0.789	0.3718

The occupancy factors for V(2) and O(8) are 0.5

RUBIDIUM VANADIUM(III) DIPHOSPHATE
RbVP$_2$O$_7$

Z. Kristallogr., 191, 137-138.

P2$_1$/c, 7.511, 10.035, 8.254, 105.74, Z = 4, R = 0.029. [Isostructural with the
Cs compound (56A, 220) and with the KAl compound (39A, 291).]

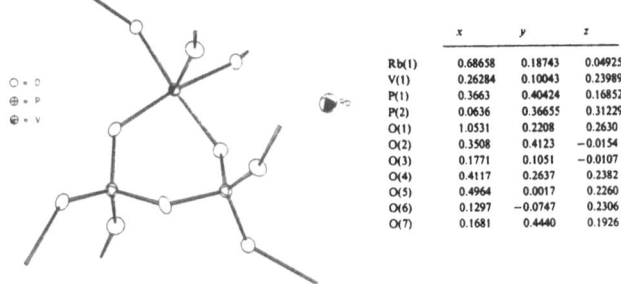

	x	y	z
Rb(1)	0.68658	0.18743	0.04925
V(1)	0.26284	0.10043	0.23989
P(1)	0.3663	0.40424	0.16852
P(2)	0.0636	0.36655	0.31229
O(1)	1.0531	0.2208	0.2630
O(2)	0.3508	0.4123	−0.0154
O(3)	0.1771	0.1051	−0.0107
O(4)	0.4117	0.2637	0.2382
O(5)	0.4964	0.0017	0.2260
O(6)	0.1297	−0.0747	0.2306
O(7)	0.1681	0.4440	0.1926

O = O
⊕ = P
◉ = V

RUBIDIUM VANADIUM(IV) PYROPHOSPHATE
$Rb_2V_3P_4O_{17}$ $Rb_2(VO)_3(P_2O_7)_2$

J. Chin. Chem. Soc., $\underline{37}$, 141-149.

Pnma, 17.502, 7.292, 11.399, Z = 4, R = 0.030. Framework of V_2O_{10} units (VO_5 square pyramid and VO_6 octahedron sharing a corner), chains of corner-sharing VO_6 octahedra, and pyrophosphate groups, with tunnels containing 9-coordinate Rb ions. One V site is disordered.

RUBIDIUM VANADIUM PYROPHOSPHATE
$RbV_3P_4O_{17.14}$

Inorg. Chem., $\underline{29}$, 3298-3301.

$P4_2/mnm$, 13.651, 7.289, Z = 4, R = 0.037. Infinite chains along **c** and finite chains along <110> of trans-corner-sharing VO_6 octahedra, linked into a three-dimensional structure by P_2O_7 groups; the finite chain contains four VO_6 sub-units. Two V sites are disordered and one O site is partially occupied. V-O = 1.54-2.51, P-O = 1.49-1.60, Rb-6 O = 3.09, 3.14 A.

	x	y	z	occ
Rb	0.19973	0.19973	0.0	1.0
V(1)	0.5	0.0	0.25	1.0
V(2)	0.6612	0.6612	0.5	0.876
V(2)'	0.3076	0.3076	0.5	0.124
V(3)	0.1281	0.1281	0.5	0.536
V(3)'	0.0815	0.0815	0.5	0.464
P(1)	0.47399	0.23352	0.8016	1.0
O(1)	0.0	0.0	0.5	1.0
O(2)	0.5	0.0	0.0	1.0
O(3)	0.1786	0.0412	0.6906	1.0
O(4)	0.5276	0.1424	0.7502	1.0
O(5)	0.2518	0.3849	0.6812	1.0
O(6)	0.4248	0.2204	0.0	1.0
O(7)	0.9186	0.0814	0.0	1.0
O(8)	0.2116	0.2116	0.5	0.64

trans-DIAMMINE-cis-DIAQUO[PYROPHOSPHATO(3-)]CHROMIUM(III) MONOHYDRATE and DIHYDRATE
$[Cr(NH_3)_2(H_2O)_2(HP_2O_7)]\cdot nH_2O$ (n = 1, 2)

Acta Cryst., $C\underline{46}$, 2369-2374.

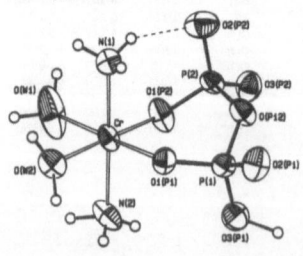 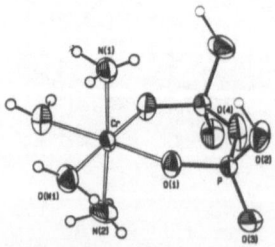

n = 1, PĪ, 7.127, 8.390, 9.619, 72.14, 98.86, 76.98, Z = 2, R = 0.057. n = 2,
C2/m, 13.118, 12.101, 7.436, 105.09, Z = 4, R = 0.043. Octahedral molecular
complexes. In the monohydrate, the six-membered chelate ring has a boat conformation with an intramolecular N-H...O hydrogen bond; in the dihydrate, the
ring has an unusual planar conformation which does not permit intramolecular
hydrogen bonding.

TRIAMMINECHROMIUM PYROPHOSPHATE HYDRATES
mer-[Cr(HP$_2$O$_7$)(NH$_3$)$_3$(H$_2$O)].2H$_2$O, fac-[Cr(HP$_2$O$_7$)(NH$_3$)$_3$]$_2$.4H$_2$O

Acta Cryst., C46, 951-957.

mer, P2$_1$/c, 7.825, 10.107, 15.322, 103.92, Z = 4, R = 0.050. fac, P2$_1$/c, 8.695,
10.327, 11.913, 97.81, Z = 2, R = 0.047. The mer-isomer is monomeric, but the
fac-isomer is a dimer with two bridging pyrophosphate groups.

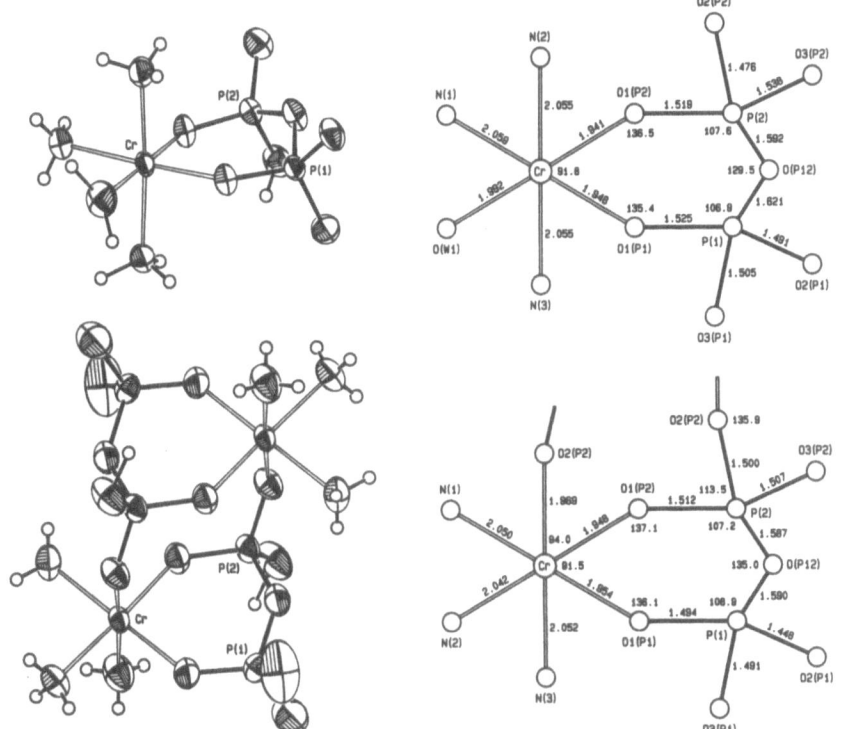

LITHIUM MOLYBDENUM PYROPHOSPHATE
LiMoP$_2$O$_7$

J. Appl. Cryst., 23, 520-525.

P2$_1$, 4.8987, 8.3912, 7.0306, 109.327, Z = 2, powder data (multiphase sample).
Isostructural with the K, Rb, and Cs compounds (54A, 238; 56A, 222), with a
framework of corner-sharing MoO$_6$ octahedra and O$_3$P-O-PO$_3$$^{4-}$ ions, with Li$^+$ ions
in tunnels.

LITHIUM IRON(III) PYROPHOSPHATE
$LiFeP_2O_7$

Mater. Res. Bull., 25, 1363-1369.

$P2_1$, 4.8229, 8.0813, 6.9419, 109.387, Z = 2, R = 0.028, previous study in 52A, 354. Isostructural with related materials (preceding report), with a framework of corner-sharing FeO_6 octahedra and tetrahedral $O_3P-O-PO_3^{4-}$ ions, with Li^+ ions in tunnels. Fe-O = 1.96-2.04, P-O = 1.50-1.52 (terminal), 1.61 (bridging), Li-4 O = 1.94-2.12 A, P-O-P = 129° (almost eclipsed conformation).

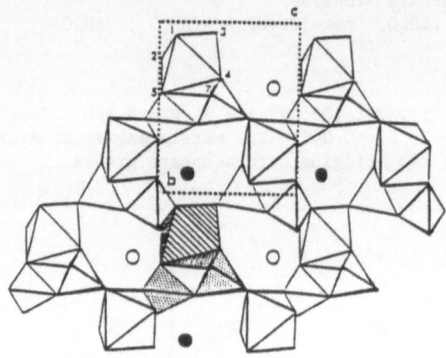

POTASSIUM COBALT(II) PYROPHOSPHATE DIHYDRATE
$K_2Co_3(P_2O_7)_2 \cdot 2H_2O$

J. Solid State Chem., 85, 275-282.

$P2_1/a$, 9.229, 8.110, 9.122, 99.31, Z = 2, R = 0.023. Chains along b of cis-edge-sharing CoO_6 octahedra linked by P_2O_7 groups into (001) layers, which are connected by 8-coordinate K ions. Co-O = 2.031-2.214, P-O = 1.504-1.527 (terminal), 1.625, 1.645 (bridging), K-O = 2.678-3.223, O-H...O = 2.701, 2.712(1) A.

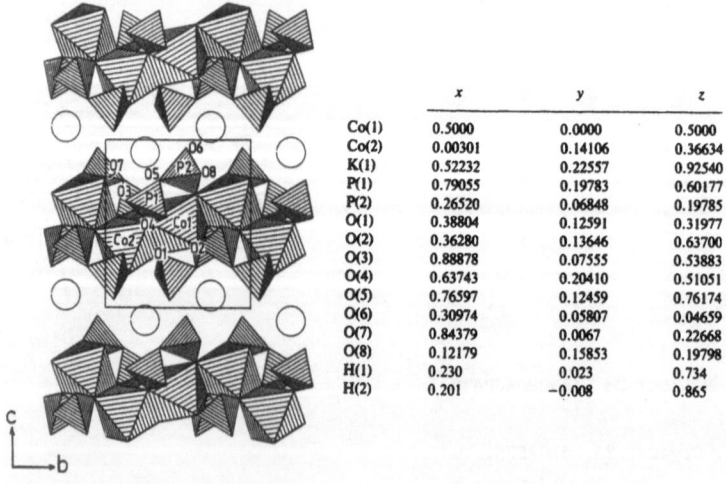

	x	y	z
Co(1)	0.5000	0.0000	0.5000
Co(2)	0.00301	0.14106	0.36634
K(1)	0.52232	0.22557	0.92540
P(1)	0.79055	0.19783	0.60177
P(2)	0.26520	0.06848	0.19785
O(1)	0.38804	0.12591	0.31977
O(2)	0.36280	0.13646	0.63700
O(3)	0.88878	0.07555	0.53883
O(4)	0.63743	0.20410	0.51051
O(5)	0.76597	0.12459	0.76174
O(6)	0.30974	0.05807	0.04659
O(7)	0.84379	0.0067	0.22668
O(8)	0.12179	0.15853	0.19798
H(1)	0.230	0.023	0.734
H(2)	0.201	−0.008	0.865

SODIUM NICKEL HYDROGEN PYROPHOSPHATE
SODIUM COBALT(II) HYDROGEN PYROPHOSPHATE
$NaNiHP_2O_7$, $NaCoHP_2O_7$

Acta Cryst., C46, 2497-2499.

The materials were previously described as $Na_2MZr(P_2O_7)_2$, M = Ni, Co, in P1
(56A, 219); refinements in P$\bar{1}$ (R = 0.033 for both compounds) now show (i) there
is no (or very little) Zr present, (ii) there is an additional H atom, forming
a very strong hydrogen bond between pyrophosphate groups, (iii) Na is split in
two sites about 0.5 Å apart.

10^4	x	y	z		x	y	z
Ni	3571	1082	3942	Co	3544	1106	3937
P(1)	−3816	3640	1987	P(1)	−3796	3640	1962
P(2)	−1410	2191	4749	P(2)	−1427	2200	4732
O(1)	−2967	2189	−588	O(6)	−3273	537	4162
O(2)	−5794	2498	2053	O(2)	−5767	2480	1986
O(3)	−3865	5975	2683	O(5)	347	1080	3297
O(4)	−2075	3757	3879	O(3)	−3854	5965	2649
O(5)	394	1040	3333	O(1)	−2900	2201	−564
O(6)	−3296	514	4139	O(7)	−894	3729	7518
O(7)	−907	3748	7536	O(4)	−2122	3794	3931
Na(1)	1141	2413	133	Na(1)	1139	2437	171
Na(2)	1615	2612	−482	Na(2)	1605	2582	−470
H	−1920	2970	−1520	H	−1896	2964	−1522

Site populations: Na(1), 0·38 ; Na(2), 0·6?   Site populations: Na(1), 0·43   ; Na(2), 0·57

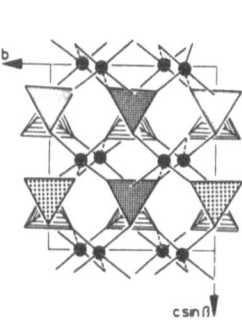

View of the short (2·44 Å) O(1)···O(7) hydrogen bond
between HP_2O_7 groups.

COPPER(II) PYROPHOSPHATE
α-$Cu_2P_2O_7$

Acta Cryst., C46, 691-692.

C2/c, 6.895, 8.113, 9.164, 109.62, Z = 4, R = 0.036. Structure as previously
described (31A, 192; 32A, 376), with $P_2O_7^{4-}$ ions linked into (001) sheets by
CuO_5 polyhedra. A high-temperature form has also been described (33A, 410).

	x	y	z
Cu	−0·01814	0·31288	0·50706
P	0·19754	0·00750	0·20566
O(1)	0	0·0466	$\frac{1}{4}$
O(2)	0·3757	0·0002	0·3614
O(3c)	0·2208	0·1559	0·1128
O(3t)	0·1785	−0·1508	0·1179

Interatomic distances (Å) and bond angles (°)

Cu—O(3t^i)	1·917 (1)	O(3t^i)—O(3c^i)	2·853 (2)	95·54 (6)
Cu—O(3c^i)	1·936 (1)	O(3t^i)—O(2^{iii})	2·871 (2)	94·88 (6)
Cu—O(2^{iv})	1·981 (1)	O(3t^i)—O(2^{iii})	3·824 (2)	155·99 (8)
Cu—O(2^{iii})	1·992 (1)	O(3c^i)—O(2^{iii})	3·882 (2)	164·73 (8)
Cu—O(3c^v)	2·327 (1)	O(3c^i)—O(2^{iii})	2·899 (2)	95·09 (5)
Cu—O(3t^{vi})	2·947 (1)	O(2^{iii})—O(2^{iii})	2·543 (3)	79·59 (7)
P—O(3t)	1·497 (1)	O(3t)—O(3c)	2·507 (2)	112·77 (8)
P—O(3c)	1·514 (1)	O(3t)—O(2)	2·508 (2)	111·30 (8)
P—O(2)	1·540 (1)	O(3t)—O(1)	2·558 (3)	112·53 (13)
P—O(1)	1·579 (1)	O(3c)—O(2)	2·509 (2)	110·47 (8)
		O(3c)—O(1)	2·443 (2)	104·35 (11)
P—O(1)—P^i	156·80 (19)	O(2)—O(1)	2·474 (1)	104·95 (5)

Symmetry code: (i) $-x, y, -z + \frac{1}{2}$; (ii) $x, -y, z + \frac{1}{2}$; (iii) $x - \frac{1}{2}, y + \frac{1}{2},$
z; (iv) $-x + \frac{1}{2}, -y + \frac{1}{2}, -z + 1$; (v) $x - \frac{1}{2}, -y + \frac{1}{2}, z + \frac{1}{2}$; (vi) $-x +$
$\frac{1}{2}, y + \frac{1}{2}, -z + \frac{1}{2}.$

CALCIUM COPPER(II) PYROPHOSPHATE
$CaCuP_2O_7$

Acta Cryst., C46, 1191-1193.

$P2_1/n$, 5.2104, 8.0574, 12.344, 91.356, Z = 4, R = 0.029. Structure closely
related to that of α-$Ca_2P_2O_7$ (33A, 407), but with different cation coordinations.

$O_3P-O-PO_3^{4-}$ anions are connected by CuO_5 square pyramids, with (6+2)-coordinated Ca ions in cages.

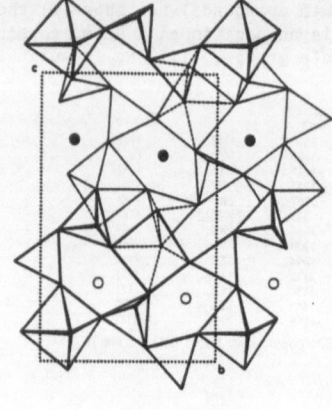

Interatomic distances (Å) and angles (°)

[P₂O₇] group

P(1)	O(1)	O(3″)	O(4″)	O(5″)
O(1)	1·510 (2)	2·451 (3)	2·490 (3)	2·534 (3)
O(3″)	107·7 (1)	1·526 (2)	2·517 (3)	2·523 (3)
O(4″)	106·7 (1)	107·6 (1)	1·593 (2)	2·512 (3)
O(5″)	114·1 (1)	112·4 (1)	108·1 (1)	1·510 (2)

P(2)	O(2″)	O(4‴)	O(6″)	O(7″)
O(2″)	1·536 (2)	2·501 (3)	2·516 (3)	2·528 (3)
O(4‴)	105·6 (1)	1·605 (2)	2·531 (3)	2·448 (3)
O(6″)	110·2 (1)	107·6 (1)	1·531 (2)	2·569 (3)
O(7″)	112·7 (1)	104·0 (1)	115·8 (1)	1·500 (2)

[CuO₅] square pyramid

Cu	O(1)	O(2ᵛⁱⁱⁱ)	O(3)	O(6″)	O(7)
O(1)	1·922 (2)	3·841 (3)	2·705 (3)	2·715 (3)	3·257 (3)
O(2ᵛⁱⁱⁱ)	160·16 (9)	1·978 (2)	2·755 (3)	2·960 (3)	3·005 (3)
O(3)	86·06 (9)	86·59 (9)	2·040 (2)	4·016 (3)	2·606 (3)
O(6″)	87·78 (9)	96·40 (9)	169·33 (8)	1·994 (2)	3·526 (3)
O(7)	104·16 (9)	91·82 (9)	75·72 (9)	114·31 (8)	2·201 (2)

Coordination of Ca

Ca—O(5ᵛ)	2·399 (2)	Ca—O(7)	2·497 (2)
Ca—O(3ᵛⁱ)	2·444 (2)	Ca—O(1)	2·498 (2)
Ca—O(5)	2·447 (2)	Ca—O(6)	2·626 (2)
Ca—O(2)	2·487 (2)	Ca—O(3)	2·846 (2)

Symmetry code: (i) $1 + x, y, z$; (ii) $x, y, 1 + z$; (iii) $x - 1, y, z$; (iv) $1/2 + x, 1/2 - y, z - 1/2$; (v) $3/2 - x, 1/2 + y, 1/2 - z$; (vi) $1/2 + x, 1/2 - y, 1/2 + z$; (vii) $x - 1/2, 1/2 - y, 1/2 + z$; (viii) $3/2 - x, y - 1/2, 1/2 - z$; (ix) $1/2 - x, 1/2 + y, 1/2 - z$.

	x	y	z
Cu	0·79798	0·14058	0·11019
Ca	0·2878	0·32276	0·27641
P(1)	0·7523	0·5280	0·16483
P(2)	0·3116	0·2020	0·98402
O(1)	0·6608	0·3508	0·1542
O(2)	0·6524	0·3979	0·4005
O(3)	0·9384	0·1207	0·2653
O(4)	0·7641	0·1067	0·4535
O(5)	0·4747	0·0450	0·2929
O(6)	0·0980	0·3338	0·4715
O(7)	0·2124	0·1867	0·0967

VANADIUM(III) DIHYDROGEN TRIPHOSPHATE
$VH_2P_3O_{10}$

Izv. Akad. Nauk SSSR, Neorg. Mater., **26**, 1064-1068.

$P2_1/n$, 7.378, 8.786, 11.702, 102.82, Z = 4, R = 0.036. Triphosphate anions of three corner-sharing PO_4 tetrahedra, linked by VO_6 octahedra.

STRONTIUM TETRATHIOCYCLOTETRAPHOSPHATE HYDRATE
$Sr_2P_4O_8S_4 \cdot 10H_2O$

Izv. Akad. Nauk SSSR, Neorg. Mater., **26**, 2194-2197.

$P\bar{1}$, 8.111, 8.631, 8.855, 116.78, 93.11, 97.35, Z = 1, R = 0.046. Cyclic anions with 8-membered P_4O_4 rings, linked by Sr ions which are coordinated to 4 O and 5 H_2O.

IRON(II) CYCLOTETRAPHOSPHATE
$Fe_2P_4O_{12}$

Z. Kristallogr., **192**, 83-90.

C2/c, 11.942, 8.370, 9.936, 118.77, Z = 4, neutron powder data. Cyclic anions, linked by two types of FeO_6 octahedra. P-O = 1.43-1.54 (terminal), 1.54-1.60(2) (bridging), Fe-O = 2.05-2.21(1) A.

POTASSIUM ERBIUM TETRACYCLOPHOSPHATE HEXAHYDRATE
$KErP_4O_{12} \cdot 6H_2O$

Ž. Neorg. Khim., <u>34</u>, 2682-2684 (1989) [Russ. J. Inorg. Chem., <u>34</u>, 1533-1534 (1989)].

C2/c, 8.643, 12.015, 14.909, 90.65, Z = 4, R = 0.055. Cyclic anions linked by $ErO_4(H_2O)_4$ and $KO_4(H_2O)_4$ polyhedra which share edges and alternate along **b**.

POTASSIUM YTTERBIUM TETRACYCLOPHOSPHATE HEXAHYDRATE
CAESIUM THULIUM TETRACYCLOPHOSPHATE HEXAHYDRATE
$KYbP_4O_{12} \cdot 6H_2O$, $CsTmP_4O_{12} \cdot 6H_2O$

Ž. Neorg. Khim., <u>35</u>, 859-862 [Russ. J. Inorg. Chem., <u>35</u>, 482-484].

C2/c, a = 8.611, 8.645, b = 11.977, 12.332, c = 14.835, 15.060 A, β = 90.64, 90.96°, Z = 4, R = 0.040, 0.029. Isostructural with the K/Er compound (preceding report). Cyclic anions are linked by LnO_8 and MO_8 polyhedra.

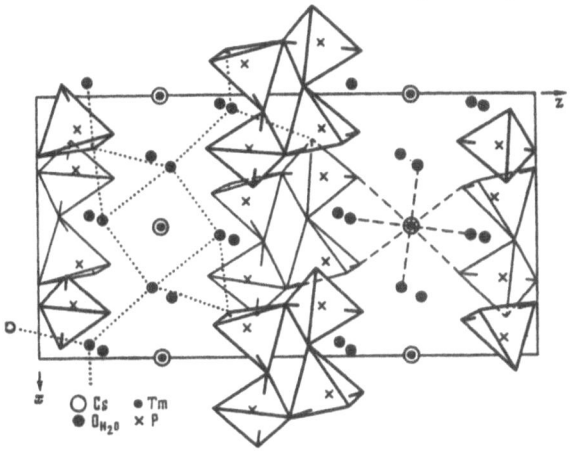

HYDROXYLAMMONIUM CYCLOHEXAPHOSPHATE TETRAHYDRATE
$(NH_3OH)_6P_6O_{18} \cdot 4H_2O$

Acta Cryst., C<u>46</u>, 2026-2028.

P$\bar{1}$, 10.365, 9.278, 7.280, 108.39, 100.30, 96.02, Z = 1, R = 0.028. Cyclic anions linked by hydrogen bonding via the cations and water molecules. P-O-P = 123-133°.

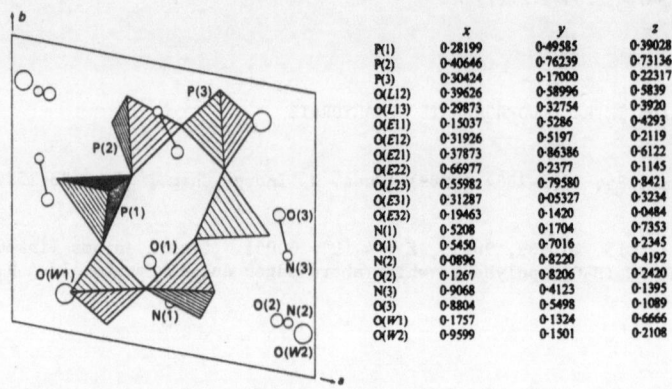

	x	y	z
P(1)	0·28199	0·49585	0·39028
P(2)	0·40646	0·76239	0·73136
P(3)	0·30424	0·17000	0·22317
O(L12)	0·39626	0·58996	0·5839
O(L13)	0·29873	0·32754	0·3920
O(E11)	0·15037	0·5286	0·4293
O(E12)	0·31926	0·5197	0·2119
O(E21)	0·37873	0·86386	0·6122
O(E22)	0·66977	0·2377	0·1145
O(L23)	0·55982	0·79580	0·8421
O(E31)	0·31287	0·05327	0·3234
O(E32)	0·19463	0·1420	0·0484
N(1)	0·5208	0·1704	0·7353
O(1)	0·5450	0·7016	0·2345
N(2)	0·0896	0·8220	0·4192
O(2)	0·1267	0·8206	0·2420
N(3)	0·9068	0·4123	0·1395
O(3)	0·8804	0·5498	0·1089
O(W1)	0·1757	0·1324	0·6666
O(W2)	0·9599	0·1501	0·2108

AMMONIUM CALCIUM CYCLOHEXAPHOSPHATE HEXAHYDRATE
$(NH_4)_2Ca_2P_6O_{18}·6H_2O$

Acta Cryst., C46, 2005–2007.

$P2_12_12$, 12.821, 12.537, 7.029, Z = 2, R = 0.035. Cyclic anions linked by 7-coordinate Ca and 8- and 12-coordinate ammonium ions. P–O = 1.474–1.486 (terminal), 1.596–1.603 (ring), Ca–O = 2.356–2.500, N–H...O = 2.881–3.085, O–H...O = 2.816–3.084 Å.

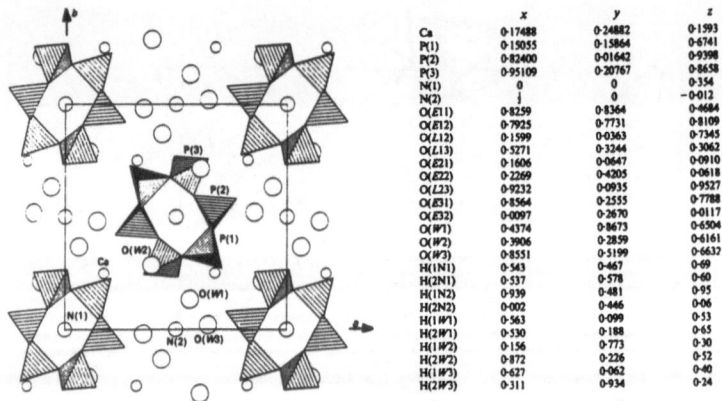

	x	y	z
Ca	0·17488	0·24882	0·1593
P(1)	0·15055	0·15864	0·6741
P(2)	0·82400	0·01642	0·9398
P(3)	0·95109	0·20767	0·8658
N(1)	0	0	0·354
N(2)	½	0	−0·012
O(E11)	0·8259	0·8364	0·4684
O(E12)	0·7925	0·7731	0·8109
O(L12)	0·1599	0·0363	0·7345
O(L13)	0·5271	0·3244	0·3062
O(E21)	0·1606	0·0647	0·0910
O(E22)	0·2269	0·4205	0·0618
O(L23)	0·9232	0·0935	0·9527
O(E31)	0·8564	0·2555	0·7788
O(E32)	0·0097	0·2670	0·0117
O(W1)	0·4374	0·8673	0·6504
O(W2)	0·3906	0·2859	0·6161
O(W3)	0·8551	0·5199	0·6632
H(1N1)	0·543	0·467	0·69
H(2N1)	0·537	0·578	0·60
H(1N2)	0·939	0·481	0·95
H(2N2)	0·002	0·446	0·06
H(1W1)	0·563	0·099	0·53
H(2W1)	0·530	0·188	0·65
H(1W2)	0·156	0·773	0·30
H(2W2)	0·872	0·226	0·52
H(1W3)	0·627	0·062	0·40
H(2W3)	0·311	0·934	0·24

LITHIUM CALCIUM CYCLOHEXAPHOSPHATE OCTAHYDRATE
$Li_2Ca_2P_6O_{18}·8H_2O$

Acta Cryst., C46, 968–970.

$P\bar{1}$, 7.767, 10.144, 7.225, 105.17, 102.76, 84.95, Z = 1, R = 0.023. (001) Layers of cyclic anions sharing corners with LiO_6 octahedra, with layers linked by 7-coordinate Ca ions and by hydrogen bonds.

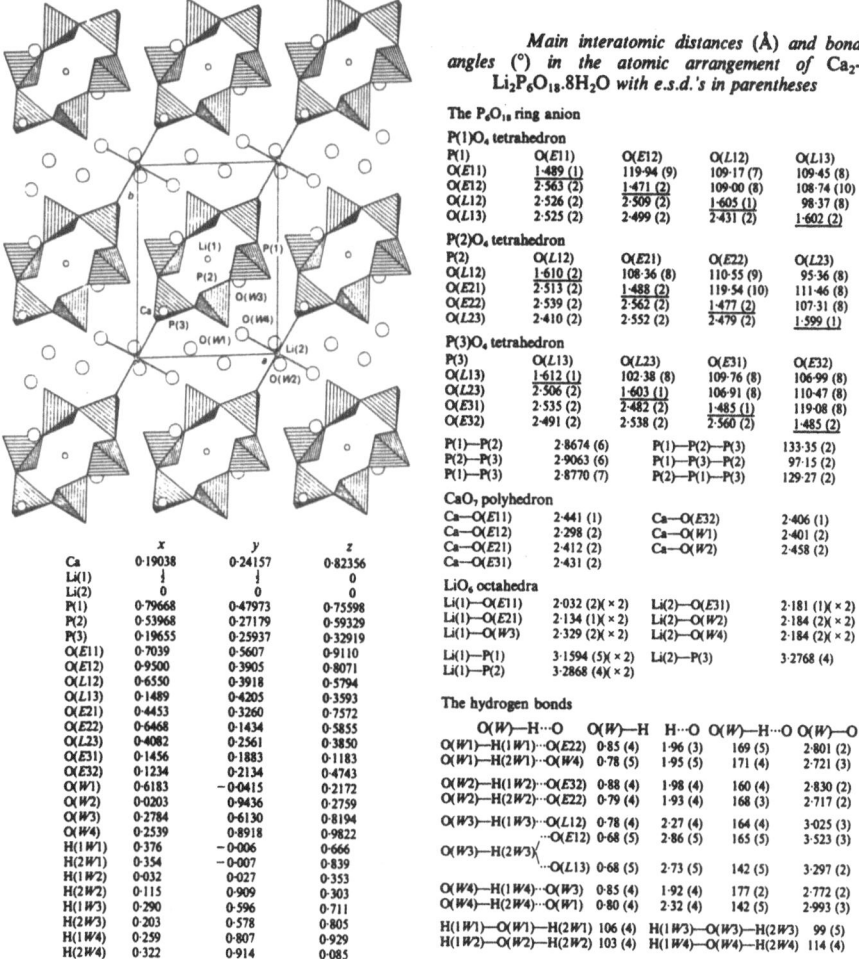

Main interatomic distances (Å) and bond angles (°) in the atomic arrangement of Ca₂-Li₂P₆O₁₈.8H₂O with e.s.d.'s in parentheses

The P$_6$O$_{18}$ ring anion

P(1)O$_4$ tetrahedron

P(1)	O(E11)	O(E12)	O(L12)	O(L13)
O(E11)	1·489 (1)	119·94 (9)	109·17 (7)	109·45 (8)
O(E12)	2·563 (2)	1·471 (2)	109·00 (8)	108·74 (10)
O(L12)	2·526 (2)	2·509 (2)	1·605 (1)	98·37 (8)
O(L13)	2·525 (2)	2·499 (2)	2·431 (2)	1·602 (2)

P(2)O$_4$ tetrahedron

P(2)	O(L12)	O(E21)	O(E22)	O(L23)
O(L12)	1·610 (2)	108·36 (8)	110·55 (9)	95·36 (8)
O(E21)	2·513 (2)	1·488 (2)	119·54 (10)	111·46 (8)
O(E22)	2·539 (2)	2·562 (2)	1·477 (2)	107·31 (8)
O(L23)	2·410 (2)	2·552 (2)	2·479 (2)	1·599 (1)

P(3)O$_4$ tetrahedron

P(3)	O(L13)	O(L23)	O(E31)	O(E32)
O(L13)	1·612 (1)	102·38 (8)	106·99 (8)	106·91 (8)
O(L23)	2·506 (2)	1·603 (1)	106·91 (8)	110·47 (8)
O(E31)	2·535 (2)	2·482 (2)	1·485 (1)	119·08 (8)
O(E32)	2·491 (2)	2·538 (2)	2·560 (2)	1·485 (2)

P(1)—P(2)	2·8674 (6)	P(1)—P(2)—P(3)	133·35 (2)
P(2)—P(3)	2·9063 (6)	P(1)—P(3)—P(2)	97·15 (2)
P(1)—P(3)	2·8770 (7)	P(2)—P(1)—P(3)	129·27 (2)

CaO$_7$ polyhedron

Ca—O(E11)	2·441 (1)	Ca—O(E32)	2·406 (1)
Ca—O(E12)	2·298 (2)	Ca—O(W1)	2·401 (2)
Ca—O(E21)	2·412 (2)	Ca—O(W2)	2·458 (2)
Ca—O(E31)	2·431 (2)		

LiO$_6$ octahedra

Li(1)—O(E11)	2·032 (2)(×2)	Li(2)—O(E31)	2·181 (1)(×2)
Li(1)—O(E21)	2·134 (1)(×2)	Li(2)—O(W2)	2·184 (2)(×2)
Li(1)—O(W3)	2·329 (2)(×2)	Li(2)—O(W4)	2·184 (2)(×2)
Li(1)—P(1)	3·1594 (5)(×2)	Li(2)—P(3)	3·2768 (4)
Li(1)—P(2)	3·2868 (4)(×2)		

The hydrogen bonds

O(W)—H···O	O(W)—H	H···O	O(W)—H···O	O(W)···O
O(W1)—H(1W1)···O(E22)	0·85 (4)	1·96 (3)	169 (5)	2·801 (2)
O(W1)—H(2W1)···O(W4)	0·78 (5)	1·95 (5)	171 (4)	2·721 (3)
O(W2)—H(1W2)···O(E32)	0·88 (4)	1·98 (4)	160 (4)	2·830 (2)
O(W2)—H(2W2)···O(E22)	0·79 (4)	1·93 (4)	168 (3)	2·717 (2)
O(W3)—H(1W3)···O(L12)	0·78 (4)	2·27 (4)	164 (4)	3·025 (3)
O(W3)—H(2W3)⟨···O(E12)	0·68 (5)	2·86 (5)	165 (5)	3·523 (3)
O(W3)—H(2W3)⟨···O(L13)	0·68 (5)	2·73 (5)	142 (5)	3·297 (2)
O(W4)—H(1W4)···O(W3)	0·85 (4)	1·92 (4)	177 (2)	2·772 (2)
O(W4)—H(2W4)···O(W1)	0·80 (4)	2·32 (4)	142 (5)	2·993 (3)

H(1W1)—O(W1)—H(2W1)	106 (4)	H(1W3)—O(W3)—H(2W3)	99 (5)
H(1W2)—O(W2)—H(2W2)	103 (4)	H(1W4)—O(W4)—H(2W4)	114 (4)

	x	y	z
Ca	0·19038	0·24157	0·82356
Li(1)	¼	¼	0
Li(2)	0	0	0
P(1)	0·79668	0·47973	0·75598
P(2)	0·53968	0·27179	0·59329
P(3)	0·19655	0·25937	0·32919
O(E11)	0·7039	0·5607	0·9110
O(E12)	0·9500	0·3905	0·8071
O(L12)	0·6550	0·3918	0·5794
O(L13)	0·1489	0·4205	0·3593
O(E21)	0·4453	0·3260	0·7572
O(E22)	0·6468	0·1434	0·5855
O(L23)	0·4082	0·2561	0·3850
O(E31)	0·1456	0·1883	0·1183
O(E32)	0·1234	0·2134	0·4743
O(W1)	0·6183	-0·0415	0·2172
O(W2)	0·0203	0·9436	0·2759
O(W3)	0·2784	0·6130	0·8194
O(W4)	0·2539	0·8918	0·9822
H(1W1)	0·376	-0·006	0·666
H(2W1)	0·354	-0·007	0·839
H(1W2)	0·032	0·027	0·353
H(2W2)	0·115	0·909	0·303
H(1W3)	0·290	0·596	0·711
H(2W3)	0·203	0·578	0·805
H(1W4)	0·259	0·807	0·929
H(2W4)	0·322	0·914	0·085

SODIUM CADMIUM CYCLOHEXAPHOSPHATE TETRADECAHYDRATE
Na$_2$Cd$_2$P$_6$O$_{18}$·14H$_2$O

Acta Cryst., C46, 10-13.

P$\bar{1}$, 7.709, 11.028, 9.231, 108.25, 110.06, 79.77, Z = 1, R = 0.019. Cyclic anions linked by MO$_6$ octahedra, with some Cd/Na disorder; one water molecule is distributed over two sites, O(W7) and O(W8). There is an extensive hydrogen bond system. Mean Cd-O = 2.29, Na-O = 2.45, Na,Cd-O = 2.36 A.

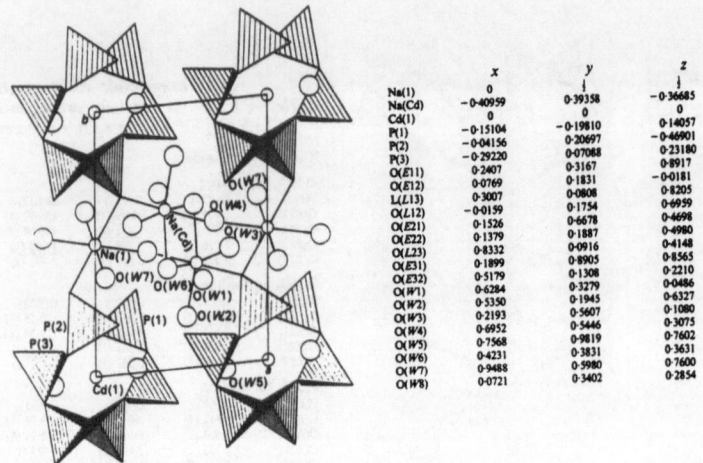

	x	y	z
Na(1)	0	0·39358	−0·36685
Na(Cd)	−0·40959	0	0
Cd(1)	0	−0·19810	0·14057
P(1)	−0·15104	−0·20697	−0·46901
P(2)	−0·04156	0·20697	0·23180
P(3)	−0·29220	−0·07088	0·8917
O(E11)	0·2407	0·3167	0·8917
O(E12)	0·0769	0·1831	−0·0181
O(E13)	0·3007	0·0808	0·8205
L(L13)			
O(L12)	−0·0159	0·1754	0·6959
O(E21)	0·1526	0·6678	0·4698
O(E22)	0·1379	0·1887	0·4980
O(L23)	0·8332	0·0916	0·4148
O(E31)	0·1899	0·8905	0·8565
O(E32)	0·5179	0·1308	0·2210
O(W1)	0·6284	0·3279	0·6327
O(W2)	0·5350	0·1945	0·1080
O(W3)	0·2193	0·5607	0·3075
O(W4)	0·6952	0·9819	0·7602
O(W5)	0·7568	0·3831	0·3631
O(W6)	0·4231	0·5980	0·7600
O(W7)	0·9488	0·5980	0·7600
O(W8)	0·0721	0·3402	0·2854

SODIUM HYDROGEN TRIARSENATE

$Na_3H_2As_3O_{10}$

Acta Cryst., C46, 1185–1188.

C2/c, 10.860, 9.323, 18.270, 103.00, Z = 8, R = 0.035. $H_2As_3O_{10}^{3-}$ anion of three corner-sharing tetrahedra, connected by hydrogen bonds into ($10\bar{1}$) layers, which are linked by octahedrally-coordinated Na^+ ions.

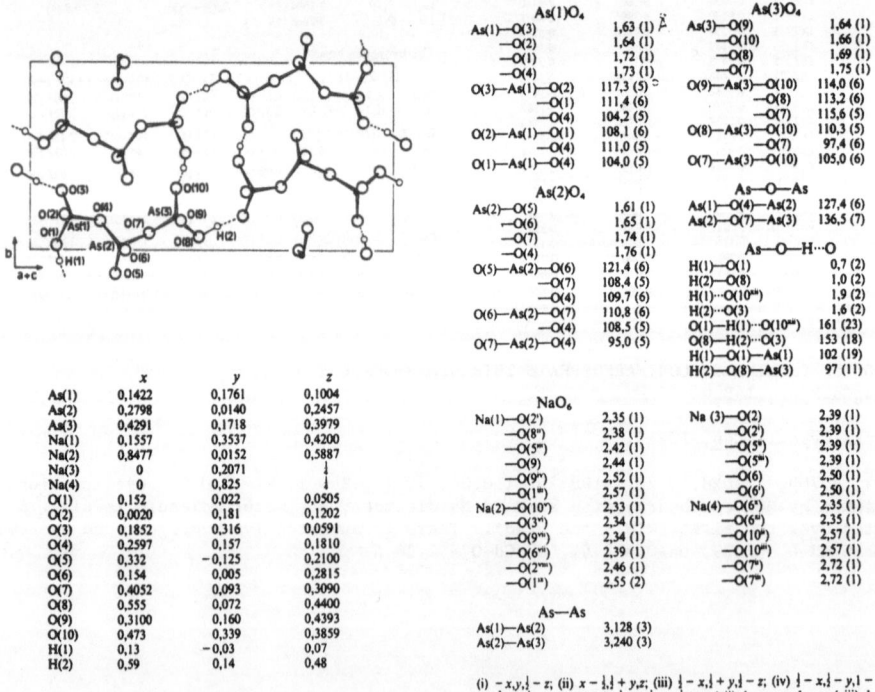

As(1)O₄

As(1)—O(3)	1.63 (1) Å
—O(2)	1.64 (1)
—O(1)	1.72 (1)
—O(4)	1.73 (1)
O(3)—As(1)—O(2)	117.3 (5)
—O(1)	111.4 (6)
—O(4)	104.2 (5)
O(2)—As(1)—O(1)	108.1 (6)
—O(4)	111.0 (6)
O(1)—As(1)—O(4)	104.0 (5)

As(3)O₄

As(3)—O(9)	1.64 (1)
—O(10)	1.66 (1)
—O(8)	1.69 (1)
—O(7)	1.75 (1)
O(9)—As(3)—O(10)	114.0 (6)
—O(8)	113.2 (6)
—O(7)	115.6 (5)
O(8)—As(3)—O(10)	110.3 (5)
—O(7)	97.4 (6)
O(7)—As(3)—O(10)	105.0 (6)

As(2)O₄

As(2)—O(5)	1.61 (1)
—O(6)	1.65 (1)
—O(7)	1.74 (1)
—O(4)	1.76 (1)
O(5)—As(2)—O(6)	121.4 (6)
—O(7)	108.4 (6)
—O(4)	109.7 (6)
O(6)—As(2)—O(7)	110.8 (6)
—O(4)	108.5 (5)
O(7)—As(2)—O(4)	95.0 (5)

As—O—As

As(1)—O(4)—As(2)	127.4 (6)
As(2)—O(7)—As(3)	136.5 (7)

As—O—H···O

H(1)—O(1)	0.7 (2)
H(2)—O(8)	1.0 (2)
H(1)···O(10^{viii})	1.9 (2)
H(2)···O(3)	1.6 (2)
O(1)—H(1)···O(10^{viii})	161 (23)
O(8)—H(2)···O(3)	153 (18)
H(1)—O(1)—As(1)	102 (19)
H(2)—O(8)—As(3)	97 (11)

NaO₆

Na(1)—O(2')	2.35 (1)	Na(3)—O(2)	2.39 (1)
—O(8'')	2.38 (1)	—O(2')	2.39 (1)
—O(5''')	2.42 (1)	—O(5')	2.39 (1)
—O(9)	2.44 (1)	—O(5'')	2.39 (1)
—O(9'')	2.52 (1)	—O(6)	2.50 (1)
—O(1''')	2.57 (1)	—O(6')	2.50 (1)
Na(2)—O(10'')	2.33 (1)	Na(4)—O(6')	2.35 (1)
—O(3'')	2.34 (1)	—O(6'')	2.35 (1)
—O(9''')	2.34 (1)	—O(10')	2.57 (1)
—O(6''')	2.39 (1)	—O(10''')	2.57 (1)
—O(2'''')	2.46 (1)	—O(7'')	2.72 (1)
—O(1'')	2.55 (2)	—O(7''')	2.72 (1)

As—As

As(1)—As(2)	3.128 (3)
As(2)—As(3)	3.240 (3)

	x	y	z
As(1)	0,1422	0,1761	0,1004
As(2)	0,2798	0,0140	0,2457
As(3)	0,4291	0,1718	0,3979
Na(1)	0,1557	0,3537	0,4200
Na(2)	0,8477	0,0152	0,5887
Na(3)	0	0,2071	¼
Na(4)	0	0,825	¼
O(1)	0,152	0,022	0,0505
O(2)	0,0020	0,181	0,1202
O(3)	0,1852	0,316	0,0591
O(4)	0,2597	0,157	0,1810
O(5)	0,332	−0,125	0,2100
O(6)	0,154	0,005	0,2815
O(7)	0,4052	0,093	0,3090
O(8)	0,555	0,072	0,4400
O(9)	0,3100	0,160	0,4393
O(10)	0,473	0,339	0,3859
H(1)	0,13	−0,03	0,07
H(2)	0,59	0,14	0,48

(i) $-x,y,\tfrac{1}{2}-z$; (ii) $x-\tfrac{1}{2},\tfrac{1}{2}+y,z$; (iii) $\tfrac{1}{2}-x,\tfrac{1}{2}+y,\tfrac{1}{2}-z$; (iv) $\tfrac{1}{2}-x,\tfrac{1}{2}-y,1-z$; (v) $\tfrac{1}{2}-x,\tfrac{1}{2}-y,1-z$; (vi) $\tfrac{1}{2}+x,\tfrac{1}{2}-y,\tfrac{1}{2}+z$; (vii) $1-x,-y,1-z$; (viii) $1+x,-y,\tfrac{1}{2}+z$; (ix) $1-x,y,\tfrac{1}{2}-z$; (x) $x,1+y,z$; (xi) $-x,1+y,\tfrac{1}{2}-z$; (xii) $\tfrac{1}{2}-x,y-\tfrac{1}{2},\tfrac{1}{2}-z$.

SODIUM MONOTHIOARSENATE DODECAHYDRATE
$Na_3AsO_3S.12H_2O$

Acta Cryst., C46, 729–732.

$P2_12_12_1$, 9.220, 12.831, 13.906, at 213K, Z = 4, R = 0.038. Units of three face-sharing $Na(H_2O)_6$ octahedra share corners to form zigzag chains along c. These chains, discrete AsO_3S^{3-} tetrahedra, and uncoordinated water molecules are linked by O–H...O and O–H...S hydrogen bonds.

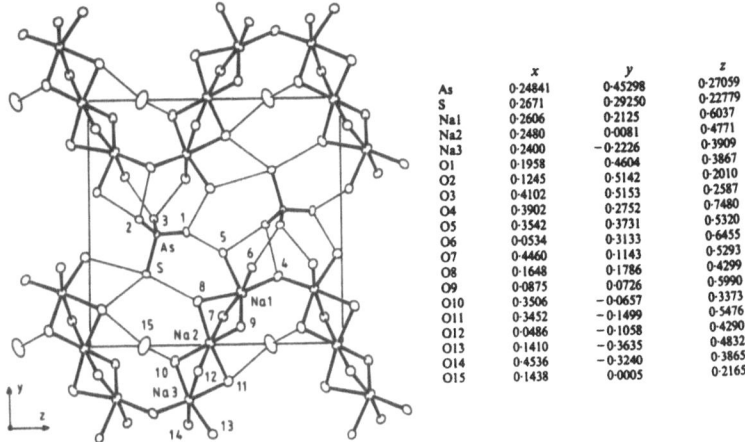

	x	y	z
As	0·24841	0·45298	0·27059
S	0·2671	0·29250	0·22779
Na1	0·2606	0·2125	0·6037
Na2	0·2480	0·0081	0·4771
Na3	0·2400	−0·2226	0·3909
O1	0·1958	0·4604	0·3867
O2	0·1245	0·5142	0·2010
O3	0·4102	0·5153	0·2587
O4	0·3902	0·2752	0·7480
O5	0·3542	0·3731	0·5320
O6	0·0534	0·3133	0·6455
O7	0·4460	0·1143	0·5293
O8	0·1648	0·1786	0·4299
O9	0·0875	0·0726	0·5990
O10	0·3506	−0·0657	0·3373
O11	0·3452	−0·1499	0·5476
O12	0·0486	−0·1058	0·4290
O13	0·1410	−0·3635	0·4832
O14	0·4536	−0·3240	0·3865
O15	0·1438	0·0005	0·2165

MIMETITE
$Pb_5(AsO_4)_3Cl$

Z. Kristallogr., 191, 125–129.

$P6_3/m$, 10.250, 7.454, Z = 2, neutron radiation, R = 0.057. Apatite-type structure as previously described (2, 101, 455).

	x	y	z
Cl	0	0	0
Pb(1)	1/3	2/3	0.0065
Pb(2)	0.2514	0.0042	1/4
As(1)	0.4096	0.3850	1/4
O(1)	0.3290	0.4937	1/4
O(2)	0.5982	0.4872	1/4
O(3)	0.3597	0.2716	0.0733

LEAD(II) HYDROGEN ARSENITE CHLORIDE
$Pb_2(AsO_2OH)Cl_2$

Acta Cryst., C46, 541–543.

$P2_1/m$, 6.410, 5.525, 9.293, 90.69, Z = 2, R = 0.044. $AsO_2(OH)^{2-}$ trigonal pyramids and rows along b of PbO_4 and PbO_2Cl pyramids, with O–H...Cl hydrogen bonds.

	x	y	z
Pb(1)	0·28878	1/4	0·44178
Pb(2)	0·52952	1/4	0·83910
As	0·7734	1/4	0·3127
O(1)	0·7415	1/4	0·1210
O(2)	0·5955	0·0144	0·3519
Cl(1)	0·2403	1/4	0·0987
Cl(2)	0·8862	1/4	0·6587

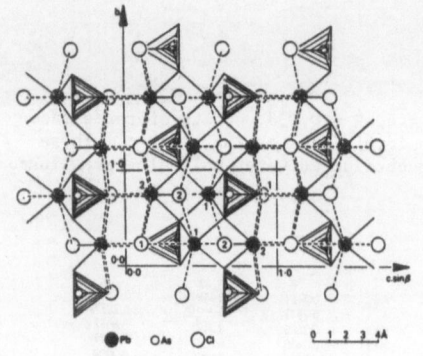

Interatomic distances (Å) and bond angles (°)

Pb(1')—O(2^{v,vi})	2·509 (7)	O(2')—Pb(1')—O(2^{vi})	62·5 (3)
Pb(1')—O(2^{v,vi})	2·515 (7)	O(2')—Pb(1')—O(2^{v})	74·2 (3)
Pb(1')—Cl(2^{vi,viii})	3·120 (2)	O(2')—Pb(1')—O(2^{vii})	109·3 (4)
Pb(1')—Cl(1')	3·200 (4)	O(2^{v})—Pb(1')—O(2')	109·3 (4)
Pb(1')—Cl(2^{v})	3·294 (4)	O(2^{vi})—Pb(1')—O(2^{viii})	74·2 (3)
		O(2^{vi})—Pb(1')—O(2^{viii})	71·0 (3)
Pb(2')—O(2^{v,vi})	2·428 (7)		
Pb(2')—Cl(2')	2·851 (4)	O(2')—Pb(2')—O(2^{viii})	74·0 (3)
Pb(2')—O(1')	2·937 (11)	O(2^{v,vii})—Pb(2')—Cl(2')	80·2 (2)
Pb(2')—Cl(1^{vii,viii})	3·182 (2)		
Pb(2')—Cl(1^{ii})	3·060 (4)		
Pb(2')—O(1^{vii,viii})	3·287 (7)	O(1')—H···Cl(1^{iii})	3·21 (1)
As'—O(2^{v,vi})	1·77 (1)		
As'—O(1')	1·79 (1)	O(2^{vi})—As'—O(1')	98·0 (6)
		O(2')—As'—O(2^{vi})	94·6 (5)

Symmetry codes: (i) x,y,z; (ii) $x,y,1+z$; (iii) $1+x,y,z$; (iv) $-1+x,y,z$; (v) $1-x,-y,1-z$; (vi) $x,0·5-y,z$; (vii) $1-x,0·5+y, 1-z$; (viii) $1-x,-0·5+y,1-z$.

ALKALI–METAL TITANYL ARSENATES
$MTiOAsO_4$ (M = K, Rb, Cs, Tl)

Mater.Res. Bull., **25**, 1193–1202.

Isostructural with $KTiOPO_4$ (**40A**, 245; **52A**, 275; this volume, p. 258), but cell parameters and atomic positions not given.

POTASSIUM ZIRCONIUM(IV) ARSENATE
$KZr_2(AsO_4)_3$

Z. anorg. Chem., **584**, 178–184.

$R\bar{3}c$, a = 9.028, c = 24.399 A, Z = 6, R = 0.041. Isostructural with $NaZr_2(PO_4)_3$ (**33A**, 297; **53A**, 207), with a framework of corner-sharing AsO_4 tetrahedra and ZrO_6 and KO_6 octahedra. As–O = 1.67, Zr–O = 2.04, 2.06 A.

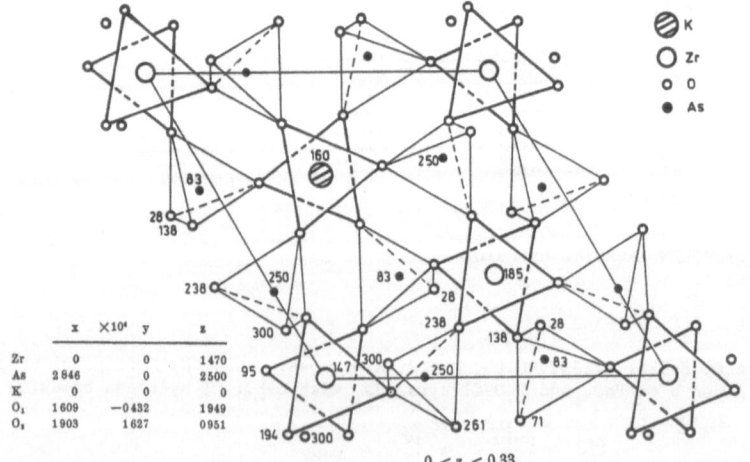

	x ×10^4	y	z
Zr	0	0	1470
As	2 846	0	2500
K	0	0	0
O$_1$	1 609	−0 432	1949
O$_2$	1 903	1 627	0951

$0 < z < 0,33$

SODIUM IRON DIARSENATE
$Na_7Fe_3(As_2O_7)_4$

Acta Cryst., C46, 1584-1587.

C2/c, 9.940, 8.5483, 28.762, 93.683, Z = 4, R = 0.033. Three-dimensional framework of corner-sharing FeO_6 octahedra and As_2O_7 groups, with Na ions in interstices (partial occupancies).

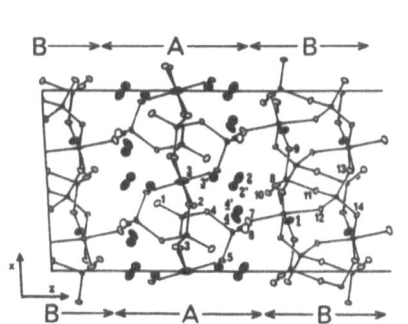

Partial view of the structure along [010]. Na sites in black.
The underlined numbers refer to Na sites; other numbers refer
to O sites.

	Occupancy	x	y	z
As(1)	1	0·29984	0·29537	0·24664
As(2)	1	0·21985	0·16184	0·33893
As(3)	1	0·51105	0·66258	0·44285
As(4)	1	0·41229	0·79584	0·53617
Fe(1)	1	0	0·1202	¼
Fe(2)	1	0·31751	0·9771	0·43648
Na(1)	1	0·2783	0·3808	0·4457
Na(2)	0·88	0·5211	0·3577	0·3561
Na(2')	0·16	0·475	0·384	0·3447
Na(3)	0·60	¼	0·0168	¼
Na(3')	0·20	0·525	0·961	0·3094
Na(4)	0·63	0·2957	0·6031	0·3461
Na(4')	0·31	0·339	0·719	0·3510
O(1)	1	0·3742	0·2214	0·2019
O(2)	1	0·3623	0·4638	0·2685
O(3)	1	0·1313	0·2994	0·2385
O(4)	1	0·3294	0·1598	0·2926
O(5)	1	0·0632	0·1199	0·3172
O(6)	1	0·2371	0·3355	0·3626
O(7)	1	0·2751	0·0041	0·3690
O(8)	1	0·4772	0·8462	0·4254
O(9)	1	0·6768	0·6382	0·4532
O(10)	1	0·4353	0·5292	0·4100
O(11)	1	0·4463	0·6451	0·4982
O(12)	1	0·3425	0·9450	0·5056
O(13)	1	0·5593	0·8364	0·5642
O(14)	1	0·3086	0·714:	0·5722

TRANSITION-METAL PYROARSENATES
β-$Co_2As_2O_7$, β-$Ni_2As_2O_7$, $Mn_2As_2O_7$

J. Solid State Chem., 86, 1-15.

C2/m, a = 6.6316, 6.5391, 6.7442, b = 8.5541, 8.5007, 8.7545, c = 4.7633, 4.7437,
4.8025, β = 103.56, 103.19, 102.76°, Z = 2, X-ray and neutron powder data. Thort-
veitite-type structures (2, 124; 55A, 295), as for $Mg_2As_2O_7$ (35A, 357). Room-
temperature α-forms of the Co and Ni compounds have space group P1, and structures
are proposed based on those of the β-forms.

β-$Co_2As_2O_7$

	x	y	z
Co	0	0.3123	0.5
As	0.2246	0	0.9047
O1	0	0	0
O2	0.3960	0	0.2252
O3	0.2310	0.1620	0.7130

β-$Ni_2As_2O_7$

	x	y	z
Ni	0	0.3136	0.5
As	0.2263	0	0.9030
O1	0	0	0
O2	0.3983	0	0.2303
O3	0.2315	0.1648	0.7129

$Mn_2As_2O_7$

	x	y	z
Mn	0	0.3110	0.5
As	0.2268	0	0.9062
O1	0	0	0
O2	0.3977	0	0.2156
O3	0.2336	0.1582	0.7107

CLINOCLASE
$Cu_3(AsO_4)(OH)_3$

Acta Cryst., C46, 2291-2294.

$P2_1/c$, 7.257, 6.457, 12.378, 99.51, Z = 4, R = 0.044. Structure as previously
described (30A, 397), with a framework of AsO_4 tetrahedra, $CuO_3(OH)_3$ distorted
octahedra, and $CuO_2(OH)_3$ square pyramids. As-O = 1.67-1.70, Cu-O = 1.90-3.00 A.

	x	y	z
Cu(1)	0·7877	0·1400	0·3294
Cu(2)	0·8153	0·3813	0·1274
Cu(3)	0·3869	0·3531	0·4126
As	0·3087	0·1499	0·1796
O(1)	0·4149	0·0710	0·0738
O(2)	0·8377	-0·1577	0·3652
O(3)	0·1798	-0·0531	0·2130
O(4)	0·4711	0·2205	0·2854
OH(1)	0·7799	0·2034	0·4780
OH(2)	0·8088	0·0943	0·1773
OH(3)	0·1808	0·1670	0·4106

FLUOROSULPHURIC ACID
$FSO_2(OH)$

Acta Cryst., C46, 319-320.

$P2_12_12_1$, 4.868, 6.736, 9.359, at 123K, Z = 4, R = 0.025. Tetrahedral molecules
linked into chains along c by O-H...O hydrogen bonds.

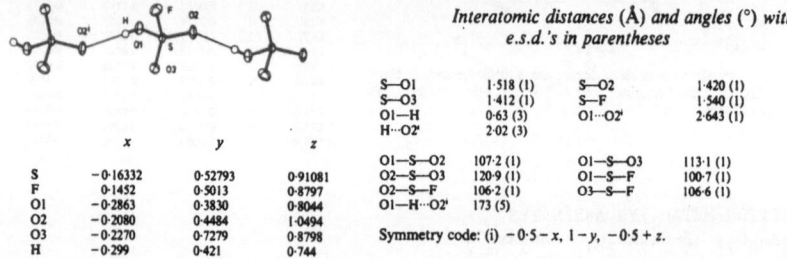

Interatomic distances (Å) and angles (°) with
e.s.d.'s in parentheses

	x	y	z
S	-0·16332	0·52793	0·91081
F	0·1452	0·5013	0·8797
O1	-0·2863	0·3830	0·8044
O2	-0·2080	0·4484	1·0494
O3	-0·2270	0·7279	0·8798
H	-0·299	0·421	0·744

S—O1	1·518 (1)	S—O2	1·420 (1)
S—O3	1·412 (1)	S—F	1·540 (1)
O1—H	0·63 (3)	O1···O2'	2·643 (1)
H···O2'	2·02 (3)		

O1—S—O2	107·2 (1)	O1—S—O3	113·1 (1)
O2—S—O3	120·9 (1)	O1—S—F	100·7 (1)
O2—S—F	106·2 (1)	O3—S—F	106·6 (1)
O1—H···O2'	173 (5)		

Symmetry code: (i) $-0·5-x$, $1-y$, $-0·5+z$.

CALCIUM CHLOROSULPHATE
$Ca(SO_3Cl)_2$

Z. anorg. Chem., 585, 204-208.

Pbca, 9.182, 17.269, 10.250, Z = 8, R = 0.048. Tetrahedral anions linked by
CaO_6 octahedra. S-Cl = 2.00, S-O = 1.41-1.45, Ca-O = 2.29-2.35(1) A.

IODINE(III) BIS(FLUOROSULPHATE) IODIDE
$I(OSO_2F)_2I$

Inorg. Chem., 29, 1527-1530.

$P2_12_12_1$, 5.511, 12.054, 13.573, Z = 4, R = 0.035. The central I(III) is bonded
to two O and one I (T-shaped geometry plus two lone pairs), with these molecules
linked by weaker I...O interactions. I-O = 2.09, 2.26(1), I-I = 2.676(1), I...O
= 2.66, 3.07(1) A.

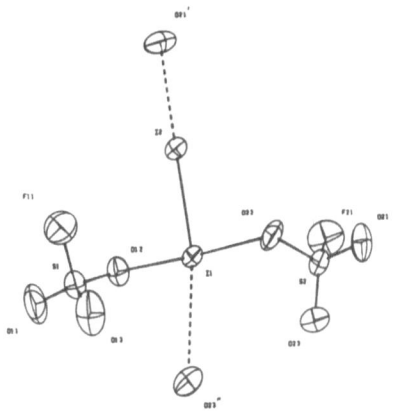

LEAD(II) SULPHITE (ORTHORHOMBIC)
$PbSO_3$

Acta Chem. Scand., <u>44</u>, 688-691.

Pnma, 7.925, 5.485, 6.816, Z = 4, neutron powder data.   Structure as previously
described (<u>50</u>A, 286).

cis-TETRAAMMINEBIS(HYDROGENSULPHITO)RUTHENIUM(II)
$Ru(SO_3H)_2(NH_3)_4$

Acta Cryst., C<u>46</u>, 889-890.

$P2_1/c$, 6.202, 7.021, 11.761, 115.30, Z = 2, R = 0.019.   Centrosymmetric molecular
complexes linked by hydrogen bonds.

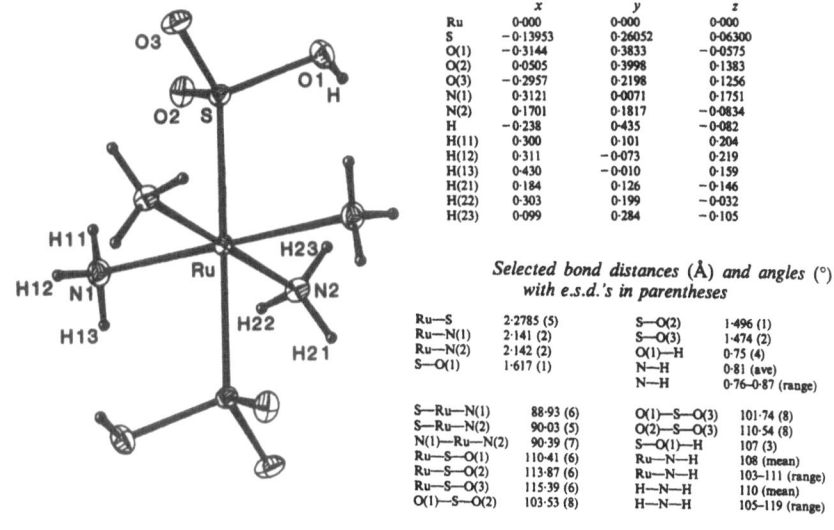

	x	y	z
Ru	0·000	0·000	0·000
S	−0·13953	0·26052	0·06300
O(1)	−0·3144	0·3833	−0·0575
O(2)	0·0505	0·3998	0·1383
O(3)	−0·2957	0·2198	0·1256
N(1)	−0·3121	0·0071	0·1751
N(2)	0·1701	0·1817	−0·0834
H	−0·238	0·435	−0·082
H(11)	0·300	0·101	0·204
H(12)	0·311	−0·073	0·219
H(13)	0·430	−0·010	0·159
H(21)	0·184	0·126	−0·146
H(22)	0·303	0·199	−0·032
H(23)	0·099	0·284	−0·105

*Selected bond distances (Å) and angles (°)
with e.s.d.'s in parentheses*

Ru—S	2·2785 (5)	S—O(2)	1·496 (1)
Ru—N(1)	2·141 (2)	S—O(3)	1·474 (2)
Ru—N(2)	2·142 (2)	O(1)—H	0·75 (4)
S—O(1)	1·617 (1)	N—H	0·81 (ave)
		N—H	0·76–0·87 (range)

S—Ru—N(1)	88·93 (6)	O(1)—S—O(3)	101·74 (8)
S—Ru—N(2)	90·03 (5)	O(2)—S—O(3)	110·54 (8)
N(1)—Ru—N(2)	90·39 (7)	S—O(1)—H	107 (3)
Ru—S—O(1)	110·41 (6)	Ru—N—H	108 (mean)
Ru—S—O(2)	113·87 (6)	Ru—N—H	103–111 (range)
Ru—S—O(3)	115·39 (6)	H—N—H	110 (mean)
O(1)—S—O(2)	103·53 (8)	H—N—H	105–119 (range)

trans-TETRAAMMINEBIS(HYDROGENSULPHITO)RUTHENIUM(II)
$Ru(SO_3H)_2(NH_3)_4$

Z. Naturforsch., 45B, 1651-1656.

$P2_1/n$, 6.194, 7.018, 10.686, 96.32, Z = 2, R = 0.040. trans-Octahedral molecules
(with S-bonded ligands) linked by O-H...O hydrogen bonds into rods along b, which
are further linked by N-H...O hydrogen bonds. Ru-S = 2.276 A (rather short),
Ru-N = 2.14 A.

POTASSIUM HYDROGEN SULPHATE
$K_3H(SO_4)_2$

J. Phys. Soc. Japan, 59, 2804-2810.

A2/a, 9.777, 5.674, 14.667, 102.97, Z = 4, R = 0.024. Isostructural with the
ammonium and rubidium salts (44A, 268; 48A, 302; 52A, 300, 302), with H disordered
in two sites in the hydrogen bond, O...O = 2.493(1) A.

$K_3D(SO_4)_2$

J. Phys. Soc. Japan, 59, 3249-3253.

A2/a, 9.787, 5.682, 14.697, 103.00, Z = 4, R = 0.023. Structure as for the H compound, with disordered D atoms; O...O = 2.519(2) A.

RUBIDIUM LITHIUM HYDROGEN SULPHATE
$Rb_4LiH_3(SO_4)_4$

Acta Cryst., C46, 1199-1202.

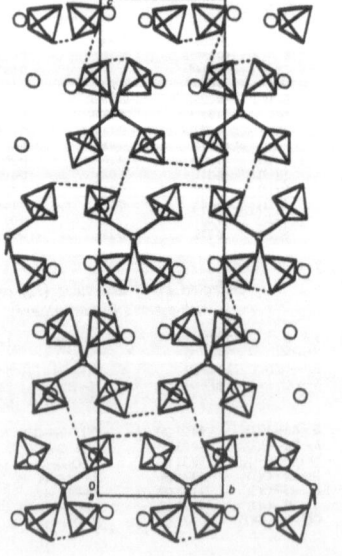

	x	y	z
Rb(1)	0·0645	0·8853	0·7180
Rb(2)	0·3652	0·3840	0·70491
Rb(3)	0·5291	0·9751	0·08343
Rb(4)	0·2271	0·4808	0·07356
S(1)	0·4620	0·1245	0·20459
O(11)	0·575	0·255	0·2250
O(12)	0·377	0·209	0·1618
O(13)	0·319	0·077	0·2332
O(14)	0·561	−0·022	0·1862
S(2)	0·4107	0·1222	0·45631
O(21)	0·341	0·253	0·4867
O(22)	0·256	0·028	0·4343
O(23)	0·510	−0·006	0·4793
O(24)	0·506	0·206	0·4197
S(3)	0·0361	−0·0053	0·08669
O(31)	0·170	0·108	0·0664
O(32)	0·130	−0·154	0·1088
O(33)	−0·059	0·087	0·1221
O(34)	−0·087	−0·068	0·0522
S(4)	0·5380	0·6973	0·33303
O(41)	0·393	0·599	0·3503
O(42)	0·489	0·845	0·3064
O(43)	0·627	0·774	0·3766
O(44)	0·668	0·588	0·3102
Li	0·279	0·869	0·2704

$P4_1$, 7.615, 29.458, Z = 4, R = 0.042 (twinned crystal containing 13% of opposite ($P4_3$) enantiomorph). Double (001) layers of sulphate tetrahedra and 8-coordinate Rb ions, linked by tetrahedrally-coordinated Li ions and by hydrogen bonds. Li-O = 1.89-1.95(2) A.

CAESIUM HYDROGENSULPHATE
$CsHSO_4$

Acta Cryst., C46, 358-361.

$P2_1/c$, 8.214, 5.809, 10.984, 119.39, Z = 4, R = 0.057; the c axis is doubled with respect to a previous incorrect description (48A, 306; see also 53A, 226; 54A, 259).

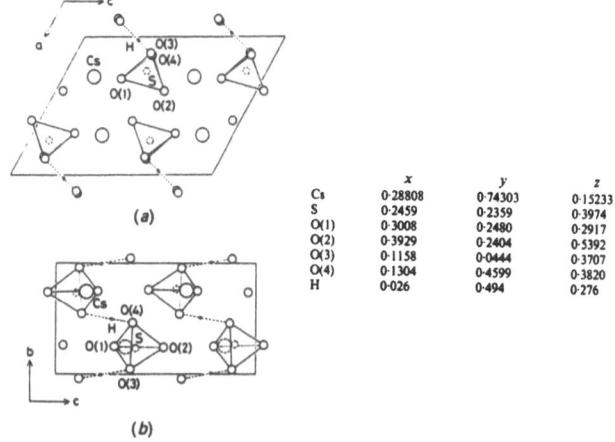

	x	y	z
Cs	0·28808	0·74303	0·15233
S	0·2459	0·2359	0·3974
O(1)	0·3008	0·2480	0·2917
O(2)	0·3929	0·2404	0·5392
O(3)	0·1158	0·0444	0·3707
O(4)	0·1304	0·4599	0·3820
H	0·026	0·494	0·276

(a) The b-axis and (b) a-axis views of the $CsHSO_4$ structure. The hydrogen bonds are drawn with broken lines.

Kristallografija, 35, 658-660 [Soviet Physics - Crystallography, 35, 383-384].

Phase I, $I4_1/amd$, 5.729, 14.21, Z = 4. Phase II, $P2_1/c$, 7.789, 8.146, 7.726, 111.04, Z = 4. Phase III, $P2_1/m$, 7.3039, 5.8099, 5.4908, 101.51, Z = 2. Neutron powder data. Structures as previously desccibed (48A, 306; 53A, 226; 54A, 259).

MAGNESIUM HYDROGEN SULPHATE
$MgH_6(SO_4)_4$

Kristallografija, 35, 852-855 [Soviet Physics - Crystallography, 35, 500-502].

$P2_1/c$, 5.091, 15.329, 7.882, 104.19, at 183K, Z = 2, R = 0.075. $SO_3(OH)^-$ and $SO_2(OH)_2$ tetrahedra linked by octahedrally coordinated Mg and by hydrogen bonds. S-OH = 1.52-1.54, S-O = 1.43-1.47, Mg-O = 2.03-2.09, O-H...O = 2.86-2.96 A.

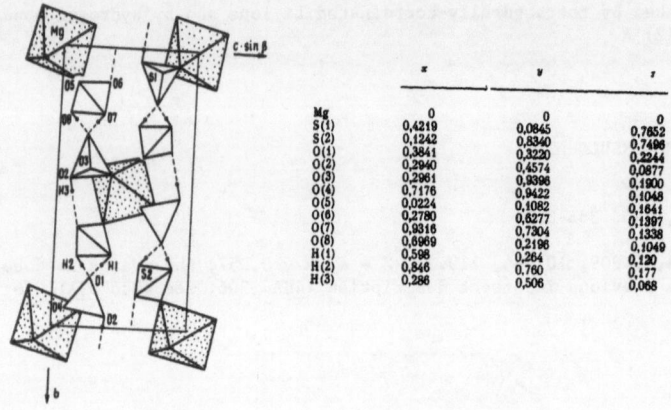

	x	y	z
Mg	0	0	0
S(1)	0,4219	0,0845	0,7652
S(2)	0,1242	0,8340	0,7496
O(1)	0,3841	0,3220	0,2244
O(2)	0,2940	0,4574	0,0677
O(3)	0,2961	0,9396	0,1900
O(4)	0,7176	0,9422	0,1048
O(5)	0,0224	0,1082	0,1641
O(6)	0,2780	0,6277	0,1397
O(7)	0,9316	0,7304	0,1338
O(8)	0,6969	0,2196	0,1049
H(1)	0,598	0,264	0,120
H(2)	0,846	0,760	0,177
H(3)	0,286	0,506	0,068

BARITE
$BaSO_4$

Z. Kristallogr., **191**, 161-171.

Pnma, a = 8.848-9.075, b = 5.441-5.582, c = 7.132-7.258 A, at 298-1308K, Z = 4, R = 0.023-0.050 at 6 temperatures. Structure as previously described (**1**, 343).

		25°C	345°C	670°C	885°C	1010°C	1035°C
Ba	x	0.18444	0.18495	0.18525	0.18536	0.1849	0.1842
	y	0.25	0.25	0.25	0.25	0.25	0.25
	z	0.15854	0.15902	0.16024	0.1606	0.1609	0.161
	B						
S	x	0.0627	0.0612	0.0602	0.0600	0.0610	0.061
	y	0.25	0.25	0.25	0.25	0.25	0.25
	z	0.6906	0.6906	0.6918	0.6923	0.687	0.685
	B						
O(1)	x	−0.0883	−0.0890	−0.090	−0.086	−0.084	−0.086
	y	0.25	0.25	0.25	0.25	0.25	0.25
	z	0.6066	0.612	0.613	0.613	0.614	0.63
	B						
O(2)	x	0.1828	0.1798(9)	0.1785(9)	0.176(1)	0.183	0.182
	y	0.25	0.25	0.25	0.25	0.25	0.25
	z	0.5488	0.551(1)	0.552(1)	0.554(2)	0.553	0.56
	B						
O(3)	x	0.0802	0.0803	0.0812	0.0815	0.079	0.080
	y	0.0301	0.029	0.032	0.036	0.041	0.028
	z	0.8116	0.8087	0.810	0.811	0.804	0.804

LEADHILLITE
$Pb_4(SO_4)(CO_3)_2(OH)_2$

Neues Jb. Miner. Mh., 255-268.

$P2_1/a$, 9.11, 20.82, 11.59, 90.46, Z = 8, R = 0.055; there is a pseudo-trigonal subcell, $P\bar{3}m1$, 5.23, 11.59, Z = 1. (001) $PbCO_3$ and $Pb(SO_4)_{0.5}OH$ sheets, with 9- and 10-coordinate Pb^{2+} ions; Pb-O = 2.2-3.5 A.

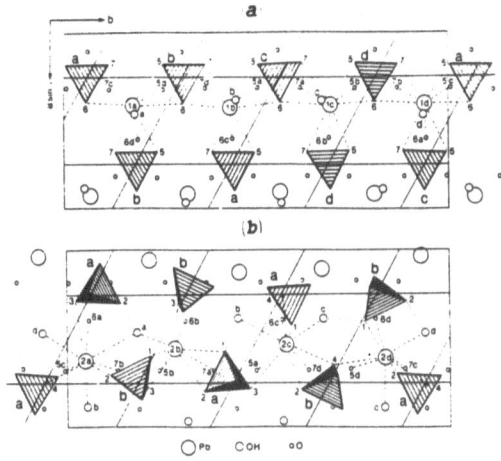

Slabs of the crystal structure of leadhillite projected along [001]: a) carbonate sheet, $0 < z < 0.30$; b) sulfate sheet, $0.28 < z < 0.62$.

VANADYL SULPHATE HEXAHYDRATE
(VO)SO$_4$.6H$_2$O

$[VO(H_2O)_5]SO_4.H_2O$

Acta Cryst., B46, 1-7.

P$\bar{1}$, 6.190, 7.471, 10.100, 92.28, 101.96, 95.35, at 120K, Z = 2, R = 0.018. Structure as previously described (46A, 354), with $[VO(H_2O)_5]^{2+}$ octahedra, sulphate tetrahedra, and one additional water molecule. Electron-density study.

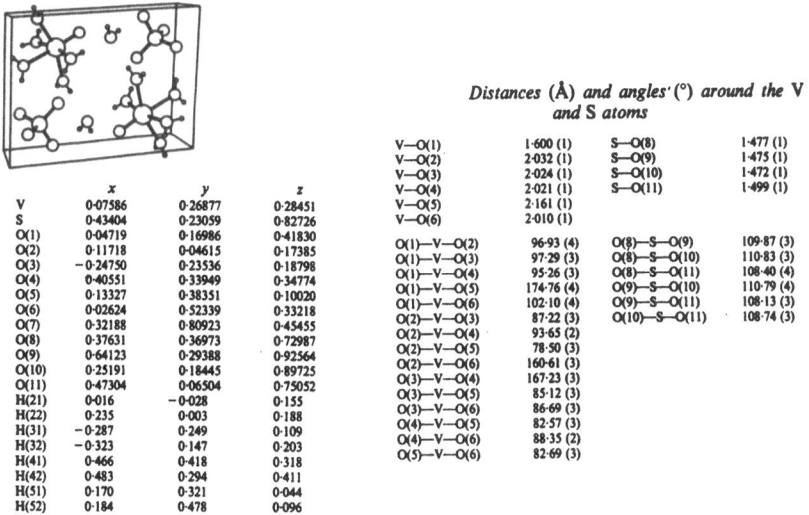

	x	y	z
V	0·07586	0·26877	0·28451
S	0·43404	0·23059	0·82726
O(1)	0·04719	0·16986	0·41830
O(2)	0·11718	0·04615	0·17385
O(3)	−0·24750	0·23536	0·18798
O(4)	0·40551	0·33949	0·34774
O(5)	0·13327	0·38351	0·10020
O(6)	0·02624	0·52339	0·33218
O(7)	0·32188	0·80923	0·45455
O(8)	0·37631	0·36973	0·72987
O(9)	0·64123	0·29388	0·92564
O(10)	0·25191	0·18445	0·89725
O(11)	0·47304	0·06504	0·75052
H(21)	0·016	−0·028	0·155
H(22)	0·235	0·003	0·188
H(31)	−0·287	0·249	0·109
H(32)	−0·323	0·147	0·203
H(41)	0·466	0·418	0·318
H(42)	0·483	0·294	0·411
H(51)	0·170	0·321	0·044
H(52)	0·184	0·478	0·096
H(61)	0·117	0·598	0·371
H(62)	−0·093	0·557	0·311
H(71)	0·380	0·853	0·403
H(72)	0·256	0·879	0·478

Distances (Å) and angles (°) around the V and S atoms

V—O(1)	1·600 (1)	S—O(8)	1·477 (1)
V—O(2)	2·032 (1)	S—O(9)	1·475 (1)
V—O(3)	2·024 (1)	S—O(10)	1·472 (1)
V—O(4)	2·021 (1)	S—O(11)	1·499 (1)
V—O(5)	2·161 (1)		
V—O(6)	2·010 (1)		
O(1)—V—O(2)	96·93 (4)	O(8)—S—O(9)	109·87 (3)
O(1)—V—O(3)	97·29 (3)	O(8)—S—O(10)	110·83 (3)
O(1)—V—O(4)	95·26 (3)	O(8)—S—O(11)	108·40 (4)
O(1)—V—O(5)	174·76 (4)	O(9)—S—O(10)	110·79 (4)
O(1)—V—O(6)	102·10 (4)	O(9)—S—O(11)	108·13 (3)
O(2)—V—O(3)	87·22 (3)	O(10)—S—O(11)	108·74 (3)
O(2)—V—O(4)	93·65 (2)		
O(2)—V—O(5)	78·50 (3)		
O(2)—V—O(6)	160·61 (3)		
O(3)—V—O(4)	167·23 (3)		
O(3)—V—O(5)	85·12 (3)		
O(3)—V—O(6)	86·69 (3)		
O(4)—V—O(5)	82·57 (3)		
O(4)—V—O(6)	88·35 (2)		
O(5)—V—O(6)	82·69 (3)		

SODIUM VANADYL SULPHATE
$Na_2VO(SO_4)_2$

Inorg. Chem., 29, 3294-3298.

$P2_12_12_1$, 6.303, 6.803, 16.682, Z = 4, R = 0.026.   Three-dimensional framework of
corner-sharing VO_6 octahedra and SO_4 tetrahedra.   V-O = 1.595, 2.016-2.068, 2.150,
S-O = 1.439-1.506, Na-6 O = 2.352-2.414(3) A.

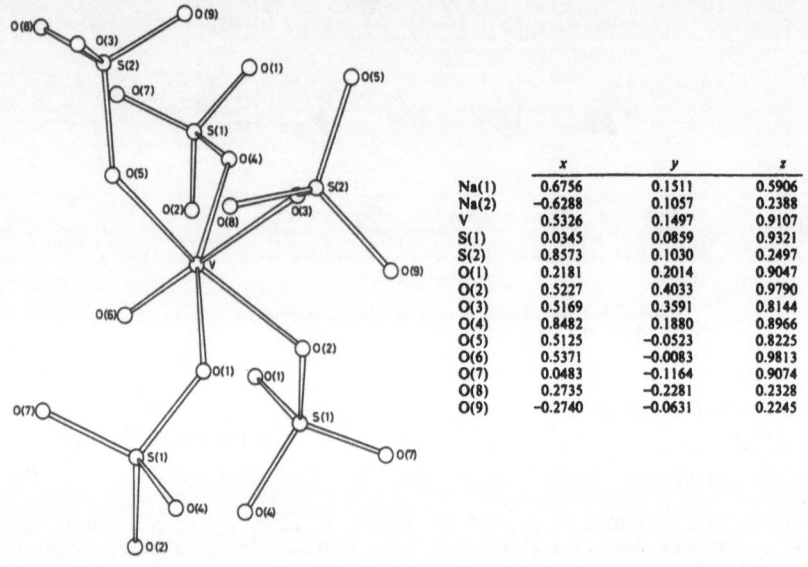

	x	y	z
Na(1)	0.6756	0.1511	0.5906
Na(2)	-0.6288	0.1057	0.2388
V	0.5326	0.1497	0.9107
S(1)	0.0345	0.0859	0.9321
S(2)	0.8573	0.1030	0.2497
O(1)	0.2181	0.2014	0.9047
O(2)	0.5227	0.4033	0.9790
O(3)	0.5169	0.3591	0.8144
O(4)	0.8482	0.1880	0.8966
O(5)	0.5125	-0.0523	0.8225
O(6)	0.5371	-0.0083	0.9813
O(7)	0.0483	-0.1164	0.9074
O(8)	0.2735	-0.2281	0.2328
O(9)	-0.2740	-0.0631	0.2245

POTASSIUM NIOBIUM SULPHATE
$K_7Nb(SO_4)_6$

POTASSIUM TANTALUM SULPHATE
$K_7Ta(SO_4)_6$

Acta Chem. Scand., 44, 328-331.

$R\bar{3}$, a = 15.106, 14.969, c = 9.254, 9.283 A, Z = 3 (rhombohedral cells, a = 9.251,
9.179 A, α = 109.46, 109.26°, Z = 1), R = 0.035, 0.027.   Complex anions with Nb
or Ta bonded octahedrally to six unidentate sulphate ligands.   Nb-O = 1.955(2),
Ta-O = 1.957(2), S-O = 1.55 (bonded to Nb or Ta), 1.43-1.44 (terminal) A,
Nb/Ta-O-S = 149.8, 148.6°.

CHROMIUM(II) SULPHATE TRIHYDRATE
$CrSO_4 \cdot 3H_2O$

Z. anorg. Chem., 580, 224; 586, 141-148.

Cc, 5.7056, 13.211, 7.485, 96.73, Z = 4, R = 0.038.   Isostructural with the Cu
compound (33A, 368), with chains of alternating corner-sharing SO_4 tetrahedra
and $CrO_3(H_2O)_3$ distorted octahedra; chains are linked via sulphate groups and
hydrogen bonds.   S-O = 1.466-1.478, Cr-O = 2.031-2.051, 2.458, 2.469 A.

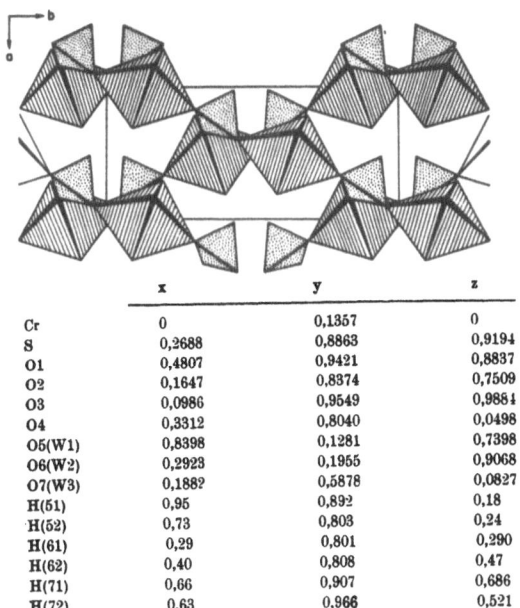

	x	y	z
Cr	0	0,1357	0
S	0,2688	0,8863	0,9194
O1	0,4807	0,9421	0,8837
O2	0,1647	0,8374	0,7509
O3	0,0986	0,9549	0,9884
O4	0,3312	0,8040	0,0498
O5(W1)	0,8398	0,1281	0,7398
O6(W2)	0,2923	0,1955	0,9068
O7(W3)	0,1882	0,5878	0,0827
H(51)	0,95	0,892	0,18
H(52)	0,73	0,803	0,24
H(61)	0,29	0,801	0,290
H(62)	0,40	0,808	0,47
H(71)	0,66	0,907	0,686
H(72)	0,63	0,966	0,521

AMMONIUM CHROMIUM(II) SULPHATE HEXAHYDRATE
$(NH_4)_2Cr(SO_4)_2 \cdot 6H_2O$

Acta Cryst., B<u>46</u>, 577-586.

$P2_1/a$, a = 9.424, 9.490, b = 12.699, 12.816, c = 6.204, 6.104 A, β = 106.60, 107.09°, at 295, 84K, Z = 2, R = 0.024, 0.030. Tutton's salt structure (<u>2</u>, 93; <u>29</u>, 354), with Jahn-Teller distorted $Cr(OH_2)_6$ octahedra: Cr-O = 2.054, 2.125, 2.323 A at 295K; 2.053, 2.079, 2.389(1) A at 84K. Electron-density study.

295 K	x	y	z	84 K	x	y	z
Cr	0	0	0	Cr	0	0	0
S	3965	1412	7426	S	38917	14293	74827
O(3)	3920	2348	5971	O(3)	37872	23646	60319
O(4)	5365	846	7723	O(4)	52962	08894	77213
O(5)	2726	707	6279	O(5)	26572	07175	63351
O(6)	3800	1741	9627	O(6)	37705	17253	97681
O(7)	1677	1071	1713	O(7)	16291	10439	17135
O(8)	-1802	1170	402	O(8)	-18323	11776	5005
O(9)	-10	-713	2967	O(9)	664	-7196	30340
N	1271	3560	3643	N	12691	36404	37701
H(11)	69	345	252	H(11)	59	351	260
H(12)	189	316	393	H(12)	189	317	412
H(13)	91	357	460	H(13)	93	373	481
H(14)	162	412	360	H(14)	167	418	357
H(15)	211	95	313	H(15)	200	93	315
H(16)	237	127	119	H(16)	230	120	117
H(17)	-263	106	-39	H(17)	-266	109	-34
H(18)	-157	178	15	H(18)	-164	178	29
H(19)	-82	-66	326	H(19)	-72	-70	337
H(20)	30	-135	323	H(20)	32	-135	322

($\times 10^4$, except for H, $\times 10^3$ at 295 K; $\times 10^5$, except H, $\times 10^3$ at 84 K)

POTASSIUM SODIUM IRON(III) SULPHATE HYDRATE
$K_{3.86}Na_{5.30}(H_3O)_{0.84}Fe_6O_2(SO_4)_{12}\cdot 17H_2O$

Z. Kristallogr., <u>190</u>, 47-62.

$P\bar{3}$, 9.611, 17.830, Z = 1, R = 0.10. Structure similar to that of related compounds (<u>46</u>A, 358; <u>48</u>A, 315; <u>50</u>A, 293; <u>52</u>A, 306). Large $[Fe_6O_2(SO_4)_{12}\cdot 6H_2O]^{4-}$ clusters are connected by K^+ and Na^+ ions into sheets, which are linked by H_3O^+ and Na^+ ions.

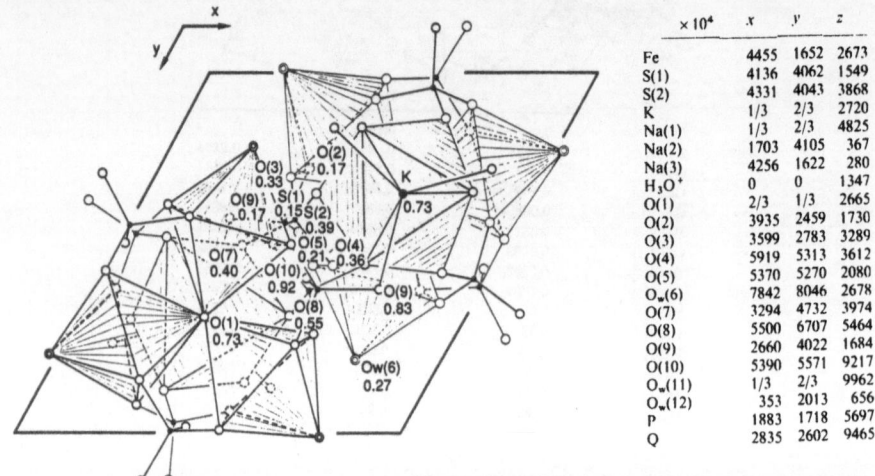

$\times 10^4$	x	y	z
Fe	4455	1652	2673
S(1)	4136	4062	1549
S(2)	4331	4043	3868
K	1/3	2/3	2720
Na(1)	1/3	2/3	4825
Na(2)	1703	4105	367
Na(3)	4256	1622	280
H_3O^+	0	0	1347
O(1)	2/3	1/3	2665
O(2)	3935	2459	1730
O(3)	3599	2783	3289
O(4)	5919	5313	3612
O(5)	5370	5270	2080
O_w(6)	7842	8046	2678
O(7)	3294	4732	3974
O(8)	5500	6707	5464
O(9)	2660	4022	1684
O(10)	5390	5571	9217
O_w(11)	1/3	2/3	9962
O_w(12)	353	2013	656
P	1883	1718	5697
Q	2835	2602	9465

RUBIDIUM AMMONIUM IRON SULPHATE HYDRATE
$Rb_{2.74}(NH_4)_{2.26}Fe_3O(SO_4)_6\cdot 7H_2O$

Acta Cryst., C<u>46</u>, 972-976.

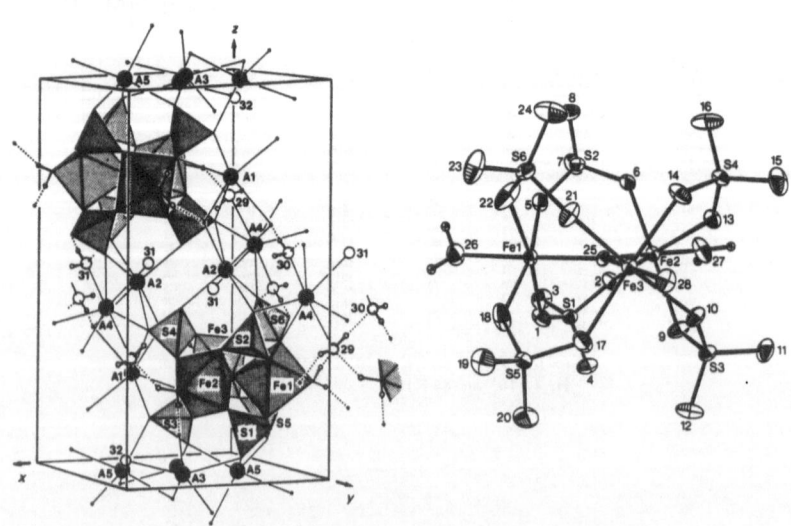

P$\bar{1}$, 9.783, 9.586, 18.389, 95.30, 93.19, 118.12, Z = 2, R = 0.033. (001) Layers of discrete trinuclear $[Fe_3(H_2O)_{30}(SO_4)_6]^{5-}$ units, linked by five Rb/NH$_4$ cations. Fe-O = 1.91-2.10 A.

CAESIUM IRON(III) SULPHATE DODECAHYDRATE
CAESIUM IRON(III) SELENATE DODECAHYDRATE
CsFe(XO$_4$)$_2$.12H$_2$O   (X = S, Se)

J. Chem. Soc., Dalton, 395-400.

Pa3, a = 12.354, 12.593 A, at 15K, Z = 4, neutron radiation, R = 0.029, 0.034. β- and α-alum structures, respectively (3, 108, 111; 48A, 308; 50A, 294). The Fe-O bonds are 19° out of the H$_2$O plane in the selenate; Cs coordinations are cubo-octahedral in the sulphate and icosahedral in the selenate.

POTASSIUM OSMIUM SULPHATE HYDRATE
K$_8$[Os$_2$O$_2$(SO$_4$)$_6$].4H$_2$O

Ž. Neorg. Khim., 35, 2250-2252 [Russ. J. Inorg. Chem., 35, 1281-1282].

P2$_1$/n, 10.288, 10.481, 14.191, 95.31, Z = 2, R = 0.058. Dimeric anion with 2 Os, 2 bridging O, 2 bridging bidentate sulphates, and four terminal unidentate sulphates. Os-Os = 2.446(1) A, indicating a metal-metal bond.

COBALT(II) SULPHATE   NICKEL SULPHATE
β-CoSO$_4$ NiSO$_4$

Z. Kristallogr., 191, 223-229.

Cmcm, a = 5.192, 5.166, b = 7.856, 7.846, c = 6.530, 6.362 A, Z = 4, R = 0.034, 0.049. Co or Ni in 4(a): 0,0,0; S in 4(c): 0,y,1/4, y = 0.3523, 0.3537; O(1) in 8(f): 0,y,z, y = 0.2513, 0.2531, z = 0.0627, 0.0579; O(2) in 8(g): x,y,1/4, x = 0.2342, 0.2344, y = 0.4633, 0.4654. Structures as previously described (21, 365; 22, 453, 454; 30A, 370). Sulphate tetrahedra linked by chains of edge-sharing MO$_6$ octahedra. S-O = 1.455-1.497, Co-O = 2.016 (x 2), 2.157(1) (x 4), Ni-O = 2.020(5) (x 2), 2.118(2) (x 4) A. Other forms (Pnma, P2$_1$/m) of the cobalt compound are known (22, 452; 30A, 370).

NICKEL SULPHATE HEXADEUTERATE (TETRAGONAL)
NiSO$_4$.6D$_2$O

Acta Cryst., B46, 27-39.

P4$_3$2$_1$2, 6.7803, 18.288, Z = 4, R = 0.014 (X-rays), 0.035 (neutrons). Structure as previously described (2, 95, 433; 31A, 200; 54A, 264; 55A, 266). Electron-density study.

POTASSIUM COPPER(II) SULPHATE HYDROXIDE HYDRATE
$KCu_2(H_3O_2)(SO_4)_2$

Acta Cryst., C46, 175–177.

C2/m, 8.955, 6.265, 7.628, 117.45, Z = 2, neutron radiation, R = 0.017.
$[HO-H...OH]^-$ groups with symmetry 2/m and disordered central H site; the two O
atoms are each bonded to two Cu, and are hydrogen bonded to sulphate oxygens.

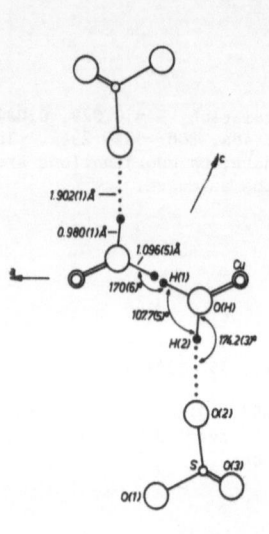

		x	$\times 10^5$ y	z
K	2(d)	0	½	½
Cu	4(f)	¼	¼	0
S	4(i)	8005	0	28776
O(1)	4(i)	18600	0	18542
O(2)	4(i)	18462	0	49968
O(3)	8(j)	−2809	19442	22737
O(H)	4(i)	15621	½	7523
H(1)	4(i)†	1876	½	1207
H(2)	4(i)	20284	½	21942

† Occupancy: ½.

Bond distances (Å) and bond angles (°) not
involved in the hydrogen-bond system

E.s.d.'s are given in parentheses.

K—O(1)	2·7284 (3) 2×	O(1)—K—O(2)	51·26 (1) 2×	128·74 (1) 2×
—O(2)	2·8231 (3) 2×	—O(3)	68·73 (1) 4×	111·27 (1) 4×
—O(3)	2·7515 (2) 4×	O(2)—K—O(3)	75·14 (1) 4×	104·86 (1) 4×
		O(3)—K—O(3)	88·18 (1) 2×	91·82 (1) 2×
Cu—O(1)	2·3515 (2) 2×	O(1)—Cu—O(3)	90·85 (1) 2×	89·15 (1) 2×
—O(3)	1·9750 (2) 2×	—O(H)	95·67 (1) 2×	84·33 (1) 2×
—O(H)	1·9830 (2) 2×	O(3)—Cu—O(H)	89·74 (1) 2×	90·26 (1) 2×
S—O(1)	1·4807 (7) 1×	O(1)—S—O(2)	110·30 (4) 1×	
—O(2)	1·4472 (7) 1×	—O(3)	108·66 (2) 2×	
—O(3)	1·4908 (4) 2×	O(2)—S—O(3)	109·81 (2) 2×	
		O(3)—S—O(3)	109·58 (5) 1×	

Bond distances (Å) and bond angles (°) in the
hydrogen-bond system

E.s.d.'s are given in parentheses.

D—H···A	D···A	D—H	H···A	∠D—H···A
O(H)—H(1)···O(H)	2·487 (1)	1·096 (5)	1·392 (5)	178 (2)
O(H)—H(2)···O(2)	2·878 (1)	0·980 (1)	1·902 (1)	174·2 (3)

H(1)—O(H)—H(2)	107·7 (5) 1×	∠S—O(2)···H(2)	173·0 (3)
H(1)—O(H)—Cu	114·4 (2) 2×		
H(2)—O(H)—Cu	107·85 (4) 2×		
Cu—O(H)—Cu	104·34 (2) 1×		

POTASSIUM CERIUM(III,IV) SULPHATE
$KCe_2(SO_4)_4$

Ž. Strukt. Khim., 31, No. 5, 14–18.

Cc, 11.734, 15.695, 9.268, 132.00, Z = 4, R = 0.086. Both Ce ions have anti-
prismatic 8-coordination, Ce(IV)–O = 2.26–2.38, Ce(III)–O = 2.39–2.56(3) A; K
sites are half-occupied and 9-coordinate.

AMMONIUM TERBIUM SULPHATE HYDRATE
$(NH_4)_6Tb_4(SO_4)_9 \cdot 2H_2O$

Ž. Neorg. Khim., 30, 1722–1726 (1985) [Russ. J. Inorg. Chem., 30, 978–981 (1985)].

C2/c, 9.008, 18.323, 21.39, 95.33, Z = 4, R = 0.058. Layers of SO_4 tetrahedra
and TbO_9 and $TbO_7(H_2O)$ polyhedra, further linked into a three-dimensional framework, with ammonium ions in holes in the framework.

POTASSIUM CAESIUM DYSPROSIUM SULPHATE DIHYDRATE
$K_{4.05}Cs_{1.95}Dy_4(SO_4)_9 \cdot 2H_2O$

Kristallografija, 35, 617-620 [Soviet Physics - Crystallography, 35, 360-362].

C2/c, 8.998, 18.256, 21.232, 95.73, Z = 4, R = 0.025. Isostructural with the
ammonium terbium analogue (preceding report), with a framework of sulphate tetrahedra linked by 8- and 9-coordinate Dy, with K/Cs in gaps in the framework.

SODIUM THORIUM(IV) SULPHATE HYDRATE
$Na_2Th(SO_4)_3 \cdot 6H_2O$ $Na_2[Th(SO_4)_3(H_2O)_3] \cdot 3H_2O$

Acta Cryst., C46, 957-960.

$P2_1/c$, 5.567, 16.81, 15.76, 91.925, Z = 4, R = 0.043. Chains along **a** of
$ThO_6(H_2O)_3$ tricapped trigonal prisms with three bridging, bidentate sulphate
groups; chains are linked by 6-coordinate Na ions and hydrogen bonds. Th-O =
2.35-2.62(1) A.

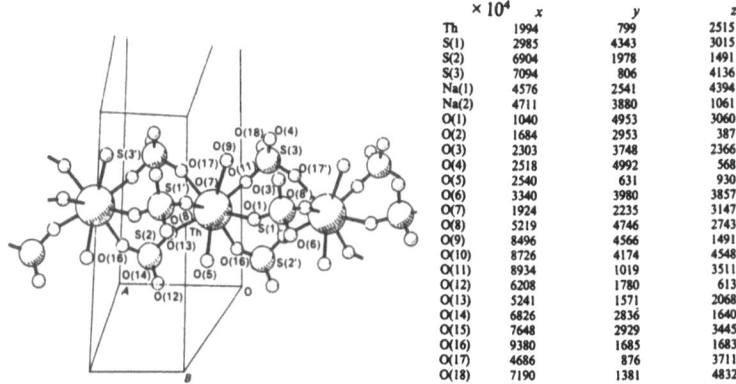

	$\times 10^4$ x	y	z
Th	1994	799	2515
S(1)	2985	4343	3015
S(2)	6904	1978	1491
S(3)	7094	806	4136
Na(1)	4576	2541	4394
Na(2)	4711	3880	1061
O(1)	1040	4953	3060
O(2)	1684	2953	387
O(3)	2303	3748	2366
O(4)	2518	4992	568
O(5)	2540	631	930
O(6)	3340	3980	3857
O(7)	1924	2235	3147
O(8)	5219	4746	2743
O(9)	8496	4566	1491
O(10)	8726	4174	4548
O(11)	8934	1019	3511
O(12)	6208	1780	613
O(13)	5241	1571	2068
O(14)	6826	2836	1640
O(15)	7648	2929	3445
O(16)	9380	1685	1683
O(17)	4686	876	3711
O(18)	7190	1381	4832

AMMONIUM DISELENITE
$(NH_4)_2Se_2O_5$

Kristallografija, 35, 889-893 [Soviet Physics - Crystallography, 35, 523-526];
Ferroelectrics, 107, 275-280.

Below 312K, $P2_12_12_1$, 7.481, 13.429, 6.991, at 293K, Z = 4; above 312K, $P2_12_12_1$,
7.546, 13.115, 7.137, at 333K, Z = 4; neutron radiation, R = 0.040, 0.044. Room-
temperature structure as previously described (46A, 365), with H atoms now located. The phase transition involves rotation of the ions and reorganization of
the hydrogen bond system.

AMMONIUM HYDROGEN SELENATE
NH_4HSeO_4

Kristallografija, 35, 647-657 [Soviet Physics - Crystallography, 35, 377-383].

B2 (c unique), a = 19.754, 19.863, b = 4.607, 4.620, c = 7.550, 7.593 A, γ = 102.59, 102.59°, at 293, 400K, Z = 6, R = 0.047, 0.054. Structure as previously described (48A, 320), with one ordered and one disordered O-H...O hydrogen bond.

SODIUM HYDROGEN SELENATE HYDRATE
$Na_5H_3(SeO_4)_4 \cdot 2H_2O$

Ž. Neorg. Khim., 35, 1363-1368 [Russ. J. Inorg. Chem., 35, 771-774].

P$\bar{1}$, 9.957, 7.254, 5.954, 100.3, 98.8 [in Abstract, 91.8 in text], 102.4, Z = 1, R = 0.042. Tetrameric $H_3(SeO_4)_4^{5-}$ units containing two types of hydrogen bond: a normal bond (2.59 A) and a shorter bond with disordered H site (2.48 A). These units are linked by NaO_6 octahedra and by hydrogen bonds via the water molecules.

CAESIUM HYDROGEN SELENATE (PHASE I)
$Cs_3H(SeO_4)_2$

Kristallografija, 35, 355-360 [Soviet Physics - Crystallography, 35, 200-203].

R$\bar{3}$m, a = 6.4260, c = 23.447 A, at 470K, Z = 3, R = 0.022. Cs(1) in 3(a): 0,0,0; Cs(2) in 6(c): 0,0,z, z = 0.20542; Se in 6(c): z = 0.41061, O(1) in 18(h): x,$\bar{x}$,z, x = 0.141, z = 0.5661; 6 O(2) in 18(h): x = 0.046, z = 0.3416. Isostructural with the Rb compound (54A, 268; 55A, 271), with selenate tetrahedra linked by an O...H...O hydrogen bond (2.71 A), and 12- and 10-coordinate Cs ions. O(2) is probably disordered. Se-O = 1.66 (x 3), 1.70, Cs-O = 3.19-3.49 A. A C2/m phase (phase III) is described in 55A, 272.

COBALTOMENITE
$CoSeO_3 \cdot 2H_2O$

AHLFELDITE
$NiSeO_3 \cdot 2H_2O$

Neues Jb. Miner. Mh., 353-362.

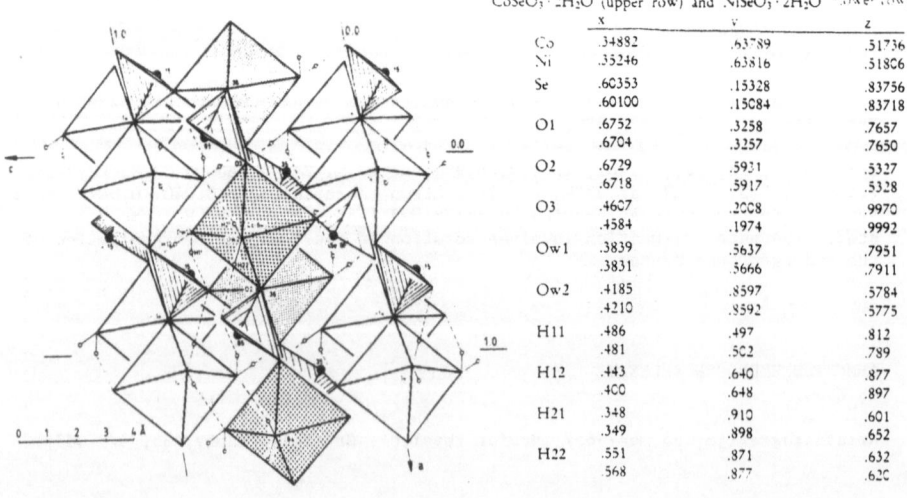

$CoSeO_3 \cdot 2H_2O$ (upper row) and $NiSeO_3 \cdot 2H_2O$ (lower row)

	x	y	z
Co	.34882	.63789	.51736
Ni	.35246	.63816	.51806
Se	.60353	.15328	.83756
	.60100	.15084	.83718
O1	.6752	.3258	.7657
	.6704	.3257	.7650
O2	.6729	.5931	.5327
	.6718	.5917	.5328
O3	.4607	.2008	.9970
	.4584	.1974	.9992
Ow1	.3839	.5637	.7951
	.3831	.5666	.7911
Ow2	.4185	.8597	.5784
	.4210	.8592	.5775
H11	.486	.497	.812
	.481	.502	.789
H12	.443	.640	.877
	.400	.648	.897
H21	.348	.910	.601
	.349	.898	.652
H22	.551	.871	.632
	.568	.877	.620

$P2_1/n$, a = 6.496, 6.441, b = 8.809, 8.746, c = 7.619, 7.522, β = 98.87, 99.00°,
Z = 4, R = 0.043, 0.069. Isostructural with the Zn compound (29, 358). Pairs
of edge-sharing $MO_4(H_2O)_2$ octahedra are linked by pyramidal selenite ions and
by hydrogen bonds. Co-O = 2.04-2.19, Ni-O = 2.02-2.13, Se-O = 1.69-1.71 A.

FRANCISITE
$Cu_3Bi(SeO_3)_2O_2Cl$

Amer. Min., 75, 1421-1425.

Pmmn, 6.354, 9.630, 7.220, Z = 2, R = 0.044. Framework of SeO_3^{2-} trigonal
pyramids, $Cu(II)O_4$ square planes, and 8-coordinate Bi(III), with tunnels along
c which contain Cl^- ions and the Se lone-pairs. Se-O = 1.70, Cu-O = 1.92-1.98,
Bi-O = 2.23-2.80 A.

SILVER AMIDOSELENATE
$\alpha-Ag_3NSeO_3$

Canad. J. Chem., 68, 1606-1610.

R3c, a = 8.462, c = 11.372 A, Z = 6, R = 0.032. Ag in 18(b): 0.26030,0.05443,0.5;
Se in 6(a): 0,0,z, z = 0.2804; N in 6(a): z = 0.4299; O in 18(b): 0.2085,0.0798,
0.2238. Se-N unit on the threefold axis, with coordination at each atom completed
by three O for Se and 3 Ag for N; these structural units are linked by Ag...O
interactions. Se-N = 1.70(2) (very short), Se-O = 1.65(1), Ag-N = 2.16(1), Ag...O
= 2.23(1) A, N-Ag-O = 175°.

AMMONIUM CADMIUM SELENATE TRIHYDRATE
$(NH_4)_2Cd_2(SeO_4)_3 \cdot 3H_2O$

J. Solid State Chem., 89, 88-93.

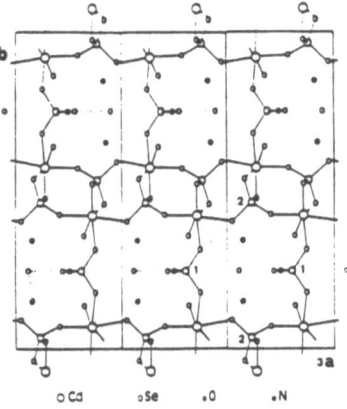

		X	Y	Z
CD(1)	10^5	27502	92539	23077
SE(1)		61730	25000	93308
SE(2)		77161	96479	26937
O(11)	10^4	5110	2500	6670
O(12)		4439	2500	11292
O(13)		7520	1820	9582
O(21)		8373	10388	1656
O(22)		2759	10233	4471
O(23)		5854	9268	1259
O(24)		9506	9060	2607
O(1)		3542	8606	5360
O(2)		1003	2500	5930
N		8532	8459	7718

$P2_1/m$, 6.836, 19.372, 5.690, 94.02, Z = 2, R = 0.043. Framework of corner-
sharing SeO_4 tetrahedra and $CdO_5(H_2O)$ octahedra with cavities, each of which
contains one water molecule and two ammonium ions. Se-O = 1.61-1.68, Cd-O =
2.18-2.43(1) A.

SODIUM YTTRIUM SELENITE
SODIUM LANTHANUM SELENITE
$NaLn(SeO_3)_2$ (Ln = Y, La)

Acta Cryst., C46, 2013-2017.

Ln = Y, $P2_1cn$, 5.397, 8.525, 12.765, Z = 4, R = 0.027. Ln = La, $P2_1/n$, 6.696,
6.761, 13.199, 101.51, Z = 4, R = 0.056. Both structures contain SeO_3 trigonal
pyramids and 5-coordinate Na: Y has 7-coordination (capped trigonal prism) and
La has 10-coordination (bicapped square antiprism).

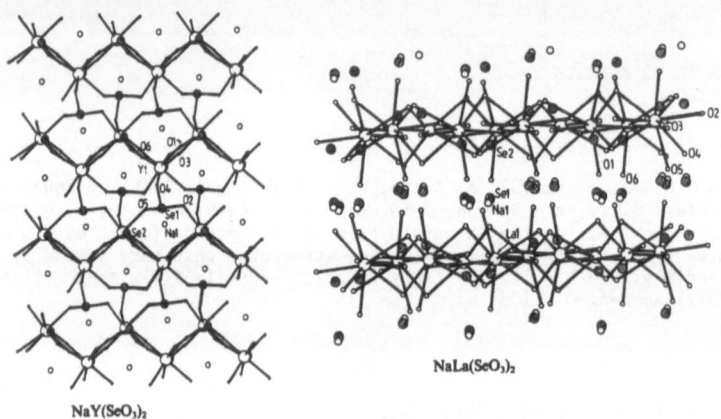

NaY(SeO₃)₂   NaLa(SeO₃)₂

CERIUM(IV) SELENATE
$Ce(SeO_4)_2$

Kristallografija, 35, 1089-1093 [Soviet Physics - Crystallography, 35, 640-643].

Pbca, 9.748, 9.174, 13.740, Z = 4, R = 0.035. Isostructural with the orthorhombic
forms of the sulphate (40A, 267) and of thorium molybdate (40A, 201; 50A, 206).
Se-4 O = 1.61-1.64, Ce-8 O = 2.27-2.42 A (square antiprism).

POTASSIUM PRASEODYMIUM SELENATE
$KPr(SeO_4)_2$

Kristallografija, 35, 1083-1088 [Soviet Physics - Crystallography, 35, 637-640].

$P2_1/c$, 8.823, 7.371, 11.139, 91.33, Z = 4, R = 0.032. Isostructural with the monoclinic form of the Nd sulphate (55A, 270).

POTASSIUM DYSPROSIUM SELENATE
$KDy(SeO_4)_2$

Kristallografija, <u>35</u>, 1099-1104 [Soviet Physics - Crystallography, <u>35</u>, 646-649].

$Pn2_1a$, 27.470, 8.989, 5.657, Z = 8, R = 0.050. Barite-related structure, with
SeO_4 tetrahedra linked by DyO_8 bicapped trigonal prisms and 9-coordinate K ions.

POTASSIUM SODIUM LITHIUM TELLURATE
$K_3Na_2LiTeO_6$

Z. anorg. Chem., <u>586</u>, 125-135.

Cc, 9.283, 11.874, 6.787, 93.8, Z = 4, R = 0.019. TeO_6 octahedra and LiO_4 tetrahedra share edges to form chains, which are linked by 6-, (6+2)-, and (4+2)-coord-
inate K and 5-coordinate Na ions. Te-O = 1.94-1.97, Li-O = 1.91-2.04 A.

	$x \quad 10^4$	$y \quad 10^4$	$z \quad 10^4$
Te (4a)	0	2430	0
K1 (4a)	7573	4606	7539
K2 (4a)	9392	9526	0744
K3 (4a)	1084	5092	2401
Na1 (4a)	8548	2841	4047
Na2 (4a)	6581	7314	6024
Li (4a)	7484	2221	7503
O1 (4a)	5557	6993	2722
O2 (4a)	4442	7894	7277
O3 (4a)	6128	3789	4010
O4 (4a)	6735	8374	0032
O5 (4a)	8880	6328	5963
O6 (4a)	8262	8525	4860

RUBIDIUM TELLURATE
$Rb_6(TeO_5)(TeO_4)$

Z. anorg. Chem., <u>584</u>, 105-113.

C2/c, 12.075, 12.663, 11.053, 123.1, Z = 4, R = 0.095. (001) Sheets of TeO_5 trigonal bipyramids, TeO_4 tetrahedra, and 6- to 9-coordinate Rb ions. Te-O = 1.80-
1.92 A.

		$x \cdot 10^4$	$y \cdot 10^4$	$z \cdot 10^4$
Te1	(4e)	5000	6459	2500
Te2	(4e)	5000	8739	7500
Rb1	(8f)	7105	3878	2392
Rb2	(8f)	8531	1463	4729
Rb3	(8f)	5995	9003	4774
O1	(4e)	5000	2716	5000
O2	(8f)	3933	1287	0433
O3	(8f)	6241	0430	2448
O4	(8f)	4514	4381	5978
O5	(8f)	3634	7303	2174

AMMONIUM CYCLOHEXAPHOSPHATE TELLURATE DIHYDRATE
$(NH_4)_6P_6O_{18} \cdot Te(OH)_6 \cdot 2H_2O$

Acta Cryst., <u>C46</u>, 179-181.

$P\bar{1}$, 9.899, 11.042, 7.632, 109.53, 106.74, 100.91, Z = 1, R = 0.018. (001) Layers
of alternating $Te(OH)_6$ octahedra and cyclic hexaphosphate anions, with layers
linked by hydrogen bonding via the ammonium ions and (twofold-disordered) water
molecules. Te-O = 1.88-1.93 A.

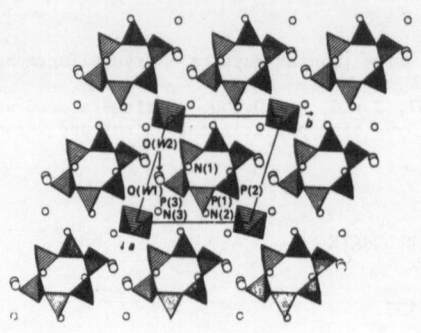

NEODYMIUM TELLURATE
$Nd_2Te_4O_{11}$

J. Solid State Chem., 85, 100–107.

C2/c, 12.635, 5.204, 16.277, 106.02, Z = 4, R = 0.023. Alternating (001) layers
of edge-sharing NdO_8 distorted square antiprisms, and corner-sharing TeO_4 polyhedra (trigonal bipyramids with equatorial lone pair), linked by sharing O atoms.
Nd–O = 2.365–2.603, Te–O = 1.830–1.989, 2.434–2.694(4) Å.

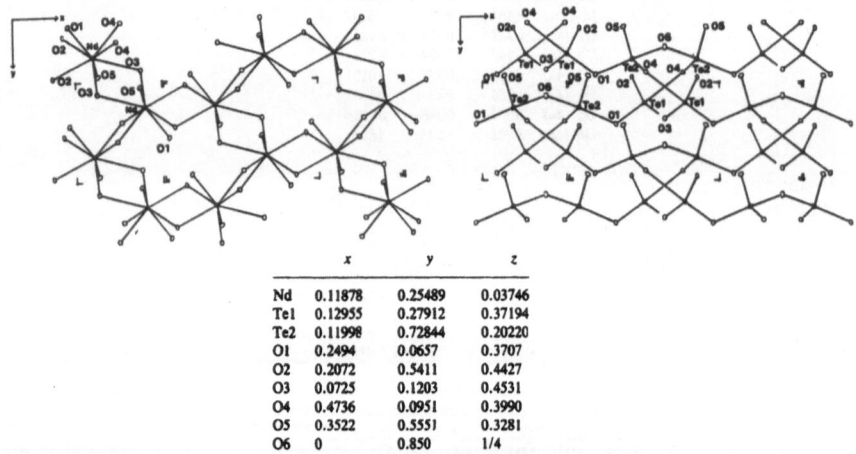

	x	y	z
Nd	0.11878	0.25489	0.03746
Te1	0.12955	0.27912	0.37194
Te2	0.11998	0.72844	0.20220
O1	0.2494	0.0657	0.3707
O2	0.2072	0.5411	0.4427
O3	0.0725	0.1203	0.4531
O4	0.4736	0.0951	0.3990
O5	0.3522	0.5551	0.3281
O6	0	0.850	1/4

LEAD CHLORITE
MAGNESIUM CHLORITE HEXAHYDRATE
SILVER CHLORITE
$Pb(ClO_2)_2$ (I), $Mg(ClO_2)_2 \cdot 6H_2O$ (II), $AgClO_2$ (III)

Acta Cryst., C46, 1755–1759.

I, Ccca, 6.004, 12.504, 6.010, Z = 4, R = 0.030. II, $P4_2mc$, 7.471, 9.980, Z = 2,
R = 0.041. III, Pcca, 6.0754, 6.6796, 6.1226, Z = 4, R = 0.029. All three
structures contain bent ClO_2^- anions. I contains (010) layers with anions linked
by 8-coordinate Pb^{2+} ions. In II [previous study in 5, 77] anions and $Mg(H_2O)_6^{2+}$
octahedra are linked by hydrogen bonds. The structure of III is as previously
described (26, 495).

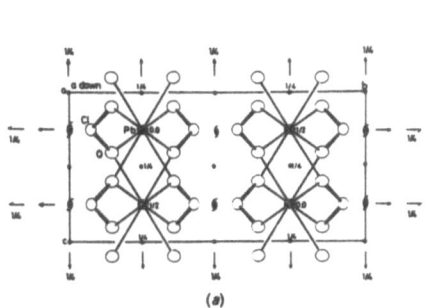

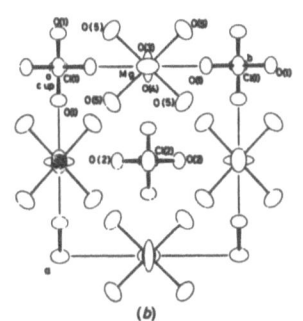

Projection of the crystal structure of (a) (I) along a and (b) (II) along c.

Positional parameters (× 10⁴; for Ag × 10⁵)

	x	y	z	B/B_{eq} (Å² × 10)
Compound (I)				
Pb	0	2500	2500	
Cl	0	−736	2500	
O	−1551	−1437	4036	
Compound (II)				
Cl(1)	0	0	44	
Cl(2)	5000	5000	−33	
Mg	0	5000	2538	
O(1)	1715	0	911	
O(2)	3260	5000	−996	
O(3)	0	5000	413	
O(4)	0	5000	4503	
O(5)	1930	2978	2454	
Compound (III)				
Ag	0	11313	25000	
Cl	0	6324	2500	
O	3509	2274	1056	

Bond distances (Å) and bond angles (°)

Compound (I)

Cl—O	1·577 (9)	O—Cl—Oⁱⁱⁱ	112·5 (5)
Pbⁱⁱ···O	2·628 (8)	Pbᵛⁱ···O	2·640 (9)

Symmetry code: (i) ½ − x, y, −½ + z; (ii) ½ − x, −y, z; (iii) −x, y, ½ − z; (iv) −½ + x, −y, ½ − z; (v) −x, −y, −z; (vi) −½ + x, y, −z; (vii) x, −y, ½ + z.

Compound (II)

Cl(1)—O(1)	1·537 (4)	O(1)—Cl(1)—O(1)ⁱⁱ	111·5 (2)
Cl(2)—O(2)	1·607 (5)	O(2)—Cl(2)—O(2)ⁱ	106·6 (3)
Mg—O(3)	2·121 (12)	O(3)—Mg—O(4)	180·0 (6)
Mg—O(4)	1·961 (11)	O(3)—Mg—O(5)	87·7 (4)
Mg—O(5)	2·071 (7)	O(4)—Mg—O(5)	92·3 (4)
O(1)···O(5)	2·694 (9)		

Symmetry code: (i) y, x, z − ½; (ii) 1 − x, y, z; (iii) −y, x, z − ½; (iv) y, −x, z − ½; (v) x, −y, z; (vi) −x, y, z; (vii) −y, −x, z − ½.

Compound (III)

Cl—Oⁱ	1·575 (2)	O—Cl—Oⁱⁱ	107·0 (1)
Ag···O	2·431 (1)	Ag···Oᵛⁱ	2·479 (2)
Ag···Oⁱⁱⁱ	2·431 (1)	Ag···Oⁱᵛ	2·603 (1)
Ag···Oⁱⁱⁱ	2·479 (2)	Ag···Oᵛ	2·603 (1)

Symmetry code: (i) ½ − x, 1 − y, z; (ii) 1 − x, y, ½ − z; (iii) −x, y, ½ − z; (iv) −½ + x, −y, ½ − z; (v) ½ − x, −y, z; (vi) −½ + x, y, −z; (vii) ½ − x, y, ½ + z; (viii) x, 1 − y, ½ + z; (ix) ½ − x, −y, z; (x) ½ + x, y, −z; (xi) ½ + x, y, 1 − z.

NICKEL CHLORATE HEXAHYDRATE
$Ni(ClO_3)_2 \cdot 6H_2O$

Acta Cryst., C<u>46</u>, 350–354.

Pa3, 10.3159, Z = 4, R = 0.026. Isostructural with zinc bromate hexahydrate (<u>1</u>, 437, 454; <u>4</u>, 63, 194), with $Ni(H_2O)_6^{2+}$ octahedra and trigonal pyramidal ClO_3^- ions, linked by hydrogen bonds. Ni-O = 2.054(1), Cl-O = 1.487(1) A, O-Cl-O = 106.5°, O-H...O = 2.838–3.226 A.

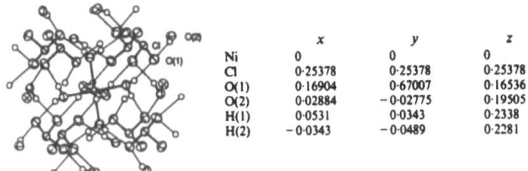

	x	y	z
Ni	0	0	0
Cl	0·25378	0·25378	0·25378
O(1)	0·16904	0·67007	0·16536
O(2)	0·02884	−0·02775	0·19505
H(1)	0·0531	0·0343	0·2338
H(2)	−0·0343	−0·0489	0·2281

STRONTIUM BROMATE MONOHYDRATE
$Sr(BrO_3)_2 \cdot H_2O$

Z. Kristallogr., <u>193</u>, 71–78.

I2/c, 8.870, 7.611, 9.369, 91.82, Z = 4, neutron radiation, R = 0.063. Structure
as previously described (53A, 245). O–H = 0.939(1) A, H–O–H = 112.9°, O...H =
1.875(1) A.

BARIUM BROMATE (FORM I)   STRONTIUM IODATE
$Ba(BrO_3)_2$ $Sr(IO_3)_2$

Z. Naturforsch., 45B, 587-592.

C2/c, a = 13.315, 12.995, b = 7.902, 7.899, c = 8.580, 8.072 A, β = 134.17,
132.62°, Z = 4, R = 0.064 for Ba compound. Isostructural with α-$Ba(IO_3)_2$ (46A,
375), with trigonal pyramidal anions, linked by 10-coordinate Ba or Sr cations.
Br–O = 1.65-1.66(1) A, O–Br–O = 101-105°. Heating produces other forms.

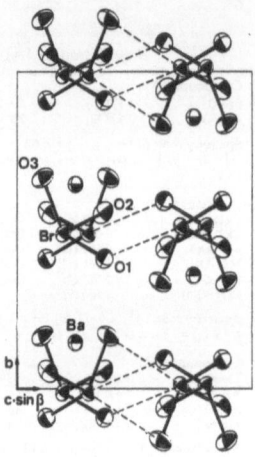

	x	y	z
$Ba(BrO_3)_2$			
Ba	0.0	0.1440	0.25
Br	0.19141	0.4893	0.1985
O1	0.1889	0.409	0.3729
O2	0.3568	0.554	0.366
O3	0.0980	0.665	0.110

COBALT(II) BROMATE HEXAHYDRATE
$Co(BrO_3)_2 \cdot 6H_2O$ $[Co(H_2O)_6](BrO_3)_2$

Acta Cryst., B46, 712-716.

Pa3, 10.3505, Z = 4, R = 0.027. Isostructural with the Ni chlorate (this volume,
p. 295), with a fluorite-type arrangement of $Co(H_2O)_6^{2+}$ octahedra and BrO_3^- trigonal pyramids, linked by hydrogen bonds. Co–O = 2.095(2), Br–O = 1.653(2),
O–H...O = 2.767-3.224(4) A, O–Br–O = 104.1°.

	x	y	z
Co	0	0	0
Br	0·26028	0·26028	0·26028
O(1)	0·15936	0·66844	0·15729
O(2)	0·03344	−0·02848	0·19762
H(1)	0·0492	0·0222	0·2319
H(2)	−0·0392	−0·0523	0·2292

NICKEL PERBROMATE HEXAHYDRATE
$Ni(BrO_4)_2 \cdot 6H_2O$

Acta Cryst., C46, 1580-1584.

P3̄, 7.874, 5.423, Z = 1, R = 0.029. The structure is similar to that at 169K
(55A, 278), with tetrahedral BrO_4^- and octahedral $Ni(H_2O)_6^{2+}$ ions linked by hydrogen
bonds. Br–O = 1.63, Ni–O = 2.06 A (libration corrected).

	x	y	z
Ni	0	0	¼
Br	⅓	⅔	0·17900
O(1)	⅓	⅔	0·88353
O(2)	0·24091	0·09702	0·27594
O(3)	0·41429	0·52839	0·28054
H(1)	0·2897	0·2113	0·2430
H(2)	0·3295	0·0681	0·2934

LITHIUM HYDROGEN IODATE
$Li_{0.67}H_{0.33}IO_3$

Solid State Comm., 75, 539-543.

$P6_3$, 5.554, 4.947, Z = 2, R = 0.032 and neutron powder data. I in 2(b): 1/3,2/3,1/2; O in 6(c): 0.3543,0.2647,0.1738; 1.33Li in 2(a): 0,0,0.440; 0.67H in6(c): 0.241,0.290,-0.032. Isostructural with α-LiIO_3 (2, 49; 50A, 307), but with the H atoms forming hydrogen bonds between oxygen atoms.

STRONTIUM IODATE MONOHYDRATE
$Sr(IO_3)_2 \cdot H_2O$

Acta Cryst., C46, 979-981.

I2/c, 8.9003, 7.748, 9.6496, 90.230, Z = 4, neutron radiation, R = 0.042. Structure as previously described (38A, 355; 53A, 245). I-3 O = 1.788-1.804(1) A, O-I-O = 99.2-99.7°, Sr-11 O = 2.52-3.27 A.

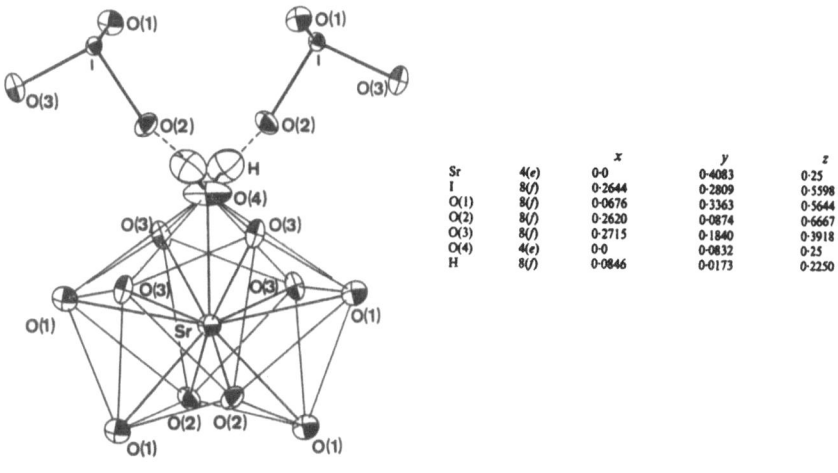

		x	y	z
Sr	4(e)	0·0	0·4083	0·25
I	8(f)	0·2644	0·2809	0·5598
O(1)	8(f)	0·0676	0·3363	0·5644
O(2)	8(f)	0·2620	0·0874	0·6667
O(3)	8(f)	0·2715	0·1840	0·3918
O(4)	4(e)	0·0	0·0832	0·25
H	8(f)	0·0846	0·0173	0·2250

STISHOVITE
SiO_2

Amer. Min., 75, 739-749.

[$P4_2/mnm$], a = 4.1801-4.1023, c = 2.6678-2.6402 A, at 0.0001-16 GPa, Z = 2, R = 0.014-0.043. [Si in 2(a): 0,0,0; O in 4(f): x,x,0], x = 0.3067-0.3053. Rutile-type structure, as previously described (27, 675; 37A, 220; 50A, 308; 54A, 283); a compresses approximately twice as much as c with increasing pressure.

ZEOLITE THETA-1
SiO$_2$

Acta Cryst., C46, 172-173.

Cmc2$_1$, 13.836, 17.415, 5.042, Z = 24, R = 0.116 (synchrotron single-crystal data). Structure approximately as previously derived from powder data (52A, 345), with revision of z-parameters by up to 1.5 A.

Selected bond lengths (Å) and angles (°)

	x	y	z
Si(1)	0·2947	0·0490	0·2500
Si(2)	0·2056	0·2116	0·329
Si(3)	0·5000	0·2257	0·725
Si(4)	0·5000	0·1234	0·225
O(11)	0·2722	0·9815	0·461
O(12)	0·2285	0·1203	0·323
O(14)	0·4059	0·0731	0·278
O(22)	0·2722	0·2580	0·118
O(23)	0·4075	0·2780	0·776
O(34)	0·5000	0·1554	0·917
O(43)	0·5000	0·1966	0·423

Bond	Length	Bond	Length
Si(1)—O(11)	1·61 (2)	Si(3)—O(23)	1·59 (2)
Si(1)—O(11ⁱⁱ)	1·58 (2)	Si(3)—O(23ⁱ)	1·59 (2)
Si(1)—O(12)	1·59 (2)	Si(3)—O(34)	1·56 (2)
Si(1)—O(14)	1·60 (2)	Si(3)—O(43)	1·61 (2)
Si(2)—O(12)	1·62 (2)	Si(4)—O(14)	1·59 (2)
Si(2)—O(22)	1·62 (2)	Si(4)—O(14ⁱⁱ)	1·65 (2)
Si(2)—O(22ⁱⁱ)	1·59 (2)	Si(4)—O(34ⁱⁱ)	1·62 (2)
Si(2)—O(23)	1·60 (2)	Si(4)—O(43)	1·65 (2)
Si(1)—O(11)—Si(1ⁱ)	144·3 (1·3)	Si(2)—O(22ⁱⁱ)—Si(2ⁱⁱ)	151·7 (1·5)
Si(1)—O(12)—Si(2)	152·2 (1·4)	Si(2ⁱⁱ)—O(23)—Si(3)	151·5 (1·4)
Si(1)—O(14)—Si(4)	156·1 (1·3)	Si(3)—O(34)—Si(4ⁱⁱ)	148·1 (1·5)
		Si(3)—O(43)—Si(4)	146·4 (1·4)

Symmetry code: (i) x, −y, ½+z; (ii) ½−x, ½−y, ½+z; (iii) x, y, 1+z; (iv) x, −y, −½+z; (v) 1−x, y, z; (vi) x, y, −1+z.

ZEOLITE H-ZSM-5
SiO$_2$

Acta Cryst., B46, 731-735.

Pnma, 20.078, 19.894, 13.372, at 350K, Z = 96, R = 0.040. Framework essentially as previously described (48A, 369; 49A, 349; 53A, 267; 54A, 304).

Zeolites, 10, 235-242.

P2$_1$/n (a unique), 20.107, 19.879, 13.369, α = 90.67°, Z = 96, R = 0.044. Complicated displacement of the framework relative to the orthorhombic phase (preceding report).

ZEOLITE ZSM-12
SiO$_2$

J. Phys. Chem., 94, 3718-3721.

C2/c, 24.863, 5.012, 24.328, 107.72, Z = 56, synchrotron powder data. Framework of corner-sharing SiO$_4$ tetrahedra with 12-ring channels (see also 52A, 345).

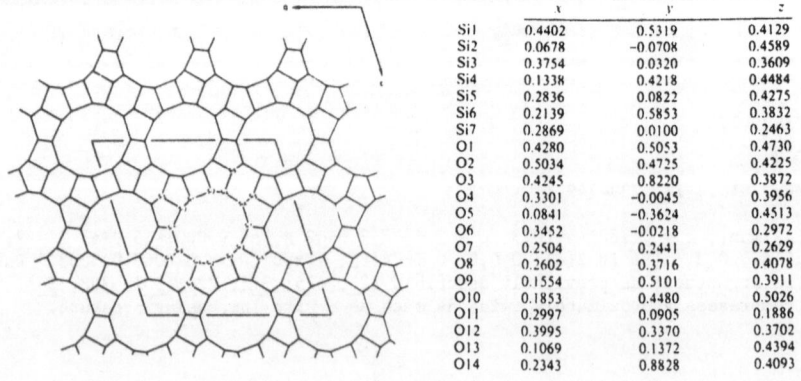

	x	y	z
Si1	0.4402	0.5319	0.4129
Si2	0.0678	-0.0708	0.4589
Si3	0.3754	0.0320	0.3609
Si4	0.1338	0.4218	0.4484
Si5	0.2836	0.0822	0.4275
Si6	0.2139	0.5853	0.3832
Si7	0.2869	0.0100	0.2463
O1	0.4280	0.5053	0.4730
O2	0.5034	0.4725	0.4225
O3	0.4245	0.8220	0.3872
O4	0.3301	-0.0045	0.3956
O5	0.0841	-0.3624	0.4513
O6	0.3452	-0.0218	0.2972
O7	0.2504	0.2441	0.2629
O8	0.2602	0.3716	0.4078
O9	0.1554	0.5101	0.3911
O10	0.1853	0.4480	0.5026
O11	0.2997	0.0905	0.1886
O12	0.3995	0.3370	0.3702
O13	0.1069	0.1372	0.4394
O14	0.2343	0.8828	0.4093

LITHIUM DISILICATE (METASTABLE FORM)
Li₂Si₂O₅

Acta Cryst., C46, 363–365.

Pbcn, 5.683, 4.784, 14.648, Z = 4, R = 0.063. Isostructural with α-Na₂Si₂O₅ (26,
506; 33A, 461), with sheets of corner-sharing SiO₄ tetrahedra, and Li ions in
tetrahedral sites between the sheets. Si–O = 1.583–1.654(4), Li–O = 1.89–2.10(1)
A. A stable form is described in 26, 506.

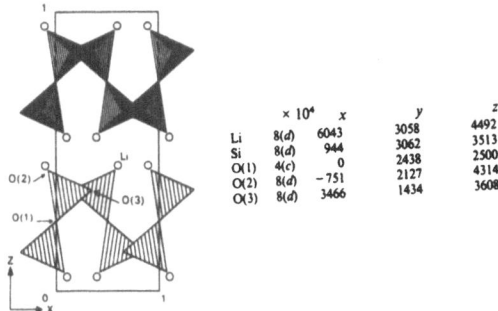

		×10⁴	x	y	z
Li	8(d)		6043	3058	4492
Si	8(d)		944	3062	3513
O(1)	4(c)		0	2438	2500
O(2)	8(d)		−751	2127	4314
O(3)	8(d)		3466	1434	3608

SODIUM HYDROGEN SILICATE MONOHYDRATE
Na₃HSiO₄.H₂O

Acta Cryst., C46, 1365–1368.

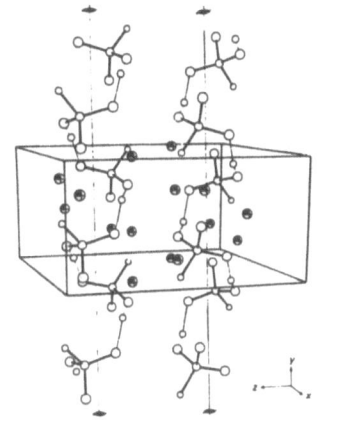

Bond lengths (Å) and angles (°) shown for the hydrogen-
bonded [SiO₃(OH)]³⁻ tetrahedra.

	x	y	z
Si	0·6907	0·2087	0·3647
Na(1)	0·2500	0·0102	0·3342
Na(2)	0·3990	0·3071	0·5667
Na(3)	0·9772	−0·2439	0·4391
O(1)	0·4902	0·2715	0·3826
O(2)	0·7694	0·0138	0·4636
O(3)	0·8098	0·4343	0·3914
O(4)	0·6631	0·1359	0·2198
O(5)	0·0739	0·3471	0·2789
H(11)	0·553	0·887	0·156
H(51)	−0·001	0·398	0·316
H(52)	0·143	0·466	0·285

*Bond lengths (Å) and angles (°) in Na₃[SiO₃-
(OH)].H₂O at 296 K with e.s.d.'s in parentheses*

The hydrogen-bonding system

The [SiO₃(OH)]³⁻ tetrahedron

H(11)—O(1)	0·83 (5)		
H(11)···O(4)	1·78 (5)	O(1)—H(11)···O(4)	172 (4)
O(1)···O(4)	2·597 (3)	Si—O(1)—H(11)	113 (3)

The H₂O molecule

H(51)—O(5)	0·87 (4)		
H(51)···O(3)	1·92 (4)	O(5)—H(51)···O(3)	166 (4)
O(5)···O(3)	2·756 (3)		
H(52)—O(5)	0·88 (4)		
H(52)···O(4)	1·85 (4)	O(5)—H(52)···O(4)	159 (4)
O(5)···O(4)	2·695 (3)	H(51)—O(5)—H(52)	100 (4)

Bond lengths in the Na–O polyhedra

Na(1)—O(1)	2·402 (2)	Na(2)—O(2)	2·302 (2)
Na(1)—O(2)	2·304 (2)	Na(2)—O(3)	2·394 (2)
Na(1)—O(3)	2·470 (2)	Na(2)—O(4)	2·338 (2)
Na(1)—O(4)	2·456 (2)	Na(3)—O(2)	2·317 (2)
Na(1)—O(5)	2·427 (2)	Na(3)—O(2')	2·428 (2)
Na(1)—O(5')	2·710 (2)	Na(3)—O(3)	2·310 (2)
Na(2)—O(1)	2·359 (2)	Na(3)—O(3')	2·446 (2)
Na(2)—O(1')	2·670 (2)	Na(3)—O(5)	2·414 (2)

$P2_1/c$, 7.898, 5.960, 11.142, 105.57, Z = 4, R = 0.019. Chains along **b** of $SiO_3(OH)^{3-}$ tetrahedra linked by strong hydrogen bonds; chains are linked by 5- and 6-coordinate Na^+ ions and by hydrogen bonds via the water molecules.

POTASSIUM SILICATE
K_4SiO_4

Z. anorg. Chem., **589**, 129-138.

$P2_1/c$, 10.370, 6.392, 10.366, 112.83, Z = 4, R = 0.080. Cs_4SnO_4-type (this volume, p. 159).

SODIUM BERYLLIUM SILICATE (ORTHORHOMBIC)
Na_2BeSiO_4

J. Solid State Chem., **86**, 64-74.

$Pca2_1$, 9.861, 4.911, 13.875, Z = 8, R = 0.049 (and R = 0.048 at 623K), twinned crystal. Cristobalite-related framework, as previously described (**56A**, 249), with chains of alternating corner-sharing SiO_4 and BeO_4 tetrahedra. Na ions are in cavities.

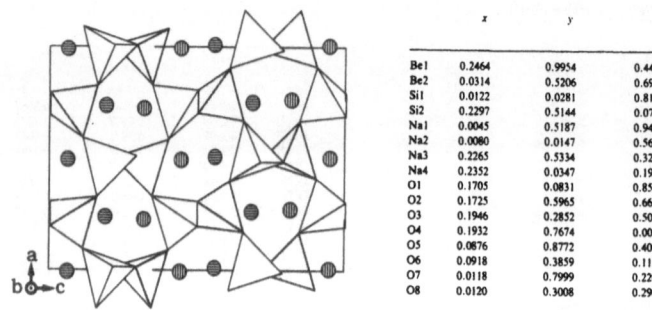

	x	y	z
Be1	0.2464	0.9954	0.4449
Be2	0.0314	0.5206	0.6928
Si1	0.0122	0.0281	0.8184
Si2	0.2297	0.5144	0.0739
Na1	0.0045	0.5187	0.9496
Na2	0.0080	0.0147	0.5664
Na3	0.2265	0.5334	0.3226
Na4	0.2352	0.0347	0.1925
O1	0.1705	0.0831	0.8512
O2	0.1725	0.5965	0.6639
O3	0.1946	0.2852	0.5060
O4	0.1932	0.7674	0.0068
O5	0.0876	0.8772	0.4065
O6	0.0918	0.3859	0.1157
O7	0.0118	0.7999	0.2209
O8	0.0120	0.3008	0.2980

$Na_{1.8}Be_{0.9}Si_{1.1}O_4$

Solid State Ionics, **44**, 51-54.

Pbca, 4.920, 9.876, 13.922, Z = 8, powder data. Structure similar to that of Na_2BeSiO_4 (preceding report), but with Be/Si disorder resulting in a higher symmetry space group. Cristobalite-related framework of corner-sharing MO_4 tetrahedra (M = Be/Si), with 4- and 5-coordinate Na ions in cavities.

SODIUM TETRABERYLLOTETRASILICATE
$Na_{10}Be_4Si_4O_{17}$

Acta Cryst., **B46**, 736-739.

$P\bar{4}3m$, 7.2811, Z = 1, R = 0.033. $Be_4Si_4O_{17}$ units of corner-sharing BeO_4 and SiO_4 tetrahedra, linked by 6-, 8-, and (4+4)-coordinated Na ions.

parameters (× 10⁴)

		x	y	z
Si	4(e)	2201	2201	2201
Be	4(e)	8688	8688	8688
Na(1)	4(e)	6589	6589	6589
Na(2)	3(d)	½	0	0
Na(3)	3(c)	0	½	½
O(1)	12(l)	2612	2612	9
O(2)	4(e)	3464	3464	3464
O(3)	1(a)	0	0	0

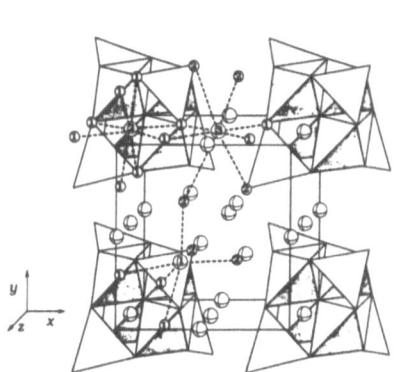

An illustration of the $Na_{10}Be_4Si_4O_{17}$ structure. Shaded tetrahedra contain Be atoms and unshaded tetrahedra Si atoms The oxygen coordination around the Na atoms is illustrated by the dashed lines

Representative bond lengths (Å) and angles (°) with e.s.d.'s in parentheses

SiO₄ tetrahedron		BeO₄ tetrahedron	
Si—O(2)	1·592 (11)	Be—O(3)	1·654 (20)
Si—O(1) 3 ×	1·652 (6)	Be—O(1) 3 ×	1·648 (8)
O(1)—O(1') 2 ×	2·680 (9)	O(1)—O(1ᵃ) 2 ×	2·698 (9)
O(1)—O(2)	2·664 (10)	O(1)—O(3)	2·689 (4)
O(1)—Si—O(1')	108·5 (6)	O(1ᵃ)—Be—O(1ᵐ)	109·9 (8)
O(1)—Si—O(2)	110·4 (4)	O(3)—Be—O(1)	109·1 (7)

Na—O polyhedra			
Na(1)—O(2) 3 ×	2·277 (7)	Na(3)—O(1) 4 ×	2·459 (4)
Na(1)—O(1) 3 ×	2·622 (8)	Na(3)—O(2) 4 ×	2·977 (1)
Na(2)—O(1) 8 ×	2·5768 (2)		

Na—Na distances			
Na(1)—Na(3) 3 ×	2·9741 (2)	Na(2)—Na(3) 6 ×	3·6405 (2)
Na(1)—Na(1') 3 ×	3·273 (8)	Na(2)—Na(1) 4 ×	3·698 (4)
Na(3)—Na(1) 4 ×	2·9741 (2)		
Na(3)—Na(2) 4 ×	3·6405 (2)		

Symmetry code: (i) y, z, x; (ii) −z, x, −y; (iii) 1−x, 1−y, 1+z; (iv) 1−y, 1+z, 1−x; (v) −y, −x, z.

SODIUM CALCIUM HEXASILICATE (HIGH–TEMPERATURE)
$Na_4Ca_4Si_6O_{18}$

Acta Cryst., B46, 125–131.

Above 748K, R3̄m, a = 7.519 A, α = 89.22°, Z = 1 (hexagonal cell, a = 10.561, c = 13.199 A, Z = 3), at 773K, R = 0.047. Isostructural with $Na_6Ca_3Si_6O_{18}$ (52A, 322), with one partly-occupied Na site. The room-temperature form (P3₁21) has been described previously (50A, 319; 53A, 248).

CALCIUM STRONTIUM SILICATE (HIGH–TEMPERATURE)
$(Ca,Sr)_2SiO_4$

Kristallografija, 35, 84–90, 91–93 [Soviet Physics – Crystallography, 35, 50–54, 54–56].

Pna2₁, 20.863, 9.5000, 5.6005, Z = 12, R = 0.038. The a cell parameter is tripled with respect to the α' phase (51A, 328). The first paper describes transitions between the various phases (43A, 299; 46A, 376; 50A, 309, 310; 51A, 328).

HYDROGEN ALUMINUM SILICATE
$HAlSi_2O_6$

Z. Kristallogr., 190, 7–18.

$P4_32_12$, a = 7.586, 7.576, c = 8.402, 8.386 A, Z = 4, R = 0.054 and neutron powder data. Keatite-type framework (23, 338) as for the Li compound (33A, 456); bands of 5-rings are stacked more closely as a result of hydrogen bonding.

		x	y	z
(Si,Al)(1)	X-ray	0.3194	0.1168	0.2656
	neutron	0.3248	0.1105	0.2695
(Si,Al)(2)		0.4041	0.4041	0
		0.4058	0.4058	0
O(1)		0.4252	0.1332	0.4375
		0.4276	0.1269	0.4333
O(2)		0.1072	0.1220	0.3170
		0.1082	0.1198	0.3190
O(3)		0.3675	0.2753	0.1485
		0.3626	0.2785	0.1474
H,D		–	–	–
		0.889	0.857	−0.059

Neues Jb. Miner. Mh., 232–240.

$P6_222$, 5.083, 5.509, Z = 1, R = 0.049, at 298K, and neutron powder data at 5K. Si/Al in 3(c): 1/2,0,0; O in 6(j): x,2x,1/2, x = 0.2080, 0.2065 (X-rays, neutrons), 1 H in 12(k): 0.1275,0.0980,0.5489. β-Quartz-type structure (1, 166), with disordered Si/Al, and H in channels.

LITHIUM ALUMINOSILICATE
$LiAlSi_2O_6$

SODIUM HYDROGEN ALUMINOSILICATE
$Na_{0.5}H_{0.5}AlSi_2O_6$

Neues. Jb. Miner. Mh., 493–503.

$P6_222$, a = 5.226, 5.150, c = 5.465, 5.458 A, Z = 1, R = 0.020, 0.087. 3 Si/Al in 3(c): 1/2,0,0; 6 O in 6(j): x,2x,1/2, x = 0.2030, 0.2079; 1 Li in 3(a): 0,0,0 or 0.5 Na in 3(b): 0,0,1/2. β-Quartz-type structure as previously described for the Li compound (37A, 324), with Li in tetrahedral sites, but the larger Na in distorted octahedral sites (H atom not located).

POTASSIUM ALUMINUM SILICATE
$K_2Al_2Si_3O_{10} \cdot KCl$

Amer. Min., 75, 947–950.

$P\bar{4}2_1m$, 9.755, 6.488, Z = 2, R = 0.056. Zeolite-type aluminosilicate framework, similar to that in edingtonite (3, 173, 529; 42A, 404), with K^+ and Cl^- ions in two types of channels.

		Site occ.	x	y	z
T(1)	2(a)	1	0	0	0
T(2)	8(f)	1	0.1524	0.1157	0.6213
Cl	2(c)	0.98	1/2	0	0.0478
K(1)	2(c)	0.98	1/2	0	0.5450
K(2)	4(e)	1.00	0.2598	0.2402	0.1226
O(1)	4(e)	1	0.2985	0.2015	0.6234
O(2)	8(f)	1	0.0445	0.1900	0.4524
O(3)	8(f)	1	0.0818	0.1123	0.8589

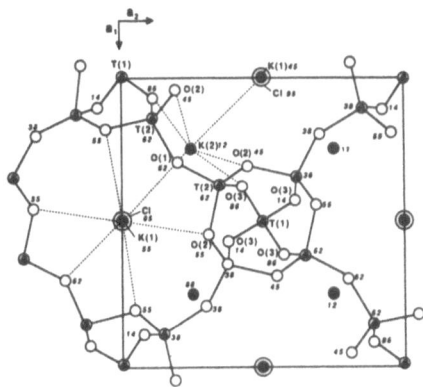

LITHIUM GALLIUM SILICATE
$Li_{3.4}Ga_{0.2}SiO_4$

J. Solid State Chem., **88**, 564-570.

$P2_1/m$, 5.1478, 6.1316, 5.2505, 89.795, Z = 2, powder data. Disordered Li_4SiO_4-type (**33A**, 467; **45A**, 360; **49A**, 321), with Ga in the Li(1) site and partial occupancy for all Li sites.

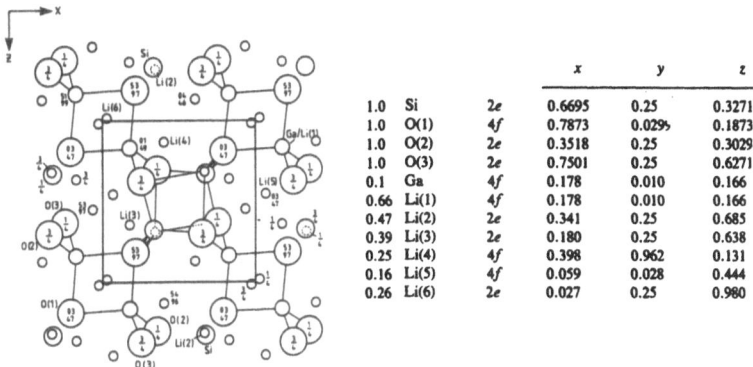

			x	y	z
1.0	Si	2e	0.6695	0.25	0.3271
1.0	O(1)	4f	0.7873	0.0295	0.1873
1.0	O(2)	2e	0.3518	0.25	0.3029
1.0	O(3)	2e	0.7501	0.25	0.6271
0.1	Ga	4f	0.178	0.010	0.166
0.66	Li(1)	4f	0.178	0.010	0.166
0.47	Li(2)	2e	0.341	0.25	0.685
0.39	Li(3)	2e	0.180	0.25	0.638
0.25	Li(4)	4f	0.398	0.962	0.131
0.16	Li(5)	4f	0.059	0.028	0.444
0.26	Li(6)	2e	0.027	0.25	0.980

CALCIUM GERMANATE SILICATE
$Ca_5Ge_{2.23}Si_{0.77}O_{11}$ (I)

$Ca_5[(Ge,Si)_2O_7][(Ge,Si)O_4]$

SODIUM CALCIUM SILICATE
$Na_2Ca_6Si_4O_{15}$ (II)

$Na_2Ca_6[Si_2O_7][SiO_4]_2$

Amer. Min., **75**, 963-969.

I, C2/m, 10.912, 8.695, 11.000, 96.87, Z = 4, R = 0.033. XO_4^{4-} and $X_2O_7^{6-}$ ions (X = Ge/Si), linked by Ca^{2+} ions which have octahedral and trigonal prismatic coordinations. Ge/Si-O = 1.69-1.81 A.

II, $P2_1/c$, 5.525, 17.413, 14.489, 90.57, Z = 4, R = 0.025. SiO_4^{4-} and $Si_2O_7^{6-}$ ions, linked by 6- and 7-coordinate cations with some Na/Ca disorder. Si-O = 1.60-1.67 A.

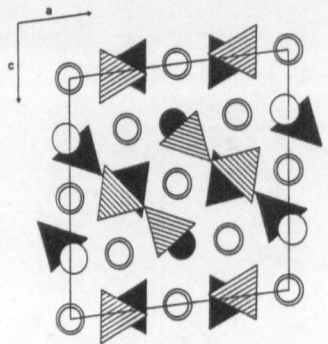

Structure of $Ca_3(Ge,Si)_3O_{11}$ projected along the b axis. Circles represent Ca positions: open circles at y = 0, 1; double circles at y = 0.18–0.32; solid circles at y = ½. Hatched tetrahedra are at y = 0, 1; solid tetrahedra at y = ½.

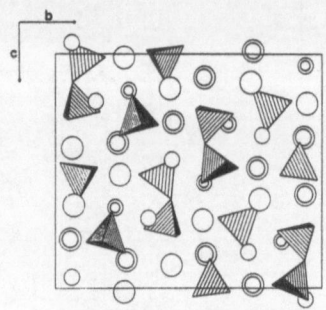

Structure of $Na_2Ca_4Si_6O_{17}$ projected along the a axis. Large circles represent Ca sites, small circles Na positions. Circles with double outlines are at x = 0.2–0.3, those with single outlines at x = 0.7–0.8. Tetrahedra are at corresponding heights.

		x	y	z
Ge1 0.73	Ge	0.5606	0	0.2709
0.27	Si			
Ge2 0.81	Ge	−0.1894	0	0.4631
0.19	Si			
Ge3 0.89	Ge	−0.2053	0	0.0177
0.31	Si			
Ca1		−0.0162	0	0.2497
Ca2		0.2701	0.1876	0.2436
Ca3		0	0.2890	0
Ca4		½	0.1768	½
O1		0.4187	0	0.3234
O2		−0.1464	0.1540	0.1035
O3		−0.1299	0.1546	0.3925
O4		−0.1430	0	−0.1205
O5		0.6490	0	0.4208
O6		0.1648	0	0.3813
O7		−0.1406	½	0.0267
O8		0.5889	−0.1719	0.2070

		x	y	z
Na1 0.63	Ca	0.7174	0.43889	0.4516
0.37	Na			
Na2 0.32	Ca	0.2481	0.2755	0.1549
0.68	Na			
Na3		0.2274	0.6503	0.2985
Ca1		0.2446	0.74221	0.98536
Ca2		0.1948	0.56255	0.09917
Ca3		0.7049	0.73337	0.12063
Ca4		0.2888	0.56823	0.85372
Ca5		0.2538	0.44438	0.29060
Si1		0.7302	0.59724	0.96371
Si2		0.2044	0.80792	0.21690
Si3		0.7257	0.57795	0.2677
Si4		0.7812	0.61555	0.4710
O1		0.6865	0.5639	0.3805
O2		0.4700	0.8346	0.0073
O3		0.4800	0.5484	0.2202
O4		−0.0622	0.6581	0.0163
O5		0.2598	0.4171	0.1272
O6		0.0556	0.8563	0.9557
O7		0.2902	0.7382	0.1509
O8		−0.0437	0.5287	0.2342
O9		0.3571	0.6132	0.6975
O10		0.2245	0.4832	0.9643
O11a 0.52	O	0.217	0.7118	0.8282
O11b 0.48	O	0.282	0.7111	0.8259
O12		0.6014	0.8128	0.9865
O13		0.9175	0.8248	0.2009
O14		0.7703	0.6680	0.2539
O15a 0.38	O	0.199	0.4506	0.4493
O15b 0.62	O	0.268	0.4492	0.4495

BARIUM TITANIUM NIOBIUM SILICATE
$Ba_3Ti_{1.2}Nb_{4.8}Si_4O_{25.4}$

Kristallografija, 35, 346-348 [Soviet Physics - Crystallography, 35, 195-196].

$P\bar{6}2m$, 9.03, 7.868, Z = 1, R = 0.040. Structure similar to that of related materials (35A, 267, 269), with triple columns along c of $(Nb,Ti)O_6$ octahedra, linked by Si_2O_7 groups (linear Si-O-Si); Ba ions are in channels.

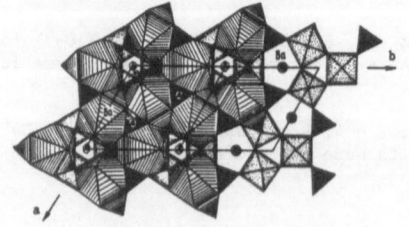

	Multiplicity of position	Degree of filling of position	x	y	z
Ba	3	1	0.6047	0	½
Nb	6	1	0.2391	0	0.2383
Si	4	1	⅓	⅓	0.196
O1	2	1	⅓	⅓	0
O2	3	1	0.288	0	0
O2	6	1	0.822	0	0.228
O4	12	1	0.487	0.180	0.293
O3	3	0.8	0.242	0	½

COBALT NICKEL ZINC METASILICATE
(Co,Ni,Zn)SiO$_3$

Acta Cryst., B46, 493-497.

Pbca, 18.209, 8.915, 5.2182, Z = 16, Mo and synchrotron radiations, R = 0.036.
Orthopyroxene structure (2, 134).

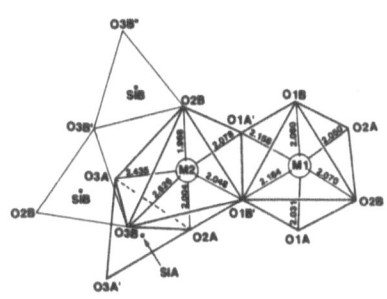

	x	y	z
M1	0·37571	0·65381	0·87979
M2	0·37654	0·49476	0·37242
SiA	0·27290	0·33981	0·06186
SiB	0·47272	0·33552	0·79094
O1A	0·18385	0·33744	0·05107
O1B	0·56211	0·33801	0·79066
O2A	0·31028	0·50181	0·06606
O2B	0·43253	0·48448	0·69441
O3A	0·30353	0·23281	−0·16765
O3B	0·44764	0·20091	0·58968

Orthopyroxene (Co,Ni,Zn)SiO$_3$: occupancy parameters

The values are expressed in three figures so that total occupancy of each site becomes unity, which makes it easy to calculate the average ionic radius.

	Co	Ni	Zn
M1	0·273	0·530	0·197
M2	0·394	0·137	0·469

BARIUM COPPER DISILICATE
BaCu$_2$Si$_2$O$_7$

Acta Cryst., C46, 1383-1385

Pnma, 6.866, 13.190, 6.909, Z = 4, R = 0.031. Isolated Si$_2$O$_7^{6-}$ anions, linked by
Cu^{2+} and Ba^{2+} ions which have irregular (4+1)- and 7-coordinations, respectively.

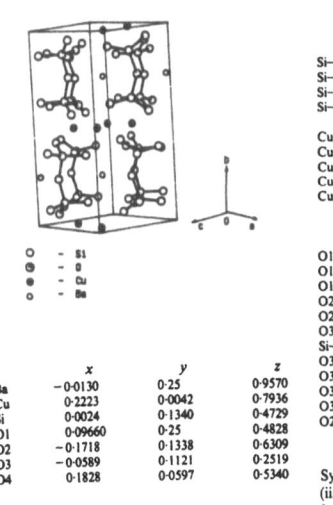

	x	y	z
Ba	−0·0130	0·25	0·9570
Cu	0·2223	0·0042	0·7936
Si	0·0024	0·1340	0·4729
O1	0·09660	0·25	0·4828
O2	−0·1718	0·1338	0·6309
O3	−0·0589	0·1121	0·2519
O4	0·1828	0·0597	0·5340

Selected interatomic distances (Å) and bond angles (°) with e.s.d.'s in parentheses

Si—O1	1·662 (2)
Si—O2	1·619 (4)
Si—O3	1·610 (4)
Si—O4	1·635 (3)
Cu—O4	1·956 (3)
Cu—O4'	1·973 (3)
Cu—O3'	2·789 (4)
Cu—O2^{ii}	1·930 (3)
Cu—O3^{iii}	1·926 (3)

O1—O2	2·605 (5)
O1—O3	2·645 (5)
O1—O4	2·603 (3)
O2—O3	2·746 (5)
O2—O4	2·708 (5)
O3—O4	2·652 (5)
Ba—O1^{vii}	2·713 (5)
Ba—O2	2·932 (3)
Ba—O2^{iv}	2·932 (3)
Ba—O3^{v}	2·749 (3)
Ba—O3^{vi}	2·749 (3)
Ba—O2^{ii}	2·863 (3)
Ba—O2^{viii}	2·863 (3)

O1—Si—O2	105·2 (2)
O1—Si—O3	107·8 (2)
O1—Si—O4	104·0 (2)
O2—Si—O3	116·4 (2)
O2—Si—O4	112·8 (2)
O3—Si—O4	109·6 (2)
Si—O1—Si	133·4 (1)
O3'—Cu—O4	103·1 (2)
O3'—Cu—O4'	65·0 (2)
O3'—Cu—O2^{ii}	101·8 (2)
O3'—Cu—O3^{iii}	91·5 (2)
O2^{ii}—Cu—O4	91·7 (2)

O2^{ii}—Cu—O4'	88·2 (2)
O3^{iii}—Cu—O4	93·9 (2)
O3^{iii}—Cu—O4'	89·3 (2)
O1^{vii}—Ba—O2^{iv}	147·3 (2)
O1^{vii}—Ba—O2^{ii}	147·3 (2)
O1^{vii}—Ba—O2	75·6 (2)
O1^{vii}—Ba—O2^{iv}	75·6 (2)
O1^{vii}—Ba—O3^{v}	76·8 (2)
O1^{vii}—Ba—O3^{vi}	76·8 (2)
O2—Ba—O3^{v}	100·5 (2)
O2—Ba—O2^{ii}	62·8 (2)
O2—Ba—O3^{vi}	150·6 (2)

Symmetry code: (i) 0·5 − x, −y, 0·5 + z; (ii) 0·5 + x, y, 1 + 0·5 − z; (iii) −x, −y, 1 − z; (iv) x, 0·5 − y, z; (v) x, 0·5 − y, 1 + z; (vi) x, 0·5 − y, 1 + z; (vii) 0·5 + x, 0·5 − y, 1·5 − z; (viii) −0·5 + x, y, 1·5 − z.

BARIUM CALCIUM COPPER HEXASILICATE
Ba$_3$CaCuSi$_6$O$_{17}$

Acta Cryst., C46, 2028-2030.

B2mb, 14.405, 16.077, 7.088, Z = 4, R = 0.052. Rings of six SiO₄ tetrahedra are linked into chains, which are connected by CuO₄ square planes and 8-coordinate Ba and Ca.

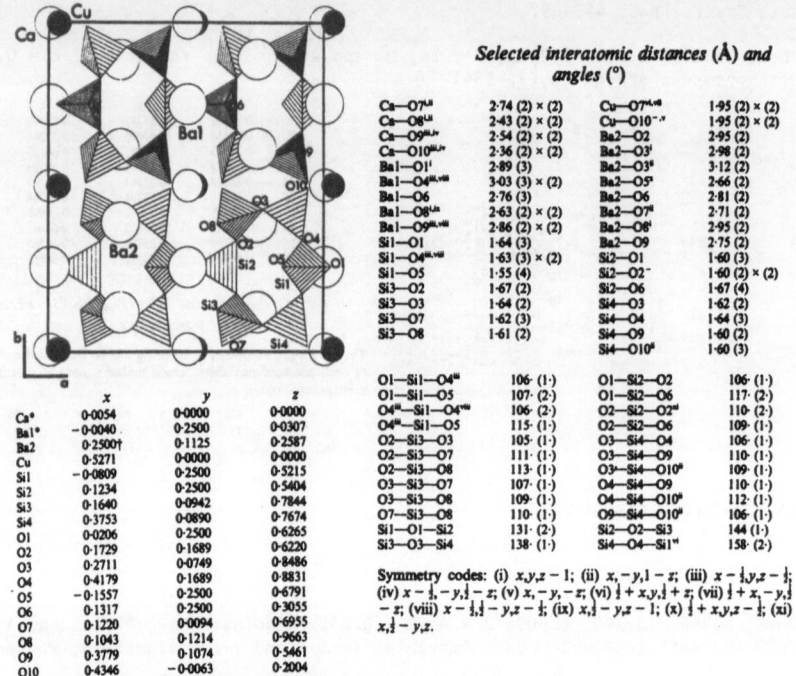

Selected interatomic distances (Å) and angles (°)

Ca—O7^{i,ii}	2·74 (2) × (2)	Cu—O7^{v,vi}	1·95 (2) × (2)
Ca—O6^{i,ii}	2·43 (2) × (2)	Cu—O10^{v,v}	1·95 (2) × (2)
Ca—O9^{iii,iv}	2·54 (2) × (2)	Ba2—O2	2·95 (2)
Ca—O10^{iii,iv}	2·36 (2) × (2)	Ba2—O3^i	2·98 (2)
Ba1—O1^i	2·89 (3)	Ba2—O3^{ii}	3·12 (2)
Ba1—O4^{iii,viii}	3·03 (3) × (2)	Ba2—O5^x	2·66 (2)
Ba1—O6	2·76 (3)	Ba2—O6	2·81 (2)
Ba1—O8^{i,ix}	2·63 (2) × (2)	Ba2—O7^{xi}	2·71 (2)
Ba1—O9^{iii,xii}	2·86 (2) × (2)	Ba2—O8^i	2·95 (2)
Si1—O1	1·64 (3)	Ba2—O9	2·75 (2)
Si1—O4^{iii,viii}	1·63 (3) × (2)	Si2—O1	1·60 (3)
Si1—O5	1·55 (4)	Si2—O2$^-$	1·60 (2) × (2)
Si3—O2	1·67 (2)	Si2—O6	1·67 (4)
Si3—O3	1·64 (2)	Si4—O3	1·62 (2)
Si3—O7	1·62 (3)	Si4—O4	1·64 (3)
Si3—O8	1·61 (2)	Si4—O9	1·60 (2)
		Si4—O10^{ii}	1·60 (3)

O1—Si1—O4^{iii}	106· (1·)	O1—Si2—O2	106· (1·)
O1—Si1—O5	107· (2·)	O1—Si2—O6	117· (2·)
O4^{iii}—Si1—O4^{viii}	106· (2·)	O2—Si2—O2^{iii}	110· (2·)
O4^{iii}—Si1—O5	115· (1·)	O2—Si2—O6	109· (1·)
O2—Si3—O3	105· (1·)	O3—Si4—O4	106· (1·)
O2—Si3—O7	111· (1·)	O3—Si4—O9	110· (1·)
O2—Si3—O8	113· (1·)	O3^v—Si4—O10^{ii}	109· (1·)
O3—Si3—O7	107· (1·)	O4—Si4—O9	110· (1·)
O3—Si3—O8	109· (1·)	O4—Si4—O10^{ii}	112· (1·)
O7—Si3—O8	110· (1·)	O9—Si4—O10^{ii}	106· (1·)
Si1—O1—Si2	131· (2·)	Si2—O2—Si3	144 (1·)
Si3—O3—Si4	138· (1·)	Si4—O4—Si1^v	158· (2·)

	x	y	z
Ca*	0·0054	0·0000	0·0000
Ba1*	−0·0040	0·2500	0·0307
Ba2	0·2500†	0·1125	0·2587
Cu	0·5271	0·0000	0·0000
Si1	−0·0809	0·2500	0·5215
Si2	0·1234	0·2500	0·5404
Si3	0·1640	0·0942	0·7844
Si4	0·3753	0·0890	0·7674
O1	0·0206	0·2500	0·6265
O2	0·1729	0·1689	0·6220
O3	0·2711	0·0749	0·8486
O4	0·4179	0·1689	0·8831
O5	−0·1557	0·2500	0·6791
O6	0·1317	0·2500	0·3055
O7	0·1220	0·0094	0·6955
O8	0·1043	0·1214	0·9663
O9	0·3779	0·1074	0·5461
O10	0·4346	−0·0063	0·2004

Symmetry codes: (i) $x,y,z-1$; (ii) $x,-y,1-z$; (iii) $x-\frac{1}{2},y,z-\frac{1}{2}$; (iv) $x-\frac{1}{2},-y,\frac{1}{2}-z$; (v) $x,-y,-z$; (vi) $\frac{1}{2}+x,y,\frac{1}{2}+z$; (vii) $\frac{1}{2}+x,-y,\frac{1}{2}-z$; (viii) $x-\frac{1}{2},\frac{1}{2}-y,z-\frac{1}{2}$; (ix) $x,\frac{1}{2}-y,z-1$; (x) $\frac{1}{2}+x,y,z-\frac{1}{2}$; (xi) $x,\frac{1}{2}-y,z$.

* Refined occupancies of these two sites are: Ca = 0·966 Ca + 0·034 Ba, Ba1 = 0·966 Ba + 0·034 Ca.
† The x coordinate of Ba2 was fixed to define the origin.

POTASSIUM ZINC TETRASILICATE
K₂ZnSi₄O₁₀

Acta Cryst., C46, 1373–1376.

P2₁2₁2₁, 10.0676, 14.047, 7.0673, Z = 4, R = 0.048. Framework of SiO₄ and ZnO₄ tetrahedra, with rings of ten SiO₄ tetrahedra; K ions have 8- and 9-coordinations. Si-O = 1.58-1.65, Zn-O = 1.93-1.98, K-O = 2.65-3.48(1) A.

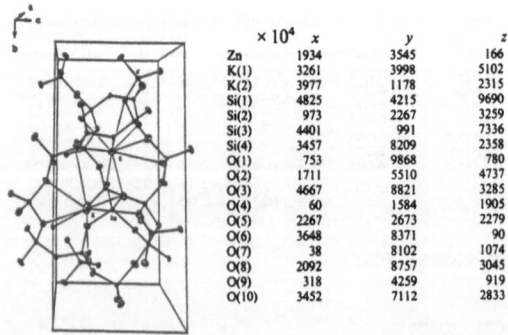

	×10⁴ x	y	z
Zn	1934	3545	166
K(1)	3261	3998	5102
K(2)	3977	1178	2315
Si(1)	4825	4215	9690
Si(2)	973	2267	3259
Si(3)	4401	991	7336
Si(4)	3457	8209	2358
O(1)	753	9868	780
O(2)	1711	5510	4737
O(3)	4667	8821	3285
O(4)	60	1584	1905
O(5)	2267	2673	2279
O(6)	3648	8371	90
O(7)	38	8102	1074
O(8)	2092	8757	3045
O(9)	318	4259	919
O(10)	3452	7112	2833

YTTRIUM PYROSILICATE
δ-$Y_2Si_2O_7$

Z. Kristallogr., 191, 117-123.

Pnam, 13.655, 5.016, 8.139, Z = 4, R = 0.077. Essentially isostructural with the Gd compound, which has been described in Pna2$_1$ (32A, 436; 35A, 301), with O_3Si-O-SiO_3^{6-} ions linked by 7-coordinate Y^{3+} ions. Si-O = 1.61-1.65 (terminal), 1.66, 1.70 (bridging), Y-O = 2.22-2.50 A.

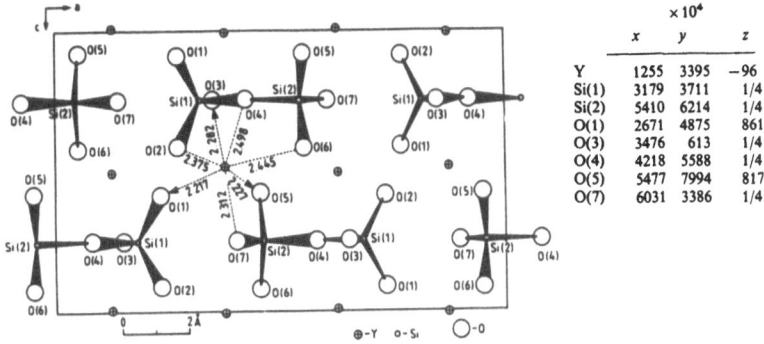

	×10⁴		
	x	y	z
Y	1255	3395	-96
Si(1)	3179	3711	1/4
Si(2)	5410	6214	1/4
O(1)	2671	4875	861
O(3)	3476	613	1/4
O(4)	4218	5588	1/4
O(5)	5477	7994	817
O(7)	6031	3386	1/4

NEODYMIUM SELENOSILICATE
Nd_2SeSiO_4

Z. Naturforsch., 45B, 465-468.

Pbcm, 6.182, 7.174, 11.024, Z = 4, R = 0.033. Alternating NdSe and NdSiO₄ sheets.

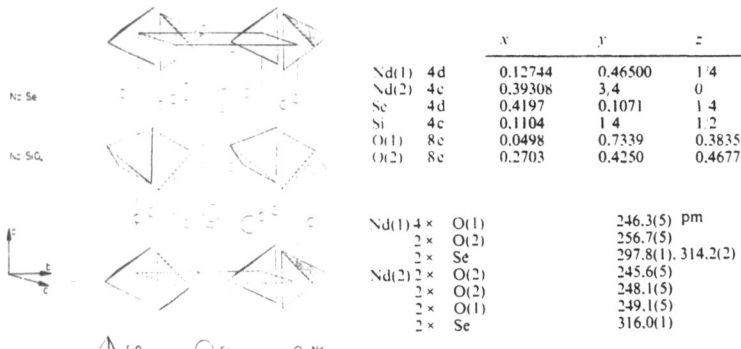

		x	y	z
Nd(1)	4d	0.12744	0.46500	1/4
Nd(2)	4c	0.39308	3/4	0
Se	4d	0.4197	0.1071	1/4
Si	4c	0.1104	1/4	1/2
O(1)	8e	0.0498	0.7339	0.3835
O(2)	8e	0.2703	0.4250	0.4677

Nd(1)	4 ×	O(1)	246.3(5) pm
	2 ×	O(2)	256.7(5)
	2 ×	Se	297.8(1), 314.2(2)
Nd(2)	2 ×	O(2)	245.6(5)
	2 ×	O(2)	248.1(5)
	2 ×	O(1)	249.1(5)
	2 ×	Se	316.0(1)

LANTHANUM NEODYMIUM SILICATE OXIDE
$La_3Nd_{11}(SiO_4)_9O_3$

Kristallografija, 35, 328-331 [Soviet Physics - Crystallography, 35, 184-186].

P6$_3$/m, 9.638, 21.350, Z = 2, R = 0.041. Apatite-type structure (e.g. 44A, 302), with tripled c-axis resulting from ordering of La and Nd.

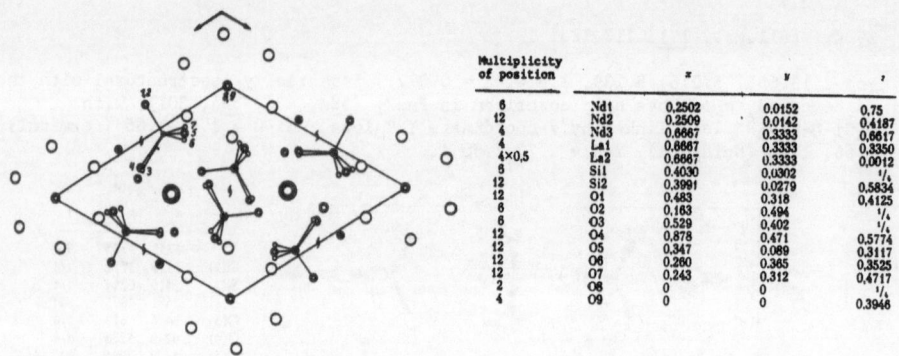

Multiplicity of position		x	y	z
6	Nd1	0,2502	0,0152	0,75
12	Nd2	0,2509	0,0142	0,4187
4	Nd3	0,6667	0,3333	0,6617
4	La1	0,6667	0,3333	0,3350
4×0,5	La2	0,6667	0,3333	0,0012
6	Si1	0,4030	0,0302	$\frac{1}{4}$
12	Si2	0,3991	0,0279	0,5834
12	O1	0,483	0,318	0,4125
6	O2	0,163	0,494	$\frac{1}{4}$
6	O3	0,529	0,402	$\frac{1}{4}$
12	O4	0,878	0,471	0,5774
12	O5	0,347	0,089	0,3117
12	O6	0,260	0,365	0,3525
12	O7	0,243	0,312	0,4717
2	O8	0	0	$\frac{1}{4}$
4	O9	0	0	0,3946

ALBITE (LOW)
$NaAlSi_3O_8$

Amer. Min., 75, 135-140, 141-149.

$C\bar{1}$, 8.137, 12.785, 7.1583, 94.26, 116.60, 87.71, Z = 4, R = 0.020. Structure as previously described (3, 164; 53A, 249).

	x	y	z
T₁O(Al)	0.00887	0.16846	0.20805
T₁m(Si)	0.00375	0.82051	0.23737
T₂O(Si)	0.69162	0.11021	0.31466
T₂m(Si)	0.68129	0.88190	0.36076
Na	0.26799	0.98865	0.1465
O_A1	0.0049	0.13103	0.9666
O_A2	0.59176	0.99756	0.2804
O_BO	0.8123	0.10966	0.1901
O_Bm	0.8200	0.85101	0.2587
O_CO	0.01288	0.30238	0.2706
O_Cm	0.02329	0.69368	0.2291
O_DO	0.20780	0.10896	0.3890
O_Dm	0.1840	0.86817	0.4362

BIOTITE
$K(Mg,Fe)_3AlSi_3O_{10}(OH)_2$

Amer. Min., 75, 305-313.

C2/m, a = 5.337-5.355, b = 9.242-9.258, c = 10.211-10.246 A, β = 100.02-100.26°, for 5 plutonic 1M biotites, Z = 2, R = 0.021-0.062. Structure as previously described (27, 696; 41A, 378).

	x	y	z
O1	0.0268	0.0	0.1690
O2	0.3192	0.2357	0.1682
O3	0.1312	0.1678	0.3912
O4	0.1286	0.5	0.3957
M2	0.0	0.3348	0.5
M1	0.0	0.0	0.5
K	0.0	0.5	0.0
T	0.0749	0.1668	0.2258

BOGGSITE
$Ca_{7.8}Na_{2.9}Al_{18.3}Si_{77.5}O_{192}\cdot70H_2O$

Amer. Min., 75, 501-507.

Imma, 20.236, 23.798, 12.798, Z = 1, R = 0.092.

The four-connected 3-D net linking the tetrahedral vertices has 4-, 5-, 6-, 10- and 12-rings. Three-connected 2-D nets of the gmelinite and ferrierite types occur respectively in the bc and ac planes. The topology of the 3-D net is obtained by replacing ⅔ of the edges of the gmelinite net by pentasil chains found in the silicalite/ZSM-5/ZSM-11 family of synthetic microporous materials. Each 12-ring channel along a has offset 10-ring windows into left and right channels along b. Correspondingly, each 10-ring channel is connected to left and right 12-ring channels to yield 3-D access. Each 12-ring is nearly circular with a free diameter between framework oxygens (assumed radius 1.35 Å) of 7.4 by 7.2 Å. Two bifurcated 4-rings reduce the free diameter of each 10-ring (5.2 by 5.1 Å) to approximately that for a near-circular 9-ring. A unique assignment of Ca, Na, and H_2O to the broad irregular peaks in the channels was not achieved. A highly disordered ionic solution that lacks systematic bonding to the silica-rich framework is indicated. The good correlation between mean T-O distance and T-O-T angle indicates Si,Al disorder over all sites.

	Population	x	y	z
Si(1)	16 × 1.0	0.18881	0.18550	0.6719
Si(2)	16 × 1.0	0.19006	0.02407	0.3297
Si(3)	16 × 1.0	0.07689	0.18517	0.8357
Si(4)	16 × 1.0	0.07768	0.02210	0.1643
Si(5)	16 × 1.0	0.22108	0.08300	0.5378
Si(6)	16 × 1.0	0.12270	0.08371	0.9656
O(1)	8 × 1.0	0.1882	¼	0.6291
O(2)	16 × 1.0	0.1194	0.1707	0.7322
O(3)	16 × 1.0	0.1950	0.1456	0.5679
O(4)	16 × 1.0	0.1900	0.0702	0.4236
O(5)	16 × 1.0	0.1194	0.0319	0.2722
O(6)	16 × 1.0	0.0893	¼	0.8731
O(7)	8 × 1.0	0	0.1738	0.8043
O(8)	8 × 1.0	0	0.0263	0.1968
O(9)	16 × 1.0	0.1944	0.0376	0.6204
O(10)	16 × 1.0	0.0978	0.1464	0.9327
O(11)	16 × 1.0	0.0959	0.0725	0.0813
O(12)	16 × 1.0	0.2007	0.0800	0.9682
O(13)	16 × 1.0	0.0949	0.0395	0.8847
O(14)	8 × 1.0	¼	0.1755	¾
O(15)	8 × 1.0	¼	−0.0375	¾
W(1)	8 × 1.85	0	0.1735	0.136
W(2)	16 × 1.35	0.1936	0.1701	0.203
W(3)	8 × 0.92	0.193	¼	0.057
W(4)	16 × 1.16	0.1074	0.1754	0.372
W(5)	8 × 3.2	0	0.175	0.570
W(6)	8 × 0.91	0	0.0284	0.429
W(7)	16 × 0.63	0.0562	0.1147	0.410
W(8)	16 × 0.77	0.0431	0.0812	0.626
W(9)	8 × 0.50	0.149	¼	0.237
W(10)	8 × 0.61	0.106	¼	0.086
W(11)	8 × 1.32	0.206	¼	0.401

BRITHOLITE

$(Na,Ca,La,Ce)_{10}[(Si,P)O_4]_6(OH,F)_2$

LESSINGITE

$Ca_4(La,Ce)_6(SiO_4)_6(OH,F)_2$

MINERAL X

$Ca_2Na(La,Ce)_7(SiO_4)_6(OH,F)_2$

Z. Kristallogr., 191, 249–263.

P6₃, a = 9.629, 9.664, 9.699, c = 7.059, 7.090, 7.125 A, Z = 1, R = 0.033, 0.039, 0.036. Apatite structures (2, 99), with oxygen atoms displaced slightly from P6₃/m sites and two Ln(1) sites.

britholite-(Ce)		Site symmetry	s.o.f.	x	y	z
REE(1)	b	3	0.771	0.33333	0.66667	−0.0028
REE(1a)	b	3	0.654	0.66667	0.33333	0.0015
REE(2)	c	1	0.789	0.23654	−0.01176	0.2500
Si	c	1	1.048	0.4014	0.3729	0.2506
O(1)	c	1		0.3254	0.4886	0.252
O(2)	c	1		0.5927	0.4697	0.2718
O(3)	c	1		0.3218	0.2461	0.422
O(3a)	c	1		0.3677	0.2651	0.0609
O(4)	a	3		0.000	0.000	0.201
lessingite-(Ce)						
REE(1)	b	3	0.732	0.33333	0.66667	−0.0047
REE(1a)	b	3	0.721	0.66667	0.33333	−0.0011
REE(2)	c	1	0.828	0.23851	−0.01142	0.2500
Si	c	1	1.004	0.4007	0.3726	0.2507
O(1)	c	1		0.3235	0.4878	0.253
O(2)	c	1		0.5942	0.4688	0.232
O(3)	c	1		0.316	0.2400	0.419
O(3a)	c	1		0.3632	0.2636	0.059
O(4)	a	3		0.000	0.000	0.205
'min X'						
REE(1)	b	3	0.876	0.33333	0.66667	−0.0035
REE(1a)	b	3	0.897	0.66667	0.33333	−0.0016
REE(2)	c	1	0.961	0.23694	−0.01118	0.2500
Si	c	1	1.038	0.4020	0.3733	0.2488
O(1)	c	1		0.3244	0.4869	0.255
O(2)	c	1		0.5952	0.4690	0.2681
O(3)	c	1		0.3178	0.2439	0.417
O(3a)	c	1		0.3626	0.2622	0.0589
O(4)	a	3		0.000	0.000	0.197

CLINOPTILOLITE (NATURAL and Cs-EXCHANGED)

$Na_{1.3}K_{1.2}Ca_{1.55}Al_{6.2}Si_{29.8}O_{72}\cdot23H_2O$, $Cs_{3.8}Ca_{1.2}Al_{6.1}Si_{29.7}O_{72}\cdot19H_2O$

Amer. Min., <u>75</u>, 522–528.

C2/m, a = 17.633, 17.692, b = 17.941, 17.945, c = 7.400, 7.404 A, β = 116.39, 116.36°, Z = 1, R = 0.062, 0.083. Heulandite-type structure, as previously described (<u>41</u>A, 380; <u>43</u>A, 358). Cs positions differ from the cation positions in the natural sample.

natural clinoptilolite

	occ(tot)	x	y	z
T(1)	1.00	0.1789	0.1705	0.0953
T(2)	1.00	0.2131	0.4104	0.5030
T(3)	1.00	0.2080	0.1907	0.7152
T(4)	1.00	0.0654	0.2989	0.4129
T(5)	1.00	0	0.2160	0
O(1)	1.00	0.1973	½	0.4571
O(2)	1.00	0.2320	0.1212	0.6138
O(3)	1.00	0.1835	0.1565	0.8859
O(4)	1.00	0.2356	0.1065	0.2518
O(5)	1.00	0.3245		½
O(6)	1.00	0.0811	0.1614	0.0570
O(7)	1.00	0.1274	0.2343	0.5484
O(8)	1.00	0.0110	0.2682	0.1857
O(9)	1.00	0.2119	0.2534	0.1830
O(10)	1.00	0.1174	0.3723	0.4079
M(1)	0.50	0.1478	0	0.6661
M(2)	0.39	0.0404	½	0.2167
M(3)	0.64	0.2344	½	0.0252
W(2)	0.98	0.0798	0	0.8531
W(3)	0.89	0.0798	0.4190	0.9655
W(4)	1.00	0	½	½
W(5)	0.36	0.0202	0.0901	0.525
W(6)	0.81	0.0865	0	0.2654

Cs-exchanged clinoptilolite

	occ(tot)	x	y	z
T(1)	1.00	0.1781	0.1699	0.0937
T(2)	1.00	0.2139	0.4107	0.5069
T(3)	1.00	0.2096	0.1903	0.7167
T(4)	1.00	0.0674	0.2976	0.4183
T(5)	1.00	0	0.2179	0
O(1)	1.00	0.1965	0.5	0.4622
O(2)	1.00	0.2359	0.1210	0.8183
O(3)	1.00	0.1867	0.1553	0.8885
O(4)	1.00	0.2298	0.1048	0.2501
O(5)	1.00	0	0.3209	½
O(6)	1.00	0.0790	0.1639	0.0456
O(7)	1.00	0.1272	0.2304	0.5487
O(8)	1.00	0.0157	0.2707	0.1868
O(9)	1.00	0.2131	0.2516	0.1843
O(10)	1.00	0.1199	0.3717	0.4279
M(1)	0.179	0.0283	0	0.1274
M(2)	0.217	−0.0116	½	0.474
M(3)	0.088	0.1913	½	−0.0327
M(4)	0.184	0.0593	0	0.2521
M(5)	0.327	0.2829	0	0.9799
W(1)	1.0	0.3809	½	0.3011
W(2)	1.0	0.4229	0.0771	0.0379
W(3)	0.64	0.5412	0	0.2141

CLINOPYROXENES
$Ca_{0.8}Mg_{1.2}Si_2O_6$

Z. Kristallogr., 192, 183-199.

C2/c, a = 9.728-9.823, b = 8.910-9.060, c = 5.244-5.287 A, β = 106.22-106.79°, at 143-1373K, Z = 4, R = 0.029-0.049. Diopside-type (2, 130), with split M2 site (Ca/Mg ordering) (49A, 334).

$Na(In,Sc)Si_2O_6$

Acta Cryst., B46, 742-747.

C2/c, a = 9.900-9.852, b = 9.131-9.070, c = 5.366-5.352 A, β = 107.23-107.17°, for 100-20% In, Z = 4, R = 0.017-0.024. Isostructural with other pyroxenes (2, 130; 32A, 458; 40A, 280).

DAVYNE
$Na_4K_2Ca_2Si_6Al_6O_{24}(SO_4)Cl_2$

Neues Jb. Miner. Mh., 97-112.

$P6_3/m$, 12.705, 5.368, Z = 1, R = 0.049. Cancrinite-type framework (3, 150; 48A, 344), with chains of Ca and chloride ions in undecahedral cages along the threefold axes, and Na, K, sulphate, and chloride ions in large channels along the sixfold axis.

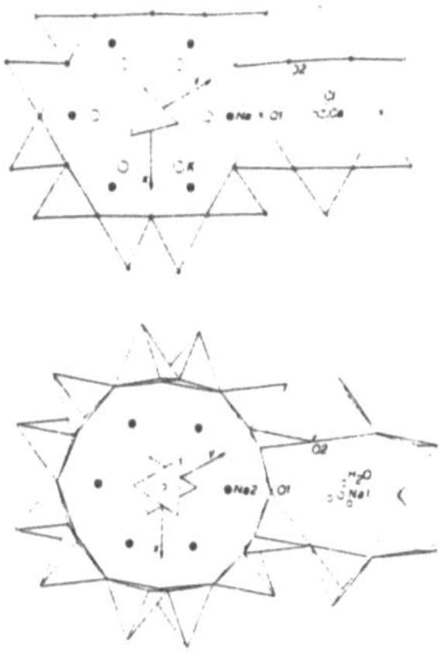

a.

b.

	M	Occupancy	x	y	z
Si	½	1 Si	0.3284	0.4093	¾
Al	½	1 Al	0.0691	0.4086	¼
O1	½	1 O	0.2145	0.4307	¾
O2	½	1 O	0.1002	0.5567	¾
O3	1	1 O	−0.0087	0.3230	1.0084
Ca	⅙	1 Ca	⅓	⅔	¾
Cl	½	⅓ Cl	0.3150	0.6372	¼
Na	½	½ Na	0.1510	0.3097	¾
K	½	0.21 Na + 0.29 K	0.2217	0.1136	¾
S	⅙	⅓ S	0	0	¼
OA	1	1/18 O	0.0400	0.0200	0.0200
OB	1	⅙ O	0.0701	0.1132	0.3746
Cl′	1	1/18 Cl	0.0490	0.0400	−0.0750

drawing of (a) davyne and (b) carbonate-rich cancrinite

Canad. Miner., 28, 341-349.

$P6_3$, a = 12.793, 12.854, c = 5.367, 5.357 A, for two samples, Z = 1, R = 0.048, 0.118. Isostructural with cancrinite (3, 150; 48A, 344), with ordered Al/Si distribution, and $[CaCl]^+$ clusters in the cage occupied by $[Na(H_2O)]^+$ in other members of the cancrinite group.

312 INORGANIC COMPOUNDS

EDGARBAILEYITE
$Hg_6Si_2O_7$

Amer. Min., **75**, 1192-1196.

C2/m, 11.755, 7.678, 5.991, 111.73, Z = 2, R = 0.058 (twinned crystal). Tetrahedral $O_3Si-O-SiO_3{}^{6-}$ ions linked by $Hg_2{}^{2+}$ pairs, each Hg having three O neighbours. Hg-Hg = 2.523(2), Hg-O = 2.12-2.86(2), Si-O = 1.59-1.63(2) A.

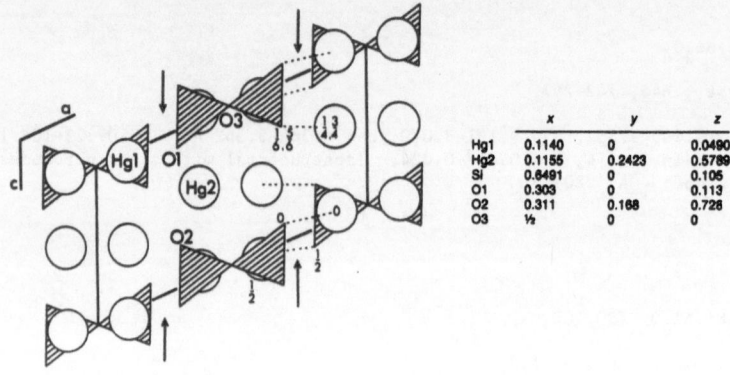

	x	y	z
Hg1	0.1140	0	0.0490
Hg2	0.1155	0.2423	0.5789
Si	0.6491	0	0.105
O1	0.303	0	0.113
O2	0.311	0.168	0.726
O3	½	0	0

EUDIALITE (POTASSIUM OXONIUM)
$(Na,K)_5(H_3O)_7Ca_6Fe_3Zr_3(Si_3O_9)_2[Si_9O_{24}(OH)_3]_2Cl_{1.5}(O,OH)_6$

Kristallografija, **35**, 1381-1387 [Soviet Physics - Crystallography, **35**, 814-817].

R3m, a = 14.245, c = 30.12 A, Z = 3, R = 0.057. Zeolite framework as previously described (**37A**, 332; **39A**, 341; **54A**, 311; **55A**, 290), with potassium and oxonium ions replacing sodium.

FELDSPARS (ANORTHITE-RICH)
$(Na,Ca)(Al,Si)_4O_8$

Amer. Min., **75**, 150-162.

Refinements of 3 P$\bar{1}$ and 17 I$\bar{1}$ samples. Structures as previously described (**27**, 679; **51A**, 348).

GMELINITE (K-RICH)
$K_{2.72}(Ca,Sr,Na,Mg)_{2.41}Al_8Si_{16}O_{48}\cdot23\cdot5H_2O$

Neues Jb. Miner. Mh., 504-516.

P6₃/mmc, 13.621, 10.254, Z = 1, R = 0.068. Structure as previously described (**31A**, 227; **49A**, 337; **51A**, 342; **53A**, 255), with low occupancy of one of the two exchangeable-cation sites, and eight partially-occupied water sites.

KAOLINITE
$Al_2Si_2O_5(OH)_4$

Cryst. Res. Technol., 25, 105-110.

C1, 5.156, 8.944, 7.405, 91.70, 104.84, 89.83, Z = 2, powder data. Structure as previously described (2, 548; 10, 161; 24, 471).

Cryst. Res. Technol., 25, 305-312.

P1, 5.158, 8.942, 7.397, 91.67, 104.86, 89.99, Z = 2, R = 0.042. Structure as previously described (2, 548; 10, 161; 24, 471).

LINTISITE
$Na_3LiTi_2(Si_2O_6)_2O_2 \cdot 2H_2O$

Z. Kristallogr., 193, 137-148.

C2/c, 28.583, 8.600, 5.219, 91.03, Z = 4, R = 0.033. The structure contains pyroxene-like tetrahedral silicate chains, chains of edge-sharing TiO_6 and $Na(O,H_2O)_6$ octahedra, columns of edge-sharing LiO_4 tetrahedra, and NaO_8 distorted cubes.

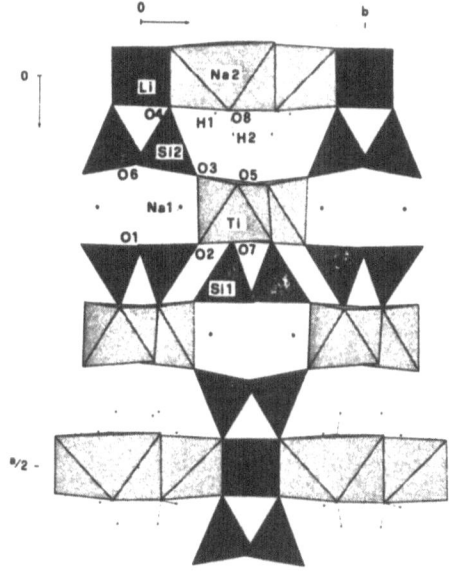

MAKAROCHKINITE
$(Ca,Na)_2(Fe,Mg)_{5.5}O_2(Si,Al,Be)_6O_{18}$

Kristallografija, 35, 1388-1394 [Soviet Physics - Crystallography, 35, 818-822].

P$\bar{1}$, 10.352, 10.744, 8.864, 105.73, 96.16, 124.91, Z = 2, R = 0.030. Isostructural with aenigmatite (37A, 327).

MONTESOMMAITE
$(K,Na)_9Al_9Si_{23}O_{64} \cdot 10H_2O$

Amer. Min., 75, 1415-1420.

Fdd2 (pseudo-I4$_1$/amd), 10.099, 10.099, 17.307, Z = 1, powder data. Substructure
proposed related to that of merlinoite (45A, 372).

PREHNITE
$Ca_2Al(Si_3AlO_{10})(OH)_2$

J. Solid State Chem., 86, 330-333.

Clinoprehnite, P2/n (a unique), 4.6314, 5.4839, 18.4887, α = 90.611°, Z = 2, R =
0.079; orthoprehnite, P2cm, 4.6260, 5.4820, 18.4826, Z = 2, R = 0.026, atomic positional parameters not listed. Predominantly ordered Al/Si distributions lower
the Pncm symmetry (24, 482; 32A, 468; see also 30A, 434).

PYROXENES
$Ca(Ni,Mg)Si_2O_6$

Amer. Min., 75, 1274-1281.

C2/c, a = 9.736-9.747, b = 8.893-8.924, c = 5.228-5.252 A, β = 105.83-105.94°,
for 5 samples with Ni content = 1.0, 0.8, 0.5, 0.25, 0, Z = 4, powder data.
Diopside structure (2, 130).

SODDYITE
$(UO_2)_2SiO_4 \cdot 2H_2O$

Kristallografija, 35, 1563-1564 [Soviet Physics - Crystallogrpahy, 35, 921-922].

Fddd, 8.297, 11.219, 18.661, Z = 8, neutron powder data. U in 16(g): 7/8,7/8,z,
z = 0.045; Si in 8(a): 7/8,7/8,7/8; O(1) in 32(h): 0.706,0.955,0.046; O(2) in 32(h):
0.022,0.045,0.063; O(3) in 16(g): z = 0.167. Previous X-ray study in 45A, 391.

SPESSARTITE
$Mn_3Al_2Si_3(O,F,OH)_{12}$ (idealized)

Amer. Min., 75, 314-318.

Ia3d, 11.628, Z = 8, R = 0.054 . Mn in 24(c): 1/8,0,1/4; Al in 16(a): 0,0,0;
Si in 24(d): 3/8,0,1/4; O in 96(h): 0.0336,0.0481,0.6520. Garnet structure (1,
363), as previously described (2, 521; 37A, 334). There is Si-deficiency, with
some F in O sites.

STAUROLITE

$(Fe,Mg)_{4.6}Al_{17.9}Si_{7.5}O_{45}(OH)_3$

Acta Cryst., B$\underline{46}$, 292–301.

C2/m, a = 7.8700, 7.8713, b = 16.6228, 16.6235, c = 5.6613, 5.6608 A, β = 90.124, 90.016°, for minerals from Heas, Pyrenees and Scaer, Brittany, Z = 1, R = 0.018, 0.026. Structure as previously described ($\underline{55}$A, 294), with threefold-disordered Fe site.

		$10^5 x$	y	z
	Heas			
0·817	Fe	39279	0	24889
0·064	U(A)	50000	0	0
0·034	U(B)	50000	0	50000
	Si	13411	16606	24900
	Al(1A)	50000	17526	0
	Al(1B)	50000	17510	50000
	Al(2)	26333	41033	25118
0·588	Al(3A)	0	0	0
0·406	Al(3B)	0	0	50000
	O(1A)	23344	0	96396
	O(1B)	23521	0	53395
	O(2A)	25533	16148	1522
	O(2B)	25470	16122	48422
	O(3)	174	8880	24692
	O(4)	2132	24917	24976
	O(5)	52712	10013	24962
	Scaer			
0·857	Fe	39191	0	24986
0·030	U(A)	50000	0	0
0·029	U(B)	50000	0	50000
	Si	13409	16603	24992
	Al(1A)	50000	17529	0
	Al(1B)	50000	17528	50000
	Al(2)	26303	41048	25009
0·507	Al(3A)	0	0	0
0·496	Al(3B)	0	0	50000
	O(1A)	23445	0	96478
	O(1B)	23467	0	53498
	O(2A)	25512	16130	1548
	O(2B)	25508	16134	48453
	O(3)	174	8876	24979
	O(4)	2143	24913	25002
	O(5)	52680	10028	25001

THOMSONITE

$Na(Ca,Sr)_2Al_5Si_5O_{20}\cdot6H_2O$

Acta Cryst., C$\underline{46}$, 1370–1373.

Pncn, 13.1043, 13.0569, 13.2463, at 13K, Z = 4, neutron radiation, R = 0.039. Structure as previously described ($\underline{3}$, 171, 529; $\underline{49}$A, 346; $\underline{52}$A, 343), with the Ca/Na site still disordered.

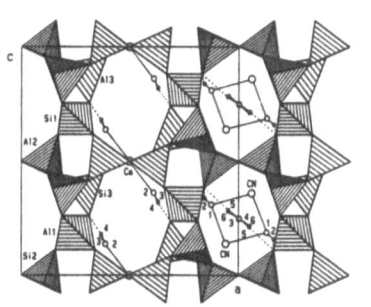

An ac projection of thomsonite at 13 K. Largest circles cations (the CaNa site is denoted CN), intermediate circles water O atoms and smallest circles H atoms.

	$\times 10^5$ x	y	z
Ca/Na	5834	50375	36020
Ca	49916	47480	49915
Si(1)	25000	25000	68808
Al(1)	25000	75000	69025
Si(2)	11302	69465	50019
Al(2)	11948	30637	49585
Si(3)	30886	38484	37739
Al(3)	31012	62393	38013
O(1)	16923	31182	61795
O(2)	15866	69043	61417
O(3)	31377	33088	75611
O(4)	31301	65759	76235
O(5)	214	63666	50126
O(6)	18383	62910	42295
O(7)	19041	38525	41545
O(8)	10428	81256	46124
O(9)	11827	18080	45269
O(10)	35702	49933	38296
OW(1)	12620	50179	18922
OW(2)	38972	49775	63828
OW(3)	0	65045	75000
OW(4)	0	34407	75000
H(1)	15862	43726	66126
H(2)	15679	55944	66093
H(3)	36905	43779	67664
H(4)	37059	55568	68072
H(5)	4677	69191	71289
H(6)	5035	30212	71813

YOSHIOKAITE
$CaAl_2SiO_6$

Amer. Min., **75**, 1186–1191.

$P\bar{3}$, 9.927, 8.220, Z = 5.3, R = 0.051 (twinned crystal gives apparent $P\bar{3}c1$ symmetry).
Stuffed derivative of high tridymite with layers having the topology of nepheline
(**11**, 478; **19**, 476; **38A**, 370); the Na site of nepheline is occupied by Ca, and the K
site is vacant.

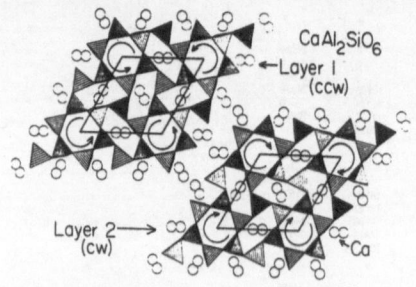

	x	y	z
T1	0.67820	0.75715	−0.06751
T2	0.75703	0.67636	0.56732
T3	1/3	2/3	0.9618
T3'	1/3	2/3	0.0421
T4	1/3	2/3	0.5407
T4'	1/3	2/3	0.4606
O1	0.2816	0.0567	0.0762
O2	0.0556	0.2614	0.4241
O3	0.5134	0.3580	0.0028
O4	0.3556	0.5132	0.4957
O5	0.6827	0.0004	0.2506
O6	1/3	2/3	0.2501
O6'	2/3	1/3	0.2573
Ca	0.44326	0.00004	0.25006

Polyhedra projection parallel to *c* of the two layers
composing yoshiokaite. Layer 1 (layer A) is similar except in
detail to the nepheline layer   Layer 2 (layer B) is
similar to the nepheline layer but with an opposite sense of rotation. Abbreviations: cw = clockwise; ccw = counter clockwise.

ZEOLITE A (NICKEL and AMMONIUM EXCHANGED)
$(NiOH)_2(NH_4)_{10}Al_{12}Si_{12}O_{48} \cdot nH_2O$

J. Phys. Chem., **94**, 7662–7665.

Pm3m, a = 12.289 A, for a partly-desolvated sample, R = 0.063. Zeolite framework
with partially-occupied nickel and ammonium sites. The hydrated structure is less
precisely established.

ZEOLITE (Ag,Tl-EXCHANGED)
$Ag_{7.5}Tl_{4.5}Al_{12}Si_{12}O_{48}$

Nonmunjip – Cheju Taehakkyo, **29**, 221–227 (1989).

Pm3m, 12.256, Z = 1, R = 0.053. Zeolite framework with Ag_6 clusters at the
centre of one-sixth of the sodalite units, Ag^+ ions in the 6-rings, and Tl^+
ions in 8-rings.

ZEOLITE A (THALLIUM(I), ZINC EXCHANGED)
$Tl_{3.4}Zn_{4.3}Al_{12}Si_{12}O_{48}$, $Tl_{5.5}Zn_{3.25}Al_{12}Si_{12}O_{48}$

Bull. Korean Chem. Soc., **11**, 150–154.

Pm3m, a = 12.100, 12.092 A, Z = 1, R = 0.075, 0.064. Zeolite structures, with Zn
coordinated in 6-ring sites to three framework oxygens, and Tl in 6-ring and 8-
ring sites.

ZEOLITE A (Cd and Rb EXCHANGED)
$Cd_xRb_{12-2x}Al_{12}Si_{12}O_{48}$ (x = 4.0, 5.0, 5.95)

Bull. Korean Chem. Soc., <u>11</u>, 328-331.

Pm3m, a = 12.204, 12.202, 12.250 A, Z = 1, R = 0.087, 0.059, 0.079. Zeolite
structures, with Cd on threefold axes at the centre of a 6-ring, and Rb preferentially in 8-rings.

ZEOLITE Na-P
$Na_4Al_4Si_{12}O_{32} \cdot 14H_2O$ (I), $Na_{3.6}Al_{3.6}Si_{12.4}O_{32} \cdot 14H_2O$ (II)

Acta Cryst., C<u>46</u>, 1361-1362, 1363-1364.

I, Pnma, 9.868, 10.082, 10.098, Z = 1, R = 0.053. II, I4$_1$/amd, 9.9989, 10.0697,
Z = 1, R = 0.037. Gismondine-type tetrahedral framework (<u>28</u>, 279; <u>37A</u>, 348; <u>53A</u>,
255), with Na ions in cavities. See also <u>38A</u>, 377.

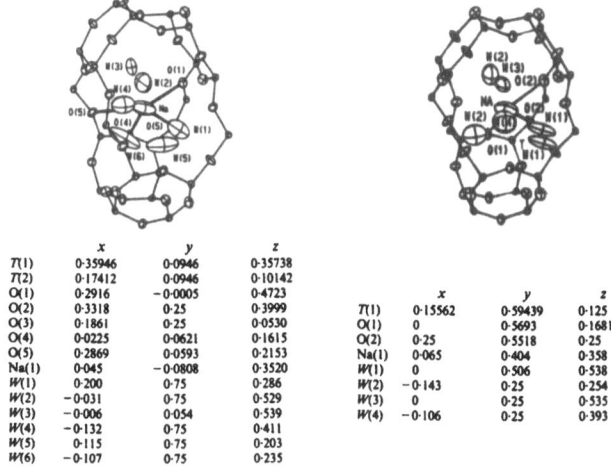

	x	y	z
T(1)	0·35946	0·0946	0·35738
T(2)	0·17412	0·0946	0·10142
O(1)	0·2916	−0·0005	0·4723
O(2)	0·3318	0·25	0·3999
O(3)	0·1861	0·25	0·0530
O(4)	0·0225	0·0621	0·1615
O(5)	0·2869	0·0593	0·2153
Na(1)	0·045	−0·0808	0·3520
W(1)	0·200	0·75	0·286
W(2)	−0·031	0·75	0·529
W(3)	−0·006	0·054	0·539
W(4)	−0·132	0·75	0·411
W(5)	0·115	0·75	0·203
W(6)	−0·107	0·75	0·235

	x	y	z
T(1)	0·15562	0·59439	0·125
O(1)	0	0·5693	0·1681
O(2)	0·25	0·5518	0·25
Na(1)	0·065	0·404	0·358
W(1)	0	0·506	0·538
W(2)	−0·143	0·25	0·254
W(3)	0	0·25	0·535
W(4)	−0·106	0·25	0·393

ZEOLITE RHO
$(Ca,ND_4)_xAl_{12}Si_{36}O_{96}$ (I), $(Ca,D)_xAl_{12}Si_{36}O_{96}$ (II)

J. Amer. Chem. Soc., <u>112</u>, 4821-4830.

I$\bar{4}$3m, a = 14.410 and 14.110 A (two phase sample for I), 13.965 A for II, Z = 1,
neutron powder data. Si/Al in 48(h): 0.2816,0.1310, 0.4345; O(1) in 24(g): x,x,z,
x = 0.2365, z = 0.4141; O(2) in 24(g): x = 0.1104, z = 0.6142; O(3) in 48(h):
0.0481,0.2032,0.3913 (parameters for II); Ca/NH$_4$ in 6(b): 1/2,0,0. Structure
essentially as previously described (<u>55A</u>, 297), with deviation from Im3m, and
Ca in the centre of the double 8-ring.

ZEOLITE X (Li-EXCHANGED)
$Li_{62}Na_{30}Al_{92}Si_{100}O_{384} \cdot nH_2O$ (n = 200 and 0)

Zeolites, <u>10</u>, 61-63.

Fd3, a = 24.821, 24.75 A, at 298, 548K, Z = 1, R = 0.067, 0.097. Framework as previously described ($\underline{24}$, 480; $\underline{50}$A, 342), with Li sites determined. There are large distortions of the framework on dehydration.

ZEOLITE NaX (Mg-EXCHANGED)
ZEOLITE CaX (Mg-EXCHANGED)
$Mg_{31}Na_{30}Al_{96}Si_{96}O_{384} \cdot nH_2O$ (n = n and 0)
$Ca_{23}Mg_{23}Al_{96}Si_{96}O_{384} \cdot nH_2O$ (n = n and 0)

Zeolites, $\underline{10}$, 32-37.

Fd3, R = 0.056, 0.062 for hydrated, 0.046, 0.057 for dehydrated forms. Zeolite X structures ($\underline{24}$, 480; $\underline{50}$A, 342), with cation sites located. Dehydration results in large distortions of the zeolite framework.

ZEOLITE ZSM-11 (GALLIUM)

Cuihua Xuebao, $\underline{11}$, 196-203.

$I\bar{4}m2$, 20.073, 13.403, R = 0.058. Zeolite framework with Ga replacing framework Si.

ZEOLITE ZSM-57

Zeolites, $\underline{10}$, 293-296.

Imm2, 7.4510, 14.1711, 18.767. Structure related to that of ferrierite, ZSM-35 ($\underline{54}$A, 293), but without the 4-rings; there is a channel system of 8- and 10-rings.

ZIRCONS
$ZrSiO_4$, $ZrSi_{0.96}V_{0.04}O_4$, $ZrSi_{0.94}V_{0.06}O_4$, $Zr_{0.94}Tb_{0.06}SiO_4$

Z. anorg. Chem., $\underline{583}$, 67-77.

$I4_1/amd$, a ∿ 6.6, c ∿ 6.0 A, Z = 4, R = 0.023, 0.023, 0.025, 0.024. Zircon structures ($\underline{1}$, 345), with V in an interstitial distorted tetrahedral site, and Tb in the Zr site.

TABLE I

Some structural information has also been given for the following materials (listed with abbreviated 1990 reference).

Compound	Structure	Reference
Al_2O_3 Cu_2O	Electron density studies	Acta Cryst., A46, 271
Zirconia (Ca-doped), $Zr_{0.85}Ca_{0.15}O_{1.85}$	Defect structure	Ibid., A46, 799
Calcium strontium bismuth cuprate, $Ca_{5.64}Sr_{4.05}Bi_{0.31}Cu_{17}O_{29}$	Superspace group description of the OD structure (55A, 198)	Ibid., B46, 39
ZrO_2	Study of monoclinic-tetragonal phase transformation	Ibid., B46, 724
Magnesium perrhenate tetrahydrate, $Mg(ReO_4)_2.4H_2O$	Space group is $P\bar{1}$, rather than P1 (53A, 170)	Ibid., C46, 349
Neodymium tantalate, $NdTaO_4$	Space group is P2/a, rather than Pa (53A, 152), with Ta and Nd on twofold axes	
$Sr_2Nb_5O_9$	$Ba_2Nb_5O_9$	Acta Chem. Scand., 44, 222
$Ti_{0.92}O$	NaCl	Ibid., 44, 851
Vanadium bronzes	Review of the crystal chemistry	Amer. Min., 75, 508
PON	High cristobalite-type, with O/N disorder	Ann. Chim., 14, 475 (1989)
$Ca_{1-x}CuO_2$	Superstructure of $NaCuO_2$-type	Chem. Mater., 2, 192
$MAlSi_2O_6$, M = Na, K	Keatite	Eur. J. Mineral., 2, 155
Magnesium aluminophosphate, $Mg_2Al_4P_6O_{24}.2(CH_3)_4N$	Sodalite framework	Ibid., 2, 787
Zeolite-L	Some cation sites determined	Ibid., 2, 851
$Y_3I_4F_7$ $Y_6O_3F_8$ $Y_7O_6F_9$	Modulated, incommensurate fluorite-type superstructures	Eur. J. Solid State Inorg. Chem., 27, 451
Potassium octagermanate, $K_2Ge_8O_{17}$	Refinement confirms Pnma (39A, 214)	Ibid., 27, 831
$RbHSeO_4$ NH_4HSeO_4	Refinements of the structures of the paraelectric and ferroelectric phases	Ferroelectrics, 170 281

K_xCrF_3	Hexagonal $BaTa_2O_6$, ortho-rhombically-distorted $BaTa_2O_6$, and orthorhombic-ally-distorted tetragonal tungsten bronze phases	Inorg. Chem., $\underline{29}$, 3037
K_2SeBr_6	K_2PtCl_6, Rb_2TeI_6, and K_2TeBr_6-type phases	J. Chem. Phys., $\underline{93}$, 8321
N_2	Study of high-pressure phases	Ibid., $\underline{93}$, 8968
$BaNpO_4$	$BaUO_4$	J. Less-Common Metals, $\underline{162}$, 323
$Ln_2Te_4O_{11}$	$Nd_2Te_4O_{11}$ [this volume, p. 294]	Ibid., $\underline{166}$, 367
Thulium phosphate, TmP_5O_{14}	HoP_5O_{14}	J. Mater, Sci. Lett., $\underline{9}$, 235
Barium sodium niobate, $Ba_{4.17}Na_{1.67}Nb_{10}O_{30}$	Quasi-commensurate structure	J. Phys.: Condens. Matter, $\underline{2}$, 25
Mo_8O_{23}	Incommensurately modulated structure	Ibid., $\underline{2}$, 45
Bi_2O_3 Ln_2O_3	Fluorite	J. Solid State Chem., $\underline{84}$, 183
Bi_2O_3-P_2O_5	Sillenite	Ibid., $\underline{85}$, 76
$Ba_3MSb_2O_9$, M = Mg, Ni, Zn	6H perovskites	Ibid., $\underline{85}$, 144
$(Fe,Ta,Zn)(F,O)_2$	Rutile	Ibid., $\underline{86}$, 41
$(Fe,M,Zn)(F,O)_2$, M = Nb, Ta	α-PbO_2	
$Sr(Co,Mn)O_3$	Perovskite	Ibid., $\underline{86}$, 75
$Bi_9Ti_6CrO_{27}$ $Bi_9Ti_6FeO_{27}$ $BaBi_8Ti_7O_{27}$	Structures proposed with perovskite layers	Ibid., $\underline{86}$, 206
$Cu(Cr,Rh)_2O_4$	Spinel	Ibid., $\underline{86}$, 286
$KBaFe_{23}O_{36}$	$NaNdAl_{23}O_{36.5}$ ($\underline{53A}$, 128)	Ibid., $\underline{87}$, 159
$Li(Al,Ti)_2O_4$ $Li(Cr,Ti)_2O_4$	Spinels, with Al and Cr in octahedral sites	Ibid., $\underline{89}$, 345
$BaCeLa(O,N)_4$ $BaCe_2(O,N)_4$	$CaFe_2O_4$	Ibid., $\underline{89}$, 366
$(M,Ln)F_{2-x}$	Fluorite	Kristallografija, $\underline{35}$, 777 [Soviet Physics - Crystallography, $\underline{35}$, 454]
$BaNd_2CuO_5$	$BaLa_2CuO_5$	Mater. Lett., $\underline{9}$, 401

$(LnM)(BiPb)O_7$, Ln = lanthanon M = Ba, Sr, Ca	Pyrochlore	Mater. Res. Bull., $\underline{25}$, 107
$Pb(Mg,Nb)O_3$	Perovskite	Ibid., $\underline{25}$, 283
$NaNbO_3$ $LiNbO_3$	Ilmenite	Nippon Seramikkusu Kyokai Gakujutsu Ronbushi, $\underline{98}$, 384
$\beta\text{-}(Mg_{0.9}Fe_{0.1})_2SiO_4$	$\beta\text{-}Mg_2SiO_4$-type ($\underline{48A}$, 336), with Fe preferentially in the smaller M1 and M3 sites	Phys. Chem. Miner., $\underline{17}$, 293
Olivines, $(Mg,Fe)_2SiO_4$	Olivine structure	Ibid., $\underline{17}$, 301
$Li_2Si_2O_5$	Structure approximately as previously described ($\underline{26}$, 506; see also this volume, p. 299)	Powder Diffr., $\underline{5}$, 137
Zeolite ZSM-18	Model for the structure, with a 12-ring channel	Science, $\underline{247}$, 1319
$(La,Ba)_2NiO_4$	La_2NiO_4	Solid State Comm., $\underline{76}$, 1327
$Bi_{12}GeO_{20}$ $Bi_{12}Ti_{0.9}O_{19.8}$	Sillenite	Vysokočist. Veščestva, No. 2, 158
$BiTa_7O_{19}$	$LnTa_7O_{19}$ ($\underline{53A}$, 151)	Z. anorg. Chem., $\underline{581}$, 183
$LnTa_3O_9$, Ln = Pr, Nd	Positions proposed for heavy atoms	Ibid., $\underline{582}$, 61
$NaZnPO_4$ $NaZnAsO_4$	Beryllonite	Ibid., $\underline{582}$, 179
$NiNb_{14}O_{35}F_2$ $CoNb_{14}O_{35}F_2$	$MgNb_{14}O_{35}F_2$	Ibid., $\underline{585}$, 113
$Cs_2Ag_2Cl_6$	$Cs_2Au_2Cl_6$	Ibid., $\underline{589}$, 7
K_2AlF_5	$Rb_2MnF_5\text{-}$ and $\alpha\text{-}(NH_4)_2FeF_5\text{-}$ types	Ibid., $\underline{589}$, 221
Li_2MXO_4 M = Mg, Mn etc X = Zr, Hf	$\alpha\text{-}LiFeO_2$	Z. Kristallogr., $\underline{190}$, 161
High-cordierite (indialite)	Possible true space groups are P6, P$\bar{6}$, or P3	Ibid., $\underline{190}$, 271
Molecular sieve VPI-5	Model for the structure	Zeolites, $\underline{10}$, 163
γ-Eucryptite, $LiAlSiO_4$	Stuffed cristobalite	Ibid., $\underline{10}$, 193
$NaLn(WO_4)_2$, Ln = La, Pr, Nd	Cubic, a = 5.349, 5.334, 5.278 A	Zhongguo Xitu Xuebao, $\underline{8}$, 37

$Ni(ReO_4)_2$ $Co(ReO_4)_2$	$Zr(MoO_4)_2$	Ž. Neorg. Khim., <u>35</u>, 545 [Russ. J. Inorg. Chem., <u>35</u>, 310]
β–Na_2TiO_3	Distorted (rhombohedral) NaCl type	Ibid., <u>35</u>, 1085 [Ibid., <u>35</u>, 613]

ELECTRON DIFFRACTION

The following compounds have been studied by electron diffraction of the vapours (listed with abbreviated 1990 reference). Bond lengths in Å, angles in degrees.

WCl_3	T-shaped W–Cl $\qquad$ 2.269	Ž. Strukt Khim., <u>31</u>, No. 1, 49 [J. Struct. Chem., <u>31</u>, 40]
WCl_4	Irregular tetrahedral W–Cl $\qquad$ 2.247	
W_2Cl_6	Two tetrahedra sharing an edge W–Cl $\qquad$ 2.270	
$InCl_3$	D_{3h} symmetry In–Cl $\qquad$ 2.262	Ibid., <u>31</u>, No. 2, 46 [Ibid., <u>31</u>, 222]
In_2Cl_6	Two tetrahedra sharing an edge In–Cl (bridging) 2.472 Cl(b)–In–Cl(b) $\quad$ 90 Cl(t)–In–Cl(t) $\quad$ 130	

PRELIMINARY NOTES

Hydrogen sulphide (low-temperature phase), H_2S	Chem. Comm., 515
Thallium titanyl phosphate, $TlTiOPO_4$	Ibid., 540
Copper oxides	Dokl. Akad. Nauk SSSR, <u>309</u>, 1165 (1989)
Tuliokite, $Na_6BaTh(CO_3)_6$	Ibid., <u>310</u>, 99
Bi_9I_2	Ibid., <u>310</u>, 117
$KFeFPO_4$	Ibid., <u>310</u>, 1129
$Ba_2Y(Cu,Al)_3O_7$	Ibid., <u>310</u>, 1407
Girvasite	Ibid., <u>311</u>, 1372
$BaNb_4O_6$ $Ba_2Nb_5O_9$	Ibid., <u>312</u>, 615

Alluaivite	Dokl. Akad. Nauk SSSR, 312, 1379
K(K,Na)$_2$Zn$_3$Mn$_{1.5}$(Si$_{12}$O$_{30}$)	Ibid., 313, 865
Hydrosodalite	Ibid., 314, 625
Sr$_6$HoSc(BO$_3$)$_6$	J. Amer. Chem., Soc., 112, 7068
La$_2$CeTaO$_6$Cl$_3$	Z. anorg. Chem., 583, 223

Reports on the following structures were prepared too late for inclusion in the main text:

SILVER TETRAAMIDO-CYCLO-TRINITRIDOSULPHODIPHOSPHATE
Ag[N$_3$P$_2$(NH$_2$)$_4$SO$_2$]

Ž. Neorg. Khim., 35, 2850-2854 [Russ. J. Inorg. Chem., 35, 1618-1621].

P2$_1$/n, 9.908, 9.306, 10.563, 108.75, Z = 4, R = 0.084. The anion contains a six-membered N-P-N-P-N-S ring. Ag is bonded to two ring N atoms of different anions (Ag-N = 2.33 A, N-Ag-N = 147°), with two longer Ag...N (2.96 A) and one Ag...H (2.87 A); there are also N-H...O hydrogen bonds.

BARIUM HEXAFLUOROANTIMONATE(V)
Ba(SbF$_6$)$_2$

Ž. Neorg. Khim., 35, 2831-2836 [Russ. J. Inorg. Chem., 35, 1608-1611].

P1, 4.885, 5.074, 8.627, 104.70, 100.25, 116.50, Z = 1, powder data. SbF$_6^-$ octahedra linked by 12-coordinate Ba.

MERCURY(II) BROMIDE
HgBr$_2$

Ž. Neorg. Khim., 35, 2476-2478 [Russ. J. Inorg. Chem., 35, 1407-1409].

Cmc2$_1$, 4.628, 6.802, 12.476, Z = 4, R = 0.08. Hg (0,0.1687,0.4959), Br(1) (1/2,0.0637,0.1267), Br(2) (1/2,0.4006,0.3654). Structure as previously described (2, 18, 250); Hg-Br = 2.445(7) A.

324

SUBJECT INDEX

This index contains the names of the substances printed at
the heads of the reports, and some additional entries.

Ahlfeldite, 290
Albite (low), 308
Alkali-metal silver oxides, 227
Alkali-metal titanyl arsenates,
276
Alluaivite, 323
Alumina (beta, barium strontium
copper), 151
Alumina (beta, silver), 152
Alumina (delta), 148
Alumina (gamma), 148
Alumina (sigma), 148
Aluminoborates, 241
Aluminum hydroxide germanate, 233
Aluminum phosphate (form CJ2),
255
Aluminum phosphate (form 25), 255
Aluminum telluroiodide, 142
Alvanite, 173
Amblygonite, 256
Amminecopper(II) tetracyano-
nickelate, 145
Ammonium bismuth chromate di-
chromate monohydrate, 185
Ammonium cadmium selenate
trihydrate, 291
Ammonium calcium cyclohexaphosph-
ate hexahydrate, 272
Ammonium chromium(II) sulphate
hexahydrate, 285
Ammonium cobalt hexaborate hydr-
ate, 241
Ammonium cyclohexaphosphate
tellurate dihydrate, 293
Ammonium decamolybdate, 187
Ammonium dihydrogen phosphate,
253
Ammonium diselenite (high-temp),
289
Ammonium diselenite, 289
Ammonium hexavanadate, 170
Ammonium hydrogen selenate
(orthorhombic), 319
Ammonium hydrogen selenate
(tetragonal), 319
Ammonium hydrogen selenate, 289
Ammonium magnesium decavanadate
hexadecahydrate, 172

Ammonium molybdenum bronze, 186
Ammonium molybdoscandouranate
hydrate, 194
Ammonium praseodymium nitrate
hydrate, 250
Ammonium terbium sulphate hydr-
ate, 288
Ammonium trifluorocadmate (ortho-
rhombic), 116
Ammonium trimolybdate(VI), 186
Anorthite (I$\bar{1}$), 312
Anorthite (P$\bar{1}$), 312
Apatite (monoclinic), 254
Apatite, 254

Barite, 282
Barium aluminum copper oxide, 215
Barium aluminum fluoride (form
I), 105
Barium bismuth cuprate, 218
Barium bismuth dysprosium oxide,
228
Barium bismuth lanthanum oxide,
218
Barium bismuth oxides, 163
Barium bismuth(III) oxychloride,
136
Barium bismuth yttrium oxide, 228
Barium bismuthate(V), 218
Barium borate (beta), 240
Barium bromate (form I), 296
Barium calcium copper hexasilic-
ate, 305
Barium calcium dysprosium
zirconium oxide, 230
Barium calcium holmium yttrium
oxide, 231
Barium calcium lanthanum cuprate,
218
Barium calcium lanthanum iron
oxide, 205
Barium calcium lanthanum yttrium
oxide, 231
Barium calcium neodymium cuprate,
218
Barium calcium neodymium ferrate,
206
Barium calcium samarium ferrate,

336

METALS FORMULA INDEX

The entries are in alphabetical order by formula.

$Ag_{0.3}Al_{3.7}Ba$, 13
$Ag_{0.7}Al_{3.3}Ca$, 13
$Ag_{0.6}Al_{3.4}Ce$, 13
$Ag_{0.7}Al_{3.3}Pr$, 13
$Ag_{0.8}Al_{3.2}Sr$, 13
$Ag_{0.5}BaGa_{3.6}$, 13
$(Ag,Cu)Bi_5Pb_3Se_{11}$, 59
$Ag_{1.5}Bi_{5.5}S_9$, 59
$Ag_{3.5}Bi_{7.5}S_{13}$, 59
AgBiSr, 13
$Ag_{0.5}CaGa_{3.5}$, 13
AgCaSb, 13
$Ag_{0.5}Ga_{3.5}Sr$, 13
$AgMoS_8$, 59
$Ag_{0.6}NbS_2$, 60
AgP_2SmZn, 45
$Ag_{0.167}S_2Ti$, 60
$AlAs_2Cs_3$, 53
$Al_4As_8K_{12}$, 45
AlAu, 3
$AlAu_4$, 3
$Al_{11}Au_6$, 3
Al_3B_xHo, 27
$Al_{3.8}BaCu_{0.2}$, 13
Al_2CdS_4, 61
$AlCe_3$, 3
$Al_{65}Co_{15}Cu_{20}$, 88
Al-Co-Ni, 89
Al_8Cr_4Er, 13
Al_8Cr_4U, 13
$Al_2Cs_6P_4$, 45
$AlCs_6Sb_3$, 14
$Al_7Cu_{16}H_3Zr_6$, 25
$Al_7Cu_{16}H_{8.2}Zr_6$, 25
$Al_{3.5}Cu_{0.5}La$, 13
Al_2CuLi, 14
$Al_7Cu_{16}Lu_6$, 14
$Al_{3.4}Cu_{0.6}Sr$, 13
Al_8DyMn_4, 13
$(Al,Fe)_{12}Y$, 89
$(Al,Ga)_2Tm$, 88
$Al_{0.33}Ge_{1.67}Hf$, 14
$Al_3Ge_2NiY_3$, 25
AlGeY, 15

Al_2HgS_4, 61
Al_8HoMn_4, 13
$Al_4K_{12}P_8$, 45
$AlMg_3Pt_2$, 15
Al-Mn, 89
Al_8Mn_4Nd, 13
Al_8Mn_4Pr, 13
Al_8Mn_4Tb, 13
Al_8Mn_4U, 13
$Al_{2.25}Nb_{0.75}Ti_3$, 15
$AmPt_5$, 88
As_2BCs_3, 47
As_2BK_3, 49
As_4BaCu_8, 46
As_2BeK_4, 47
As_2CaCu_2, 53
AsCoS, 77
As_2Cs_3Ga, 55
As_2Cs_2Pd, 55
As_2CsRh_2, 90
As_2CsRu_2, 90
As_2DyNi_4, 48
$AsFe_3Ga$, 56
$As_{0.3}Fe_3Ga_{1.7}$, 56
$As_4Ga_3K_3$, 56
$As_{12.66}Ga_6K_{20}$, 20
As_2GaK_2Na, 57
$As_{0.065}GaP_{0.935}$, 90
$As_{4.5}Ga_{4.5}Pd_{25}$, 89
As_3HfNa_5, 57
As_2K_2Ni, 50
As_2Mn_3, 57
$AsNa_3$, 58
$AsNi_2Si$, 49
$As_2Ni_{1.6}U$, 58
As_4NiU_3, 58
$Au_{0.8}CaGa_{3.2}$, 13
AuKTe, 84
Au_3LiNa_2, 16
AuSbYb, 13
$Au_{51}U_{14}$, 4

B_2CCe, 27
$B_2Ce_3N_4$, 41

$BCoFe_{13}Nd_2$, 28
$B_{40}Co_{40}Sm_{11}$, 28
BCs_3P_2, 47
B_3ErMo, 29
B_2ErRh_3, 89
$B_4Eu_2Rh_5$, 30
$B_6Eu_3Rh_5$, 30
$BFe_{14}Ho_2$, 89
$B_{10}Ho_3Ni_{19}$, 29
$B_2Ir_5Mg_2P_{0.7}$, 30
$B_2Ir_5Mg_2Si$, 30
BK_3P_2, 49
B_6LaNi_{12}, 30
$B_4La_2Rh_5$, 30
BN, 39
BN_2Na_3, 44
BNa_3P_2, 50
B_6Nb_5, 88
$B_{12}Pr_8Re_{12.62}$, 31
B_5Sm_2, 31
B_6Ta_5, 31
$BaCu_2P_4$, 45
$BaCu_8P_4$, 46
$Ba_2Cu_3P_4$, 53
$BaCu_{5.65}S_{4.5}$, 61
BaCuSb, 13
$Ba_2In_2S_5$, 68
$Ba_2In_2Se_5$, 68
$BaMo_6S_8$, 65
BaNNi, 39
$BaNi_9P_5$, 46
Ba_2S_5Ta, 61
$Ba_{16.5}S_{39}Ta_9$, 62
BeK_4P_2, 47
$Be_{13}U$, 4
BiCaCu, 13
$(Bi,Sb)CuNiS_3$, 78
Bi_2PbS_4, 78
$BkPt_5$, 88

C_3CrSc_2, 32
CDy_2Fe_{17}, 33
$C_4Er_4Ni_{13}$, 32
C_2Er_2Re, 33

CORRIGENDA

Change required on page	From	To
31A, 25, Ba_5Si_3	16 Ba(2) in 14(g)	16(g)
38A, 113, $Fe_{9.5}Lu$ and $Fe_{9.6}Lu$	Z = 8	Z = 4
40A, 137, $KAsF_6$	Z = 2	Z = 3
43A, 47, middle report	K. KATO and I. KAWADA	K. KATO, I. KAWADA and T. TAKAHASHI
390	Add	Takahashi, T., 47
51A, 127, line 4	Strukturbericht, **3**, 122, 479	Structure Reports, **9**, 196
249, lines 7/8	Z = 3	Z = **2**
53A, 145, line 3	Ba	Sr
55A, 309	CAESIUM OZONIDE, 301	302
56A, 85, bottom report, line 4 of text	LiO_6	LiF_6
187 and 287	THALLIUM OXIDE TELLURIDE	THORIUM(IV) OXIDE TELLURIDE